Scaling Phenomena in Disordered Systems

NATO ASI Series

Advanced Science Institutes Series

A series presenting the results of activities sponsored by the NATO Science Committee, which aims at the dissemination of advanced scientific and technological knowledge, with a view to strengthening links between scientific communities.

The series is published by an international board of publishers in conjunction with the NATO Scientific Affairs Division

A Life Sciences **B Physics**	Plenum Publishing Corporation New York and London
C Mathematical and Physical Sciences	D. Reidel Publishing Company Dordrecht, Boston, and Lancaster
D Behavioral and Social Sciences **E Engineering and Materials Sciences**	Martinus Nijhoff Publishers The Hague, Boston, and Lancaster
F Computer and Systems Sciences **G Ecological Sciences**	Springer-Verlag Berlin, Heidelberg, New York, and Tokyo

Recent Volumes in this Series

Volume 127—Phenomena Induced by Intermolecular Interactions
edited by G. Birnbaum

Volume 128—Techniques and Concepts of High-Energy Physics III
edited by Thomas Ferbel

Volume 129—Transport in Nonstoichiometric Compounds
edited by George Simkovich and Vladimir S. Stubican

Volume 130—Heavy Ion Collisions: *Cargèse* 1984
edited by P. Bonche, Maurice Lévy, Philippe Quentin, and Dominique Vautherin

Volume 131—Physics of Plasma-Wall Interactions in Controlled Fusion
edited by D. E. Post and R. Behrisch

Volume 132—Physics of New Laser Sources
edited by Neal B. Abraham, F. T. Arecchi, Aram Mooradian, and Alberto Sona

Volume 133—Scaling Phenomena in Disordered Systems
edited by Roger Pynn and Arne Skjeltorp

Volume 134—Fundamental Processes in Atomic Collision Physics
edited by H. Kleinpoppen, J. S. Briggs, and H. O. Lutz

Series B: Physics

Scaling Phenomena in Disordered Systems

Edited by

Roger Pynn
Laue-Langevin Institute
Grenoble, France

and

Arne Skjeltorp

Institute for Energy Technology
Kjeller, Norway

Plenum Press
New York and London
Published in cooperation with NATO Scientific Affairs Division

Proceedings of a NATO Advanced Study Institute,
held April 8–19, 1985,
in Geilo, Norwav

Library of Congress Cataloging in Publication Data

Main entry under title:

Scaling phenomena in disordered systems.

(NATO ASI series. Series B, Physics; vol. 133)
"Published in cooperation with NATO Scientific Affairs Division."
"Proceedings of a NATO Advanced Study Institute held April 8–19, 1985"—T.p. verso.
Includes bibliographies and index.
1. Orders-disorder models. 2. Scaling laws (Nuclear physics) 3. Phase transformations (Statistical physics) 4. Particles. 5. Chemistry, Physical and theoretical. I. Pynn, R. II. Skjeltrop, A. III. North Atlantic Treaty Organization. Scientific Affairs Division. IV. NATO Advanced Study Institute (1985: Geilo, Norway) V. Series: NATO ASI series. Series B, Physics; v. 133.
QC173.4.O73S28 1985 539′.1 85-24472
ISBN 0-306-42112-7

A Division of Plenum Publishing Corporation
233 Spring Street, New York, N.Y. 10013

Printed in the United States of America

Fig. 1. An exactly self-similar fractal: the Serpinski Gasket. It is generated recursively by replacing a filled equilateral triangle by three smaller triangles as shown in the upper left-hand corner. The large figure shows the result of 6 stages of iteration. (R. F. Voss)

Fig. 2. Sample aggregate formed by 2D diffusion-limited aggregation (DLA) as discused in the article by L. M. Sander in this volume. Individual particles diffuse until reaching a sticky site at the center. When they touch, they become part of the immobile sticky aggregate. The different colors of the particles indicate the order in which they were attached to the aggregate. (R. F. Voss)

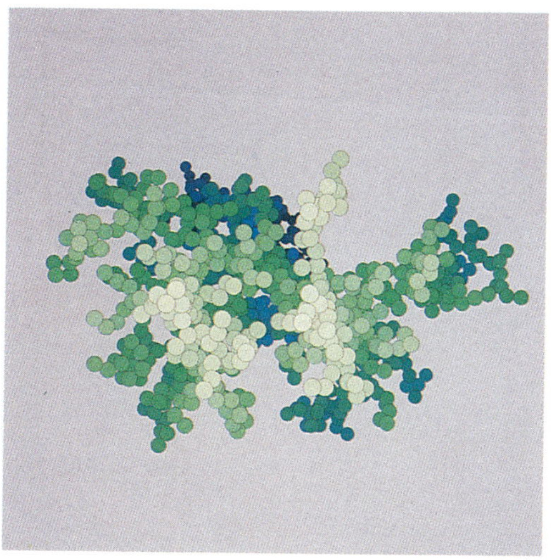

Fig. 3. Cluster of 1024 particles obtained from diffusion-limited particle aggregation in three dimensions. In this process, particles diffuse at very low concentration toward a center of growth, situated at the origin of the aggregate. They stick rigidly to the aggregate at the first contact with the cluster. The resulting aggregate, shown in projection, is very ramified in three dimensions and can be characterized by a fractal dimension $D = 2.5$. As the fractal dimension is larger than 2, the projection appears compact. (M. Kolb et al.).

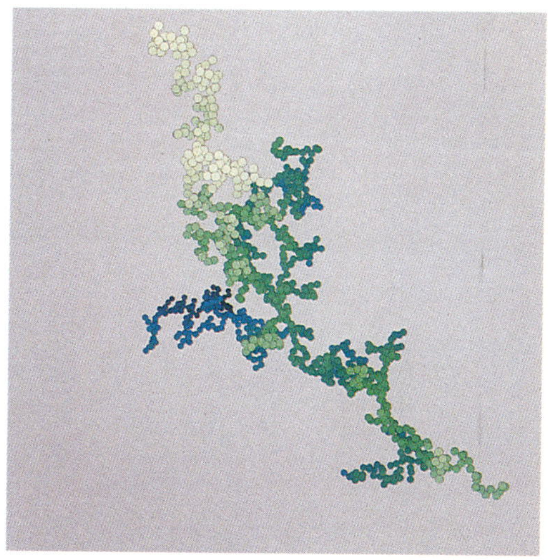

Fig. 4. Cluster of 1024 particles from diffusion-limited cluster aggregation in three dimensions. Here, clusters of comparable size diffuse and stick together when they touch each other. Clusters grown in this hierarchical fashion are distinctly more ramified than the ones when accreting single particles. The fractal dimension is $D = 1.8$ in three dimensions. As D is less than 2, the clusters also appear fractal, in projection, with the same D. The difference in D between particle and cluster aggregation can be seen as follows: the two clusters shown have about the same overall size, and the same number of particles. The size of the individual particles, however, is much larger for PA than for CA. For details, see the article "Scaling properties of cluster and particle aggregation." (M. Kolb et al.)

Fig. 5. 1D cellular automata vs time (vertical). (S. Wolfram)

Fig. 6. 1D cellular automata vs time (vertical). (S. Wolfram)

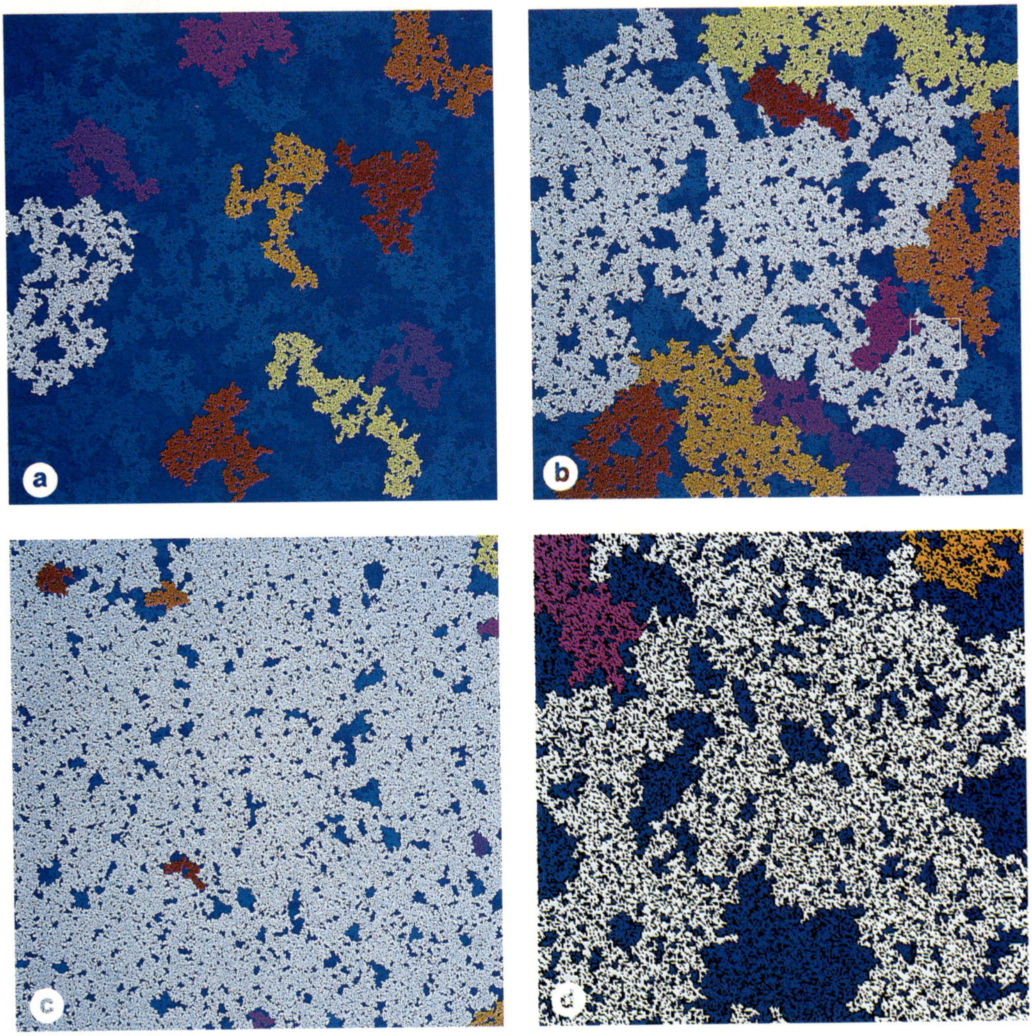

Fig. 7. 2D site percolation on a 3000 by 3000 square lattice showing how the cluster geometry changes as the site occupation probability, p, varies from below to above the critical probability, p_c. (See articles by Voss, Stinchcombe, et al. in this volume). In each sample, the largest clusters are shown in bright colors (the white cluster is the largest) while the smaller clusters are in various shades of blue. (a) $p = 0.587$ below p_c. The largest clusters are finite with a size comparable to the correlation length. (b) $p = 0.593$ close to the expected value of $p_c = 0.59277$, where the largest cluster just spans the sample and has a statistically self-similar fractal geometry on all length scales. (c) $p = 0.599$ above p_c, where the largest cluster spans the sample in all directions. In this case, the "holes" in the largest cluster are roughly the size of the correlation length. (d) A blowup of the 300 by 300 area in the white box in (b) emphasizing the self-similarity of the largest cluster at p_c. (R. V. Voss)

PREFACE

This volume comprises the proceedings of a NATO Advanced Study Institute held in Geilo, Norway, between 8-19 April 1985. Although the principal support for the meeting was provided by the NATO Committee for Scientific Affairs, a number of additional sponsors also contributed, allowing the assembly of an unusually large number of internationally recognized speakers. Additional funds were received from:

EXXON Research and Engineering Co.
IBM (Europe)
Institutt for energiteknikk (NorwaY)
Institut Lauge-Langevin (France)
The Norwegian Research Council for Science and Humanities
NORDITA (Denmark)
The Norwegian Foreign Office
The U.S. Army Research, Development and Standardization Group (Europe)
The U.S. National Science Foundation
The Norwegian Council for Science and Letters

The organizing committee would like to take this opportunity to thank these contributors for their help in promoting a most exciting rewarding meeting.

This Study Institute was the eighth of a series of meetings held in Geilo on subjects related to phase transitions. In contrast to previous meetings which were principally concerned with transitions in ordered systems, this school addressed the problems which arise when structural order is absent. The unifying feature among the subjects discussed at the school and the link to themes of earlier meetings was the concept of scaling. How much can be learned about the behaviour of apparently disordered systems by extrapolating between short-range behaviour and macroscopic properties ? In this context the concept of self-similarity under dilation and of fractal geometry play important roles and were recurring themes of the meeting. Inevitably these relatively new ideas have produced a plethora of theoretical speculations and calculations in the research literature. Experimental investigations of model systems (the "fruit-flies" of disordered media) are few and their relationship to theoretical conjecture is not always clear. This bias is to some extent reflected by the present volume. However, we believe that the lectures presented here provide a useful compendium of current views; they certainly give a window on several fascinating subjects and may help to stimulate the experimental activity which is necessary if this field is to flourish.

At the beginning of the school, the power of Mandelbrot's development of fractal geometry was demonstrated by a talk on fractal forgeries. The sensory impression left by computer imitations of landscapes, archipelagos, galaxies and one-over-f music is strong and the obvious extrapolations to description of our environment and to visual art and music are extremely tempting. For obvious reasons these visual effects cannot be fully reproduced here. However, several of the "forgeries" presented at Geilo are contained in Mandelbrot's book "The fractal geometry of nature".

The connection between the "traditional" studies of phase transitions and the more recent applications of scaling to disordered media was provided by several speakers. Concepts such as real-space renormalisation and decimation were linked to the idea of self-similarity under dilation in one of the first lectures. Percolation emerged as the archetypical example of both approaches and was considered by several authors. Both regular fractals and statistically self-similar structures were discussed in this context. Although many of the results presented were exact (or at least accepted by the establishment) others were preliminary and heuristic, giving an accurate impression of the state of controversy and discussion which is an inevitable and exciting aspect of a subject in a state of active growth. Nowhere was this transient nature more obvious than in the alphabet soup of exponents and (fractal) dimensions. In many cases the proponents of different notations are still arguing the relative merits of their contributions to the new vocabulary. No attempt to artificially terminate this debate and impose a coherent notation has been made in editing this volume; rather a partial table of symbols and definitions is given on page 50.

Aggregates of various types were widely discussed at the school and a computer-generated film (produced by a group from Orsay in France) nicely contrasted the particle dynamics involved in diffusion limited aggregation (DLA) of particles and clusters with that of reaction-limited aggregation. In these two cases experiments on the aggregation of colloidal gold particles and of immunoglobulin were available for comparison with some aspects of theory. Further proof that computer simulation of DLA is relevant to real systems was provided by experiments on viscous fingering in a Hele-Shaw cell and on electrolytic deposition of copper. Even so, the search for the essential ingredients of growth models is largely empirical at present. Phenomena which lead either to regular patterns or apparently disordered, self-similar structures during growth are not yet fully catalogued. A demonstration that regularity of growth may be far from trivial was provided by an explanation of diffraction patterns of quasi crystals displaying pentagonal symmetry.

The use of light, X-ray and neutron scattering techniques for the determination of the fractal dimensions of aggregates (volume fractals) and interfaces (surface fractals) was described in some detail. In many cases, however, the interpretation of the results of such studies in terms of fractal objects is difficult. The identification of the length scales over which objects may be considered fractal and the possibility that these scales may change with time during growth introduce ambiguities in the interpretation of such measurements.

A relatively large section of the program of the school was devoted to transport properties in disordered media. In this context, "walks on fractals" appeared as a theorist's dream and an experimentalist's nightmare. The ingenuity (or stupidity !) of the "ant in a labyrinth" or of its near relatives "the termite" and "the parasite" (a new term coined for either an ant or a termite diffusing on a lattice animal) have led theorists to a plethora of predictions. Yet the relationship to real systems is often not clear and the imagination of theorists has clearly outstripped the ingenuity of experimentalists. The whole subject of the dynamics of fractal objects is still fraught with questions. Has the fracton edge been observed or has it not ? What does the density of fracton states really look like : different calculations give different answers. Are localised phonons included correctly in the fracton calculations and does bond-bending matter ? These questions and many others are addressed in this volume together with the theoretical constructs on which they are based.

Many of the self-similar systems considered at the Study Institute were of technological interest; polymers, gels and porous media are obvious examples. In the former, scaling theory has been widely used to describe both the individual and collective motion of polymer chains. The confrontation between theory and experiment has, in this case, reached relative maturity. Nevertheless, although techniques such as neutron scattering have been successful in vindicating many theoretical constructs, the observation of repetition, which is related to the rich and varied rheological properties of polymers, remains elusive. In at least one area of polymer physics, the adsorption of polymers at surfaces, theory is either ahead of or asking different questions than experiment.

Deterministic systems which appear to display many of the phenomena usually associated with stochastic processes were the subject of a series of lectures. These cellular automata, which evolve under the application of simple mathematical rules generate complex behaviour from simple components. Homogenous states, limit cycles, strange attractors and complicated localised structures can all be generated in this way. Indeed the study of cellular automata leads to a number of fundamental questions which are of relevance in solid-state physics. The definition of randomness, the classification of complexity and the possibility (or otherwise) of predicting the behaviour of a system without simulating it are examples among those which were described at the school.

The material contained in this volume is derived from many subdisciplines of physics. Examples not already mentioned in this preface include random-field systems such as diluted magnets and binary liquid mixtures in gels, temporal chaos in liquid and solid systems, the mechanical properties of disordered materials, domain growth at first-order transitions, branched polymers, and many others. The common features are the disorder of the systems considered and the applicability of scaling arguments to their description. One of the goals of this school was to assemble experts and students interested in many different areas in the hope that an exchange of ideas would prove fruitful and might even lead to some consensus. Although there was general agreement among participants that the school was stimulating and useful there was certainly no consensus, as befits a subject which is growing daily.

R.A. Cowley
R. Pynn
S.K. Sinha
A.T. Skjeltorp
H. Thomas

CONTENTS

RANDOM FRACTALS: characterization and measurement

Richard F. Voss
IBM Thomas J. Watson Research Center
Yorktown Heights, NY 10598
and
Harvard University, Division of Applied Science
Cambridge, MA 01238

ABSTRACT

Mandelbrot's fractal geometry provides both a description and a mathematical model for many of the seemingly complex shapes found in nature. Such shapes often possess a remarkable invariance under changes of magnification. This statistical self-similarity may be characterized by a fractal dimension D, a number that agrees with our intuitive notion of dimension but need not be an integer. A brief mathematical characterization of random fractals is presented with emphasis on variations of Mandelbrot's fractional Brownian motion. The important concepts of fractal dimension and exact and statistical self-similarity and self-affinity will be reviewed. The various methods and difficulties of estimating the fractal dimension and lacunarity from experimental images or point sets are summarized.

random fractals: an introduction

Mandelbrot's fractal geometry [1] has revolutionized the application of geometrical constructs to the natural sciences. Fractals provide the proper mathematical framework for a treatment of the irregular, seemingly-complex shapes found in nature from the small scale structure of disordered systems and percolation clusters to coastlines, mountain ranges, clouds, and the distribution of stars in the night sky. Some of the building blocks of fractal geometry originated in the exactly self-similar mathematical "monsters" (such as the Koch curve and Seirpinski gasket) of the early 1900's. Although such exact deterministic constructs serve as useful tools in building understanding and intuition about scaling properties, the fractal shapes found in nature possess a statistical rather than exact self-similarity. The following sections present an expository summary of the major mathematical definitions and relations used in the characterization and measurement of random or statistical fractals as condensed from Mandelbrot[1]. A detailed discussion of the algorithmic considerations in generating or simulating such random fractals is found in [2].

fractional Brownian motion

One of the most useful mathematical models for the random fractals found in nature (such as mountainous terrain and clouds) has been the *fractional Brownian motion* (fBm) of Mandelbrot and Wallis[1,3]. It is an extension of the central concept of *Brownian motion* that has played an important role in both physics and mathematics. Sample *traces* of fBm are shown in Fig. 1. Almost all computer graphics fractal simulations[2] are based on an extension of fBm to higher dimensions such as the fractional Brownian landscape of Fig. 2. Fractional Brownian motion is also a good starting point for understanding anomalous diffusion and random walks on fractals.

A fractional Brownian motion, $V_H(t)$, is a single valued function of one variable, t (usually time). Its increments $V_H(t_2)$-$V_H(t_1)$ have a Gaussian distribution with variance

$$<|V_H(t_2) - V_H(t_1)|^2> \propto |t_2 - t_1|^{2H}, \tag{1}$$

where the brackets < and > denote averages over many samples of $V_H(t)$ and the parameter H has a value $0<H<1$. Such a function is both stationary and isotropic. Its mean square increments depend only on the time difference t_2-t_1 and all t's are statistically equivalent. The special value $H=1/2$ gives the familiar Brownian motion with $\Delta V^2 \propto \Delta t$.

As with the usual Brownian motion, although $V_H(t)$ is continuous, it is nowhere differentiable. Nevertheless, many constructs have been developed (and are relevant to the problem of light scattering from fractals) to give meaning to "derivative of fractional Brownian motion" as *fractional Gaussian noises*[1,3]. Such constructs are usually based on averages of $V_H(t)$ over decreasing scales. The derivative of normal Brownian motion, $H=1/2$, corresponds to uncorrelated *Gaussian white noise*, and Brownian motion is said to have *independent increments.* Formally, for any three times such that $t_1<t<t_2$, $V_H(t)$-$V_H(t_1)$ is statistically independent of $V_H(t_2)$-$V_H(t)$ for $H=1/2$. For $H>1/2$ there is a positive correlation both for the increments of $V_H(t)$ and its derivative fractional Gaussian noise. For $H<1/2$ the increments are negatively correlated. Such correlations extend to arbitrarily long time scales and have a large effect on the visual appearance of the fBm traces as shown in Fig. 1.

$V_H(t)$ shows a statistical scaling behavior. If the time scale t is changed by the factor r, then the increments ΔV_H change by a factor r^H. Formally,

$$<\Delta V_H(rt)^2> \propto r^{2H}<\Delta V_H(t)^2>. \tag{2}$$

Unlike statistically self-similar curves (such as the coastlines in Fig. 2), a $V_H(t)$ trace requires *different* scaling factors in the two coordinates (r for t but r^H for V_H) reflecting the special status of the t coordinate. Each t can correspond to only one value of V_H but any specific V_H may occur at multiple t's. Such non-uniform scaling is known as *self-affinity* rather than self-similarity.

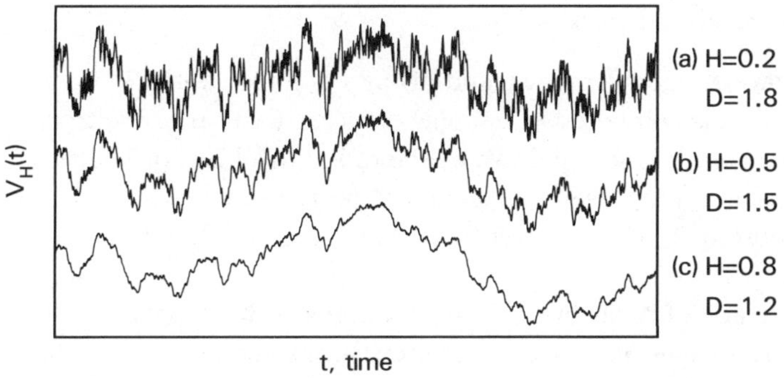

Figure 1. Sample traces of statistically self-affine fractional Brownian motion. (a) Increments are negatively correlated. (b) Normal Brownian motion with uncorrelated increments. (c) Increments are positively correlated.

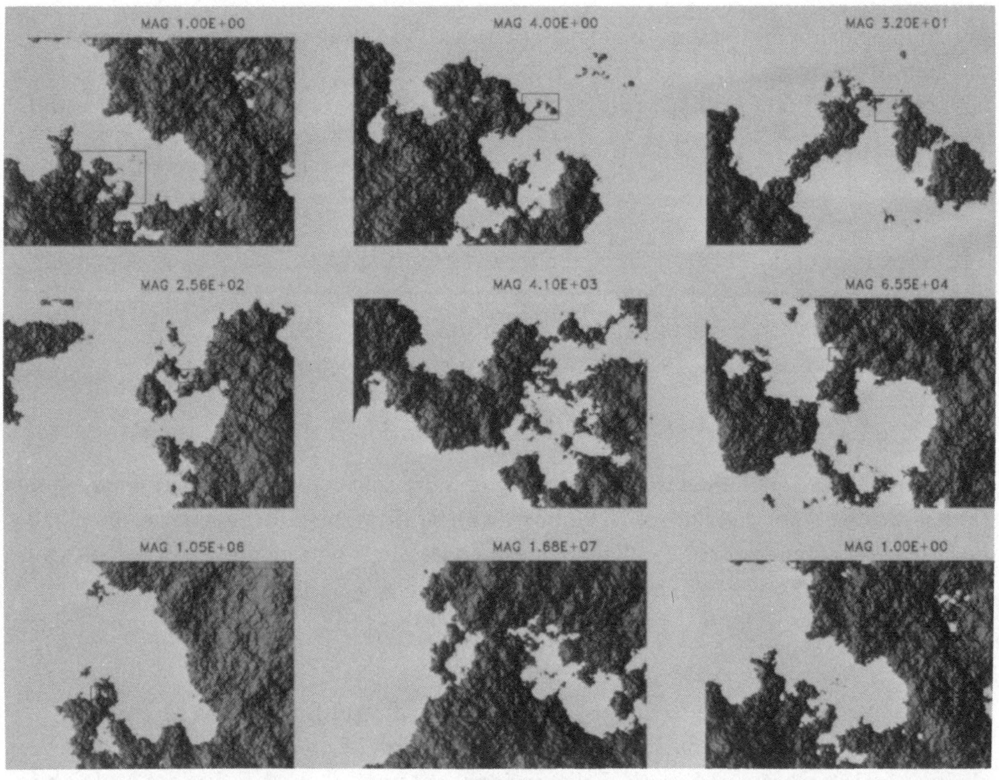

Figure 2. Computer realization of a statistically self-affine fractional Brownian landscape with H=0.8 D=2.2 at various magnifications. The altitude variations along any straight line path give a fBm trace such as Fig. 1(c). Successive frames show magnified view of the black outine from the previous frame and demonstrate that the landscape's coastline is statistically self-similar.

self-similar vs self-affine fractals.

The distinction between similarity and affinity is important. By way of summary, a *self-similar* object is composed of N copies of itself (with possible translations and rotations) each of which is scaled down by the ratio r in all E coordinates from the whole. More formally, consider a set S of points at positions $\vec{x} = (x_1, \ldots, x_E)$ in Euclidean space of dimension E. Under a *similarity* transform with real scaling ratio $0<r<1$, the set S becomes rS with points at $r\vec{x} = (rx_1, \ldots, rx_E)$. A bounded set S is *self-similar* when S is the union of N distinct (non-overlapping) subsets each of which is congruent to rS. *Congruent* means identical under translations and rotations. The fractal or *similarity dimension* of S is then given by

$$1 = Nr^D \quad \text{or} \quad D = \frac{\log N}{\log 1/r}. \tag{3}$$

This relation leads to several important methods of *estimating* D for a given set S.

For topologically one-dimensional fractal "curves", the apparent "length", varies with the measuring ruler size. If the entire self-similar curve is of maximum size L_{max}, then, on a smaller scale $L = rL_{max}$, the curve consists of $N = 1/r^D$ segments of length L. Thus, for $D \geq 1$

$$\text{LENGTH} = L \times N = L \times (L_{max}/L)^D \propto 1/L^{D-1}. \tag{4}$$

The fractal dimension D also characterizes the covering of the set S by E-dimensional "boxes" of linear size L. If the entire S is contained within one box of size L_{max}, then each of the $N=1/r^D$ subsets will fall within one box of size $L = rL_{max}$. Thus, the number of boxes of size L, $N_{box}(L)$, needed to cover S is given by

$$N_{box}(L) = (L_{max}/L)^D \quad \text{or} \quad N_{box}(L) \propto 1/L^D. \tag{5}$$

This definition of *box dimension* is one of the most useful methods for estimating the fractal dimension of a given set. The box dimension can be conveniently estimated by dividing the E-dimensional Euclidean space containing the set into a grid of boxes of size L^E and counting the number of such boxes $N_{box}(L)$ that are non-empty. It is useful to examine as large a range of L as possible and to average over various origins for the boxes.

One can also estimate the "volume" or "mass" of the set S by a covering with boxes of linear size L. If one considers only distances of order L *about a given point in S*, one finds a single box of size L with E-dimensional volume L^E. If the distance scale about the same point is increased to $L_{max}=L/r$, one now finds a total of $N=1/r^D=(L_{max}/L)^D$ boxes of mass L^E covering the set. Thus, the mass within a distance L_{max} of some point in S, $M(L_{max}) = N \times M(L) = M(L) \times (L_{max}/L)^D$ or

$$M(L) \propto L^D. \tag{6}$$

The fractal dimension D, thus, also corresponds to the commonly used *mass dimension* in physics. Mass dimension also has a strong connection with intuitive notions of dimension. The amount of material within a distance L of a point in a one-dimensional object increases as L^1. For an E-dimensional object it varies as L^E. The mass dimension is another extremely useful method for estimating the fractal dimension of a given object.

The set S is also *self-similar* if each of the N subsets is scaled down from the whole by a different similarity ratio r_n. In this case, D is given implicitly by

$$1 = \sum_{n=1}^{N} r_n^D , \tag{7}$$

which reduces to the familiar result in Eq. (3) when all of the r_n are equal.

The set S is *statistically self-similar* if it is composed of N distinct subsets each of which is scaled down by the ratio r from the original and is identical in all statistical respects to rS. The similarity dimension is again given by Eq. (3). In practice, it is impossible to verify that all moments of the distributions are identical, and claims of statistical self-similarity are usually based on only a few moments. Moreover, a sample of a random set (such as a coastline) is often statistically self-similar for all scaling ratios r. Its fractal dimension is usually estimated from the dependence of box coverings $N_{box}(L)$ or mass M(L) on varying L as in Eqs. (5) and (6).

Under an *affine* transform, on the other hand, each of the E coordinates of $\vec{x}$ may be scaled by a different ratio ($r_1, \ldots, r_E$). Thus, the set S is transformed to r(S) with points at $r(\vec{x}) = (r_1x_1, \ldots, r_Ex_E)$. A bounded set S is *self-affine* when S is the union of N distinct (non-overlapping) subsets each of which is congruent to r(S). Similarly, S is *statistically self-affine* when S is the union of N distinct subsets each of which is congruent *in distribution* to r(S). The fractal dimension D, however, is not as easily defined as with self-similarity.

the relation of D to H for self-affine fractional Brownian motion

The assignment of a fractal dimension D to a self-affine set can be illustrated with a trace of fractional Brownian motion $V_H(t)$ from above. Consider, for convenience, a trace of $V_H(t)$ covering a time span $\Delta t=1$ and a vertical range $\Delta V_H=1$. $V_H(t)$ is statistically self-affine when t is scaled by r and V_H is scaled by r^H. Suppose the time span is divided into N equal intervals each with $\Delta t=1/N$. Each of these intervals will contain one portion of $V_H(t)$ with vertical range $\Delta V_H = \Delta t^H = 1/N^H$. Since $0<H<1$ each of these new sections will have a large vertical to horizontal size ratio and the occupied portion of each interval will be covered by $\Delta V_H/\Delta t = (1/N^H)/(1/N) = N/N^H$ square boxes of linear scale $L=1/N$. In terms of box dimension, as t is scaled down by a ratio $r=1/N$ the number of square boxes covering the trace goes from 1 to N(L) = number of intervals × boxes per interval or

$$N(L) = N \times N/N^H = N^{2-H} = 1/L^{2-H}. \tag{8}$$

Thus, by comparison with Eq. (5),

$$D = 2 - H \quad \text{for a trace of } V_H(t). \tag{9}$$

Consequently, the trace of normal Brownian motion has $D = 1.5$.

It is important to note that the association of a similarity dimension D with a self-affine fractal such as fBm is implicitly fixing a scaling between the (otherwise independent) coordinates. The result depends strongly on whether one is looking at scales large or small compared to this (artificially introduced) characteristic length. The difference is particularly clear when one attempts to estimate D for a trace of fBm from Eq. (4). As above, one can divide the t axis into N segments of size $\Delta t = 1/N$. For each of these segments the typical V variation will be $\Delta V = \Delta t^H$. The length *along* each segment is typically $\ell = (\Delta t^2 + \Delta V^2)^{1/2}$ and the total

$$\text{LENGTH} = N \times \ell \propto (1 + \Delta V^2/\Delta t^2)^{1/2} \propto (1 + 1/\Delta t^{2-2H})^{1/2}. \qquad (12)$$

On small scales with $\Delta t << 1$, the second term dominates and LENGTH $\propto 1/\Delta t^{1-H}$ so D = 2-H by comparison with Eq. (4) and in agreement with Eq. (9). On the other hand, on large scales with $\Delta t >> 1$, LENGTH is independent of Δt and D = 1.

The *zeroset* of fBm is the the intersection of the trace of $V_H(t)$ with the t axis, the set of all points such that $V_H(t) = 0$. The zeroset is a disconnected set of points with topological dimension zero and a fractal dimension D_0 = D-1 = 1-H that is less than 1 but greater than 0. Although the trace of $V_H(t)$ is self-affine, its zeroset is self-similar.

trails of fBm

Consider a particle undergoing a fractional Brownian motion or random walk in E dimensions where each of the coordinates is tracing out an independent fBm in time. Over an interval Δt each coordinate will vary by typically $\Delta L = \Delta t^H$. If overlap can be neglected, the "mass" of the trail $M \propto \Delta t \propto L^{1/H}$. In comparison with Eq. (6), the trail of fBm has a fractal dimension

$$D = 1/H \qquad (13)$$

provided 1/H < E. Normal Brownian motion with H=0.5 has D=2. 1/H is known as the *latent* fractal dimension[4,5] of a *trail* of fBm. When 1/H > E overlap cannot be neglected and the actual D = E. When 1/H = E the trail is *critical* and most quantities will have important logarithmic corrections to the usual fractal power laws. Note that although the *trail* of a fBm in E dimensions is self-similar, each of the E coordinates has a self-affine *trace* vs time.

self-affinity in higher dimensions: Mandelbrot landscapes and clouds

The traces of fBm, particularly Fig. 1(c) with H=0.8, bear a striking resemblance to a mountainous horizon. The modelling of the irregular Earth's surface as a generalization of traces of fBm was first proposed by Mandelbrot. The single variable t can be replaced by coordinates x and y in the plane to give $V_H(x,y)$ as the surface altitude at position x,y as shown in Fig. 2. In this case, the altitude variations of a hiker following any straight line path at constant speed in the xy plane is a fractional Brownian motion. In analogy with Eq. (1),

$$< | V_H(x_2, y_2) - V_H(x_1, y_1) |^2 > \propto [(x_2 - x_1)^2 + (y_2 - y_1)^2]^H. \qquad (10)$$

Once again, the fractal dimension D must be greater than the topological dimension 2 of the surface. Here,

$$D = 3 - H \quad \text{for a fractal landscape, } V_H(x,y). \tag{11}$$

The intersection of a vertical plane with the surface $V_H(x,y)$ is a self-affine fBm trace with D=2-H, smaller by one than the value of Eq. (11). Similarly, the zeroset of $V_H(x,y)$, its intersection with a horizontal plane, also has a fractal dimension D_0=2-H. This intersection, which produces a family of (possibly disconnected) curves, forms the coastlines of the $V_H(x,y)$ landscape. Since the two coordinates x and y are, however, equivalent, the coastlines of $V_H(x,y)$ are self-similar, not self-affine. Figure 2 demonstrates how the coastline remains statistically invariant under changes of magnification.

This generalization of fBm can continue to still higher dimensions to produce, for example, a self-affine fractal temperature or density distribution $V_H(x,y,z)$. Here, the variations of an observer moving at constant speed along any straight line path in space generate a fBm and the fractal dimension

$$D = 4 - H \quad \text{for a fractal cloud } V_H(x,y,z). \tag{12}$$

The zeroset $V_H(x,y,z)$ = constant now gives a self-similar fractal with D_0=3-H.

To summarize, a statistically self-affine fractional Brownian function, V_H of $\vec{x} = (x_1, \ldots, x_E)$ in E Euclidean dimensions satisfies

$$< | V_H(\vec{x}_2) - V_H(\vec{x}_1) |^2 > \; \propto \; | \vec{x}_2 - \vec{x}_1 |^{2H}. \tag{13}$$

and has a fractal dimension

$$D = E + 1 - H. \tag{14}$$

The zerosets of $V_H(\vec{x})$ form a statistically self-similar fractal with dimension D_0 = E-H.

perimeter vs area scaling

The fractal dimension of the coastlines of $V_H(x,y)$ can also be estimated from the perimeter area scaling of the islands. Eq. (4) gives the LENGTH or perimeter of a fractal object of size L_{max} when "measured" with a ruler of size L. Consider a population of such islands with the same coastline D and all measured with the same ruler L. For a given island, perimeter $P \propto L^D_{max}$ from Eq. (4). Provided the coastline $D < 2$, each island will have a well-defined area $A \propto L^2_{max}$ and

$$P \propto A^{D/2}. \tag{15}$$

A study of P vs A scaling is a useful method of estimating a coastline D for a population of fractal objects.

spectral densities for fBm and the spectral exponent β

Random functions in time V(t) are often characterized[6,7] by their *spectral densities* $S_V(f)$. If V(t) is the input to a narrow bandpass filter at frequency f and bandwidth

Δf, then $S_V(f)$ is the mean square output $V(f)$ divided by Δf, $S_V(f) = |V(f)|^2/\Delta f$. $S_V(f)$ gives information about the time correlations of $V(t)$. When $S_V(f)$ increases steeply at low f, $V(t)$ varies more slowly. If one defines $V(f,T)$ as the Fourier transform of a specific sample of $V(t)$ for $0<t<T$,

$$V(f,T) = \frac{1}{T}\int_0^T V(t)e^{2\pi ift}dt, \text{ then } S_V(f) \propto T\,|V(f,T)|^2 \text{ as } T \to \infty. \tag{16}$$

An alternate characterization of the time correlations of $V(t)$ is given by the *2 point autocorrelation function*

$$G_V(\tau) = \langle V(t)V(t+\tau)\rangle - \langle V(t)\rangle^2.$$

$G_V(\tau)$ provides a measure of how the fluctuations at two times separated by τ are related. $G_V(\tau)$ and $S_V(f)$ are not independent. In many cases they are related by the Wiener-Khintchine relation[6,7]

$$G_V(\tau) = \int_0^\infty S_V(f)\cos(2\pi f\tau)\,df. \tag{17}$$

For a Gaussian white noise $S_V(f)$ = constant and $G_V(\tau) = \Delta V^2\delta(\tau)$ is completely uncorrelated. For certain simple power laws for $S_V(f)$, $G_V(\tau)$ can be calculated exactly. Thus, for

$$S_V(f) \propto 1/f^{\beta}, \text{ with } 0<\beta<1, \quad G_V(\tau) \propto \tau^{\beta-1}. \tag{18}$$

Moreover, $G_V(\tau)$ is directly related to the mean square increments of fBm,

$$\langle |V(t+\tau) - V(t)|^2\rangle = 2[\langle V^2\rangle - G_V(\tau)]. \tag{19}$$

Roughly speaking, $S_V(f) \propto 1/f^{\beta}$ corresponds to $G_V(\tau) \propto \tau^{1-\beta}$ and a fBm with $2H=\beta$-1 from Eqs. (1) and (19). Thus, the statistically self-affine fractional Brownian function, $V_H(\vec{x})$, with $\vec{x}$ in an E-dimensional Euclidean space, has a fractal dimension D and spectral density $S_V(f) \propto 1/f^{\beta}$, for the fluctuations along a straight line path in any direction in E-space with

$$D = E + 1 - H = E + \frac{3-\beta}{2} \tag{20}$$

This result agrees with other "extensions" of the concepts of *spectral density* and *Wiener-Khintchine relation* to *non-stationary* noises where some moments may be undefined. Moreover, it provides an extremely useful connection between D, H and β for finite simulations. For H in the range $0<H<1$, $E<D<E+1$, and $1<\beta<3$. The value $H \simeq 0.8$ is a good choice for many natural phenomena.

Although the formal definition of fBm restricts H to the range $0<H<1$, it is often useful to consider integration and an appropriate definition of "derivative" as extending the range of H. Thus, integration of a fBm produces a new fBm with H increased by 1, while "differentiation" reduces H by 1. When $H \to 1$, the derivative of fBm looks like a fBm with $H \to 0$. In terms of spectral density, if $V(t)$ has $S_V(f) \propto 1/f^{\beta}$ then its derivative dV/dt has spectral density $f^2/f^\beta = 1/f^{\beta-2}$. In terms of Eq. (20), differentiation decreases β by 2 and decreases H by 1.

measuring fractal dimensions: Mandelbrot measures

Numerical simulation or experimental image analysis often produces a geometrical object defined by a set S of points at positions $\vec{x} = (x_1, \ldots, x_E)$ in an E-dimensional Euclidean space. Lacking other information, all of the points may be assumed to be equivalent and all points are equally probable origins for analysis. The spatial arrangement of the points determines P(m,L). P(m,L) *is the probability that there are m points within an E-cube (or sphere) of size L centered about an arbitrary point in S.* P(m,L) is normalized

$$\sum_{m=1}^{N} P(m,L) = 1 \quad \text{for all L} \tag{21}$$

P(m,L) is directly related to other probability measures as used by Mandelbrot[1,8], Hentschel and Procaccia[9], and others. This particular definition of P(m,L) is, however, particularly efficient to implement on a computer.

The usual quantities of interest are derived from the moments of P(m,L). The *mass dimension*,

$$M(L) = \sum_{m=1}^{N} m\, P(m,L). \tag{22}$$

The number of boxes of size L needed to cover S,

$$N_{box}(L) = \sum_{m=1}^{N} \frac{1}{m} P(m,L). \tag{23}$$

The configurational entropy when space is divided into cubes of size L

$$S(L) = \sum_{m=1}^{N} \log m\;\; P(m,L). \tag{24}$$

For a fractal set $M(L) \propto L^D$, $N_{box}(L) \propto 1/L^D$, and $e^{S(L)} \propto L^D$.

In fact, one can define all moments

$$M^q(L) = \sum_{m=1}^{N} m^q\, P(m,L), \tag{25}$$

for $q \neq 0$ (the $q=0$ case is given by Eq.(24) above) and one can then estimate D from the logarithmic derivatives

$$D = \frac{1}{q} \left\langle \frac{\partial \log M_q(L)}{\partial \log L} \right\rangle \quad \text{for } q \neq 0, \tag{26}$$

and

$$D = \frac{1}{q}\left\langle\frac{\partial S(L)}{\partial \log L}\right\rangle \quad \text{for } q = 0. \tag{27}$$

A double logarithmic plot of $M^q(L)$ vs L is an essential tool in verifying whether a fractal interpretation is valid for S. One expects a fractal to have power-law behavior of $M^q(L)$ over a wide range of L.

For a uniform fractal (fractal set) as the number of points examined, $N \to \infty$ the distribution is expected to take the scaling form $P(m,L) \to f(m/L^D)$ and all moments give the same D. For a non-uniform fractal (a fractal measure such as a Poincare map) the moments satisfy the relation[9] that $D_{q_1} \leq D_{q_2}$ for $q_1 < q_2$.

lacunarity

It is obvious from the above discussion that the fractal dimension D characterizes only part of the information in the distribution P(m,L). Different fractal sets may share the same D but have different appearances or *textures* corresponding to different P(m,L). As an initial step toward quantifying texture, Mandelbrot [1, chapter 34] has introduced the parameter *lacunarity*, Λ (*lacuna* is Latin for gap). Although the qualitative visual effect of changing lacunarity at fixed D is quite striking [1], to date there have been no quantitative measurements of the lacunarity of various random fractals and Mandelbrot [1] offers several alternative definitions. The most useful derives from the width of the distribution P(m,L) at fixed L. Given $M^q(L)$ as defined by Eq. (25),

$$\Lambda(L) = \frac{\langle M^2(L)\rangle - \langle M(L)\rangle^2}{\langle M(L)\rangle^2}. \tag{28}$$

When P(m,L) takes the scaling form $f(m/L^D)$, Λ is just the relative mean square width of the distribution f, $\Lambda = \langle \Delta f^2\rangle / \langle f\rangle^2$.

acknowledgement

This work was made possible by the help, encouragement, and inspiration of Benoit Mandelbrot.

References

1. Mandelbrot, B.B. *The Fractal Geometry of Nature*, (Freeman, New York) 1982 and references therein. See also, *Fractals: Form, Chance, and Dimension,* W. H. Freeman and Co., San Francisco (1977).

2. Voss, R.F, "Random Fractal Forgeries:", *Proc. NATO A.S.I on Fundamental Algorithms in Computer Graphics,* Ilkley, Yorkshire, UK (1985).

3. Mandelbrot, B.B. and Wallis, J. W. "Fractional Brownian motions, fractional noises, and applications", *SIAM review, 10* (1968) 422-437.

4. Mandelbrot, B.B. "Fractals in Physics: Squig Clusters, Diffusions, Fractal Measures, and the Unicity of Fractal Dimensionality", *J. Stat. Phys. 34* (1984) 895-930.

5. Mandelbrot, B.B., Gefen, Y., Aharony, A., and Peyeiere, J. "Fractals, their transfer matrices and their eigen-dimensional sequences", *J. Phys. A 18* (1985) 335-354.

6. A good discussion is found in Reif, F. *Statistical and Thermal Physics,* McGraw-Hill Book Co., New York, (1965), Chapter 15, "Irreversible Processes and Fluctuations."

7. For example see: Freeman, J. J. *Principles of Noise,* John Wiley & Sons, Inc., New York, (1958), Chapter 1, "Fourier Series and Integrals." or Robinson, F.N.H. *Noise and Fluctuations,* Clarendon Press, Oxford, (1974).

8. Mandelbrot, B.B. "Intermittent turbulence in self-similar cascades: divergence of higher moments and dimension of the carrier", *J. Fluid Mech. 62* (1974) 331-358.

9. Hentschel, H.G.E. and Procaccia, I. "The infinite number of generalized dimensions of fractals and strange attractors", *Physica 8D* (1983) 435-444.

INTRODUCTION TO SCALING CONCEPTS

R.B. Stinchcombe

Department of Theoretical Physics
1 Keble Road
Oxford, OX1 3NP

PHASE TRANSITIONS; GEOMETRICAL VIEWPOINTS

Spin interactions cause long range order in a magnet below its transition temperature T_c. In this phase transition the order parameter (spontaneous magnetisation) goes continuously to zero as T_c is approached from below, and the susceptibility diverges at T_c while other quantities such as specific heat are singular. This is the behaviour of a continuous thermal phase transition[1]. A typical spin configuration (in which up and down spin regimes are distinguished) is shown in Figure 1 for an Ising model magnet very close to its transition.

Another continuous phase transition which is not thermal but just geometrical is the percolation transition[2,3] occuring when a lattice is diluted by random removal of bonds or sites. A bond diluted square lattice is shown for two values of the bond concentration p in Figure 2. The two concentrations lie on either side of a critical value p_c which is the least concentration for the occurrence of the "infinite cluster", required to provide a connected path across an infinite lattice. Figure 3 shows the 'backbone' of the infinite cluster[4] (with dangling ends removed) at $p = p_c$. It is clearly difficult to distinguish any scale characterising the features in this diagram.

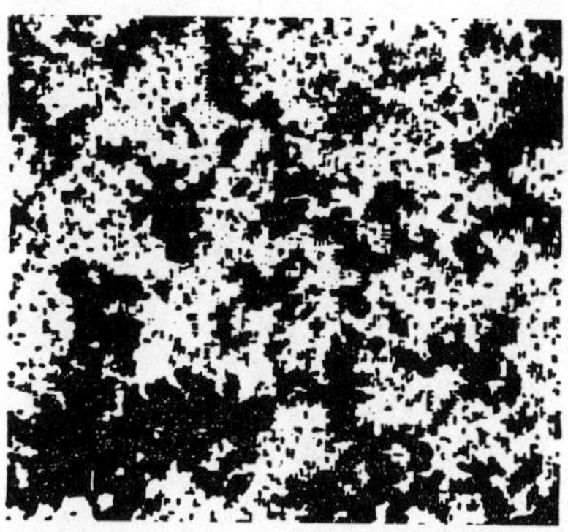

Fig. 1. Spin configuration of an Ising model at its transition.

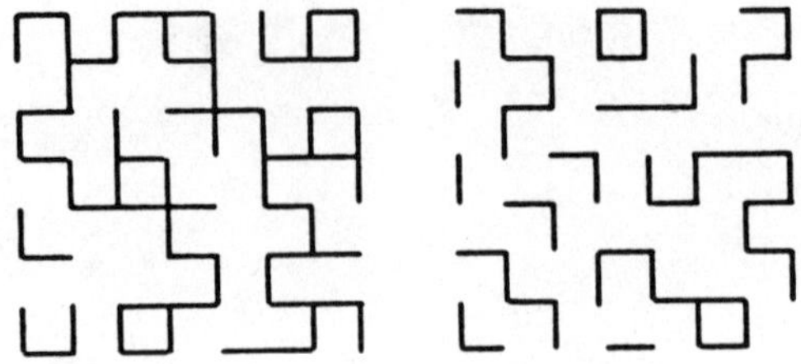

Fig. 2. Bond-diluted square lattice as percolation model.

The large range of void sizes evident in the picture will later be discussed in terms of "scale invariance" ideas (§2) which underlie all the critical effects and suggest the scaling hypothesis[1] (§3) and length scaling approaches[5,6] (§4) needed to treat them, as well as relationships to fractal viewpoints (§5).

A large finite cluster of a site percolation model is shown[8] in Figure 4, generated by using $p \sim p_c$. The shape is extremely irregular, the surface showing undulations on many length scales. A cluster of the Witten-Sander aggregation model[9] is shown in Figure 5. Again the variety of scales is apparent. However, there is obviously a largest size to the clusters in each of Figures 4, 5, unlike that in the system of which Figures 2, 3 show parts. But, as $p \to p_c$ in the site percolation model, or as the growth process is continued for longer in the aggregation model, finite clusters of larger and larger size are probable, and in the limits all scales of length are present: the self-similar property of (random) fractals is occuring.

2. SCALE INVARIANCE AND CRITICALITY

In order to make the concept of scale invariance (or self-similarity) more precise, we return to the infinite cluster backbone (Fig. 3). Suppose we construct two figures (Fig. 6(a), 6(b)) by choosing at random equal sized parts from two different reductions of Figure 3, in each case sufficient that the lattice spacing is no longer resolvable. It is not possible to tell by looking at them which figure involves the largest

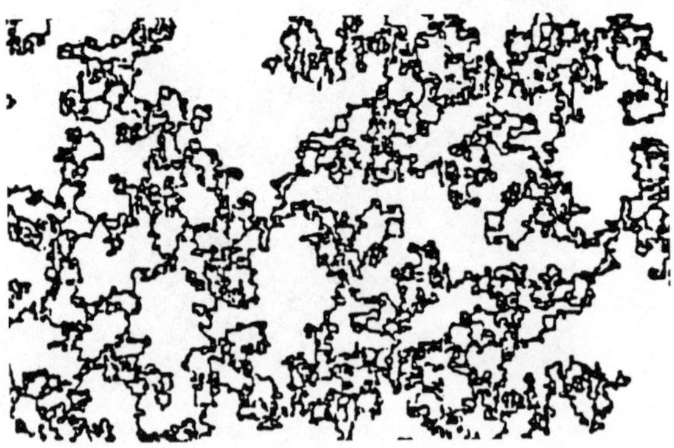

Fig. 3. Infinite percolation cluster backbone at p_c.

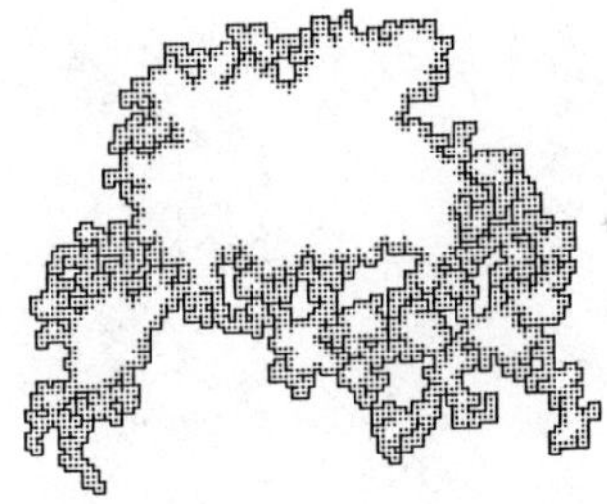

Fig. 4. Large cluster in site percolation.

reduction, since they have (statistically) similar appearances. This means that the original figure was scale invariant, or (equivalently) self similar (including parts very like itself on <u>any</u> larger scale). An alternative statement is that apart from the lattice spacing (a, say) there is no further length in the problem. Obviously, pure regular lattices have this property. But in general the configurations of Ising or percolation systems do not, since the up and down spin clusters of the Ising model or the finite clusters of the percolation model normally have a well defined average characteristic size. In such problems it is very useful to introduce a second length, the correlation length $\xi(T)$ or $\xi(p)$ to characterise the spatial extent of correlations and (in simple cases) other space dependent properties. In more complicated examples (e.g. a magnet in non-zero field) further independent characteristic lengths may be required.

The discussion, immediately above, of the scale invariance of the infinite cluster backbone, and the related discussion of finite clusters at the end of §1, suggest that the correlation length $\xi(p)$ for percolation diverges as $p \to p_c$. It does so in the following way[3]

$$\frac{\xi(p)}{a} = C|p - p_c|^{-\nu}$$

where C is a constant. A similar behaviour, with p, p_c replaced by T, T_c, and a different "critical exponent" ν, has been obtained[10] from Onsager's solution of the Ising model in two dimensions (d=2). The divergence of the correlation length in magnets at T_c is experimentally observable using neutron diffraction[11,12], as shown in Figures 7, 8. As will be discussed in §3, its divergence induces singular behaviour, e.g. in the free energy and its derivatives in the case of a thermal transition.

Fig. 5. Witten-Sander aggregation cluster

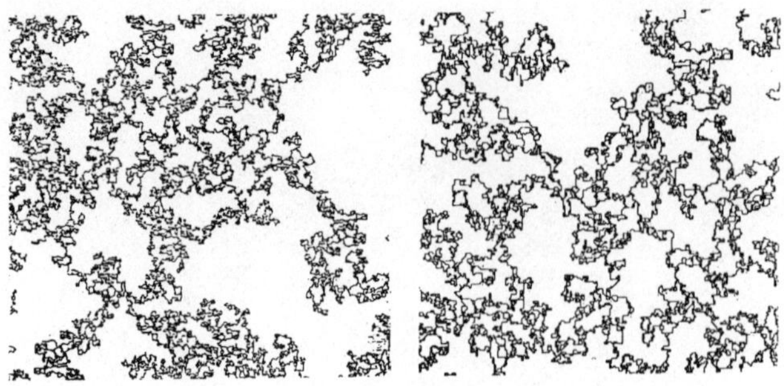

(a) (b)

Fig. 6. Scale invariance of the infinite cluster backbone of Fig. 3.

To relate self-similarity of the infinite percolation cluster backbone to fractals, we consider the following construction[4]. Take a square, divide into four equal squares, and discard one chosen at random. Repeat with the remaining three squares, and so on indefinitely. Figure 9 was obtained in this way, and is a random fractal (self-similar) because of its recursive generation. Its similarity to Figure 3 is quite striking.

The non-random fractal shown in Figure 10 (a triangular Sierpinsky gasket) is generated in a similar way by discarding the middle triangle of four into which an equilateral triangle is subdivided at each stage[7]. A fractal (i.e. self-similar) is strictly only produced by such recursive processes if carried on indefinitely. The length scales in Figure 9 range from the smallest ("lattice") spacing to the side-length L of the largest triangle, so it could be used to represent a system with correlation length $\xi = L$. Alternately, we could think of L as a length arising from finite size effects in a system with $\xi = \infty$.

Such finite size effects are crucial in discussing critical phenomena in small samples, and even more so for discussing walks or growth processes, since the number N of steps, or growth time, is one of the most important parameters for consideration. So, for a random walk of N steps, the end to end distance R, which is rather like a correlation range in other problems, depends on step length (a) and N (like $R \propto aN^{\frac{1}{2}}$ at large N). In an interacting walk model in which successive steps tend to be parallel yet another length enters, namely that up to which the walk looks more or less straight. The controlling or competing role played by such lengths is covered by a scaling hypothesis introduced in the next section.

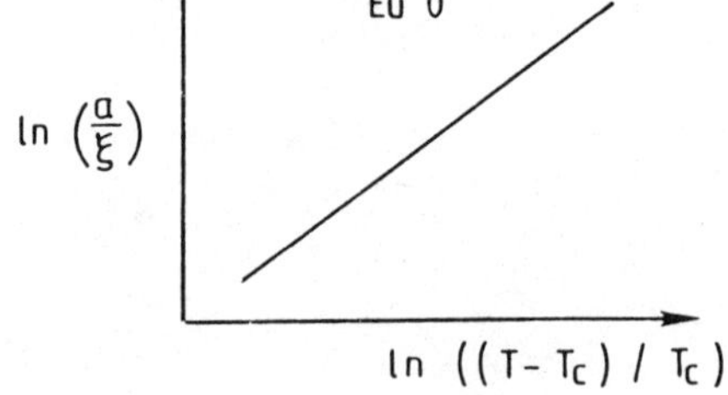

Fig. 7. Critical behaviour of correlation length of Eu 0.

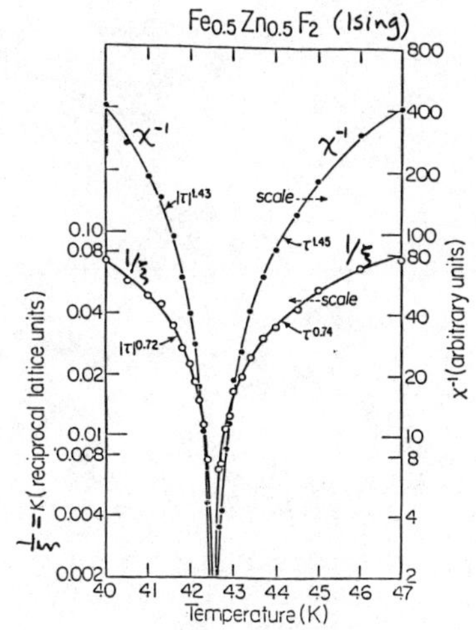

Fig. 8. Critical behaviour of inverse correlation length and inverse susceptibility of Ising magnet $Fe_{0.5}Zn_{0.5}F_2$

3. SCALING HYPOTHESIS AND CROSSOVER

To see how the diverging correlation length $\xi(T)$ causes critical behaviour in a thermal example, consider the free energy F in the combination F/k_BT occurring in the usual Boltzmann factors. This combination is dimensionless and extensive, and therefore (the singular part) is proportional to (Volume/$\xi(T)^d$), where d is the dimensionality, if $\xi(T)$ is the only relevant length for critical behaviour. Then, using the thermal equivalent of (1), with thermal exponent ν_T,

$$F/k_BT \propto [T - T_c]^{d\nu_T}. \tag{2}$$

Two derivatives with respect to temperature then produce the singular part of the specific heat in the form $[T - T_c]^{-\alpha}$ where

$$\alpha = 2 - d\nu_T. \tag{3}$$

(3) is the so-called hyper-scaling relation[1] between critical exponents.

This concept of ξ as a controlling length is not in itself enough. To extend an argument like that above to the critical behaviour of the magnetisation we would need to know its length-dependence. This is not so trivial as for the free energy because the net magnetisation is proportional to the difference between up and down spin regions. This difference is self similar at T_c (cf. Figure 1) and, like the percolation cluster of Figure 2, has a scaling dimension ("dimensionality") less than the space dimension. Without a knowledge of this "anomalous" dimension, which will be discussed in sections 4,5, the treatment is incomplete.

A second point is that ξ often occurs in competition with other long lengths. For example, the correlation function for spins separated by a large distance r in a zero field magnet near T_c depends on r and $\xi(T)$. The

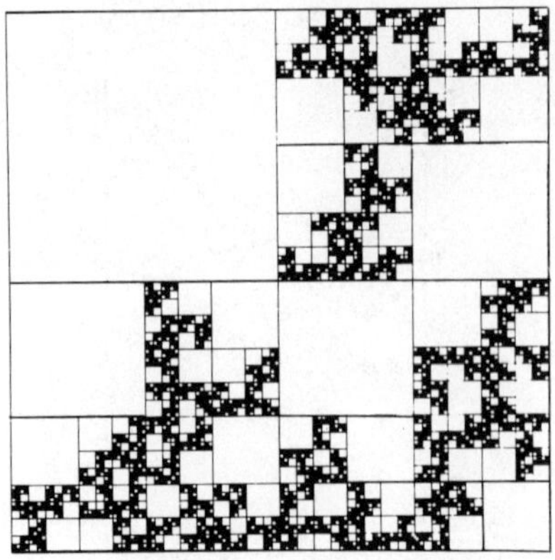

Fig. 9. Recursive generation of random self-similar structure.

scale of the r-dependence is set by ξ (by definition) and in addition there can be powers of r (or ξ) arising from sources like anomalous dimensions. In a similar way, the number n of finite clusters of s sites in percolation[3] might be expected to take the following (scaling) form for large s, ξ

$$n = s^{-\tau}\phi(s/\xi^{d_1}) \tag{4}$$

which involves the ratio of s to a typical cluster size. The exponent d_1 is a measure of the dimensionality of the clusters at p_c, and ξ controls the dependence of n on $(p - p_c)$.

Another example relates to the frequency ω of acoustic lattice vibrations of small wave vector k. On a lattice in which the only length is the lattice spacing a, a continuum picture is adequate for $ka \ll 1$, resulting in $\omega \propto k$. If the lattice is diluted to $p \sim p_c$ so that all lengths from a to a large correlation length ξ occur, the continuum result will break down for $k\xi \gg 1$, which is always achieved at any fixed small k as $p \to p_c$. According to a dynamic scaling hypothesis of the type introduced for thermal transitions[14], the frequency takes the form

$$\omega = k^z\ f(k\ ) \tag{5}$$

going over to the normal behaviour $\omega \propto k$ for $k\xi \ll 1$ and to an anomalous dependence $\omega \propto k^z$ for $k\xi \gg 1$ provided f(x) goes to x^{1-z} or to a constant for x small or large respectively. This is an example of crossover behaviour, and the relationship between the dependences on either side of

Fig. 10. Construction of a Sierpinsky gasket fractal.

the crossover actually implies mode softening (if the dynamic exponent z is greater than 1) since x^{1-z} contains a factor $(p - p_c)^{(z-1)\nu}$.

A scaling form sufficient to discuss the behaviour of the interacting walk model introduced at the end of §2 is (see §4), for large N, L,

$$\frac{R}{a} = N^{\frac{1}{2}}F(N/L) \tag{6}$$

where L is the length up to which the walk looks straight, and the other variables are as before. If $F(x) \propto x^{\frac{1}{2}}$ at small x, and $F(\infty)$ = constant, the form (6) describes the crossover from a straight walk for $N \lesssim L$ to a random walk for $N \gtrsim L$. Such scaling forms (e.g. (4), (5), (6)) suggest "universal plots" e.g. of ωk^{-z} versus $k\xi$, which reduce a dependence on two variables to a dependence on one.

A number of critical exponents have occurred in the above discussions. Such exponents, and also the scaling functions, are universal in the critical phenomena sense. They can be calculated by length scaling methods (next section) which also justify the scaling hypotheses used above and provide methods of dealing with scale-invariant situations (criticality, fractals).

4. LENGTH SCALING BY DECIMATION AND BY BLOCKING; PERCOLATION, SPIN SYSTEMS, AND WALKS; EXPONENTS

Length scaling typically involves exploring the consequences of changing some reference length in the system, while maintaining the properties on some gross scale.

The most common method is to increase the lattice constant, a, by some dilatation factor b, while maintaining the absolute correlation length ξ. For a percolation model p must then change, by virtue of (1) (similarly T has to change in a thermal example). Conversely, if we know the change of p required to maintain ξ while increasing a to a' = ba in a percolation model, we can hope to derive relationships like (1) (all our statements hold equally in percolation, or thermal, or other sorts of transition: and also walks if we for example replace a by step length, ξ by end to end distance and p by number of steps: and so on).

Another way of looking at such a length scaling is as a process which can be applied recursively until the lattice spacing has been increased to about the correlation length (assumed initially large but finite), at which stage the initial difficulties arising from the large range of length scales have been eliminated. This procedure actually implies successive changes in the parameter (p, or T) of the problem, moving it out of the critical region to a region which is simple to treat.

One particular value of the parameter which does not change under such length scalings is $p = p_c$ (or $T = T_c$), as can be seen from (1): if $p = p_c$, ξ is infinite and the scaling of a still leaves the left hand side infinite. This is clearly the self similar situation discussed in §§1,2, and is equivalent to the statement that the system does not notice the change in a if the only other length (ξ) with which to compare a has become infinite.

The length scaling process can be carried out in position space or in k-space. The k-space approach is used in the most controlled (field theoretic) methods[15] but we concentrate here on the simpler position space approach[3,16-19], which is most suitable for geometrically based problems and fractals.

The simplest method is decimation[17-19] and the simplest system on which to illustrate it is the one-dimensional bond percolation model[19] shown in Figure 11. Decimation consists of eliminating, say, every other site to arrive at a lattice with twice the lattice spacing ($a \to a' = ba$, $b = 2$), as shown in the figure. The bond concentration p is the same as the probability that nearest neighbour sites of the original lattice are connected. The probability, p', that nearest neighbour sites (e.g. 1,3) of the dilated lattice are connected is the same as the probability p^2 that those sites are connected in the original lattice, if we maintain the absolute correlation length. Thus as $a \to a' = 2a$, $\xi \to \xi' = \xi$,

$$p \to p' = p^2 \tag{7}$$

where the prime denotes quantities on the rescaled lattice. If we suppose that $\xi/a = f(p)$, (7) requires that $2f(p^2) = f(p)$, from which we obtain the exact result

$$\xi/a = f(p) = \frac{A}{\ln p} \tag{8}$$

where A could be any periodic function of $\ln\ln p$ with period $\ln b (= \ln 2)$. Since the scaling could have been done for any integer b (resulting in $p' = p^b$), A has to be a (negative) constant. It is more usual to use linearising approximation on scaling transformations of the type (7) to obtain results valid only in the critical region. We reserve this procedure for the next example, which is less trivial.

The application of decimation to bond percolation on the square lattice is illustrated in Figure 12, in which a), b) are specific configurations, for $p > p_c$, $p < p_c$, of a lattice whose scaled equivalents are shown in A), B). These scaled configurations were obtained by decimating half the sites to leave those on a sublattice with lattice spacing $b = \sqrt{2}$ greater than the spacing of the original lattice (and rotated 45° from it). The new (renormalised) bonds are thus diagonals of the square cells of the original lattice and a renormalised bond is drawn (present) if a path of bonds crosses the original square cell it replaced. Figure 13 shows the configurations of a square cell leading to a present renormalised bond along the diagonal. (All other configurations of the cell give "no bond"). The probability p' of such a bond being present is therefore given by summing the probabilities of the configurations in Figure 13, which yields

$$p' = 2p^2 - p^4 \equiv R(p) \qquad (b = \sqrt{2}) \; . \tag{9}$$

Though a simple approximate procedure was used to obtain the rescaled lattice configurations, it can be seen that they represent quite well all the gross features of the original configurations on an absolute scale: in particular, ξ is quite closely preserved.

The transformation (9) is exploited as follows.

(i) At the transition ($p = p_c$), $\xi = \infty$. So, as discussed above, the system is scale invariant and no change of the parameter p accompanies the length scaling (of a). This situation therefore arises at a "fixed point" of the transformation (9), i.e., a value p^* of p which transforms into itself:

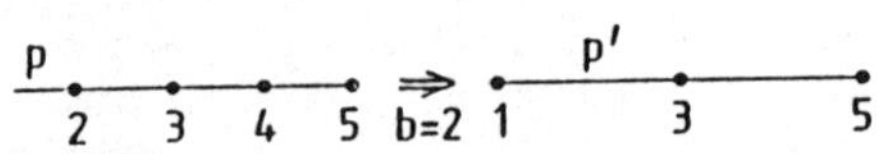

Fig. 11. Decimation of a linear chain.

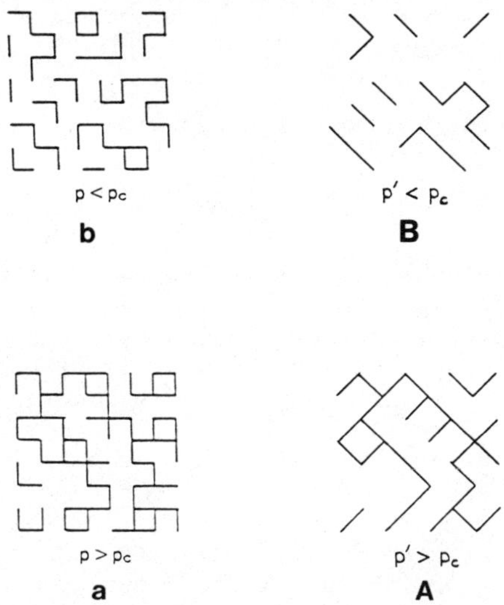

Fig. 12. Original and decimated configurations (a,A) of a dilute lattice above the threshold and below (b,B).

$$p^* = R(p^*) \ . \tag{10}$$

($p_c \equiv p^* \sim 0.618$). Locating the fixed point locates the transition.
(ii) Near the transition. Here $p \sim p_c$ so $\delta p \equiv p - p^*$ is small and so is its scaled value $\delta p'$. So the transformation (9) can be linearised in δp, $\delta p'$ to yield

$$\delta p' = \lambda \ \delta p \tag{11}$$

where $\lambda \equiv dR/dp$ at p^*. Now (1) is of the form $\xi/a = c\ \delta p^{-\nu}$, with the required exponent ν. The ratio of this equation to a corresponding equation for the scaled system is

$$\frac{\xi/a}{\xi'/a'} = \left(\frac{\delta p}{\delta p'}\right)^{-\nu} \ . \tag{12}$$

However, the scaling procedure (11) was such that $\xi' = \xi$, $a' = ba$ with $b = \surd 2$. So (12) becomes

$$b = \lambda^{\nu} \ . \tag{13}$$

Since the eigenvalue λ is known from (9) for the dilatation $b = \surd 2$, (13) yields the exponent ν ($= \ell n b / \ell n \lambda$).

This is the linearisation procedure referred to after (8), which gives exponents in terms of eigenvalues. Complete analytic solutions, such as (8), not limited to the critical region are rare, though numerical iterations of the transformation equations can give useful results not limited to that neighbourhood.

Generalised decimation techniques, such as the Migdal bond moving scheme, are available. But one of the great values of simple decimation is its direct applicability to fractals (§5).

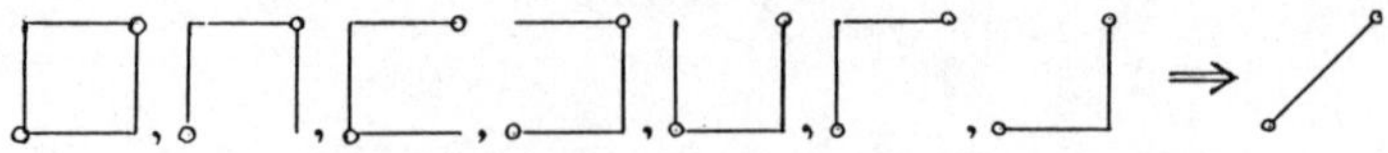

Fig. 13. Configurations of a square cell leading to a bond in the rescaled lattice.

Another very useful position space length scaling technique is blocking. This is an implementation of the original Kadanoff scaling idea[20], which again we illustrate for percolation, but in this case site percolation. Figure 14(a) shows a site-diluted triangular lattice on which the open circles denote vacant sites, and any two neighbouring occupied sites (black circles) are joined by a bond. The blocking procedure used in this illustration consists of grouping sites into threes indicated by the shaded triangles. A block then becomes a site of the rescaled lattice whose lattice spacing is $b = \sqrt{3}$ times bigger than originally. Such a "renormalised" site is taken as occupied if a majority (2 or 3) of the original three sites grouped to form the block are occupied. The resulting configuration on the rescaled lattice is shown in Figure 14(b). Again it represents the gross features of the original configuration on the same absolute scale, i.e., it preserves the correlation length. The probability that the renormalised site is occupied is

$$p' = p^3 + 3p^2 (1 - p) = 3p^2 - 2p^3 \equiv R(p). \quad (b = \sqrt{3}) . \tag{14}$$

The procedures used previously (steps (i), (ii)) may now be applied to this new transformation, yielding

(i) a fixed point at $p_c = \frac{1}{2}$ (the exact value)

(ii) a linearised equation with eigenvalue $\lambda = [dR/dp]_{p*} = 3/2$, leading to

$$\nu = \frac{\ell n \sqrt{3}}{\ell n 3/2} = 1.35 . \tag{15}$$

This is fortuitously close to the exact result, which is believed to be 4/3.

Both the decimation[19] and the blocking[21] method can be made much more reliable by the use of larger clusters and increased parameter spaces, to allow for correlations and further neighbour linkages which develop under scaling.

The treatment of a thermal problem by the scaling method is next considered. The philosophy is exactly as before, only the correlations

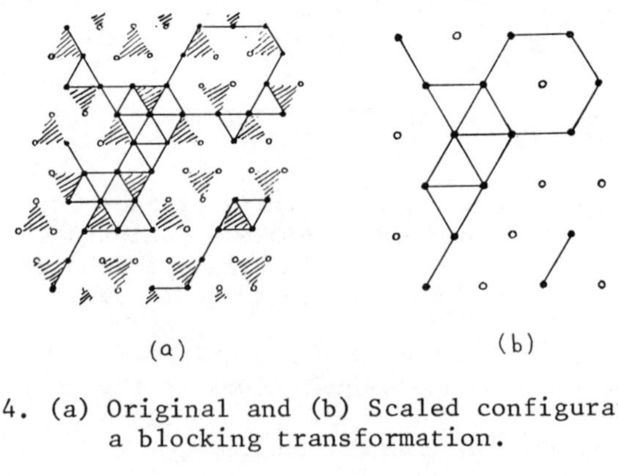

(a) (b)

Fig. 14. (a) Original and (b) Scaled configurations in a blocking transformation.

being preserved are now determined by the temperature parameter through Boltzmann weight factors and it is the temperature that transforms. The system to be treated is the Ising chain shown in Figure 15. Decimation or blocking can be applied; for simplicity we use decimation of alternate sites, i.e., dilatation by b = 2. In the original system, the probability of adjacent spins being parallel or antiparallel is $ce^{\pm K}$ where $K \equiv J/K_BT$ and c is a normalisation constant. Nearest neighbour spins of the rescaled lattice, e.g., spins 1, 3, are parallel with probability $c^2[c^{2K} + e^{-2K}]$, from the probability of the two events "123 all parallel" and "13 parallel with 2 antiparallel". Similarly 1, 3 are antiparallel with probability $2c^2$. If we again maintain correlations on an absolute scale, the ratio of these two probabilities is the ratio of the transformed values of $ce^{\pm K}$, which we denote by primed variables. Hence $e^{2K'} = \cosh 2K$ or, equivalently,

$$t' = t^2 \tag{16}$$

where $t \equiv \tanh K$. This (exact) transformation has only the fixed points $t^* = 0, 1$, corresponding to infinite temperature ($\xi = 0$) or zero temperature ($\xi = \infty$). There is therefore no finite temperature transition, as is well known for the Ising chain. ξ does however diverge as $T \to 0$, in a manner ($\xi \propto (1-t)^{-1}$) which can be obtained by linearising around the unstable fixed point $t^* = 1$. Alternatively, a complete result of the form (8) (with p replaced by t) can be obtained from (16) by the same methods used to discuss (7).

A very simple interacting walk model or linear "polymer" related to the system just discussed is shown in Figure 16. Each step of the walk can be right or left, and successive steps are in the same or opposite directions with probabilities proportional to $e^{\pm K}$. If steps to right or left are represented by spin up or down, this model maps into an Ising chain. If K is positive, there is a tendency for the equivalent spins to align , i.e., for the walk to be straight, for any sequence of up to $N_o \equiv \xi/a$ steps, where ξ is the correlation length (treated above) of the Ising chain and a the step length or lattice spacing. We consider now how this, and other properties of the (repulsive) walk arise in decimation. A (b=2) decimation again results in (16), if applied to the equivalent spin model (preserving ξ and doubling a). However, in order to describe the walk it is also necessary to consider the number N of steps (which is analogous to a finite size parameter in the Ising chain) and the mean square end to end distance R^2. The decimation groups two steps into one so N transforms according to

$$N' = N/2 \ . \tag{17}$$

At the same time the ratio $S \equiv R^2/a^2$ transforms approximately as

$$S' = 2S(1 + t) \tag{18}$$

(from averaging the contributions 4S, 0 from the parallel or antiparallel possibilities arising in the grouping of the two bonds). The scaling equations (16), (17) show that $N \ln 1/t$ (that is, N/N_o) is invariant under

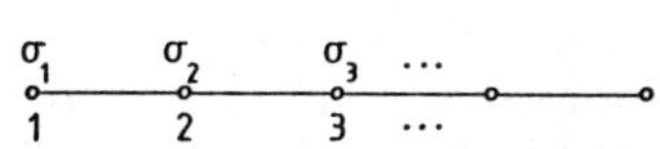

Fig. 15. Ising spin chain.

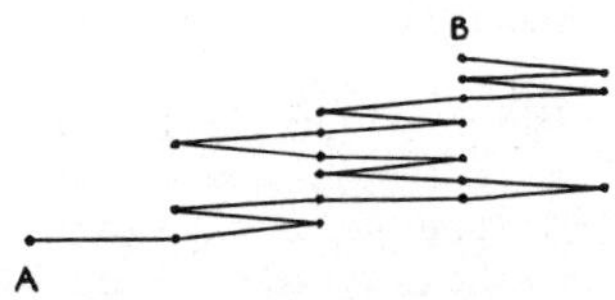

Fig. 16. A simple interacting walk model.

the scaling, so S must involve a (crossover) function of N/N_o. The procedures of §3 would suggest the form (cf. (6)).

$$S^{\frac{1}{2}} \equiv \frac{R}{a} = N^{\nu} f\left(\frac{N}{N_o}\right) . \tag{19}$$

Now, in the two limits $N_o \to 0, \infty$ (i.e. $t \to 0, 1$), (18) becomes $S' = \lambda S$, where $\lambda = 2, 4$ respectively. Comparison with (17) thus shows that $S \propto N$, and $S \propto N^2$ in the two limits $N_o \to 0$ and ∞. These are the random and (straight) self avoiding walk limits ($R \propto N^{\frac{1}{2}}$, $R \propto N$). The scaling form describes the crossover between these two limiting behaviours, with $\nu = \frac{1}{2}$ and $f(\infty) = 0$, $f(x) \sim x^{\frac{1}{2}}$ for small x. In a kinetic generalisation of such a model (N replaced by time) the crossover is between diffusion and ballistic motion.

The final example of this section is the Ising model on a square lattice. The same $b = \sqrt{2}$ decimation that was used to obtain (9) (i.e. renormalised bonds as diagonals of square cells of the original lattice) leads to the transformation[17,18]

$$t' = \frac{2t^2}{1+t^4} \qquad (b = \sqrt{2}) \tag{20}$$

where $t \equiv \tanh K \equiv \tanh (J/K_BT)$. In the equivalent form $t' = \tanh 2\overline{K}$, where $\tanh \overline{K} = t^2$, the result (20) can be seen to arise from adding two effective exchange couplings $\overline{K}$ coming from the two two-bond "chains" which join diagonally opposite sites on the square-cell cluster. For related reasons, the result (20) is the same as given by a simple series - parallel procedure in the Migdal-Kadanoff bond moving approach[22].

The relevant (unstable) fixed point of (12) is $t^* = 0.54$ (cf. the Onsager exact result $\sqrt{2} - 1$) and linearisation of (2) about this fixed point leads to an eigenvalue λ from which a (poor) approximation for the exponent ν can be arrived at by using (13) in the standard way.

It is hoped that the examples of this section demonstrate the main point: length scaling procedures can probe the divergent controlling length and deal with self-similarity. The resulting transformations of parameters yield critical conditions from fixed points and exponents from eigenvalues. Crossover is also available when more than one parameter is scaled. It should be remarked, however, that though the philosophy is always similar one does not need to proceed in the way followed above. If one has changed the length scale and maintained one thing by scaling another, one could have (at least in principle) scaled the first to maintain the other. Or one could dilate something (e.g. a time) other than a length. And so on.

5. RECURSIVE ASPECTS OF FRACTALS; DIMENSIONS

The purpose of this section is to show how the recursive nature of fractals allows them to be treated by decimation procedures which in a sense reverse the recursion generation of the fractal. Relationships between the dimensionalities of the fractals and critical exponents then become apparent.

Figure 17 shows the generation of a fractal (actually a Berker lattice[23]) by an iterative process. Such iterative processes are responsible for the self-similarity of fractals[5]. The middle and left of the diagram shows the replacement of a single bond by a diamond of original single bonds. This is equivalent to the approximate construction of the renormalised bond in the decimation of a square lattice by $b = \sqrt{2}$ (§4),

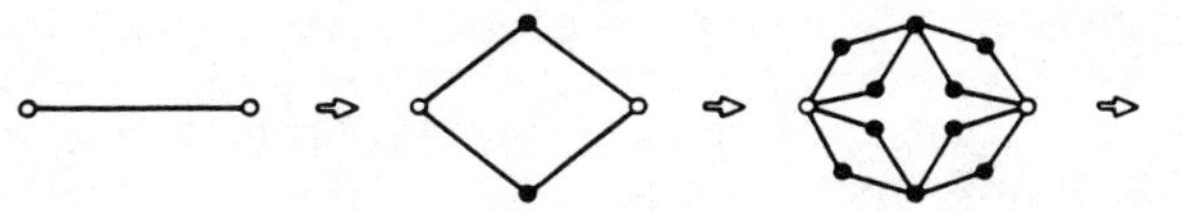

Fig. 17. Recursive construction of a Berker lattice fractal.

where the bond replaces the square cell of which it is a diagonal. So, that decimation procedure reverses the procedure used to construct this particular fractal, and in both cases the transformation is applied recursively. Thus decimation is an appropriate procedure for dealing with fractals, on which it is usually exact (e.g. results (9), (20) are exact for the fractal of Figure (17)).

Of course, various quantities may be transformed in the decimation on the fractal, with various results. This accounts for the many different dimensionalities which can be used to characterise different processes on a fractal. These are related to the exponents of the different properties which can become "critical" because of the self-similarity of the fractal.

Figure 18 shows two successive stages in the construction of a Sierpinsky gasket fractal. The arrow should be read to the left for the construction process, and to the right for a decimation reduction in which the vertices of the smallesttriangles are eliminated.

To illustrate how decimation or the construction process itself yields dimensionalities of fractals[24], consider for the Sierpinsky gasket of fixed side length L, the scaling of the "mass" M_n (number of vertices) remaining after discarding, at each of n successive steps, the central sub-triangle of each triangle divided into four. Clearly $M_n = 3 + \sum_{r=1}^{n-1} 3^r$; so, for large n,

$$M_{n+1} = 3\,M_n\ , \tag{21}$$

while the basic length a_n (side of smallest triangle) scales like

$$a_{n+1} = \tfrac{1}{2}\,a_n\ . \tag{22}$$

The Mass-length relationship is therefore (cf. (11), (12), (13))

$$M \propto \left(\frac{L}{a}\right)^{d_f}\ , \qquad d_f = \ell n\ 3/\ell n\ 2. \tag{23}$$

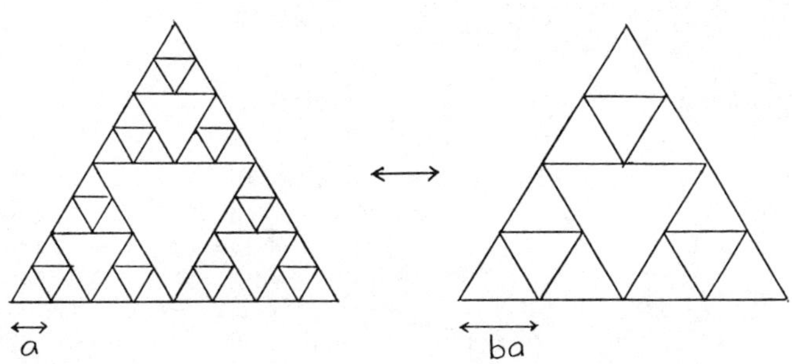

Fig. 18. Construction or decimation of a Sierpinsky gasket fractal.

Here, d_f is the "fractal dimension" replacing the usual Euclidean dimension d. In this case $1 < d_f < d = 2$. Thus, though the fractal (like the percolation infinite cluster at p_c and the difference between up and down spin regions in an Ising model at T_c) exists in a space of d dimensions, its dimensionality d_f is lower than d. The difference is an "anomalous" dimension like that introduced into the discussion of §3 to explain why the magnetisation scaling is non-trivial. (This anomalous dimension can be obtained directly for the fractal, from the scaling of the fraction of the volume remaining (i.e., $(3/4)^n$) at successive stages of the construction process).

The type of dimension d_f just discussed, thus in some way characterises the fractals occurring in critical phenomena, e.g., the infinite percolation cluster at p_c. In such a context its relationship to critical exponents can be discussed as follows. The critical exponents usually specify the singular dependences of properties on parameters like $p - p_c$ and these parameters do not occur in the fractal which, if self-similar, represents a situation occurring only at p_c. However, the dependence on $(p - p_c)$ can be replaced by a dependence on length using (1) (just as in finite size scaling[13]) and then a connection to the fractal dimensionality can be made. Specifically, consider the "Percolation" probability P(p) that an arbitrarily selected site lies on the infinite cluster. The exponent β is defined by the relation $P(p) \propto (p - p_c)^\beta$ which, in terms of ξ, becomes

$$P(p) \propto \xi^{-\beta/\nu} . \tag{24}$$

Since P(p) is essentially the mass density of the infinite cluster it scales with length like $L^{d_f - d}$, where d_f is defined as in (23), so we obtain the relationship[25] between the fractal dimensionality of the infinte cluster at p_c and percolation critical exponents:

$$d_f = d - \frac{\beta}{\nu} . \tag{25}$$

An alternative viewpoint is provided by introducing the characteristic length $\xi(p)$ into the "fractal" by terminating the recursive construction at some stage such that the ratio of largest to smallest side lengths in the "fractal" is ξ/a. This again results in relationships like (25).

Another dimension (of a fractal process) can be introduced by considering not the mass, but say the conductivity on a fractal lattice. For the Sierpinsky gasket, the scaling of the conductivity[25] is like (21), but with a different eigenvalue from 3. Thus its length-scaling exponent, i.e. dimensionality (cf. (23)), is different from d_f. If we denote this new dimensionality by d_σ an argument like that above relates d_σ to t/ν where t is the usual exponent (§6) for the percolation conductivity just above the threshold:

$$\sigma(p) = \sigma_o (p - p_c)^t . \tag{26}$$

Again, because of the conversion to a dependence on length through ξ, the relationship to the dimensionality involves a ratio of an exponent to ν. This feature is not always necessary: in some cases the exponent is defined as a power of a length, or inverse length. An example is the dynamic exponent z occuring in (5) or an exponent, related to it by the dimensionality d_f, for the density of states[26,27,28]. The exponent z can be calculated directly by decimation on the fractal[26,27,29].

It turns out[7,25,29] that the mass, conductivity, and dynamic processes on the triangular Sierpinsky gasket fractal have dimensionalities d_f, d_σ

and z reasonably close to those of the corresponding processes on the two dimensional infinite cluster backbone and the gasket fractal is often taken as a model for this. This should be done with great caution however, since an infinity of dimensionalities can be defined on a given fractal depending on what property the dimension characterises, and though a number of them may be shared with another fractal they need not all be. So for example, scaling of the Ising thermal parameter t on the Sierpinsky gasket is quite anamalous and does not lead to a behaviour, let alone a dimension, at all like that on the percolation backbone.

6. SCALING TREATMENT OF STATICS OF DILUTE SYSTEMS

To provide further selected examples of scaling, especially in the form developed in §4, a very brief account is given here of some applications to static properties of dilute networks and dilute magnets,[3,4,6,30].

(i) Dilute Networks

The conductivity $\sigma(p)$ of a diluted resistor network at concentration p vanishes at the threshold p_c like (26). Length scaling methods such as decimation can be used to obtain the behaviour of $\sigma(p)$ and in particular the exponent t. The method relies on scaling the distribution of the random conductance variable σ_i associated with an arbitrary bond i. This distribution is characterised by two parameters: p, the probability of the non-vanishing values of σ_i; and σ_o, the average of these non-zero values. The original way of carrying out the scaling[31] was to use decimation to transfrom p, σ_o in such a way as to maintain the correlation length ξ, and the average conductivity $\sigma(p)$ between two undecimated sites of the network. This results in a scaling of the form

$$p' = R(p) \tag{27}$$

$$\sigma_o' = S(p, \sigma_o) \ . \tag{28}$$

Linearising at the fixed point $(p^*, \sigma_o^*) = (p_c, 0)$ gives eigenvalues $\lambda = dR/dp$, $\Lambda = \delta S/\delta\sigma_o$ which determine the exponents ν,t (by use of (13) and an analogous form $\Lambda = \lambda^{-t}$ resulting from (26) by equating the original and scaled average conductances). A discussion of the scope of such methods is given at the end of the section.

(ii) Dilute magnets[30]

Dilute magnets may be treated by a generalisation of the method described above for dilute networks. Again a random variable occurs, in this case the exchange coupling J_i for an arbitrary bond. The combination J_i/kT (or, better, for the Ising case $t_i \equiv \tanh J_i/kT$) appears in the Boltzmann factors weighting the spin configurations and the decimation procedure, e.g. in the form of a partial trace over spins on a sublattice, leads to the scaling of the distribution of the variables t_i required to maintain the absolute correlation length $\xi(p,T)$. Extracting the scaling of the weight p and average size t of the non-zero values of t_i again leads to a scaling of the form (27), (28) but with σ_o replaced by t and a new function in place of S.

Two fixed points are now of interest: $(p^*, t^*) = (p_c, 1), (1, t_c)$. The first is the zero temperature percolation point, and the second is the pure transition point. (A new, "random" fixed point would not be located by a two variable description of the type being discussed). Under the renormalisation group transformation, the point (p, t) transforms along a "flow" line in the (p, t) plane through the initial point. All along the special flow line joining the two fixed points the correlation length $\xi(p, T)$ diverges, since all points on this flow line transform into one of the fixed points (the pure one). This line is therefore the critical line. The critical line calculated in such a way for a dilute Ising magnet is given in Figure 19, together with experimental results[32] for the transition temperature. This phase diagram has the usual interpretations (no order below p_c because of the absence of the infinite cluster, etc.). Percolation-thermal crossover occurs at the percolation fixed point, and is in this description characterised by unit crossover exponent, which is an exact result for dilute Ising systems.

In addition to the correlation length behaviour, and the critical condition (in this case a critical line) the scaling method can give all the static critical behaviour. To do this, in addition to p, t in the present case it is neccesary to scale a field variable and the free energy, which occurs through a normalisation constant in the Boltzmann distribution [5,30,33]. The results of such a calculation for a bond-diluted d = 2 Ising model[33] include the specific heat shown in Figure 20. For comparison, Figure 21 gives experimental results[34] for the specific heat of a particular (site-) diluted magnet.

We conclude by commenting very briefly on the nature of the approximations in these applications, first with relation to the dilute network. There, for a full discussion, the conductance distribution should really be allowed to achieve its invariant shape under scaling, as was done originally. This is especially important in extending such methods to more subtle problems like the dilute Ising random fixed point[35] or the spin glass[36,37], since e.g., approximating the evolving distribution in a spin glass scaling by a plus-minus distribution suppresses some of the fluctuations which prevent spin glass order for the d = 2 short range Ising spin glass. In addition, small cluster approximations involving few parameters and a neglect of correlations etc., are inadequate for the subtle problems, and make exponents inaccurate even in the simpler cases. However, for these simpler cases (dilute networks and most aspects of dilute magnetism, etc.) even the crudest scaling approaches are usually capable of providing a qualitatively correct description of most features.

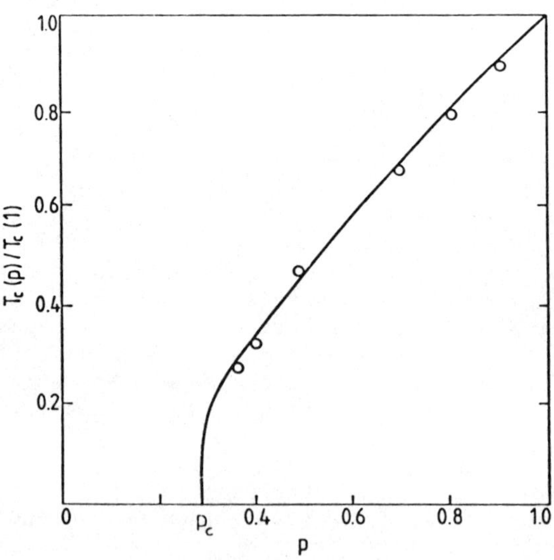

Fig. 19. Phase diagram of a diluted Ising magnet.

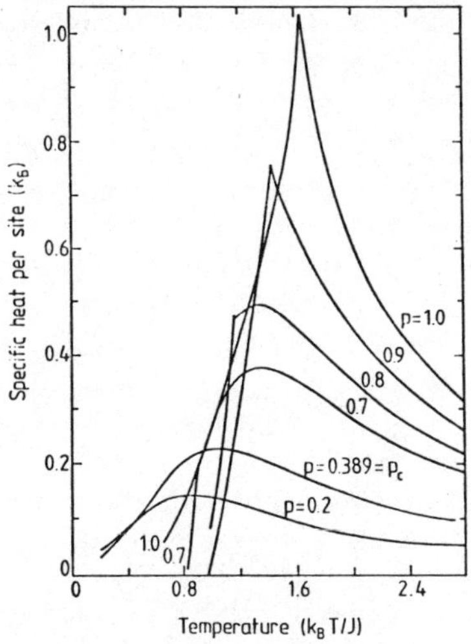

Fig. 20. Scaling results for specific heat of a dilute Ising model.

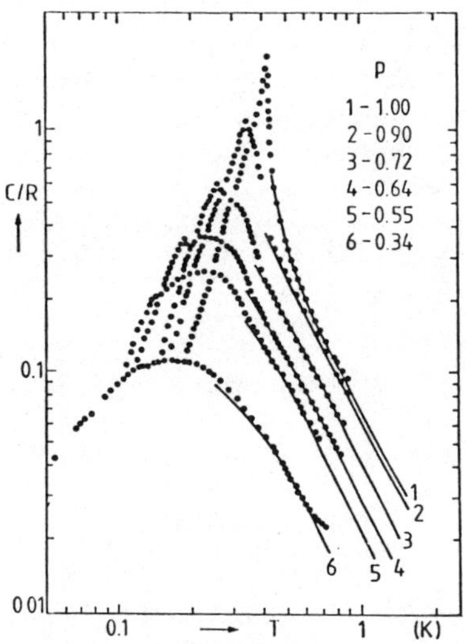

Fig. 21. Experimental results for specific heat of a dilute magnet.

REFERENCES

1. H. E. Stanley, "Introduction to Phase Transitions and Critical Phenomena", Oxford (1971).
2. S. R. Broadbent and J. M. Hammersley, Proc. Camb. Phil. Soc. 53: 629 (1957).
3. J. W. Essam, Rept. Prog. Phys. 43: 833 (1980).
4. S. Kirkpatrick, in: "Ill-Condensed Matter", ed., R. Balian, R. Maynard and G. Toulouse, North Holland (1979).
5. T. W. Burkhardt and J. M. J. van Leeuwen, "Real Space Renormalisation", Springer - Topics in Current Physics, Vol. 30 (1982).

6. R. B. Stinchcombe, in "Highlights of Condensed Matter Physics", ed., F. Bassani, F. Fumi and M. P. Tosi, Proceedings of the 1983 Varenna Summer School, North Holland (1985).
7. B. Mandelbrot, "Fractals", Freeman (1977).
8. P. L. Leath and G. R. Reich, J. Phys. C11: 4017 (1978).
9. T. A. Witten and L. M. Sander, Phys. Rev. B27: 5606 (1983).
10. T. T. Wu, Phys. Rev. 149: 380 (1966).
11. J. Als-Nielsen, O. W. Dietrich and L. Passell, Phys. Rev. B14: 4908 (1977).
12. R. J. Birgeneau, R. A. Cowley, G. Shirane, H. Yoshizawa, D. P. Belanger, A. R. King and V. Jaccarino, Phys. Rev. B27: 6747 (1983).
13. M. E. Fisher, in: "Proceedings of the Enrico Fermi International School of Physics, Varenna, Italy" ed., M. S. Green (Academic Press (1971).
14. B. I. Halperin and P. C. Hohenberg, Phys. Rev. 117: 952 (1969).
15. K. G. Wilson and J. Kogut, Phys. Rev. 12C: 75 (1974).
16. Th. Niemeyer and J. M. J. van Leeuwen, in "Phase Transitions and Critical Phenomena", ed., C. Domb and M. S. Green, Academic Press, Vol. 6: (1976).
17. M. N. Barber, J. Phys. C8: L203 (1975).
18. L. P. Kadanoff and A. Houghton, Phys. Rev. B11: 377 (1975).
19. A. P. Young and R. B. Stinchcombe, J. Phys. C8: L535 (1975); C9: L643 (1976).
20. L. P. Kadanoff, Physics 2: 263 (1966).
21. P. J. Reynolds, H. E. Stanley and W. Klein, J. Phys. A10: L203 (1977); J. Phys. A11: L199 (1978).
22. A. A. Migdal, Sov. Phys. JETP 42: 743 (1976); L. P. Kadanoff, Ann. Phys., NY, 100: 559 (1976).
23. A. N. Berker and S. Ostlund, J. Phys. C12: 4961 (1979).
24. See also R. Voss, Geilo lecture.
25. Y. Gefen, A. Aharony, B. Mandelbrot and S. Kirkpatrick, in "Disordered systems and Localisation", ed., C. Castellani, C. di Castro and L. Peliti; Springer (1981).
26. S. Alexander and R. Orbach, J. Phys. Lett., Paris, 43: 625 (1982).
27. R. Rammal and G. Toulouse, J. Phys. Lett., Paris, 44: L13 (1983).
28. S. J. Lewis and R. B. Stinchcombe, Phys. Rev. Lett., 52: 1021 (1984).
29. C. K. Harris and R. B. Stinchcombe, Phys. Rev. Lett. 50: 1399 (1983).
30. R. B. Stinchcombe, "Dilute Magnetism", in: "Phase Transitions an Critical Phenomena", ed., C. Domb and J. L. Lebowitz, Academic Press (1983).
31. R. B. Stinchcombe and B. P. Watson, J. Phys., C9: 3221 (1976).
32. E. Lagendijk and W. J. Huiskamp, Physica 62: 444 (1972).
33. C. Jayaprakash, E. F. Riedel and M. Wortis, Phys. Rev. B18: 2244 (1978).
34. H. A. Algra, L. J. de Jongh, W. J. Huiskamp and J. Reedijk, Physica 86-88B: 737 (1977).
35. D. Andelman and A. N. Berker, Phys. Rev. B29: 2630 (1984).
36. A. P. Young and R. B. Stinchcombe, J. Phys. C9: 4419 (1976).
37. A. J. Bray and M. A. Moore, J. Phys. C17: L463 (1984), and to be published.

GROWTH BY PARTICLE AGGREGATION

Leonard M. Sander

Physics Department
University of Michigan
Ann Arbor, MI

INTRODUCTION

The phenomenon of disorderly nonreversible growth is very common in nature: dust and ash, coral, trees and cities grow into diverse forms far from equilibrium. In many cases a kind of a vague similarity seems to characterize the overall shape of these structures. This review is concerned with some recent attempts to give a description of such growth via simple models. We will present some glimpses of unifying features in their morphology.

The particular class of processes we are interested in is the following: Consider the formation of a large aggregated structure by the repeated addition of identical subunits, each obeying the same dynamics. We will ask the question of whether we can predict anything useful about the aggregate when it has grown to be very large compared to the individual constituents. For definiteness the reader can think of the growth of, say, a chemical precipitate or an electrolytic deposit from solution. As the precipitate or deposit grows it can sometimes take on a sprawling, tenuous shape. This sort of growth is our subject here.

In fact, in a rough way, we can classify processes of growth by aggregation into three regimes. If the growing object is always near equilibrium despite the addition of new material then its internal structure and external shape will be describable by ordinary equilibrium considerations. In our examples, we may expect to find single crystals with the well-known equilibrium crystal shape.

As we drive the system away from equilibrium we often find a new regime of morphology, that of pattern formation. In this case, new length scales associated with the steady-state growth give rise to intricate patterns which can persist even for large objects. Examples of these are the beautiful, feathery, dendritic shapes of snowflake growth (Langer, 1980, Ben-Jacob et al., 1983, Brower et al., 1983).

Even farther from equilibrium a whole new regime can appear--that of disorderly growth with no length scales at all. The remarkable fact about this regime is that in some cases it is not merely amorphous growth (though that can happen, as we will see) but growth of scale-invariant fractals (see Mandelbrot, 1982). The fact that this can occur was first demonstrated experimentally for aggregated metal smoke particles (Forrest and Witten, 1979) and later for electrolytic deposition (Brady and Ball, 1984, Matshushita et al., 1984). Theoretical study was initiated by Witten and Sander (1981).

It will be useful for the rest of these lectures to set down some ways in which a scale-invariant fractal can be characterized. The simplest is to note that from the definition of a fractal the number of subunits (particles), N, within radius, R, of a point is given by

$$N \sim R^D \tag{1}$$

with D the fractal dimension. For a growth process we can take R to be some characteristic size of the object (like its radius of gyration) and interpret Eq. (1) as a law relating mass to size in various stages of growth. Note that if the object is scale-invariant, all characteristic sizes are proportional. For example, Figure 1 is an example of a fractal growth produced in a way we will discuss shortly. In this case D is 1.7; See Figure 2. There is another, more local, characterization. Consider the two-point correlation function:

$$c(x) = \langle \rho(\vec{r}+x)\rho(\vec{r})\rangle \ , \tag{2}$$

where the angular brackets indicate the average over the object. Now given a particle at $\vec{r}$, the density at a distance x away is (from Eq. (1)):

$$(dN(x)/dx)\ x^{-(d-1)} \sim x^{D-d} \ . \tag{3}$$

Here d is the dimension of space. This quality is exactly c(x). We expect a power-law behavior of the correlations in this case. This is quite different from the case of crystalline or amorphous matter where c(x) approaches a constant for distances much larger than a correlation length.

Finally, we should note that if we consider growth from a nucleation center, then the spherically-averaged density profile itself will exhibit a power-law behavior,

$$\rho(r) \sim 1/r^{d-D} \ . \tag{4}$$

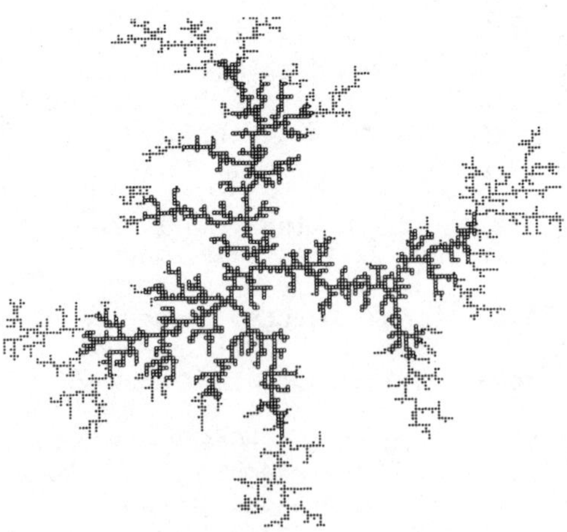

Figure 1

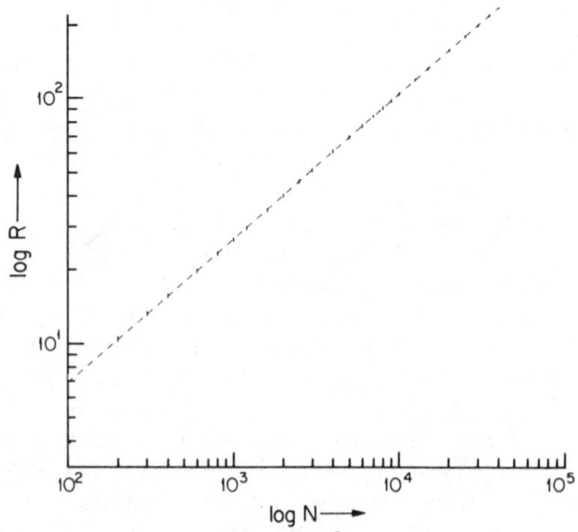

Figure 2

Since the fractal dimension is usually less than the dimension of space, these objects are tenuous and wispy: they do not fill a finite fraction of space.

DIFFUSION-LIMITED AGGREGATION

The first explicit demonstration of a process which exhibits disorderly irreversible growth together with scaling was the work of Witten and Sander (1981). This model, called diffusion-limited aggregation (DLA), has received a good deal of attention; it is probably the best understood member of the class of particle aggregation models that we will discuss.

The intention of the authors was to invent a simple, physically motivated model which could describe growth in a tractable way. The process proposed was as follows: Consider growth from a nucleation site. Assume a single particle approaches performing a free random walk from a distant point. When it comes into contact with the nucleation center (within a distance, a, say, which may be thought of as the particle diameter) it sticks. Then another random walker is released and can stick to the center or to the previous particle. After many such cycles a pattern, such as that of Figure 1, is formed. As first pointed by Brady and Ball (1984) electrodeposition may be thought of in this way; if the random walkers (the ions in solution) are at low enough concentrations they may be considered to arrive one at a time. The situation at larger concentration will be discussed below.

Simulations using this model gave the rather startling result that patterns like Figure 1 are fractal (as in Eq. (1)) and have power-law correlations. The results were extended and refined by Meakin (1983a) and by Witten and Sander (1983). Meakin's work was particularly extensive. He showed that for d=2 to 6, D ranges from about 1.7 to 5.1. In three dimensions, D=2.4, which agrees quite well with experiments of Brady and Ball. R.C. Ball will discuss these experiments in more detail later in this conference.

These results are worth detailed discussion. One's first reaction on seeing a pattern like Figure 1 is to be puzzled at the large holes in the aggregate. In fact, if the object is a fractal it must have holes on the order of the size of the object itself. But how do these arise?

Qualitatively, the holes arise from a sort of screening. Each random walker is prevented from penetrating into the interior by the extended arms already present. The walker cannot easily penetrate the long fiords. The quantitative discussion of this screening effect has led to considerable controversy. We believe that the tentative resolution, which we will now present, illustrates a subtle aspect of the scale-invariance of DLA aggregates.

A first step towards understanding the screening is to note that the probability distribution of the random walkers, $u(\vec{r},t)$, obeys, in the continuum limit:

$$\partial u/\partial t = B\nabla^2 u \ , \tag{5a}$$

where B is a diffusion constant. The boundary conditions are:

$$u = u_\infty \ , \ r \to \infty \ , \tag{5b}$$

$$u = 0, \ \text{surface} \ . \tag{5c}$$

The last condition is the usual absorbing boundary condition. As $u_\infty \to 0$ (few walkers arriving at once) the time scale for growth lengthens. In this limit the random walkers see practically a steady state. Thus we can set:

$$\nabla^2 u = 0 , \tag{6}$$

outside the aggregate.

Now we can estimate the penetration depth, ξ, of a random walker into an aggregate by noting that the absorbing boundary condition, (5c), can be incorporated into Eq. (6) by setting

$$\nabla^2 u \approx Z\rho u , \tag{7}$$

where ρ is the spherical average of the density, as above and Z is a constant. Clearly, from Eq. (3) we have:

$$\xi^{-2} = \nabla^2 u/u = 1/R^{d-D} ,$$

$$\xi \sim R^p \tag{8a}$$

$$p = (d-D)/2 . \tag{8b}$$

This equation undoubtedly underestimates ξ because it neglects the anticorrelations of u and the density. Certainly it does imply that if $(d-D)/2>1$, so that ξ grows faster than R, the aggregate would be transparent. However, this never happens for DLA.

Taken literally, Eq. (8) is quite misleading. It would imply that ξ is an independent diverging length scale in DLA, i.e., that DLA is not scale-invariant. Further, the exponent, p, in Eq. (8b) is not at all near 1 ($p \simeq 0.15$ for $d=2$) so that the violations of scaling with a single length would be very large. Plischke and Racz (1984) and Sander (1984) both pointed out the error by examination of actual simulation results. The question may be stated in several ways: since a DLA aggregate grows whenever a random walker is captured, the length, ξ, also characterizes the growing part of the density profile. We ask, then, if the density profile scales in a way beyond what is demanded by Eq. (3): does the

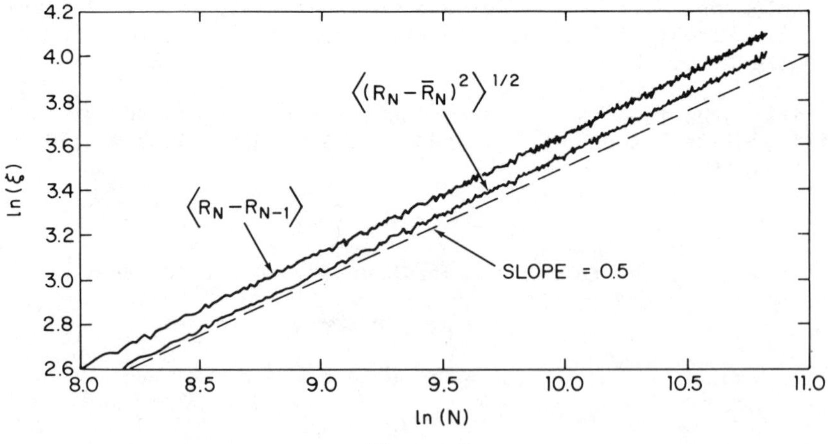

Figure 3

interface grow more slowly than R, so that ξ becomes very small compared to R as R increases, (as Eq. (8) implies) or does the interface itself scale with R.

On the basis of preliminary data which showed that, in any case, p is much larger than 0.15 for d=2, the present author (Sander, 1984) speculated that the entire profile should scale. It is hard to see how DLA could grow fractals otherwise: holes on the scale of R are formed, and random walkers should penetrate some fraction of that distance. That is, p should be 1. However, Plischke and Racz (1984) showed that for clusters up to about 2500 particles ξ appeared to grow slightly more slowly than R:

$$R \sim N^{\nu} \tag{9a}$$

$$\xi \sim N^{\overline{\nu}} \tag{9b}$$

(Of course, $\nu = 1/D$). For the small clusters of Plischke and Racz, $\overline{\nu}<\nu$. In fact, they found $\overline{\nu}/\nu=p \simeq 0.83$, violating scaling.

Meakin and Sander (1985) have reexamined this situation and have, we believe, resolved it. We have grown very large aggregates (N = 50,000) and carefully analyzed the behavior of $\xi(N)$, which we interpret as the variance of the sticking site of the Nth particle. A sample result is shown in Fig. 3. Note that the slope of the log-log plot changes with N. We find that the effective exponent p for N in the range 25,000 to 50,000 is 0.93 ± 0.1. There is now every reason to think that the asymptotic limit of p is 1.

A technical point should be added here: our large clusters were grown off-lattice. Thus any anisotropy induced by the lattice is irrelevant to the present discussion.

VARIATIONS ON THE THEME

In the previous section we demonstrated that there is at least one non-trivial growth model with interesting scaling properties. However, though the model itself is very simple, this sort of physics is sufficiently unfamiliar that it is far from clear what is essential and what incidental in the formulation. We can ask various questions: Is it necessary that the particles diffuse? What is the role of the particle diameter, a? What if we change the sticking rule? What if we relax the restriction of a single nucleation center? And, fundamentally, can we understand properties like D, which characterize long-range scaling properties, without doing a simulation every time. The answer, or rather the lack of one to the last question has determined our approach to the others. There is, as yet, no satisfactory theoretical understanding of DLA and similar processes. The community has been led to do an essentially experimental search (via numerical and real experiments) for the answers to the first questions. A limited body of theoretical understanding has emerged in the course of this process. We will return to a tentative picture, in the last section, below. It is useful to present one tool that the author has found useful in such a search, namely the analysis of growth instabilities.

Stability Analysis

A basic problem in looking at different aggregation processis is in a kind of classification: Can we say, without doing a simulation, which processes are likely to yield scale-invariant growth, and which are not. The present author (Sander, 1984) has proposed a partial answer which serves as a rough guide.

Qualitatively, we want to know when particle aggregation will not give rise to compact objects, at least. Suppose we start with an initially smooth surface. The arrival of a particle can be thought of as a small perturbation. If many more particles now arrive, we can ask whether the perturbation will be amplified or suppressed. In the latter case, certainly we have compact growth.

However, we need more: the perturbations which grow must grow on all length scales up to the size of the object in order to give rise to a fractal. Tentatively, then, we identify scale-invariant growth as having instabilities on all length scales. In what follows we will illustrate the usefulness of this idea.

Variants of DLA

It is easy to apply the instability analysis to DLA, because it amounts to a repetition of the standard treatment of Mullins and Sekerka (1963). We can see this from a continuum treatment. Let us adjoin to Eq. (6), which governs u outside the aggregate, a rule for the average growth. Let v be the velocity of the surface. On the average the surface moves with velocity proportional to the incident flux of walkers:

$$v \sim \hat{n} \cdot \nabla u|_S \quad . \tag{10}$$

This equation, together with Eqs. (5b), (5c) and (6), form a deterministic system to be solved for the growth which does exhibit an instability, as Mullins and Sekerka (1963) first showed. In fact, for a two-dimensional system (Witten and Sander, 1983) if the initial shape is slightly perturbed from being round:

$$r = R + \delta \cos m\theta \quad . \tag{11}$$

then,

$$d\delta/dt = A\delta(m-1)/R^2 \tag{12}$$

where A is a constant. Thus, for any m larger than one, the perturbation grows. Now the scale of the perturbation is R/m, so any long-wavelength perturbation grows implying the possibility of fractal growth.

We now turn to variants of DLA. We will try to use stability analysis as a guide.

The first such variant to be investigated was the role of the sticking rule (Meakin, 1983, Witten and Sander, 1983). In fact, it is clear that in a real situation we expect particles to bounce off instead of sticking. Further, they should stick more easily where there are many neighbors, and less easily on a tip. That is, partial sticking conteracts the effect of screening which makes tips grow more easily than fiords.

This effect is even more interesting because it is clearly related to surface tension: what we have given is a microscopic description of surface tension. Now we can apply a stability analysis again. Surface tension changes the boundary condition on the surface:

$$u \sim \kappa \tag{13}$$

Here κ is the curvature. But in this case, the analysis above can be repeated with the well known result that only perturbations whose wavelength is much longer than a characteristic length (the capillary length) are unstable.

However, it is the long-wavelength perturbations which are of interest in the scaling regime. Thus, we expect fractal growth for large enough aggregates.

The result of a simulation with modified sticking is shown here (Fig. 4). The branches of the aggregate have thickened. However, when the object grows large enough, it becomes a fractal again, and D is unchanged (Meakin, 1983a, Witten and Sander, 1983). Another version of surface tension has recently been investigated by Vicsek (1984). In all these cases the qualitative effect of surface tension is to replace the thickness of the arms (essentially the particle diameter in the original model) with a larger length, the capillary length.

There is another case in which we can apply known results to get an appreciation of the nature of the growth. Suppose we start the system with a finite density of walkers a mean distance, L, apart. If we let them walk all at once, what is the nature of the growth? The result of simulations (Witten and Meakin, 1983, Voss, 1984) is that a compact, amorphous object grows as soon as the size exceeds L. Qualitatively, the density cannot fall below L^{-d}.

We can see this result in terms of stability analysis. Consider the diffusion equation, again, and suppose that velocity of growth of a tip is v. Then,

$$\nabla^2 u = \frac{1}{B}\frac{\partial u}{\partial t} \sim \frac{v}{B}\frac{\partial u}{\partial r} \quad . \tag{14}$$

The characteristic length $l = v/B$ is called the diffusion length. When particles arrive one at a time, v and also l are small, so we can neglect $\partial u/\partial t$, as above. But when we have a finite density of walkers, v is finite. In this case l has the significance of a boundary layer thickness over which the diffusion field feels the effect of the boundary. Now it is known (Langer, 1980) that if the scale of the perturbation greatly exceeds l the instability goes away. This is easy to understand: points on the interface farther than l cannot communicate, so that a hole larger than l cannot grow. This analysis was done in more detail by Nauenberg, Richter and Sander, (1983).

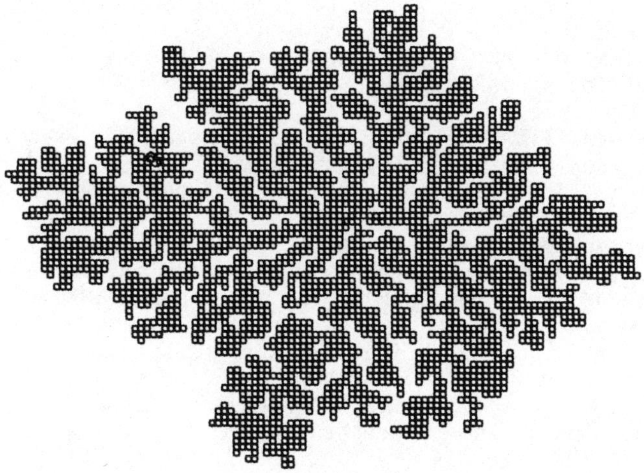

Figure 4

Dielectric Breakdown Models

As we pointed out above Eq. (5b) (5c) and (6) govern the probability distribution of random walkers. Thus if these equations are solved directly and a particle added to the aggregate with probability given by Eq. (10) we should recover DLA. This was done by Niemeyer, et al. (1984). They did, in fact, produce aggregates whose fractal dimension is close to ordinary DLA simulations. However, these workers went further and pointed out that the equations should describe dielectric breakdown. We will return to this subject, below. They were motivated in this way to introduce non-linear effects, namely to have a growth probability that replaces Eq. (10) with

$$(n\cdot\nabla u|_s)^\eta \tag{15}$$

When the growth exponent, η, is unity, DLA is recovered. the remarkable fact about these models is that D seems to depend continuously on η, and becomes equal to d (compact growth) for $\eta=0$. The last result can be produced from a stability analysis, as we will see shortly. For $\eta\neq0$, there is a Mullins-Sekerka instability.

Compact Aggregation: Ballistic Aggregation and the Eden Model

We now ask the question of what happens when the particles added do not diffuse. For example, suppose we simply add particles at random to perimeter sites. This might model the growth of a cell culture, for example. It was posed in this context by Eden (1961). Extensive numerical work (Peters, et al., 1979, Meakin, 1983b) has shown by simulation that this is compact aggregation, D=d. To our knowledge, even for this very simple case, there is no convincing proof of this statement. Stability analysis does indicate the result, however. Suppose we consider a disk with a small perturbation (a bump) and let it grow according to the Eden rules. The shape will have a uniform "skin" of new particles. Thus the bump does not grow (or shrink)--it simply remains constant. Of course, the relative size of the bump compared with R decreases in time. There is certainly no instability.

In a similar way, suppose particles do not diffuse but "rain" on a seed. That is, they move in straight lines. This is called ballistic aggregation. It is believed, on the basis of numerical results that D=d also for this case (Meakin, 1983b). And this is one of the few results that can be understood analytically in terms of the so-called causality bound (Ball and Witten, 1984a). We will return to this point below.

However, a stability analysis also works here. In exactly the same way as the Eden model, this model acquires a skin of new particles on top of a perturbation. Thus small perturbations do not grow. However, a sufficiently large perturbation will cast a geometric shadow. Thus, this sort of aggregation is compact but has a non-linear instability which gives rise to a fascinating pattern-formation problem. Some of the patterns have been sorted out by Ramanlal and Sander (1985).

Meakin has investigated various kinds of cross-overs between ballistic and diffusion-limited aggregation. In one case (Meakin, 1983b), drift was superimposed on diffusion. The objects formed were DLA-like for small sizes, but compact as R increased. More intriguing is a simulation using Levy flight trajectories (Meakin, 1984). A Levy flight is a random walk with variable step lengths, x. The probability, P, of a given step length is given by

$$P(x \geqslant u) = u^{-f} \tag{16a}$$

$$P(x<1) = 0 \quad . \tag{16b}$$

In this case D seems to depend on f. For $f \to 0$ (straight lines) the result D=d is recovered.

Cluster-Cluster Aggregation

The sort of clusters which have been studied the longest is undoubtedly colloids and related objects like coagulated aerosols (see Friedlander, 1977). The DLA model was in fact invented to study the morphology of a cluster of this type. However, it did not successfully do so. The strongest evidence for this is probably D itself. For d=3, DLA has a fractal dimension of about 2.4. But experiments (Forrest and Witten, 1979, Weitz and Olivera, 1984) give a much smaller value, $D \simeq 1.8$.

The resolution of this difficulty is due to Meakin (1983c) and Kolb, et al (1983) who showed that what is left out of DLA is the aggregation of aggregates. That is, in an atmosphere of coagulating particles we expect dimers to form and then move, so that dimers stick to dimers and trimers, and so on. The dominant combinations always seems to be that of clusters of comparable sizes.

Quite a lot is now known about this model. For example, the morphology of cluster-cluster aggregates is far more open than that of DLA. This is clear when one considers that a large cluster is far less likely to penetrate the openings in another than a single particle is. For example, for d=3, $D \simeq 1.8$, in agreement with experiments.

The kinetics of this process has been studied using the Smoluchowski equation (Botet and Jullien, 1984). It is found that this kinetics fits simulation results very well if the kernel in the equation is assumed to depend on the fractal dimension.

All of these matters could be elaborated at great length. However, they form a somewhat separate subject from particle aggregation which is the focus of these lectures.

EXPERIMENTS

The models described above are so simple that one is entitled to wonder whether they are merely computer games rather than descriptions of physical reality. For the case of cluster-cluster aggregation such doubts are easily laid to rest because of many years of experience with well-characterized colloidal and aerosol systems whose kinetics have received a very great deal of attention. The new feature in recent years has been the study of the relationship of kinetics to morphology.

For particle aggregation the situation is not so simple, thus it has been particularly exciting that DLA-like models have been successfully used for a number of physical systems that appear very different from each other. By comparing these experiments we can also try to gain insights into what is essentially necessary for fractal growth.

The experiments we will discuss fall in two categories. There are those which are more or less literal realizations of the model, such as electrodeposition. In others the relationship is more formal. In each case we will try to identify what we believe to be the essential ingredients: a diffusing field feeding unstable growth, a short-range cutoff, and the lack of a long-range cutoff.

Electrolytic Deposition

As we have already mentioned, when ions are deposited on an electrode their motion can be arranged in such a way that it is essentially diffusive. In that case we expect the deposit to be DLA-like (Brady and Ball, 1984, Matsushita, et al., 1984). For a polycrystalline deposit, the short-range cutoff can be considered to be the crystallite size, which seems to be more or less uniform. The maximum size for a fractal deposit is given by the diffusion length which can be fairly large. In a uniform drift due to convection is set up, the deposit should be compact.

Dielectric Breakdown

Niemeyer, et al., (1984) and Witten and Sander (1983) pointed out how dielectric breakdown might be represented by a DLA model. The reasoning is rather simple. Suppose we have a potential, ϕ, in a dielectric, and a part of the breakdown already formed. Then

$$\nabla^2\phi = 0 \; , \tag{17a}$$

$$\phi|_s = 0 \; . \tag{17b}$$

The last equation is a statement that the breakdown channel is a good conductor. The growth law is modelled by assuming that the probability of growth of a tip is proportional to $E = -\nabla\phi$, the electric field, i.e., Eq. (10). (Non-linearities were discussed above).

The measurements of Niemeyer, et al. on two-dimensional discharges (Lichtenberg figures) find D=1.7, as expected. The short-range cutoff is probably some characteristic length of channel in the material. In a good dielectric there should be no long-range cutoff.

Viscous Fingering

The previous two systems had a natural "graininess"--they seemed to be very naturally describably by a particle aggregation model. We now consider two systems which are not of this nature. It is a remarkable fact that they both may also show fractal growth.

Paterson (1984) was the first to point out the relationship between the classic effect of Saffman-Taylor instabilities and DLA. The reasoning is as follows (see also the lectures of Stanley in this conference.)

Consider a fluid of small viscosity pumped into a viscous fluid. For the first, the pressure, P_1, is more or less constant. For the second, in a porous medium, for example, or in motion between closely spaced plates (Hele-Shaw geometry):

$$\vec{v} = -Q\nabla P \tag{18a}$$

where Q involves the viscosity, and P is the pressure in the second fluid. Then if we set $u=P-P_1$,

$$0 = \nabla\cdot v \sim \nabla^2 u \; , \tag{18b}$$

$$u|_s = P-P|_s = -d_o\kappa \; . \tag{18c}$$

$$u_\infty = P_\infty - P_1 \tag{18d}$$

The second condition is the boundary condition at the surface with d_0 the capillary length. These equations are exactly the ones we have been discussing.

Now as Nittmann, et al. (1985) showed (see elsewhere in this conference) a DLA-like pattern can be produced in the Hele-Shaw geometry if the capillary length is made small. This is a very significant result because it seems to mean that DLA does not require large shot noise. What we have been calling the average growth, Eq. (10), seems to accurately describe the process. Presumably the variety of patterns which can be produced result from sensitivity to initial conditions. We will return to this point below.

There is a fly in the ointment, however. In order to reduce d_0 Nittman, et al. were led to use a non-Newtonian fluid. It is possible to wonder whether the shear thinning of this polymer solution led to some kind of effective length scale and, perhaps an analogue of the particle diameter.

The present author and collaborators (Ben-Jacob, et al., 1985) have been doing a set of experiments with Hele-Shaw cells containing Newtonian fluids (air and glycerine). Our approach was to modify the usual Hele-Shaw geometry in order to reduce boundary effects. We did this by injecting air into the center of the cell. Thus with relatively large d_0 we were able to produce patterns with many branches reminiscent of Fig. 1. We have not yet been able to convincingly show that we can make a fractal, however. The major point of the experiment was to study the relation to pattern formation. We will return to this below.

It is interesting to point out here, however, that we have also produced an experimental analogue of biased diffusion of the sort considered by Meakin (1983). We do this by merely lifting one end of the Hele-Shaw plates. Then, the quantity, Q, above becomes a function of position:

$$Q = b^2(x)/12\zeta \quad , \tag{19}$$

where b is the plate separation and ζ the viscosity. Then for small gradients of b, the Laplace equation for u is replaced by:

$$\nabla^2 u + k\partial u/\partial x = 0 \ . \tag{20}$$

With k a constant. This is the equation for biased diffusion.

Solidification

Everyone has seen pictures of snowflakes. Their complex branching patterns seem to remind us of Fig. 1, but it is clear that there are great differences, as well. to be precise, very large dendritic crystals show a pattern: they are not random and chaotic as Fig. 1 is. Further, they grow in a steady state with constant growth velocity. It is natural to ask what the relationship to Fig. 1 is, if any.

If one examines the literature about non-equilibrium solidification (Langer, 1980) one would conclude that solidification is a diffusion-limited problem. In most cases solidification in a supercooled fluid is assumed to be limited by the diffusion of latent heat away from the crystal. Setting $u = T_m - T$, where T_m is the equilibrium melting temperature, we have:

$$\partial u/\partial t = K\nabla^2 u \ , \qquad (21)$$

where K is the thermal conductivity. Once again, we have boundary and growth conditions like those above. What then is the difference?

In view of the above, we might be tempted to conclude that the pattern is stabilized by a diffusion length and a capillary length that are not very far apart. This may be true in some cases, but typically the diffusion length is larger than the regular micro-structure of a growing tip.

Very recent work (Ben-Jacob, et al., 1983, Brower, et al., 1983) has cast light on this. The key to snowflake patterns is the essential role of crystalline anisotropy: the patterns form, of course, in single crystals. The anisotropy enters the equations (among other ways) in an angular dependence of d_o. The qualitative result is that stable growing tips form only for a finite anisotropy. For small anisotropy the tips split.

These theoretical developments were done in a particular approximation, however, that of vanishing diffusion length. (The full problem is too difficult analytically or even numerically.) Thus they beg the question we are posing of the crossover to disorderly scaling growth.

We (Ben-Jacob, et al., 1985) have attacked this problem by returning to the Hele-Shaw experiment. We introduced anisotropy in a very simple way by milling grooves into the plates between which the fluid is confined. By changing the plate spacing we can demonstrate the same crossover from orderly growth to tip-splitting as was theoretically predicted. However, in this case, the effective diffusion length is always infinite so that we may be observing the crossover to fractal growth.

THEORY

In view of the extraordinary progress made in simulation of particle aggregation and experiments on related systems, it is remarkable that very little soid theoretical understanding has emerged. We have no way to calculate the "critical exponent" D or any other property, or even an idea of how to begin to do so. This is in stark contrast to scaling properties of equilibrium systems for which the formalism of field theory and the renormalization group is available. To be sure, estimates of various kinds have appeared based on quasi-equilibrium considerations, but these do not seem to yield much insight. In this section we will review some work which does seem solidly based and which may lead further.

Causality Bound

The only solid result, about the fractal dimension, that this author is aware of was given by Ball and Witten (1984a). It provides a bound on D.

The reasoning is rather simple. Suppose in the case of DLA we consider u far away from the aggregate. The solution to Eq. (6) is well known in this case:

$$u = u_\infty(1-(R/r)^{d-2}) \qquad (22)$$

where, as above R is some characteristic size of the aggregate. Then the rate at which particles are added is:

$$dN/dt \sim \int ds\ \hat{n}\cdot\nabla u \sim R^{d-2} , \tag{23}$$

On the other hand, from Eq. (1)

$$dN/dt \sim R^{D-1}v , \tag{24}$$

If we now assume that v, the characteristic tip velocity, is bounded, we have:

$$v \geqslant TR^{d-D-1} , \tag{25}$$

$$d-1 \leqslant D \leqslant d . \tag{26}$$

Here T is a constant.

The right hand part of the inequality follows from the definition of D. All the simulation results satisfy this bound.

Note that D never stops increasing with the dimension of space. This is quite unlike the case of a critical percolation cluster, for example, which has a constant D above a certain upper critical dimension (6 for percolation). In this sense DLA does not have an upper critical dimension. Cluster-cluster aggregation probably does have an upper critical dimension (Ball and Witten, 1984b, Kolb, 1984).

The causality bound can be reformulated for ballistic aggregation with a very interesting result. We expect (from scaling) that the inward particle flux is

$$dN/dt \sim R^{d-1} \tag{27}$$

i.e., it depends on the cross-section of the cluster. Then we have,

$$v \geqslant TR^{d-D} \tag{28}$$

$$D = d . \tag{29}$$

Mean Field Theory

Since DLA is a scaling object it is natural to attempt to describe its largest scale features alone without inquiring into details. Thus one is led to try to describe an aggregate by a coarse-grained description. Thus the density (which we have taken to be constant inside the aggregate and zero outside, thus far, becomes a continuous variable. An equation for this variable $\rho(\vec{r})$, was first posed by Witten and Sander (1983). The most successful use of this point of view, however, takes a further step. Suppose we try to cook up an equation for the ensemble average of ρ, i.e., a mean-field theory. Clearly, this quantity, ρ_m, will be spherically symmetric.

The set of equations proposed are the following:

$$\nabla^2 u = \partial\rho_m/\partial t = Zu(\rho_m + a^2\nabla^2\rho_m) . \tag{30}$$

These represent a conservation law (subtracting from the average diffusion field adds to the aggregate) and a law for the average penetration into the aggregate. The spirit is similar to that of Eq. (7), above, with the additional feature that the diffusing field sees a particle in the aggregate when it is a distance a away.

These equations have been solved (Nauenberg, 1983, Ball, Nauenberg and Witten, 1984). The solution has some reasonable and some unreasonable features, as might be expected in a mean-field treatment. The essential result is that we can write:

$$\rho \sim f(z)/r \tag{31a}$$

$$z = (r-R)/\xi \tag{31b}$$

$$R = vt;\ \xi \sim \sqrt{R}\ . \tag{31c}$$

Here the kink function f is unity for large negative arguments and zero for large positive arguments.

There are several interesting features to note here. By comparison with Eq. (4), above, we see that D=d-1 in this approximation. Also the exponent p is exactly at its naive value (d=D)/2=1/2. At most, this is a crude representation of reality. However, Eq. (30) has proved useful, for example, in discussing the crossover to compact aggregation (Nauenberg, Richter and Sander, 1983).

How can we improve on this treatment? One approach which has not yet been tried, would be to try to estimate the correlations between u and ρ which have been left out so far. We could write, for the last term of Eq. (30), something more complicated involving a correlation function that we might estimate self-consistently.

Figure 5

More attention has been given to reinterpreting Eq. (30) as describing not an ensemble average, but just a local coarse-graining. Then we try to replace the model with Eq. (30).

A first step in this direction was taken by Nauenberg and Sander (1984). We showed that, at least, the continuum equations have a surface instability of the Mullins-Sekerka type. Closer inspection seems to show that this instability is hard to excite: it has a small basin of attraction. It is the author's view that Eq. (30) does not represent DLA. Evidence for this has been presented (Kantor, et al., 1985).

Interface Dynamics

How do we resolve the difficulty presented above that the continuum equations do not seem to produce fractal growth? There seem to be two ways. One is to point out that we have left out the stochastic nature of the process in writing deterministic equations like Eq. (30). In this view, constantly feeding the system with shot noise is necessary to make fractal growth.

However, we have adduced experimental evidence (above) to the effect that continuum flows with no large noise source (surely, the discreteness of the molecules in a fluid has no effect!) becomes fractal. We can then ask whether it is the coarse-graining that we did above which is at fault.

We propose, then, that a sharp interface developing according to Eq. (5), and (6) and (10) may be enough to give fractal growth by sensitivity to initial conditions. To this end, the author and collaborators (Sander, Ramanlal, and Ben-Jacob, 1985) have attempted to solve this set of equations numerically. We present here (Fig. 5) a very preliminary theoretical result. This picture is not a fully developed fractal, but it seems to be developing in that direction via repeated tip-splittings.

Acknowledgements: I would like to thank my colleagues E. Ben-Jacob, P. Garik, J. Hautman and P. Ramanlal for many discussions. Supported by NSF Grant 82-03698 and ARO Grant DAAG-29-83-K0131.

REFERENCES

Ball, R. and Witten, T., 1984a, Phys. Rev. A 29, 2966.

Ball, R. and Witten, T., 1984b, proceedings of The Third Conference on Gaithersburg, Md.

Ben-Jacob, E., Goldenfeld, N., Langer, J.S., and Schön, G., 1983 Phys. Rev. A. Lett. 51, 30.

Ben-Jacob, E., Godbey, R., Goldenfeld, N., Koplik, J., Levine, H., Mueller, T., and Sander, L., 1985, Submitted.

Botet, R., and Jullien, R., 1984, J. Phys. A 17, 2517.

Brady, R. and Ball, R.C., 1984, Nature, 309, 225.

Brower, R., Kessler, D., Koplik, J., and Levine H., 1983, Phys. Rev. Lett. 51, 1111.

Friedlander, S., 1977, "Smoke, Dust, and Haze", Wiley, New York.

Kantor, Y., Ball, R., and Witten, T., 1984, unpublished.

Kolb, M., Botet, R., and Jullien, R., 1983, Phys. Rev. Lett. 51, 1123.

Kolb, M., 1984, Physique Lett. 45, L211.

Langer, J.S., 1980, Rev. Mod. Phys. 52, 1.

Mandelbrot, B., 1982, "The Fractal Geometry of Nature", Freeman, New York.

Matsushita, M., Sano, M., Hayakawa, Y., Honjo, H., and Sawada, Y., 1984, Phys. Rev. Lett. 53, 286.

Meakin, P., 1983a, Phys. Rev. A 27, 604; 27, 1495.

Meakin, P., 1983b, Phys. Rev. B 28, 5221.

Meakin, P., 1983c, Phys. Rev. Lett. 51, 1119.

Meakin, P., 1984, Phys. Rev. B 29, 3722.

Meakin, P. and Sander, L., 1985, Phys. Rev. Lett., Comments, accepted.

Mullins, W. and Sekerka, R., 1963, J. Appl. Phys. 34, 323.

Nauenberg, M., 1983, Phys. Rev. B 28, 449.

Nauenberg, M., Richter, R., and Sander, L., 1983, Phys. Rev. B 28, 1649.

Nauenberg, M. and Sander, L., 1984, Physica, 123A, 360.

Niemeyer, L., Pietronero, L. and Wiesman, H., Phys. Rev. Lett. 52, 1033.

Nittman, J., Daccord, G., and Stanley, H., Nature 314, 141.

Paterson, L., 1984, Phys. Rev. Lett. 52, 1621.

Plischke, M. and Racz, Z., 1984, Phys. Rev. Lett. 53, 415.

Sander, L., 1984, in "Kinetics of Aggregation and Gelation", F. Family and D.P. Landau, ed., North-Holland, New York, p. 13.

Sander, L., Ramanlal, P., and Ben-Jacob, E., 1985, to be published.

Vicsek, T., 1984, Phys. Rev. Lett. 53, 2281.

Voss, R., 1984, Phys. Rev. B 30, 334.

Weitz, D. and Olivera, M., 1984, Phys. Rev. Lett. 52, 1433.

Witten, T. and Sander, L., 1981, Phys. Rev. Lett. 47, 1400.

Witten, T. and Sander, L., 1983, Phys. Rev. B 27, 5686.

FRACTAL CONCEPTS FOR DISORDERED SYSTEMS:

THE INTERPLAY OF PHYSICS AND GEOMETRY

H. Eugene Stanley

Center for Polymer Studies and Department of Physics
Boston University
Boston, Massachusetts 02215

INTRODUCTION

One purpose of these three talks is to address the question of how fractal concepts provide a natural framework within which to discuss a range of phenomena occurring in disordered systems. A second purpose is to exemplify the degree to which the language of fractals motivates one to translate physical phenomena into geometric terms, thereby rendering complex problems more tractable.

To achieve these goals, we shall emphasize that measurement of "the" fractal dimension is generally not sufficient to capture the subtle range of behavior of a given system, just as a measurement of a single critical exponent is not sufficient, generally, to describe the nature of a critical point.

Accordingly, we shall systematically introduce ten separate critical exponents, all of which have the geometric interpretation as fractal dimensions. Some of these appear to be completely independent of the others, though it is quite possible that many will ultimately be found to be related, just as one has found the various critical exponents are related to one another.

ORGANIZATION AND NOTATION

First a word on notation. There are presently almost as many notations as there are workers in the field. This situation cannot be avoided: it is easier to get a colleague to use your toothbrush then to use your notation. An abbreviated "Rosetta stone" connecting the notations of Professors Aharony, Orbach, and myself appears as Table 1. The notation we use has two advantages: (i) there are no tildes, bars, or double bars to confuse the myopic reader and (ii) the subscript serves to denote which fractal dimension is intended, so one is not left with the feeling "If I were able to remember all those symbols I would have become a botanist" (to paraphrase E. Fermi on his exasperation with the many symbols for elementary particles).

Second, we remark that not all the material presented in these three lectures will fit in the allotted space. Accordingly, we concentrate in

TABLE 1. "Rosetta Stone" connecting notation used here with that used by Professors Aharony and Orbach in their lectures. In the final column are relations among the 10 fractal dimensions; those relations that are in the conjecture stage are indicated by the symbol $\stackrel{?}{=}$. The author wishes to thank Drs. R. Pynn and A. Skjeltorp for suggesting the inclusion of this table, and Professors A. Aharony and R. Orbach for helpful discussions on the subject.

	This lecture	Prof. Aharony	Prof. Orbach	Relations
Lecture 1:				
$N_f \sim L^x$	$x = d_f$	$x = D$	$x = \bar{d}$	(a) $d_f = y_h$
$N_w \sim L^x$	$x = d_w$	$x = 2 + \theta$	$x = d_w$	(b) $d_w \stackrel{?}{=} \frac{3}{2} d_f$
				(c) $d_w \stackrel{?}{=} d_f + 1$
$N_{min} \sim L^x$	$x = d_{min}$	$x = d_{min}$	$x = d_{min}$	(d) $d_{min} \stackrel{?}{=} d_f - d_{red}$
Lecture 2:				
$N_{resistors} \sim L^x$	$x = d_R$			(e) $d_R = d_w - d_f = d_w^{BB} - d_f^{BB}$
$N_{hull} \sim L^x$	$x = d_h$			(f) $d_h \stackrel{?}{=} 1 + d_{red}$
$N_{unscreened} \sim L^x$	$x = d_u$			(g) $d_u \stackrel{?}{=} (d_f - 1) + (d - d_f)/d_w$
$N_{growth} \sim L^x$	$x = d_g$			(h) $d_g = 2d_f - d_w$
Lecture 3:				
$N_{backbone} \sim L^x$	$x = d_{BB}$	$x = d_{BB}$		
$N_{red} \sim L^x$	$x = d_{red}$	$x = d_{red}$		(i) $d_{red} = y_T$
$N_{elastic} \sim L^x$	$x = d_e$			(j) $d_e \stackrel{?}{=} d_{min}$
Intrinsic Exponents				
$N_w \sim (N_f)^x$	$x = d_s$		$\bar{\bar{d}}$	$d_s \equiv 2d_f/d_w$
$N_f \sim (N_{min})^x$	$x = d_\ell$		$\hat{d}$	$d_\ell \equiv d_f/d_{min}$

(a) Stanley (1977)

(b) Alexander and Orbach (1982)

(c) Aharony and Stauffer (1984)

(d) Havlin and Nossal (1984)

(e) Stanley and Coniglio (1984)

(f) Sapoval et al (1984)

(g) Coniglio and Stanley (1984)

(h) Stanley (1984a,b)

(i) Coniglio (1981)

(j) Herrmann and Stanley (1985)

these notes on the new material and refer the interested reader to earlier lecture notes and reviews for background material (Stanley 1981, 1982a,b,c, 1983, 1984a,b, 1985; Stanley and Coniglio 1983).

We will organize this first talk around three questions, questions that could be used to organize an introduction to almost any subject.

QUESTION ONE: WHAT HAPPENS? WHAT ARE THE PHENOMENA?

Suppose instead of coming to Norway you took a train ride across my home state of Oklahoma. Unlike Norway, Oklahoma is flat. After a very short time, you are bored. You walk to the last car, look out the back, and take a photograph of this boring scenery (Fig. 1a) to show your friends who went to Norway what they are missing. Later you develop your film, and it occurs to you that maybe there was a buffalo or something interesting on the landscape. So you enlarge a piece of the negative, say the rectangular box of Fig. 1a. But being a physicist you drop the original print (Fig. 1a) and the enlargement of the inset (Fig. 1b). You find to your excitement that you cannot tell the original from the enlargement.

You have just re-discovered a symmetry which holds not only for Oklahoma but also for many other interesting materials in nature. These materials are the object of the present talk. The symmetry of Oklahoma is the basis of much of large-scale construction. One builds, for example, a small-scale model of an airplane. When the small-scale model does not crash, one proceeds to build a full-scale model on the assumption that it will not crash.

FIG. 1: A somewhat artificial scale-invariant photograph. The inset in (a) is enlarged in (b). Part (a) was originally drawn by S. Ma.

Regular Fractals

Now consider another two-dimensional object, the Sierpinski gasket (Fig. 2). Here we take a photograph of a triangle. On a large length scale L it appears to be solid but as we enlarge the negative (examining it on finer length scales), we detect the fact that it has holes. Its density thus decreases when the length scale changes.

This behavior is certainly very much against our normal ideas of Euclidean geometry. Objects can be either solid (density one) or full of holes (density below one), but in both cases the density is independent of the length scale on which it is measured. If a truck full of donuts has

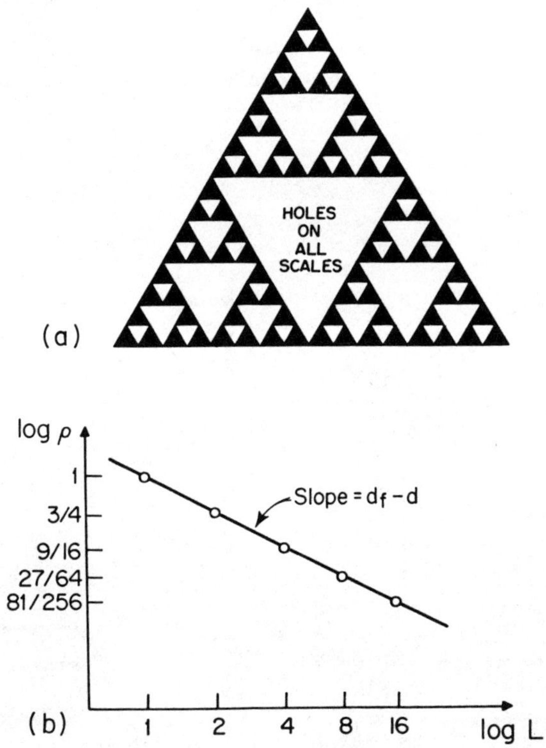

FIG. 2: The prototype "regular" or "exact" fractal, the Sierpinski gasket, shown here for a system of Euclidean dimension 2. On each iteration the density decreases, a generic feature of fractal objects.

mass M, then a truck whose cargo space is increased in edge from L to 2L will carry eight times the mass:

$$M(2L) = 8M(L) = 2^3 M(L). \tag{1a}$$

In general, for any positive number λ, we can write

$$M(\lambda L) = \lambda^d M(L). \tag{1b}$$

Equation (1a) is a functional equation. The solution is obtained by setting $\lambda = 1/L$:

$$M \sim L^d. \tag{1c}$$

The density $\rho = M/L^d$ hence scales as

$$\rho \sim L^o. \tag{2}$$

The lower is the density, the smaller are the amplitudes that appear implicitly in (1c) and (2). However regardless of how light is the material transported the exponent in (1b) and (1c) is always the Euclidean dimension d.

The above results are so familiar that we do not need to use the formal language of functional equations to understand them. However it is exactly this language that is needed to describe fractal objects. For the Sierpinski gasket example of Fig. 2, the mass obeys the functional equation

$$M(2L) = 3M(L) = 2^{d_f} M(L) \qquad \text{with} \qquad d_f = \ln 3/\ln 2 = \log_2 3. \tag{3}$$

Similarly, the density obeys

$$\rho(2L) = (3/4)\,\rho(L) = 2^{d_f - d}\,\rho(L), \tag{4}$$

where $d - d_f$ is termed the co-dimension.

In general, fractal objects obey functional equations of the form

$$M(\lambda L) = \lambda^{d_f} M(L) \qquad [d_f < d]. \tag{5a}$$

The solution of the general functional equation (5a) is

$$M(L) = L^{d_f} \text{ [FRACTAL DIMENSION No. 1]}. \tag{5b}$$

For this exact fractal there are no "correction to scaling terms"--the leading "scaling" term suffices for all values of L.

Statistical Fractals

The simplest example is an N-step random walk on, say, a square lattice (Fig. 3a). The walker, a Polya drunk, takes one step per time unit, so that N=t. Since there are 4 choices for the drunk for each next step, after a time t there are a total of 4^t distinct configurations. Most configurations are not fractals, yet certain average quantities obey functional equations identical to those obeyed by the

Sierpinski gasket. To see this, let us use for the length scale L the "range" of the walk--the Pythagorean distance from the origin after t steps, averaged over the ensemble of all t step walks. Then $L(t=1) = 1$ for the four 1-step walks, $L(2) = 2^{1/2}$ for the 16 2-step walks, $L(3) = 3^{1/2}$ for the 64 3-step walks, etc. In general,

$$L(t) = t^{1/2}. \tag{6a}$$

The proof that (6a) holds for walks on any d-dimensional hypercubic lattice is an exercise for the student.

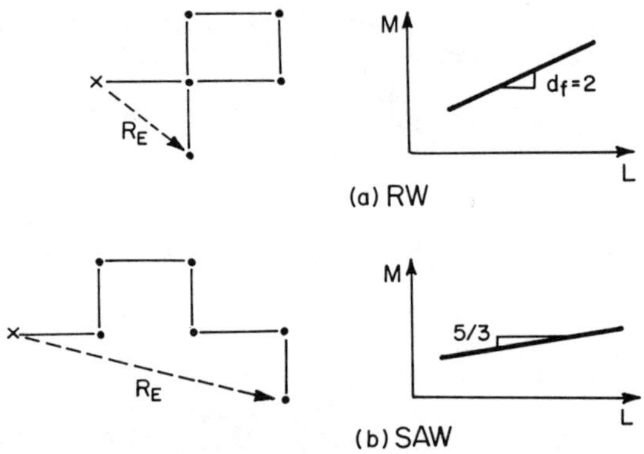

FIG. 3: Schematic illustration of (a) a random walk and (b) a self-avoiding walk, each of $M = t = 6$ steps. We show just <u>one</u> of the 4^6 possible walks, most of which have zero weight for the SAW case. Shown also are log-log plots of the relation between the characteristic length scale L (the mean end-to-end distance R_E) and the variable which formally plays the role of the mass M, namely the number of steps N or the time t.

Equation (6a) so resembles (5b) that it is tempting to use the same language to describe both phenomena, even though they are technically quite different: (5b) holds exactly and describes a <u>regular object</u> while (6a) describes a <u>statistical ensemble of objects</u> (t-step random walks).

To make the formal correspondence stronger, we can imagine that after each step is placed a monomer, and the steps correspond to chemical bonds between monomers. Then the random walk traces out a "free-flight" polymer which ignores the Archimedes principle that two objects cannot occupy the same point of space. The total polymer mass M after t steps is proportional to t, so the mean Pythagorean distance from beginning to end of a polymer with N monomers is related to total polymer mass by

$$M(L) = L^2. \tag{6b}$$

Clearly (6b) is identical in structure to (5b), suggesting that we may speak of a fractal dimension of a random walk. Hence, we write

$$M(L) \sim L^{d_w} \quad \text{[FRACTAL DIMENSION No. 2]}, \tag{7a}$$

with

$$d_w = 2 \qquad \text{[RANDOM WALK]}. \tag{7b}$$

There is an important difference between (6b) and (5b), one which often confuses newcomers to the field of fractals. Equation (6b) applies to an "EXACT" fractal: it holds exactly, provided that L is a power of two. Equation (6b) applies to a "STATISTICAL" fractal: it applies not to a single object like a Sierpinski gasket, but to a statistical *ensemble* of z^t objects (here z is the lattice coordination number and $M=t$ is the number of steps or "monomers" in the random walk). Most members of this ensemble are anything but fractal: For example, z walks are linear straight lines and hence are strictly *one-dimensional*, while z walks are blinkers oscillating from the origin to one of the z nearest neighbor sites, and so are *zero-dimensional*. However the statistical properties of the ensemble can be calculated exactly, as noted above, and one finds a power law relation between the mass scale and the length scale. Hence the exponent d_w is of use: it does describe the statistical properties of the ensemble--even though it does not describe the behavior of the individual members of the ensemble. A real random system in nature can only manifest itself statistically, so this statistical property is the most we could really use.

As an aside, we note that generally for statistical fractals the equality sign in equations such as (7a) must be replaced by a tilde sign. The tilde means that the relation holds "asymptotically," for a range of L that is much larger than the microscopic scale $L=1$ and much smaller than any "correlation lengths" ξ that serve to set an upper limit on the domain of self-similarity. For the random walk on an infinite lattice, $\xi = \infty$, but for a random walk on a large DLA cluster ξ is finite, scaling as the diameter of the fractal substrate. For random walks, the lower limit of L is generally many lattice spacings. An exception is L^2 for a random walk, for which (6b) holds for L down to $L=1$. Henceforth, we will focus on statistical fractals and the tilde symbol means that a relation holds approximately over a range of values of L . . . typically for $[1 \ll L \ll \xi]$.

QUESTION TWO: WHY DO WE CARE? WHY ARE WE INTERESTED IN FRACTALS?

"*First Answer*"

There are many answers to this question. The first answer is that *fractals occur in nature*; recognizing that an object is fractal gives one predictive power if you know the fractal parameters characterizing it. The number of such objects is vast, and it is tempting to "pull out the family album" and show you pictures of every fractal that I have come to know and love.

Many of my favorites will be treated by other lecturers at this school. Hence I will limit myself instead to one example, the discovery of which I was associated with. Suppose a low-viscosity liquid like water is forced at high pressure through a high viscosity liquid--e.g., through a thick syrupy polysaccharide solution. This delicate experiment was recently carried out (Nittmann et al 1985) using a Hele Shaw cell. This is a quasi-two-dimensional "cheese sandwich" structure in which the role of the bread is played by two transparent parallel plexiglas sheets, about a meter long and 20 cm wide. The space, typically 1 mm or less, between the bread is occupied by a "cheese" consisting of high-viscosity polysaccharide solution. Water enters under pressure from the left end of this cell. The resulting pattern, called a viscous finger, is shown for successive time intervals in Fig. 4. When we first saw this highly-ramified filamentary structure, we immediately thought that this might be an example of a statistical fractal--analogous in some ways to the random walk example discussed above but of course this is a branched object. We measured the fractal dimension and found reproducible numbers from finger to finger. The details of our analysis will be presented in the next lecture.

"<u>Second Answer</u>"

A second reason for studying something is that <u>one achieves some practical advantage</u>. For example, if we know the mass of a fractal object of edge L, we can predict, statistically speaking, the mass of another fractal object of edge L. We mentioned above the application of such scaling concepts to the trivial example of a donut truck. Another less trivial application will be described next.

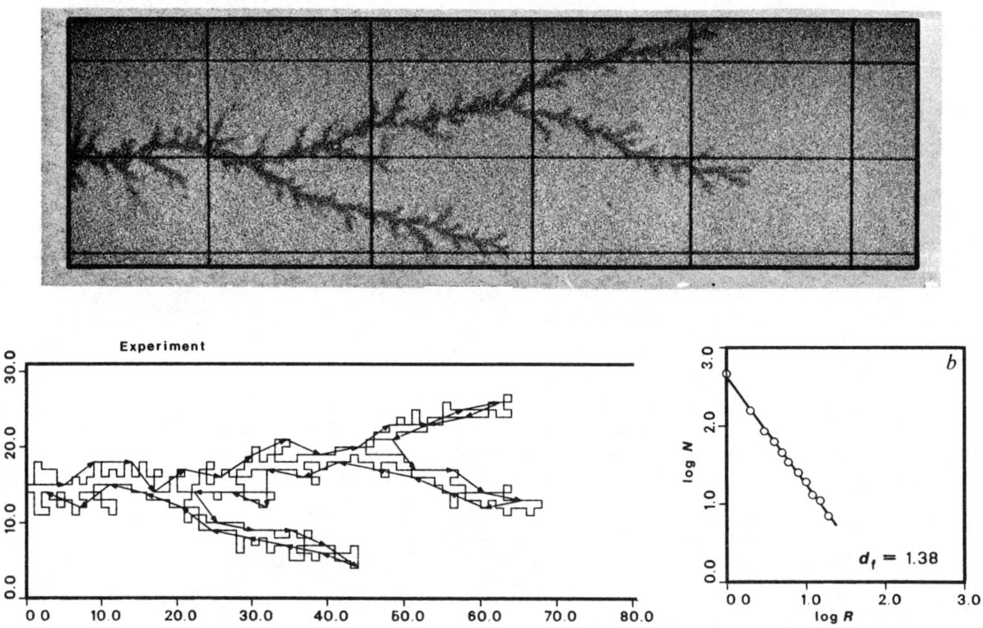

FIG. 4: Typical viscous finger created by water advancing into a Hele Shaw cell filled with a non-Newtonian fluid with zero interfacial tension. The grid rectangles have dimensions 3 cm by 5 cm. Dyed water is injected under force from a single inlet (on the left wall) into an aqueous polymer solution whose viscosity varies between 10^3 and 10^4 depending on the water velocity. Below the experimental plot is the fractal analysis. On the left is a digitized version of the finger, showing a walker taking steps of length R=5. On the right are the data for several R values, and from the slope we can calculate the fractal dimension. After Nittmann et al (1985).

Recently, it has become possible to make fractal viscous fingers unencumbered by the boundary conditions of the lateral walls by using a radial geometry (Daccord et al 1985a). The rectangular Hele Shaw cell is replaced by a round picnic table that is covered with about 1 mm of syrup; a plexiglass plate is used to cover the whole mess. Water is forced into a hole in the center of the table and the resulting patterns have no walls. We find $d_f = 1.7$ (Daccord et al 1985b). Suppose we wish to scale this system to the size of a typical oilfield configuration, in which water is pumped under pressure down a single well in the center of an oilfield. The water forces oil ahead of it which is collected at oil wells typically 1-10 km distant from the center "water well." To the extent that viscous fingers under these laboratory conditions may be expected to have fractal properties roughly resembleing those in the porous sandstone of a quasi two-dimensional oil formation. Hence if the length scales increase by 10^3, we might anticipate that the viscous finger mass increases by 10^5, which is much less than what one would obtain, 10^6, if the fingers were non-fractal.

The list of practical systems could go one and on. Another example is Cab-o-Sil, which appears to be fractal on short length scales. This extremely low density material is used as an additive to a variety of ice cream, paint, and other systems. Its fractal properties will be discussed elsewhere in this meeting.

"Third Answer"

A third reason for studying fractals is that through their study we can gain theoretical insight into the workings of nature. For example, consideration of fractals permits us to assign a reproducible parameter, d_f, to the system. The utility of being able to quantitatively characterize a random system cannot be understated. The first reaction of some to the subject of random systems is that they are random, and that is all there is to it. The second reaction of many is that all random systems are the same so what is there to be done. The truth is more complex: the set of random systems in nature appears to fall into a large but denumerable number of distinct "universality classes" just as the set of systems displaying critical points appear to have a finite number of universality classes.

A simple illustration of how the fractal dimension serves to quantitatively distinguish two different random walks is provided by comparing the random walk (RW) with the self-avoiding walk (SAW), which is identical to the random walk except for the constraint that the walk cannot visit more than once the same site--see Fig. 3b. The motivation for introducing this variation of a RW is to incorporate the "excluded volume" interaction into the model of a polymer: by forbidding two monomers to occupy the same point in space the SAW respects Archimedes. Intuitively, we would expect that the SAW constraint would have the effect of weighting the more expanded configurations at the expense of the more compact configurations. For example, the blinker configuration and a large fraction of the 4^t compact RW configurations have weight zero for a SAW since the same site is visited more than once. Accordingly, we might expect that the fractal dimension for the SAW to be smaller than for the RW. It is . . . provided the embedding space has a Euclidean dimension d that is below a "critical upper marginal dimension" $d_+ = 4$. Moreover, the difference between d_w for the SAW and the RW is greater as d decreases, since the Archimedes constraint assigns zero weight to a larger fraction of configurations.

"Fourth Answer"

A fourth reason for studying fractals is the possibility of gaining insight into extremely basic questions regarding the present theory of critical points. We illustrate this fact by citing two of the many examples where studying geometric phase transitions has given us insight into the nature of thermal phase transitions.

(i) Percolation. There is compelling numerical evidence that percolation obeys Widom-Kadanoff scaling (Nakanishi and Stanley 1980,1981 and references therein). That is, the "Gibbs potential" $G(h,\epsilon) = \sum_s n(s,\epsilon)\exp(-sh)$ is a generalized homogeneous function, obeying a functional equation that for two variables is an obvious generalization of the one-variable homogeneity relation (5a):

$$G(\lambda^{y_h} h, \lambda^{y_T}\epsilon) = \lambda^d G(h,\epsilon).$$

Here $n(s,\epsilon)$ is the number of s-site clusters per lattice site, and the two numbers y_h and y_T are called scaling powers. All critical exponents in percolation can be expressed in terms of these two scaling powers, so it would be nice if they had a simple geometric interpretation. The first paper introducing fractals into percolation noted in passing that

$$y_h = d_f,$$

and wondered "what cluster property could be related to y_T" (Stanley 1977). Four years later it was found that

$$y_T = d_{red},$$

using both Monte Carlo calculations for $d=2$ (Pike and Stanley 1981) and exact arguments valid for all d (Coniglio 1981). Here d_{red} is the fractal dimension for the singly-connected bonds (see lecture 3). Thus the Kadanoff scaling powers are simply given in terms of fractal dimensions, at least for percolation.

(ii) Polymers. Consider first the Ising model of a binary mixture or liquid-gas critical point. It is well known that the critical exponents of the Ising model depend strongly on d up to a certain "upper marginal dimension" $d_+ = 4$. For $d > 4$, the exponents stick at their mean field values. Why is there an upper critical dimension? Why does it take on the value 4? These questions are nicely answered by the Wilson renormalization group, but the degree of physical insight thereby gained depends sensitively on the prior training of the student in field theory. There "should" be a basic geometric reason underlying this fact, for this same feature is present for geometric critical points like those displayed by topologically linear structures (such as the RW and SAW) and branched structures (such as lattice animals and percolation). Indeed this is one of the reasons for the basic faith that our understanding of statistical physics is closely coupled to our understanding of geometry.

It is thus natural to seek a geometric explanation for the occurrence of an upper marginal dimension d_+. Consider, then, an infinitely long RW embedded in a d-dimensional space. Color the first half of the walk white and the second black. Each half is a semi-infinite RW and has $d_w = 2$. These two 2-dimensional objects will not intersect each other if $d > 4$, while if $d < 4$ they are certain to intersect. The intersection question is marginal at $d=4$ (Fig. 5a). The same argument works for lattice animals, for which $d_f = 4$ and hence $d_+ = 8$ (Fig. 5b). Stated more precisely, the set of intersection points of two fractals has a co-dimension that is the sum of the co-dimensions of the two fractals ($d - d_f$ is the co-dimension of a fractal set).

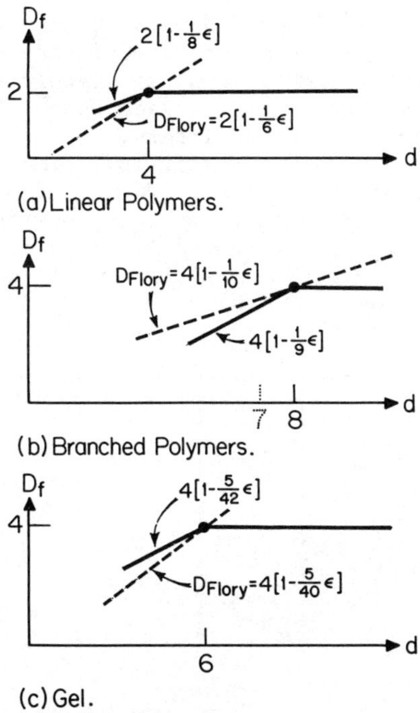

FIG. 5: Dependence on the Euclidean embedding dimension d of the fractal dimension d_f for (a) the self-avoiding walk (SAW) used to model linear polymers, (b) the random lattice animal, used to model branched polymers, and (c) percolation, used to model gelation of polyfunctional monomers. The results are based upon renormalization group calculations, which are valid only in the immediate vicinity of the upper marginal dimension d_+. Also shown are the predictions of Flory theory in the vicinity of d_+. Here $\epsilon = d_+ - d$. After Stanley (1983).

WHAT DO WE ACTUALLY DO?

A simplistic answer to this question is as follows. If you are an experimentalist, you try to measure the fractal dimension. If you are a theorist, you try to calculate the fractal dimension of suitably-chosen models. If there is no agreement then you try another model.

In reality, this simplistic answer describes research in many fields. In the present subject, however, we can be usefully guided by the previous example of other scale-invariant phenomena, namely the study of critical phenomena. There the first task was the recognition that there were a denumerable number of scale-invariant quantities, so that a separate critical exponent could be assigned to each. That the existence of a critical exponent is connected with "scale free" or "scale-invariant" behavior is made clear in Fig. 6, where we show successive re-scalings of the temperature axis for the specific heat data on ^{4}He near its point. Fig. 6 should be compared with the "scale-free" behavior of Fig. 1a and 1b. The analog of the critical exponent is clearly the fractal dimension d_f, and for this reason the similar principles may underly both disciplines.

We shall see that "the" fractal dimension d_f is not sufficient to describe these phenomena. For example, percolation and DLA both have $d_f = 2.5$ when $d=3$, yet any child can immediately see that these two fractals look completely different from each other [Fig. 7]. Hence there must be some fractal parameter that differs. Later in this lecture we shall see that we must introduce a new fractal dimension, d_{min}, to describe the tortuosity of the fractal. We will see that $d_{min} = 1$ for DLA and $d_{min} > 1$ for percolation. In fact, by the time these three lectures have ended, we shall have found it necessary to introduce a total of ten distinct fractal dimensions. Although this may sound overwhelming

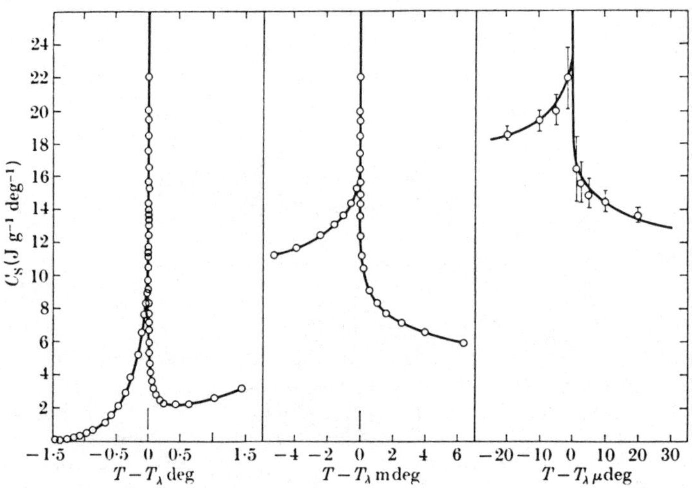

FIG. 6: Dependence of specific heat near the λ transition of ^{4}He on temperature, showing three successive "re-scaling" of the temperature axis. If the three frames were mixed up, one would be faced with the same problem that occurred when the two frames of Fig. 1 were dropped. We say that this behavior is "scale-free" or "scale-invariant" since from the figure we cannot guess the true scale. After Buckingham and Fairbank (1961).

at the present time, we shall see that these arise in as natural a fashion as the ten or so distinct critical point exponents that most students know and use regularly. Just as the critical point exponents were found to be not all independent of one another, so also the ten fractal dimensions are not all independent quantities. Rather, they are related by simple relations not altogether unlike the scaling laws relating critical exponents.

Since one can introduce fractal dimensions that play the role of critical exponents, and since relations among the dimensions play the role of scaling laws, it is natural to ask "what about renormalization group?" It turns out that one can develop a very successful renormalization group for geometrical objects. No Hamiltonian appears, yet the essential idea of "renormalization" of critical parameters upon successive re-scalings still holds. In these lectures we will not have time to go into the renormalization group work in detail, so we refer the interested reader to a recent review (Stanley et al 1982).

Our first task is to clearly state how we shall use the term fractal dimension. We shall use a very crude operational definition. Suppose we plot on log-log paper against a characteristic length scale L some quantity q that can be interpreted as a "number of objects" (equivalently, a "mass"). If there is an asymptotic (large-L) region in which this plot becomes straight, then the slope is termed the fractal dimension d_q characterizing that quantity; the notation d_q uniquely associates the the quantity under discussion. Toulouse has suggested that these be called "extrinsic" fractal dimensions, in order that the term "intrinsic" fractal dimensions can be reserved for systems for the ratio of two extrinsic dimensions.

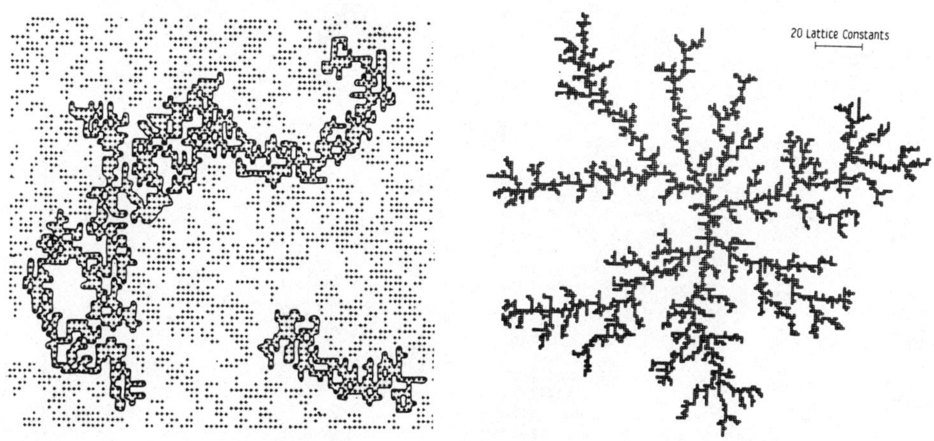

FIG. 7: Comparison between two $d=2$ aggregates with different fractal dimensions d_f: (a) site percolation cluster just below p_c (after Stanley et al 1976) for which $d_f = 91/48 \cong 1.896$, and (b) DLA aggregate (after Witten and Sander 1981) for which $d_f = 1.71$ (Meakin 1985). For $d=3$, the two fractal structures retain their characteristic appearance, yet their respective values of d_f are identical to within present calculational accuracy ($d_f \cong 2.5$). This shows d_f alone is not sufficient to characterize a fractal. In fact, other fractal dimensions such as d_{min} are quite different for percolation and DLA (see Fig. 8b).

We follow Mandelbrot and reserve the term dimension for quantities that <u>are</u> dimensions. We use the word "exponent" for ratios of dimensions, which are often called <u>intrinsic</u> fractal dimensions. Our preference is not mere pedantry, for if we insist on thinking of each "masslike" quantity M_q has having its own characteristic scaling dimension d_q, then we can write down any exponent by inspection. If, e.g., we wish to know the exponent characterizing the dependence of M_q on a different "masslike" quantity $M_{q'}$, then we simply write $d_q/d_{q'}$—no chance for error:

$$V \sim L^3 \quad \text{and} \quad A \sim L^2. \tag{8a}$$

Then

$$V \sim A^{3/2} \quad \text{or} \quad A \sim V^{2/3}. \tag{8b}$$

The quantities 3/2 or 2/3 in (8b) are not dimensions while 3 and 2 in (8b) are.

A similar situation occurs in the application of fractals to realistic physics problems. For example, suppose we wish to know the mean number of sites visited when a random walker stumbles around randomly on a fractal substrate. De Gennes (1976) has termed this problem the "ant in a labyrinth" but a "drunk in a gulag" might seem more descriptive of the actual picture [Fig. 8]. A random walker (drunken ant or drunken prisoner)

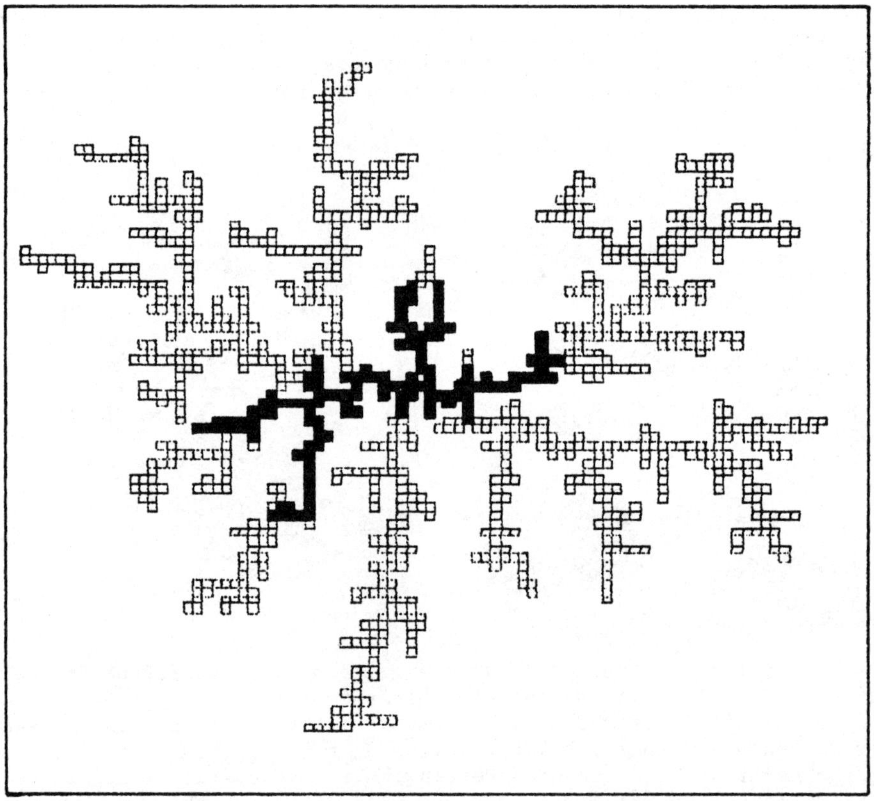

FIG. 8: A random walk of 2500 steps on a DLA fractal substrate with 1000 sites. The sites visited by this walk have been indicated by solid squares. This is a visualization of the de Gennes "ant in a labyrinth" problem. After Meakin and Stanley (1983).

has been parachuted onto a randomly-chosen site, which is then colored black. At each successive time step, the ant moves to a neighboring fractal site, and that becomes black also. At each time step t, the ant (<u>la fourmi</u>) calculates her range L, the rms displacement from the local origin where the parachute landed. Recall from (6b) that the fractal dimension d_w determines how t scales with L. We anticipate that d_w is considerably <u>larger</u> than 2 since many of the neighbors at each step are unavailable to the ant so she is obliged to return in the direction of her parachute point. Hence L increases much less fast with t than for an unconstrained RW (Fig. 9).

Now let us illustrate the analog of (8a) by asking how the number of black visited sites (one "mass") scales with the time (a different "mass"). The ant fills compactly a region of the fractal, so the first, "black" mass scales with fractal dimension d_f. The second scales with fractal dimension d_w. Hence that the number of visited sites scales with time as the <u>ratio</u> of the two fractal dimensions,

$$\langle s \rangle \sim t^{d_f/d_w}. \tag{9a}$$

Note that one additional advantage of our proposed notation is that one sees by inspection that the critical exponent in (9a) is just the ratio of the fractal dimension of the two variables s and t that appear in (9a). In short, it is possible to "write down exponents" by inspection, and virtually impossible to make mistakes.

The quantity d_f/d_w was first introduced by Alexander and Orbach (1982) (henceforth AO) in connection with studying the density of states for phonon-like excitations on a fractal substrate. The quantity that appears is

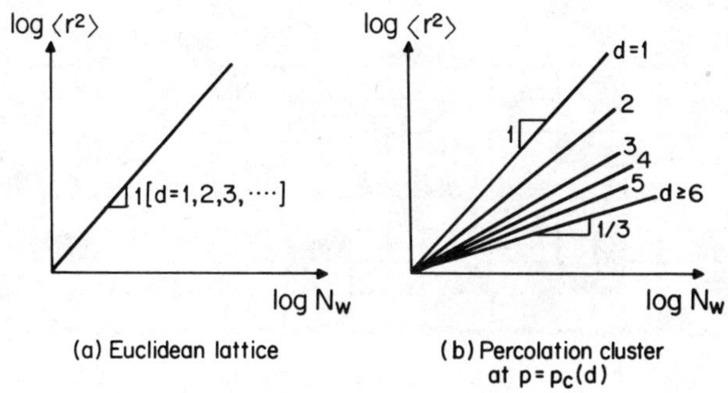

FIG. 9: Schematic illustration of the dependence of the mean square displacement on the number of steps of a random walk, plotted double logarithmically, for (a) a Euclidean lattice, and (b) a percolation cluster at the percolation threshold. For d above 6, the slope sticks at the $d=6$ value of 1/3--so that the fractal dimension of the walk sticks at $d_w = 6$.

$$d_s = 2d_f/d_w \qquad \text{[INTRINSIC FRACTAL No. 1]}, \tag{9b}$$

and is called the fracton or spectral exponent. AO tabulated the quantities d_f, d_w and d_s for percolation fractals on a d-dimensional Euclidean lattice. They noted that while d_f and d_w change dramatically with d (below $d_+ = 6$), d_s does not. Hence they made the now-famous "AO conjecture" that for percolation,

$$d_s = 4/3 \qquad [2 \leq d \leq 6]. \tag{10}$$

The phrase hyper-universal or super-universal was coined to express the possibility that an exponent could be independent not only of lattice details but also of d itself. Shortly after the AO conjecture was published, several studies were published that seemed to confirm it to higher accuracy (Pandey and Stauffer 1983; Havlin and Ben-Avraham 1983). Other studies tested whether the AO conjecture might hold for other random phenomena, such as random walks on DLA clusters (Meakin and Stanley 1983) or diffusive annihilation between random walkers who are constrained to move on a percolation fractal substrate (Meakin and Stanley 1984).

In the past few months, four independent groups have decided to put the AO conjecture to a very severe test. Using completely different computational methods all four groups concluded that d_s is about 2% larger than 4/3.

In a way, it is a pity the AO idea does not work, since if it did we would have a relation between two fractal dimensions, one of which describes "statics" (d_f) and the other of which describes "dynamics" (d_w). Accordingly, several workers have attempted to find another relation to replace (10). Aspects of this work will be discussed in Lecture 2.

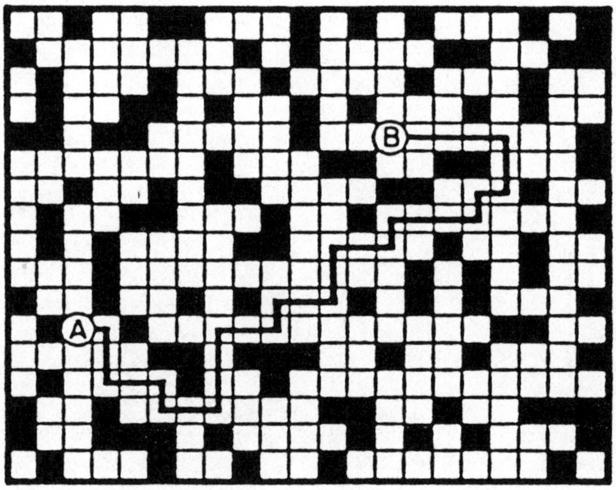

FIG. 10: Schematic illustration of the minimum path followed by a soldier in carrying an important message from General X (at position A) to General Y (at position B) in a battlefield that has been randomly infested by mines in sufficient concentration that the pair connectedness length ξ is larger than the Pythagorean distance L between A and B. This figure is adapted from Ritzenberg and Cohen (1984), where it was used to model the spread of electrical excitations in the heart tissue.

You are a soldier in a mine-infested battlefield, say Lebanon. Your boss, General X, wants you to carry a message to his boss, General Y, who is 10 km. distant (Fig. 10). You do this, and note that it takes one hour. General Y asks you to carry the same message to his boss, General Z, who is 20 km. away. It is critical that the message be delivered within two hours. You tell the impatient General Y that it will certainly take more than two hours to travel twice the distance. More precisely, you tell him that the actual minimum path ℓ obeys the functional equation

$$\ell(\lambda L) = \lambda^{d_{min}} \ell(L) \quad \text{[FRACTAL DIMENSION No. 3].} \tag{11a}$$

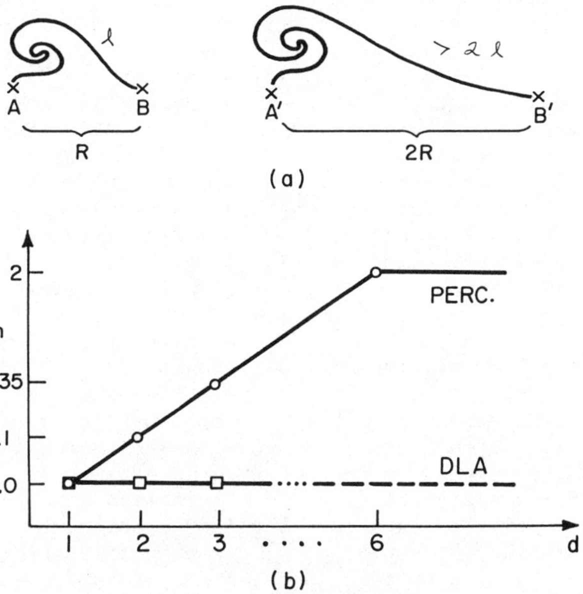

FIG. 11: (a) Schematic illustration of the significance of the fact that d_{min} is above 1: If the distance between points A and B is doubled, then the shortest path length increases by a factor $2^{d_{min}}$, which is larger than 2. (b) Numerical values of d_{min} for percolation (top curve) and DLA (bottom curve). The fact that d_{min} sticks at 2 above the upper marginal dimensions d_c means that the shortest path between two is a RW. The fact that d_{min} does not seem to approach 2 for DLA means that there is no d_+ for DLA (this figure is based on data in Meakin et al 1984).

Herrmann et al (1984) find $d_{min} \simeq 1.1, 1.3$ for $d=2,3$ percolation fractals. The solution to (11a) is found in setting $\lambda = 1/L$,

$$\ell \sim L^{d_{min}}. \tag{11b}$$

You explain that when land mines are placed at random (by the presumably drunken enemy troops), they form clusters of all possible sizes. Hence the minimum path between General Y and General Z is a convoluted path since it must avoid these clusters. If the Pythagorean distance is short, there is little probability of encountering a huge cluster L. Since General Y and General Z are separated by 2L, it is possible that you will encounter much larger mine clusters and hence your path length will more than double.

Can one relate the exponent d_{min} to the other exponents? To do this, it is first important to develop some feel for this quantity. To this end, we have embarked on a program of calculating d_{min} for various fractal substrates, including percolation, lattice animals, DLA, and cluster-cluster aggregation (see Meakin et al 1984 and references therein). We find that in all cases except DLA (Fig. 11), d_{min} increases monotonically with d up to the critical dimension. At and above d_+,

$$d_{min} = 2 \qquad [d \geq d_+]; \tag{11c}$$

this means that the minimum path between two points in a fractal above the critical dimension has the <u>same statistics as a random walk</u> or Gaussian chain. Thus we can understand the critical dimension in a different fashion: all fractals have different values of d_f, d_w, etc. above d_+ but all have $d_{min} = 2$. As d is decreased, this continues to hold until at $d = d_+$, the shortest path between two points is "straighter" than a random walk, and $d_{min} < 2$. That $d_{min} = 1$ for DLA (Meakin et al 1984) is consistent with the idea that DLA has no upper critical dimension.

The utility of d_f in providing a quantitative characterization of a fractal form was discussed above--so also the utility of d_w in characterizing diffusion ("random walks") on a fractal substrate. What is the utility of d_{min}? Clearly d_{min} applies to physical phenomena that propagate <u>efficiently</u> from site to site, not re-visiting previous sites. In a rough sense, we can say that d_w describes random walk propagation where re-visiting fractal sites is possible while d_{min} describes a sort of self-avoiding walk where re-visits are not. Our "runner" finds the minimum path from General X to General Y by considering all possible self-avoiding walks from X to Y and choosing the shortest. Physically, this is the same as imagining that every cluster site of the fractal had a tree on it. If the system is near the percolation threshold, then the trees form a self-similar substrate for length scales up to the connectedness length $\xi = (p - p_c)^{-\nu}$. Now at time $t=1$ let us ignite the trees on the site occupied by General X. At $t=2$, we ignite the trees on the neighboring sites of X, and so forth until after some time delay the site occupied by General Y is ignited. This time delay is of course simply the minimum path ℓ between X and Y, and we have described a very elaborate procedure of determining ℓ which consists of igniting successive "chemical shells" around General X.

Having described the forest fire mechanism, we can now introduce a new intrinsic fractal dimension. Suppose we ask for the total mass of burning trees at time t. From the definition of the fractal dimension d_f, we have

$$M(t) \sim L^{d_f}, \tag{12}$$

where $L = L(t)$ is the radius of gyration (or any length such as the caliper diameter, that scales with the radius of gyration). We asked for the dependence upon t of $M(t)$, where $t = \ell$ is the minimum path length between the newest shell of burning and the origin of the fire. Combining (11b) and (12), it follows that

$$M(\ell) \sim \ell^{d_\ell} \qquad [\ell \ll L^{d_{min}}]. \tag{13a}$$

Here

$$d_\ell = d_f/d_{min} \qquad \text{[INTRINSIC FRACTAL No. 2]}, \tag{13b}$$

is called the chemical dimension; Toulouse recently proposed that d_ℓ be termed the spreading dimension because it governs the spreading of something (e.g., a fire) through a fractal substarate. Toulouse and his colleagues use the symbol $\hat{d}$.

What is the velocity with which the fire front is propagated? Below p_c the fire is localized to finite clusters of burning trees but above p_c it can spread without limit. Hence the velocity is zero below p_c but rises rapidly above p_c (Fig. 12) with an exponent β defined through

$$v \sim (p - p_c)^{\beta}. \tag{14}$$

We have used the symbol β to suggest that v plays the role of an order parameter in this problem. Clearly the velocity is given by

$$v = dR/dt = dR/d\ell = (d\ell/dR)^{-1} = R^{1-d_{min}} = (p - p_c)^{\nu(d_{min}-1)}. \tag{15}$$

Combining (14) and (15), one finds (Grassberger 1985, Barma 1985)

$$\tilde{\beta} \equiv \beta/\nu = d_{min} - 1. \tag{16}$$

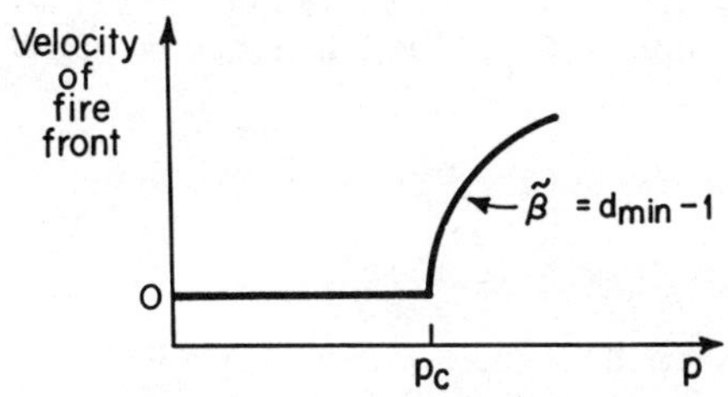

FIG. 12: The velocity of propagation of a fire front, v, as a function of p, the concentration of flammable material ("trees"). The singular behavior near the percolation p_c is reminiscent of the order parameter behavior, suggesting that v plays the role of the order parameter for dynamic phenomena on percolation clusters.

For $d=2$, $d_{min} = 1.10$, very close to unity. Hence the increase in v just above p_c is remarkably steep: a fire which fails to propagate at all just below p_c will propagate extremely fast just above.

An important application of d_ℓ and d_{min} to disordered solids will be discussed in the lecture of Professor Orbach, so these exponents will not be discussed further here.

ACKNOWLEDGEMENTS

First and foremost, I wish to thank Dr. Roger Pynn and Dr. Arne Skjeltorp for having invited me here again; it has been 14 years since I had the pleasure of lecturing at the first Geilo NATO ASI, and my joy of returning to Norway cannot be expressed in words. I also wish to thank Gerd Jarrett for having made this the most stimulating--and happy--of any conference I ever attended. Last and not least, I wish to thank my graduate students and colleagues, without whose help I would not have arrived at my present appreciation for fractal materials.

REFERENCES

Aharony A and Stauffer D 1984 Phys Rev Lett 52 2368

Alexander S and Orbach R 1982 J de Physique 43 L625

Barma M 1985 J Phys A 18 L277

Buckingham MJ and Fairbank WM 1961 In Prog Low Temp Phys Vol 3 (ed CJ Gorter) North-Holland, Amsterdam, p80

Coniglio A 1981 Phys Rev Lett 46 250

Coniglio A and Stanley HE 1984 Phys Rev Lett 52 1068

Daccord G, Nittmann J and Stanley HE 1985a in On Growth and Form [Proc 1985 Cargèse Nato ASI] eds HE Stanley and N Ostrowsky (Martinus Nijhoff, Amsterdam)

Daccord G, Nittmann J and Stanley HE 1985b Proc Les Houches Conference on Finely Divided Metter eds M Daoud and N Boccara (Springer Verlag, Heidelberg)

de Gennes PG 1976 La Recherche 7 919

Grassberger P 1985 J Phys A 18 L215

Havlin S and Ben-Avraham D 1983 J Phys A 16 L483

Havlin S and Nossal R 1984 J Phys A 17 L427

Herrmann HJ, Hong DC and Stanley HE 1984 J Phys A 17 L261

Herrmann HJ and Stanley HE 1985 to be submitted

Meakin P 1985 in On Growth and Form [Proc 1985 Cargèse Nato ASI] eds HE Stanley and N Ostrowsky (Martinus Nijhoff, Amsterdam)

Meakin P, Majid I, Havlin S and Stanley HE 1984 J Phys A 17 L975

Meakin P and Stanley HE 1983 Phys Rev Lett 51 1457

Meakin P and Stanley HE 1984 J Phys A 17 L173

Nakanishi H and Stanley HE 1980 Phys Rev B 22 2466

Nakanishi H and Stanley HE 1981 J Phys A 14 693

Nittmann J, Daccord G and Stanley HE 1985 Nature 314 141

Pandey RB and Stauffer D 1983 Phys Rev Lett 51 527

Pike R and Stanley HE 1981 J Phys A 14 L169

Ritzenberg AL and Cohen RJ 1984 Phys Rev B 30 4038

Sapoval B, Rosso M and Gouyet JF 1985 J Physique Lett 46 L149

Stanley HE 1977 J Phys A 10 L211

Stanley HE 1981 in Int Conf on Disordered Systems and Localization eds C Castellani, C DiCastro and L Peliti (Springer Verlag, Heidelberg)

Stanley HE 1982a in Proc NATO Advanced Study Institute on Structural Elements in Statistical Mechanics and Particle Physics eds K Fredenhagen and J Honerkamp (Plenum Press, New York)

Stanley HE 1982b in Physics as Natural Philosophy: Festschrift in Honor of Laszlo Tisza eds A Shimony and H Feshbach (MIT Press, Cambridge)

Stanley HE 1982c Prog Physics (Beijing) 30 95

Stanley HE 1983 J Phys Soc Japan Suppl 52 151

Stanley HE 1984a in Kinetics of Aggregation and Gelation eds F Family and D Landau (North Holland, Amsterdam)

Stanley HE 1984b J Stat Phys 36 843

Stanley HE 1985 in Ency on Polymer Science (Wiley, New York)

Stanley HE, Birgeneau RJ, Reynolds PJ and Nicoll JF 1976 J Phys C 9 L553

Stanley HE and Coniglio A 1983 in Percolation Structures and Processes eds G Deutscher, R Zallen and J Adler (Adam Hilger, Bristol)

Stanley HE and Coniglio A 1984 Phys Rev B 29 522

Stanley HE, Reynolds PJ, Redner S and Family F 1982 in Real-Space Renormalization eds TW Burkhardt and JMJ van Leeuwen (Springer Verlag, Heidelberg)

Witten TA and Sander LM 1981 Phys Rev Lett 47 1499

SCALING PROPERTIES OF CLUSTER AND PARTICLE AGGREGATION

M. Kolb[*+], R. Jullien[*] and R. Botet[*]

[*]Laboratoire de Physique des Solides, Bât. 510

Université Paris-Sud, 91405 Orsay, France

[+]Freie Universität Berlin, Institute for Theoretical Physics

D-1000 Berlin 33, W-Germany

ABSTRACT

The model of cluster aggregation and several variants of it serve to describe different disordered aggregation processes and their scaling properties. Here, the basic model is introduced and its properties are briefly discussed. Some more recent results are then presented in more detail : simplifying features in high dimension, gelation in cluster aggregation, the role of reaction kinetics, the problem of cluster readjustment and the internal structure of clusters in relation to the growth process. Finally a film, produced to illustrate how clusters aggregate, is described.

INTRODUCTION

There are many models for irreversible growth which by means of extremely simple operational procedures allow one to describe rather complex structures with non trivial fractal or scaling behaviour. This is relevant to a number of experimental situations, notably fracture, dielectric breakdown, electrodeposition, aerosols, colloid aggregation, etc...[1]. While scaling properties in experiments are observed over a limited range only, it is useful to study, classify and link theoretically various growth processes according to their asymptotic scaling behaviour. Here, we introduce the model of cluster aggregation (CA) or clustering of clusters and discuss its basic properties. Then the following topics are covered : 1) to better understand the growth mechanism, the model is treated in high dimension (transparent clusters), 2) at high cluster concentration, CA describes gelation, with different fractal properties, 3) another situation that leads to new fractal exponents is chemical or reaction limited clustering, 4) in experiments, the question of restructuring ot the clusters arises frequently. A simple model taking into account the readjustment of clusters will be outlined, 5) analysing the internal structure of cluster and particle aggregates, it is shown that the growth mechanism makes particle aggregates anisotropic and cluster aggregates isotropic.

A film has been produced to show the different aggregation processes and the resulting fractal structures. It is based on the visualisation of simulations for the growth models presented here. It will be explained, how the film was produced and what can be learned from this "real time" experiment.

CLUSTERING OF CLUSTERS

The model of CA^2 discussed here is defined as follows : N_o identical particles are placed in a volume V_o. The concentration $\rho_o = N_o/V_o$ is chosen such that the average distance between the particles is much larger than their radius ($\rho_o << 1$). The particles move independently oc each other in a completely random fashion (brownian motion). Occasionally, two particles are sufficiently close that the strong range attraction causes them to stick to each other, permanently. Such clusters of two particles continue, of course, to move with the same diffusive motion as single particles (possibly with a different diffusion constant). When these clusters encounter other clusters (or particles), they again stick together. The growth process defined this way continues indefinitely, as we suppose that the clusters never break up again. Thus, small clusters form when particles collide, larger clusters appear when small clusters (and particles) collide, etc... In a finite system the growth stops when all the particles belong to one large cluster.

In a first analysis, the fractal properties of large clusters (with several thousand particles) are determined. To this end, numerical simulations were performed by placing the particles in a box of linear size L in d dimensions (Volume $V_o = L^d$) or by placing them on a hypercubic lattice. The system is periodically bounded in all directions. Working on a lattice or off lattice does not change the measured fractal properties of large clusters.

The fractal analysis is suggested by the stringy appearance of the clusters, as shown in Fig. 1. They become less and less dense with increasing size. A quantitative measure is the radius of gyration

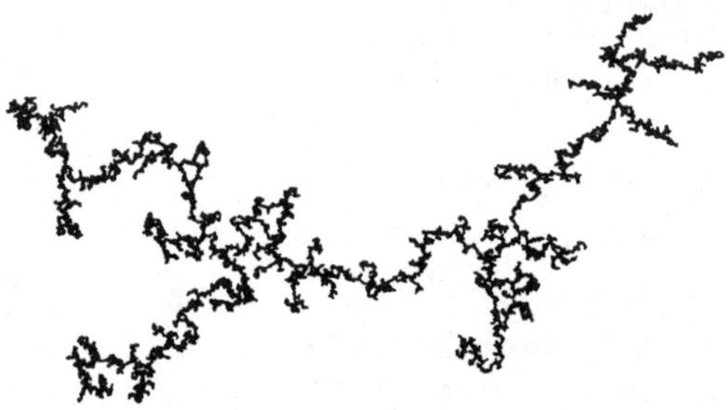

Fig. 1 : A cluster of 16384 particles obtained by diffusive cluster aggregation in two dimensions.

$R^2(m) = \frac{1}{2m^2} \sum_{i,j=1}^{m} (\vec{R}_i - \vec{R}_j)^2$, where the summation is over the m particles of the cluster, with coordinates $\vec{R}_i$. One can also study the density correlation function C(r) inside the cluster. The clusters are said to be fractal (without mathematical rigor) if

$$(1) \quad R(m) \sim m^{1/D}, \qquad m \to \infty$$

or equivalently

$$(2) \quad C(r) \sim r^{-(d-D)}, \qquad 1 << r << R$$

with $D < d$. The equivalence of (1) and (2) can be verified by integrating C(r) over r. D is the fractal dimension. In order to calculate D numerically, one takes an average over a large number of independently grown clusters. Fig. 2 shows a log-log plot of R versus m, from which the exponent D is estimated to be D = 1.43 ± 0.03 in d = 2 dimensions. Similar calculations

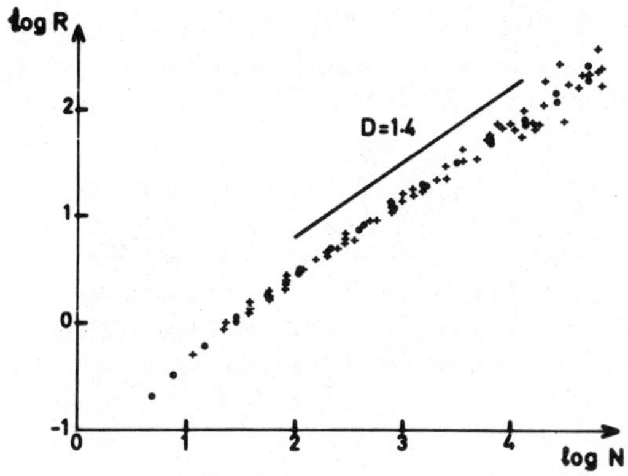

Fig. 2 : log-log plot of Radius R versus mass m of 2-dimensional aggregates grown by brownian CA. The slope determines the fractal dimension D.

for higher dimensions are quoted in Table 1 (the dimension of a brownian trajectory is two, $d_w = 2$). Experimentally, this growth process is realised in aerosols and colloids[3] where a fractal dimension D = 1.75 is measured in d = 3 dimensions, in good agreement with the theoretical value quoted in Table 1. So far, the dependence of the scaling properties on the

	D		
d	$d_w = 0$	$d_w = 1$	$d_w = 2$
2	1.55 ± 0.03	1.51 ± 0.03	1.42 ± 0.03
3	2.00 ± 0.06	1.91 ± 0.03	1.78 ± 0.05
4	2.32 ± 0.10	2.22 ± 0.04	2.04 ± 0.08
5	-	2.47 ± 0.05	2.30 ± 0.20
6	-	2.7 ± 0.3	2.60 ± 0.30

Table 1
Fractal dimensions D of CA as a function of the spatial dimension d and the dimension of the trajectories d_w ($d_w = 0$ independent of trajectory, $d_w = 1$ linear trajectory, $d_w = 2$ brownian trajectory)[4-6]

diffusion constant has not been considered. In order to take it into account we suppose that the diffusing velocity of a cluster depends on its mass as

$$(3) \qquad v(m) = m^{\alpha}$$

where α is a parameter that can be chosen to suit the experiments. In the simulations, this corresponds to moving each cluster by one lattice spacing with probability v(m) per unit time. To within numerical accuracy, one finds that D is independent of α, as long as $\alpha \lesssim 0$.

In contrast to particle aggregation[7], time has a direct physical meaning in the cluster aggregation process. Quantities such as the average radius R or mass m are time dependent and can be included in the scaling analysis,

$$(4) \qquad R \sim t^{1/z}, \qquad m \sim t^{\Theta}, \quad \Theta = D/z$$

where z is the dynamic critical exponent. In contrast to D, z depends on the exponent α, eq. (3). In fact, for $d \geqslant 2$, mean field arguments can be used to relate z to α and D, namely[8]

$$(5) \qquad z = D(1 - \alpha) - (d - 2)$$

This relation is quite well satisfied for $\alpha \lesssim 0$ and $2 < d$. Another quantity which can be expressed in scaling form, is the cluster size distribution. The number of clusters of mass m at time t, N(m, t) can be written as $N(m, t) = \frac{N(t)}{\bar{m}(t)} \; p(m/\bar{m}(t))$ with $N(t) = \sum_m N(m, t)$ and $\bar{m}(t) = \frac{N_o}{N(t)}$. p(x) then is time independent for $1 \ll N(t) \ll N_o$. Again, the simulations confirm that p(x) is indeed a scaling function and determine it as a function of α.

Is there a relation between the cluster size distribution of CA and the mean field description of cluster growth using the Smoluchowski equation ? The scaling solutions of the Smoluchowski equation depend on the degree of homogeneity of the reaction kernel just as p(x) depends on α in the simulations. If one matches these parameters such that the dynamics exponent (say for $\bar{m}(t)$) is the same, then also the scaling function has the same form[8, 9]. This shows that while the static aspects, i. e. the fractal dimension, has to be calculated numerically, the dynamics is mean-field.

Another aspect is suggested by the comparison with the Smoluchowski equation : two qualitatively different regimes can be distinguished, dependent on whether gelation (formation of an infinite cluster) occurs in a finite time or not. Indeed this distinction is also valid for CA[10]. For $\alpha > \alpha_c = 1 - \frac{d-2}{D}$ the growth process becomes that of particle aggregation (as seen explicitly for $\alpha \to \infty$, when only the largest cluster moves).

There are many generalisations possible for the CA model. The particle (and cluster) trajectories play an important role in the growth process. Instead of brownian trajectories one can consider linear trajectories ($d_w = 1$) or completely eliminate the trajectories ($d_w = 0$). The fractal dimension clearly depends on the choice of d_w, as illustrated in Table 1.

Let us now present some more recent and more detailed results of cluster aggregation and related growth processes.

CLUSTER AGGREGATION IN HIGH DIMENSION

For CA, there exists an upper critical dimension, above which the clusters can freely interpenetrate and their fractal dimension becomes independant of the spatial dimension d [11]. If one wishes to include the reaction kinetics, it is appropriate to use the Smoluchowski equation, which describes the cluster-size distribution correctly down to dimensions d = 2,3. The degree of homogeneity ω of the reaction kernel K(i, j) that governs the collision of an i-mer with a j-mer, $K(\lambda_i, \lambda_j) = \lambda^{2\omega} K(i,j)$ is the most relevant parameter for a large class of kernels ($\omega < 0$). Using the scaling function that results for large clusters at long times (depending on ω) a generalised aggregation process can be described, that includes effects such as a change over to a particle aggregation process - just as observed in the simulations[10]. The results are obtained analytically[12] and indicate that $D_c(\omega)$, the fractal dimension as a function of the parameter ω, behaves as $D_c(\omega \to 1/2) \to \infty$. The fractal dimension stays practically constant for all negative values of ω, rising from $D_c = 3.4$ to $D_c = 3.6$ when ω varies from $-\infty$ to 0. For $\omega > 1/2$, the upper critical dimension is infinite, anticipated because particle aggregation is believed to have $d_c = \infty$.

The existence of a finite or infinite upper critical dimension is illustrated clearly for CA with linear trajectories ($d_w = 1$) with and without impact parameter. In the first case (I) the clusters collide randomly, in the second case (II) the trajectories go through the centers of mass of both clusters. Fig. 3 compares the numerical results, suggesting that model II has the same upper critical dimension as the $d_w = 2$ model, whereas $D \to \infty$ as $d \to \infty$ for model I[13]. For model I, the transparency condition simply is never satisfied. This condition is required for the mean-field theory to be valid.

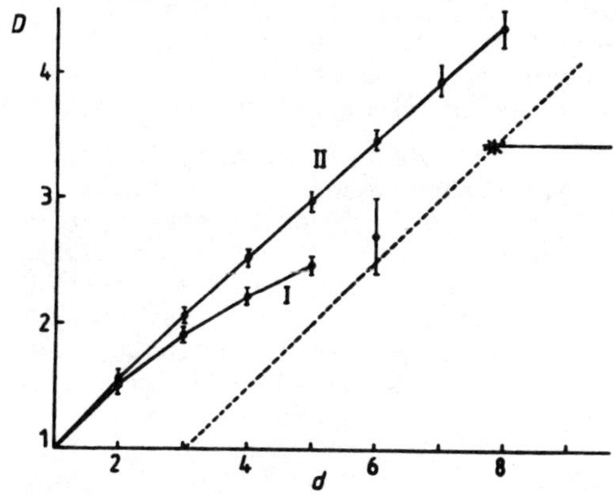

Fig. 3 : Fractal dimension D as a function of the spatial dimension d for CA with linear trajectories (d_w=1). Model I (II) corresponds to trajectories with (without) impact parameter, leading to a finite (infinite) upper critical dimension[13].

GELATION MODELLED BY CLUSTER AGGREGATION

The CA model studied so far was for low cluster concentration, precisely for $\rho_o = \frac{N_o}{V_o} \to 0$. Experimentally, there have been studies at finite cluster concentrations[14,15], finding a fractal dimension D which is larger than the value found for $\rho_o \to 0$. The reason for the increase of D is that at higher concentration the clusters interpenetrate, thus reducing the screening that makes aggregation clusters so stringy for brownian trajectories. At high concentration, the scaling behaviour occurs close to the gel point, when an infinite spanning cluster appears. Thus CA provides a mechanism to describe kinetic gel formation[16]. This aspect has been studied as a function of the cluster mobility exponent α, with the conclusion that the qualitative picture based on the Smoluchowski equation also applies to CA at high concentration. Of course, the relation between α and ω is different from the case when $\rho_o \to 0$. Varying α, one distinguishes between a finite and an infinite gel time, having different scaling properties. In

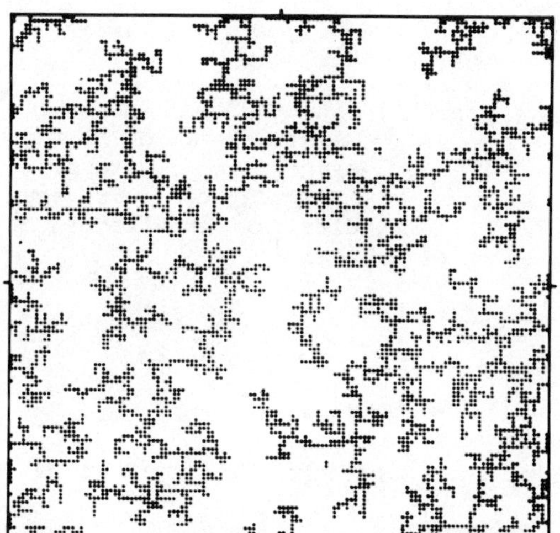

Fig. 4 : Simulation of CA in two dimensions at high concentration[16].

Fig. 4, CA at high concentration in two dimensions is shown. Its fractal dimension in two dimensions is $D = 1.75 \pm 0.07$ for $\alpha \lesssim -0.40$.

REACTION LIMITED CLUSTERING

The aggregation process considered so far is idealised in the sense that two clusters stick together at the first contact. In the experiments with gold colloids[9], it is possible to vary the reactivity, obtaining different fractal properties in the diffusion limited and the reaction limited case (when the clusters must collide many times before they stick). The bonds - once formed - are also irreversible in the reaction limited case. This new model (the $d_w = 0$ version of CA) has been studied theoretically[6] by introducing a sticking probability $0 \lesssim p \lesssim 1$. Every time two clusters touch a bond is formed with probability p. If they don't stick together they continue to diffuse independently, respecting nevertheless the steric constraints. $p = 1$ ($p = 0$) corresponds to diffusion limited (chemically limited) aggregation, $d_w = 2$ ($d_w = 0$). For $0 < P << 1$ there is a crossover from chemical to diffusive aggregation. Asymptotically, the clusters have the same properties as diffusive CA. $p = 0$ is best investigated using the hierarchical version of CA. The fractal dimension obtained from simulations, shown in table 1, agree well with the results from the experiments.

READJUSTMENT OF THE CLUSTERS

In the experiment by Allain and Jouhier[14], it has been observed, that there is a restructuring of the clusters, which changes (compactifies) their appearance, seen most clearly on the scale of a few particle radii. Notably, the clusters rotate until they form loops, which strengthen their mechanical stability. This effect has been modelled by letting the clusters in CA rotate around their point of first contact until they form a loop[17]. The simulations have been done as follows : when two clusters touch at a point, they are rotated instantly, either in the direction of the smaller, the larger or both (at random) angles, until a loop forms (sometimes two rotations are required). In the hierarchical version simulated, the choice of how one rotes does not change the results. The measured fractal dimension D in two dimensions is slightly larger when restructuring (1.49 instead of 1.42) though one cannot exclude that they are asymptotically the same (the average angle of rotation may tend to zero for large clusters). The very pronounced difference of the clusters can be seen in Fig. 5 where the

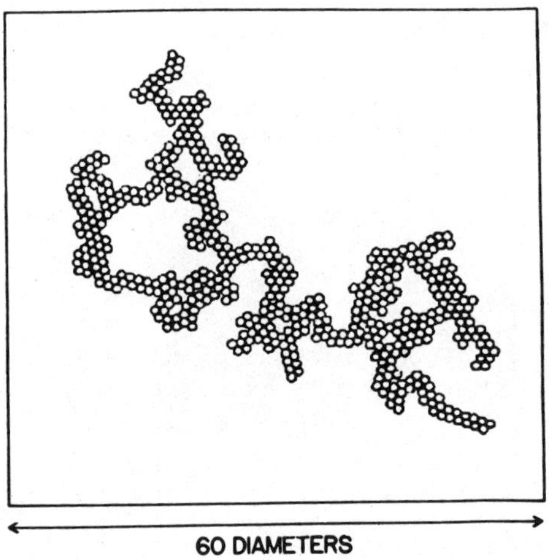

Fig. 5 : Cluster from hierarchical CA with rotation of the aggregating clusters. The structures look very different on a short scale[17].

particles form practically a crystalline structure on a very short scale.

INTERNAL STRUCTURE OF AGGREGATION CLUSTERS

While the fractal concept has been applied to many growth processes with success, no relation between the growth mechanism and the structure has been established so far. Particle and cluster aggregates do grow in a very different way : particle aggregation (PA) has a distinct center (the seed particle), cluster aggregation does not. In order to characterise the internal structure and quantify the distinction between these growth processes, angular density correlations were considered[18]. They show, that PA grows out of its center, whereas CA is isotropic. In the scaling region deep inside the cluster, the correlations $c(r) \sim r^{-A}$ parallel and perpendicular to the direction of growth scale with different exponents, for PA. The difference is small, $A_{\parallel} - A_{\perp} = \Delta A = 0.16 \pm 0.05$, which explains why the PA clusters appear isotropic on a small scale. The clusters for CA are fully isotropic on all scales. In fig. 6, the ratio of correlations

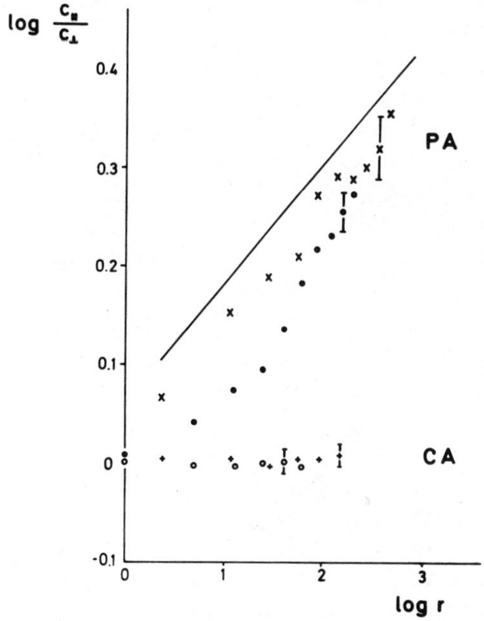

Fig. 6 : Parallel ($c_{\parallel}$) and perpendicular ($c_{\perp}$) density correlations inside clusters for PA and CA. The ratio $\frac{c_{\parallel}}{c_{\perp}}$ shows that, for PA only, the correlations scale differently in the two directions ; this leads to corrections to scaling effects.

parallel and perpendicular to the center of the cluster is shown, illustrating a marked difference for PA and CA. On a lattice, there is an additional anisotropy in the growth due to the preferred directions of the lattice. This, however only changes the amplitudes of the correlation functions.

FILM ON AGGREGATION PROCESSES

The various aggregation mechanisms presented above can be visualised directly by using the numerical simulations of these models to make a 'real time experiment' - a film.

This idea has been realised to show the growing structures in particle and cluster aggregation both with diffusion and chemically limited reaction kinetics, and at varying cluster concentrations.

The resulting movie is about 25 minutes long and was produced with the aid of the ZEAM, FU Berlin. The following technique was used : during a standard simulation of a growth process on a computer, a series of successive configurations are displayed visually on a color screen from which they are fotografed - one by one - by a 16 mm movie camera. This tedious process is required, as the frames cannot be generated at a sufficiently high rate to actually film the pictures off the screen. At 25 frames per second, the time of generating and displaying a configuration and exposing the film far exceeds the available 0.04 sec. Exposing the frames has nevertheless been mechanized by releasing the camera automatically from the screen, whenever a new drawing has been completed.

The color has been utilised to indicate the different stages of the growth - for PA succeeding layers of accreting particles are distinguished this way, for CA clusters of different sizes have different colors which illustrates the dispersity of the cluster size distribution.
The models illustrated are
1) The Eden model, a reaction limited particle aggregation model that models surface growth. The growth visibly leads to compact clusters, with a fairly narrow but rough surface.
2) The Witten-Sander model (PA) accretes particles diffusively. The screening by the tips prevents the incoming particles to penetrate into the fjords. The resulting cluster is fractal.
3) Cluster aggregation (CA) leads to clusters that are even more ramified than PA. The process is displayed exactly as described in the first paragraph above. The mobility of the clusters (eq. 3) manifests itself very clearly : for $\alpha \ll 0$ the cluster size distribution is monodisperse, for $\alpha \gg 0$, a crossover from CA to PA is visible (initially all clusters move, later on only the largest cluster moves. The final cluster visibly has the structure of a cluster aggregate on a short scale and a particle aggregate on a large scale, see above).
4) Introducing a sticking probability, $p \ll 1$, (see above) changes the aggregation process and results in more compact (though still fractal) clusters.
5) At a high concentration of clusters, CA describes gel formation. After a finite time (which depends on how the clusters move) the largest, fractal cluster spans the box ; it encloses many small clusters which through their diffusive motion gradually attach to the gelling network.

We acknowledge the collaboration and interesting discussions with H. Herrmann and P. Meakin. This work has been partially supported by the DFG, by an ATP of the C. N. R. S. and by the CCVR (Palaiseau).

REFERENCES

1. An overview of both theoretical and experimental results can be found in the abstracts of the workshop on "Kinetic models for cluster formation' (september 17-28, 1984, CECAM, Orsay) ; R. Jullien, M. Kolb, H. Herrmann and J. Vannimenus, eds. J. Stat. Phys. (April 1985).
2. P. Meakin, Phys. Rev. Lett. 51, 1119 (1983)
M. Kolb, R. Botet and R. Jullien, Phys. Rev. Lett. 51, 1123 (1983)
3. S. R. Forrest and T. A. Witten, J. Phys. A 12, 1109 (1979)
D. A. Weitz and M. Oliveria, Phys. Rev. Lett. 52, 1433 (1984)
D. A. Weitz, M. Y. Lin and C. J. Sandroff, Surf. Sci. to appear (1985)
4. R. Jullien, M. Kolb and R. Botet, J. Physique Lett. 45, L 211 (1984)
5. R. Jullien, J. Phys. A 17, L 771 (1984)
6. R. Jullien and M. Kolb, J. Phys. A 17, L 639 (1984)
M. Kolb and R. Jullien, J. de Physique Lett. 45, L 977 (1984)
7. T. A. Witten and L. M. Sander, Phys. Rev. Lett. 47, 1400 (1981)
8. M. Kolb, Phys. Rev. Lett. 53, 1653 (1984)
9. R. Botet and R. Jullien, J. Phys. A 17, 2517 (1984)
10. R. Botet, R. Jullien and M. Kolb, Phys. Rev. A 30, 2150 (1984)
11. R. Ball and T. A. Witten, Conference on Fractals ; Gaithersburg Md. (Nov. 1983)
12. R. Botet, J. Phys. A, 18, 847 (1985)
13. R. Ball and R. Jullien, J. de Physique Lett. 45, L 1031 (1984)
14. C. Allain and B. Jouhier, J. de Physique Lett. 44, L 421 (1983)
15. P. Ricetti, J. Prost and P. Barois, J. de Physique Lett., L 1137 (1984)
16. M. Kolb and H. Herrmann, J. Phys. A Lett., to appear
17. R. Jullien and P. Meakin, submitted for publication
18. M. Kolb, submitted for publication

A FRACTAL MODEL FOR CHARGE DIFFUSION ACROSS A ROUGH INTERFACE*

L. J. Gray

Engineering Physics and Mathematics Division
Oak Ridge National Laboratory
Oak Ridge, TN 37831

S. H. Liu and T. Kaplan

Solid State Division
Oak Ridge National Laboratory
Oak Ridge, TN 37831

It would appear to be a simple matter to describe the process of charge transfer across the interface between two substances (e.g., electrode and electrolyte): there is ohmic resistance in each material and a capacitance across the interface. However, experiments[1,2] indicate that, as the angular frequency $\omega \to 0$, the impedance $Z(\omega)$ behaves as

$$Z(\omega) \sim (j\omega)^{-\eta} \quad , \tag{1}$$

where $0 < \eta < 1$. This is contrary to the expected $\eta = 1$ one would get from the above simple RC circuit. Furthermore, the observed values of η depend on the roughness of the surface, with $\eta \to 1$ as the surface becomes smoother.

Since nonintegral exponents often arise from scaling, it is reasonable to examine a model wherein the interface is assumed to have a fractal structure.[3] A cross section of a simple but nonetheless illuminating model is depicted in Fig. 1; the surface is assumed to be grooved, with each groove itself similarly grooved, but scaled down in size by the factor $a > 2$ at each stage.[4] This model is related to the Cantor bar[5] which has fractal dimension $d = \ln(2)/\ln(a) < 1$, and the fractal dimension of the surface is $2 + d = d_s$. If each groove divides into N branches and $a > N$ then $d = \ln(N)/\ln(a)$. The corresponding electric circuit for this interface is shown in Fig. 2; the resistance increases by the factor a at each stage because of the reduction in cross-sectional area of the branch and the capacitance C represents the effect of the two lateral faces (the capacitance contributed by the dip between branches goes to zero at higher stages and has no effect on the asymptotic behavior of $Z(\omega)$). The impedance is given by

$$Z(\omega) = R + \cfrac{1}{jC\omega + \cfrac{2}{aR + \cfrac{1}{jC\omega + \cfrac{2}{a^2R + \cdots}}}} \tag{2}$$

Substituting ω/a for ω in the continued fraction and doing the necessary algebra to remove the 1/a factor from ω yields

$$Z(\omega/a) = R + \frac{aZ(\omega)}{2+jC\omega Z(\omega)} \quad . \tag{3}$$

For $\omega \to 0$ (assuming $\omega Z(\omega) \to 0$) this reduces to the scaling relation

$$Z(\omega/a) = (a/2)Z(\omega) \quad , \tag{4}$$

which has $Z(\omega) = A\omega^{-\eta}$, $\eta = 1 - \ln(2)/\ln(a)$, as a solution. Since $d_s = 2 + \ln(2)/\ln(a)$, it follows that η and d_s are related by $\eta = 3 - d_s$. Note that for a smooth surface $d_s \to 2$ and thus $\eta \to 1$ as required.

While this fractal model exhibits the experimentally observed behavior, it is clearly not meant to be an accurate picture of a rough interface. In a real material the rough interface is actually a disordered structure. A more realistic model would naturally allow the scaling and branching in the fractal to vary from stage to stage; this can be accomplished by assuming that the scaling and branching are random variables. In addition, the surface should be viewed as being composed of many fractal regions, each with a different fractal structure. Since these regions are connected in parallel, and since in parallel the admittances 1/Z add, the relevant physical quantity is the average admittance $\langle 1/Z(\omega)\rangle$.

For the model considered herein, the impedance is given by

$$Z(\omega) = R + \cfrac{1}{jC\omega + \cfrac{N_1}{a_1R + \cfrac{1}{jC\omega + \cfrac{N_2}{a_1a_2R + \cfrac{1}{\cdots\cdot}}}}} \quad . \tag{5}$$

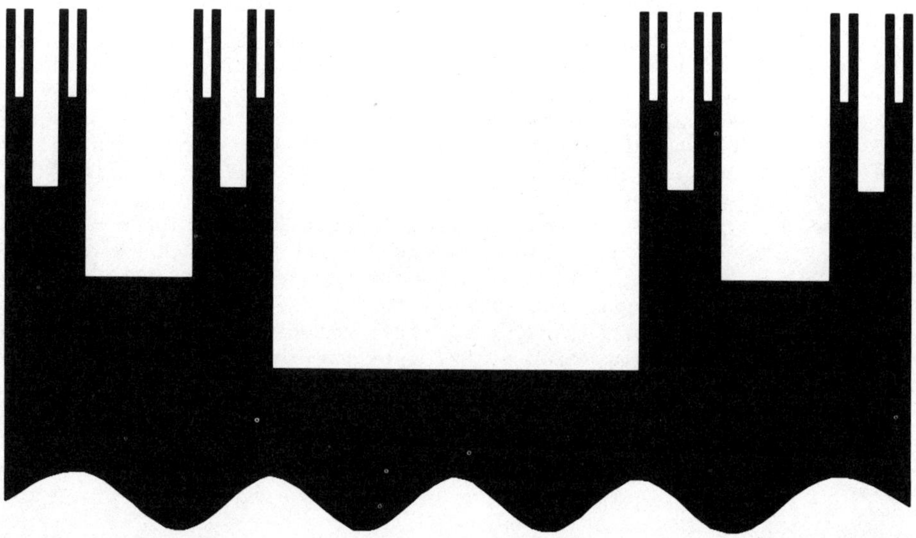

Fig. 1. Fractal model of a rough interface between an electrolyte (black) and an electrode (white). Two grooves each with four stages of branching are shown.

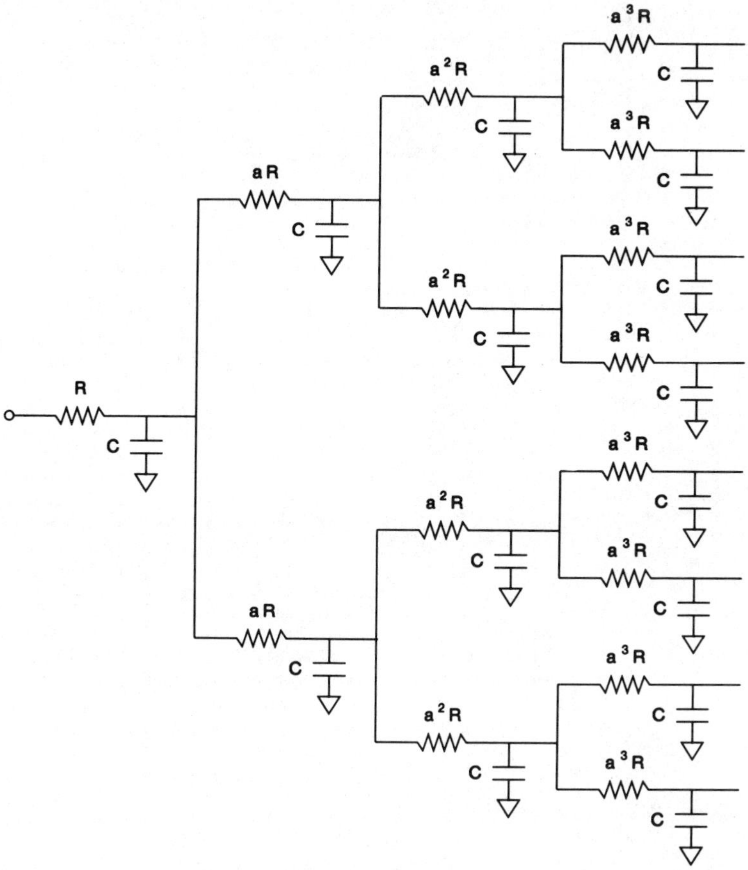

Fig. 2. The equivalent circuit model for a groove of the rough interface model in Fig. 1.

The branchings N_k are independent random variables with probability distribution $p_N(x)$, whereas the scaling variables $\{a_k\}$ are also independent variables (and independent of $\{N_k\}$) with density $p(x)$. Thus, at the k-th stage, N_k new branches appear, each of these new branches scaling according to a_k. A more complicated model, one which allows each of the new branches to scale independently, can also be analyzed;[6] the results however are the same as derived below. Returning to Eq. (5), note that

$$1/Z(\omega/a_1) = \left[R + \cfrac{1}{jC\frac{\omega}{a_1} + \cfrac{N_1}{a_1 R + \cfrac{1}{jC\frac{\omega}{a_1} + \cfrac{N_2}{a_1 a_2 R + \cdots}}}} \right]^{-1}$$

$$= \left[R + \cfrac{a_1}{jC\omega + \cfrac{N_1}{R + \cfrac{1}{jC\omega + \cfrac{N_2}{a_2 R + \cdots}}}} \right]^{-1}$$

$$= \left[R + \cfrac{a_1}{jC\omega + \cfrac{N_1}{Z_1(\omega)}} \right]^{-1} = \left[R + \frac{a_1 Z_1(\omega)}{jC\omega Z_1(\omega)+N_1} \right]^{-1}$$

$$= \left(1/Z_1(\omega)\right) \left[R/Z_1(\omega) + \frac{a_1}{jC\omega Z_1(\omega)+N_1} \right]^{-1} \quad . \tag{6}$$

Note that $Z_1(\omega)$ is precisely $Z(\omega)$ with the first stage deleted. Therefore $\langle 1/Z(\omega)\rangle = \langle 1/Z_1(\omega)\rangle_1$, where $\langle\cdot\cdot\rangle_1$ indicates averaging over all random variables except a_1 and N_1. Now for $\omega \to 0$, one expects, as before, that $1/Z(\omega) \to 0$ and $\omega Z(\omega) \to 0$, and thus in this limit Eq. (6) becomes

$$1/Z(\omega) = \frac{N_1}{a_1}\, 1/Z_1(a_1\omega) \quad . \tag{7}$$

Averaging this equation yields

$$\langle 1/Z(\omega)\rangle = \int \frac{N_1}{a_1} \langle 1/Z_1(a_1\omega)\rangle_1 p(a_1) p_N(N_1) da_1 dN_1$$

$$= \langle N\rangle \int \frac{1}{a_1} \langle 1/Z(a_1\omega)\rangle p(a_1) da_1 \quad . \tag{8}$$

If it is assumed that $\langle 1/Z(\omega)\rangle \sim \omega^{\eta}$, then this equation becomes

$$\omega^{\eta} = \langle N\rangle \int a_1^{\eta-1} \omega^{\eta} p(a_1) da_1 \quad ; \tag{9}$$

i.e., the frequency exponent η satisfies the equation

$$1 = \langle N\rangle\langle a^{\eta-1}\rangle \quad . \tag{10}$$

(It can be shown that there is a unique solution to this equation in the interval (0,1) provided the physically reasonable inequalities

$$\langle a^{-1}\rangle < 1/\langle N\rangle < 1 \tag{11}$$

are true).

In order to relate η to the geometry of the interface, as was done for the non-random model, it is necessary to calculate the fractal dimension of this random surface. This is accomplished in Ref. 6, where it is shown that

$$1 = \langle N \rangle \langle a^{2-d_s} \rangle \quad . \tag{12}$$

Thus $\eta = 3 - d_s$, which is the same relationship predicted by the non-random fractal model and the more complicated model mentioned above. Since η and d_s can be measured independently, this result can be tested experimentally. To our knowledge, Eq. (12) provides the first determination of the fractal dimension of a Cantor bar for which the scaling is random.

REFERENCES

*Research sponsored by the Division of Materials Sciences and Applied Mathematical Sciences Subprogram, U. S. Department of Energy under contract DE-AC05-84OR21400 with Martin Marietta Energy Systems, Inc.

1. See the review by P. H. Bottelberghs, in "Solid Electrolytes," edited by P. Hagenmuller and W. Van Gool, Academic Press (New York, 1978), Chap. 10, and references cited therein.
2. J. B. Bates, private communication.
3. A. Le Mehaute and G. Crepy, Solid State Ionics 9 and 10:17 (1983).
4. S. H. Liu, submitted for publication.
5. B. B. Mandelbrot, "The Fractal Geometry of Nature," Freeman (New York, 1983).
6. T. Kaplan and L. J. Gray, submitted for publication.

APPLICATIONS OF SCALING AND DISORDERLY GROWTH PHENOMENA TO OIL RECOVERY

H.E.Stanley,* G.Daccord,** H.J.Herrmann+ and J.Nittmann**

*Center for Polymer Studies and Department of Physics
Boston University, Boston, MA 02215

**Etude et Fabrications, Dowell Schlumberger
42003 St. Etienne, France

+Service Physique Theorique, CEN Saclay
91191 Gif-sur-Yvette, France

INTRODUCTION

Students often like to see some connection between abstract topics like "scaling in disorderly growth systems" and problems confronting practical people who live in the real world. For this reason, I shall devote my third and final lecture to describing two somewhat "novel" fractal growth structures, each of which has some indirect relevance to the oil recovery problem.

NOVEL FRACTAL 1. FRACTAL GEOMETRY OF THE CRITICAL PATH: "VOLATILE FRACTALS" [with H. J. Herrmann]

When a fluid just breaks through a random material, the critical path it takes is a fractal object. If the breakthrough point is like the percolation threshold, then the critical path is termed the backbone. Recently it has been demonstrated that the backbone consists of multiply connected blobs connected by singly connected links. The blobs themselves form a "volatile fractal" whose identity depends on the box size L used to embed them. Scaling for volatile fractals is described, as well as the exponential law obeyed by the strands of links.

NOVEL FRACTAL 2. VISCOUS FINGERS IN HELE-SHAW CELLS: "CROSSOVER FRACTALS" [with G. Daccord and J. Nittmann]

When one attempts to use a low viscosity fluid to push a high viscosity fluid, there arises a fluid instability phenomenon termed the "viscous fingering instability" in which the low viscosity fluid forms characteristic fingers extending well into the high viscosity fluid. Under suitable conditions, the fingers can form fractal structures whose fractal dimension crosses over continuously as a function of the width of the cell to the DLA value, $d_f = 1.71$ (Nittmann et al 1985).

Before beginning, it is appropriate to remark that, in addition to my collaborators who are joining me as co-authors, several others kindly offered their advice and criticism: A. Coniglio, R. Lemanzyk, R. Pike, P. Rossi, J. D. Sherwood and D. Stauffer.

PART 1. THE CRITICAL PATH: A "VOLATILE FRACTAL"

How does oil flow through randomly porous material just at that point where it can "break through" and reach the surface? How does electricity flow through a random resistor network where the fraction of intact resistors is close to the percolation threshold p_c? How does one describe the shape of the incipient infinite cluster that appears near p_c when one considers the polyfunctional condensation of monomers? There are but three of a host of questions that one can ask that are of some practical interest and require for their answer a clear specification of cluster structure.

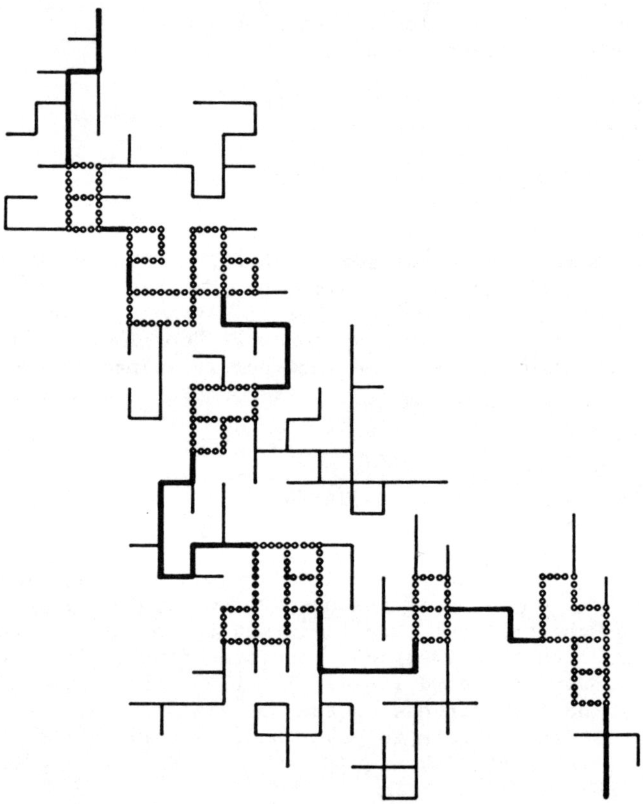

FIG. 1: A large machine-generated finite cluster just below the percolation threshold of the square lattice bond problem. The bonds of this cluster are of three sorts: dangling ends (light), singly-connected "red" bonds (heavy), and multiply-connected "blue" bonds (dotted). The red and blue bonds form the backbone, which (a) carries stress if the two ends of the cluster are pulled, (b) carries current if a batter y is applied across the cluster, (c) carries fluid if the bottom of the cluster is oil and the top is on the ground, if the cluster represents a model of a randomly-porous material with oil at the bottom, (d) propagates spin order, if we view this as a large cluster of Ising spins just below p_c. From Stanley (1977).

It has long been recognized (Kirkpatrick 1978, Shlifer et al 1979) that the dangling ends, shown as lines in Fig. 1, do not contribute to transport. Hence people began to focus on the problem obtained by decapitating all the dangling ends. The resulting structure is termed the backbone (Fig. 2). Is the backbone self-similar up to length scales L of the order of the connectedness length ξ ? If so, what is its fractal dimension d_{BB}? Many workers have verified that the backbone is indeed a fractal object, with

$$N_{BB} \sim L^{d_{BB}} \qquad \text{[FRACTAL DIMENSION No. 8]} \qquad (1.1)$$

for length scales L in the range $1 \ll L \ll \xi$. The fractal dimension d_{BB} sticks at the value 2 for $d > d_c = 6$, and then decreases "slowly" as d decreases below 6. By $d=2$, d_{BB} has decreased to about 1.65 (different workers obtain different values, since d_{BB} is quite difficult to estimate accurately). One of the problems is associated with obtaining a fast algorithm suitable for very large systems. It appears that Herrmann's burning algorithm (Herrmann et al 1984) is quite efficient and makes possible studying two-dimensional backbones in a box of size up to $(600)^2$ and $d=3$ backbones up to $(60)^3$.

Some time ago, it was noted (Stanley 1977) that there are two kinds of bonds in the backbone:
(i) singly-connected ("red") bonds, which have the property that if they are cut, the backbone ceases to conduct, and
(ii) multiply-connected ("blue") bonds, which have the property that they can be cut without interrupting the flow.

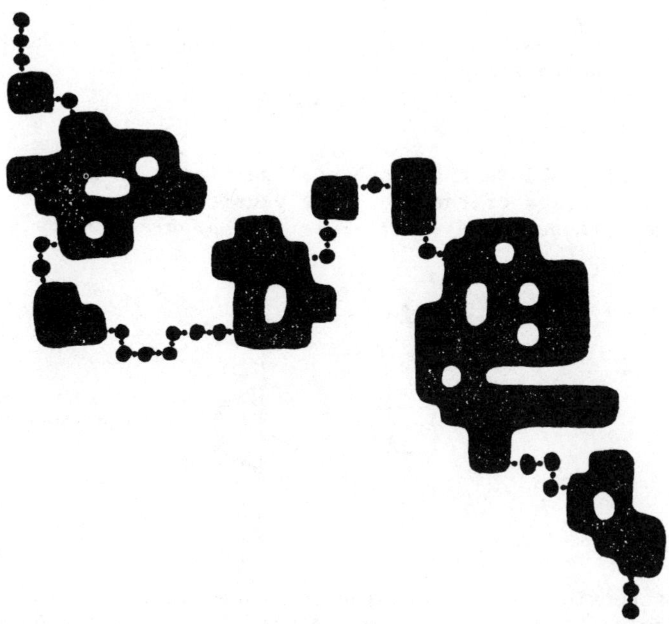

FIG. 2: Actual simulation of a backbone in site percolation just below p_c for a square lattice. The decomposition into blobs of all sizes form 1 to oo is apparent. After Herrmann and Stanley (1984).

The rationale for the terms red and blue is that the red bonds carry all the current from one bus bar to the other, and so are "hotter" than the blue bonds which "share" the current with the other blue bonds. In 1977, no one believed that the red bonds were a critical quantity. At the annual Canadian Physics Association meeting in November 1977, I posed the "red-blue" problem and an undergraduate member of the audience, Rob Pike, came up afterward to announce that he could solve the problem. In the ensuing 3 years, Pike succeeded in developing a computer algorithm that partitions bonds of a cluster into three separate colors: red, blue, and "yellow" (the dangling ends).

One of the worries that clouded the Pike project was the possibility that as p approached p_c, the mean number of red bonds would approach zero as the dangling ends found each other, transforming everything in between into a large blue blob. Fortunately, Pike's data showed clearly that all three functions, N_{red}, N_{blue} and N_{yellow} diverge (Pike and Stanley 1981):

$$N_{red} \sim L^{d_{red}} \qquad \text{[FRACTAL DIMENSION No. 9]} \tag{1.2a}$$

$$N_{blue} \sim L^{d_{BB}} \tag{1.2b}$$

$$N_{yellow} \sim L^{d_f}, \tag{2c}$$

where it is understood that $1 << L << \xi$. The same year, Coniglio (1981) provided a rigorous argument that for all d,

$$d_{red} = 1/\nu = y_T, \tag{1.3}$$

where $\xi \sim |\epsilon|^{-\nu}$ with $\epsilon = (p_c - p)/p_c$ and y_T is the thermal-like scaling field in percolation. Since

$$y_h = d_f \tag{1.4}$$

(Stanley 1977), it follows that both scaling fields in percolation are equal to geometric properties of the incipient cluster! What about d_{BB}? To what is this fractal dimension related? This is a problem we are working on just now.

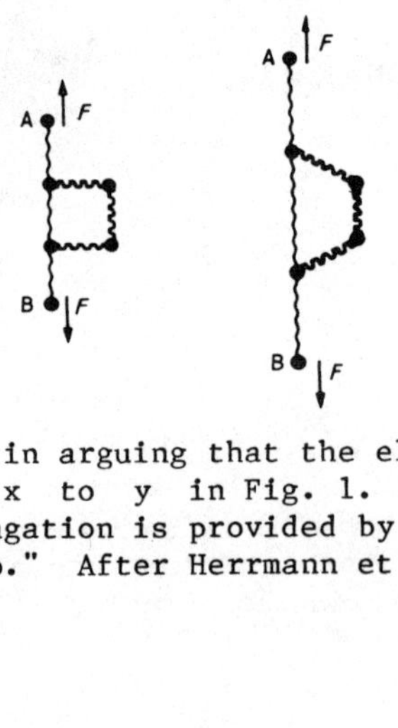

FIG. 3: Construction used in arguing that the elastic backbone is the union of all minimum paths from x to y in Fig. 1. If each bond is a spring, then the resistance to elongation is provided by only one of the four bonds comprising the 4-bond "blob." After Herrmann et al (1984).

ELASTIC BACKBONE

Suppose we regard each of the bonds in Fig. 1 as being a tiny spring (Fig. 3). Then if we pull on the two extremeties of the IIC, we find that only a subset of the backbone bonds, called the <u>elastic backbone</u>, carries stress. The fractal dimension of the elastic backbone is defined by

$$N_{Elastic} \sim L^{d_E} \quad \text{[FRACTAL DIMENSION No. 10].} \tag{1.5a}$$

Herrmann et al (1984) and Herrmann and Stanley (unpublished) have carried out extensive calculations of this quantity for $d=2,3$. It appears that within the 1-2% accuracy of the calculations,

$$d_E = d_{min}. \tag{1.5b}$$

Here d_{min} is the fractal dimension of the minimum path joining two points (see lecture 1). This is not altogether unreasonable, since the elastic backbone is the union of all minimum paths--just as the backbone between two points is the union of all self-avoiding walks between the two points. If true, Eq. (1.5b) means that the sort of loops which contribute to the elastic backbone do not occur on all length scales.

VERIFICATION OF RED-BLUE STRUCTURE OF BACKBONE: VOLATILE FRACTALS

Is there direct evidence that the red/blue (or, to use the term coined by Coniglio (1982), "links" and "blobs") picture of the backbone is valid? This is the question addressed recently by Herrmann and Stanley (1984). To this end, they studied the "blob size distribution function" for site--as opposed to bond--percolation (cf. Fig. 4). They found the probability P_s that a randomly-chosen site in the backbone belongs to an s-site blob scales, for fixed box size, as

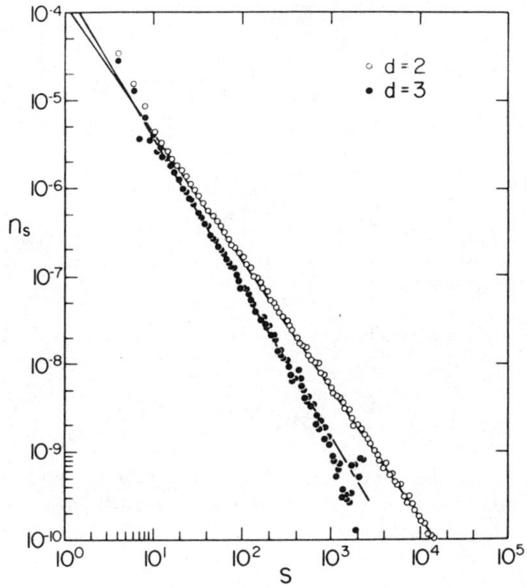

FIG. 4: Blob size distribution function for $d=2$ and 3 for $L = 600$ ($d=2$) and for $L = 60$ ($d=3$). Note that the data are linear over many decades. After Herrmann and Stanley.

"cluster size distribution scaling analysis" cluster-cluster aggregation case. The formalisms, though developed independently, are completely identical under the transformation

$$L \longleftrightarrow t. \tag{1.7}$$

From Fig. 2 it appears that the backbone is a pearl necklace, some of the pearls being swollen into a myriad of gargantuan shapes. Presumably, the backbone should be a string of pearls each of which is strung together by a drunken assembly line worker whose job is only to randomly choose a pearl from the pearl size distribution function P_s. To test this appealingly-simple possibility, we considered the statistical distribution of strings of "red sites" (1-site blobs). Specifically, we made a histogram of the number of strings with m sites against m (Fig. 6). We found that this histogram decayed exponentially with m--one of the few examples in fractals where one must use semi-log as opposed to log-log paper! The parameters entering this exponential decay are calculated in terms of the total number of red sites and the total number of blobs.

We conclude this section by making connection with the problem of anomalous diffusion and transport mentioned in the previous lecture. There we emphasized the fundamental importance of the Nernst-Einstein relation connecting a transport quantity, the macroscopic conductivity, to the diffusion constant which measures microscopic motion. Applied to the full cluster, we obtained the relation $d_R = d_w - d_f$ of Eq. (18). What happens if we chop off all the dangling ends? Clearly the conductivity cannot change, so that d_R is unaffected. However d_f is changed to d_{BB} and d_w

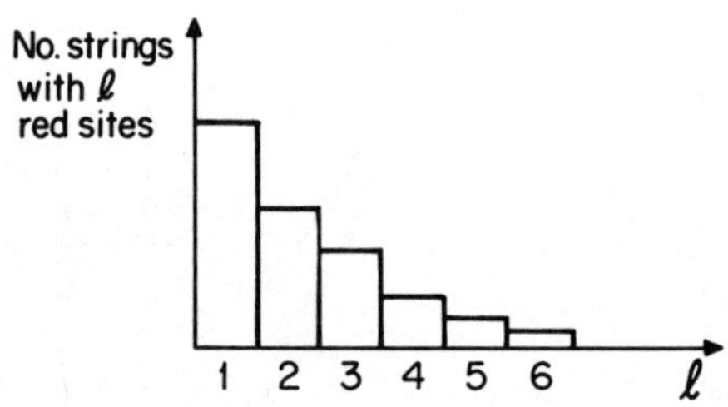

FIG. 6: Schematic illustration of histogram giving number of strings found with red (singly-connected) sites in each. Actual data for d=2,3 are in Fig. 4 of Herrmann and Stanley (1984).

$$P_s = sn_s \sim s^{-d/d_f} \sim s^{-\tau+1}. \tag{1.6}$$

Now suppose we double L. What happens to the backbone? As Fig. 5a suggests, the actual identity of the backbone changes, as tiny strands of red sites become part of a big blob. In general, the little blobs cascade into big blobs and the entire distribution shifts "downward" (Fig. 5b). We say that the backbone is a volatile fractal, in that its identity depends on the box size.

Another volatile fractal is cluster-cluster aggregation, which was treated in the seminar by M. Kolb. Here the time plays the same role as the box edge L. If we double t, then little clusters get eaten up, and become part of bigger clusters. The detailed "blob size distribution scaling analysis" for volatile fractals (Fig. 4) is presented by Herrmann and Stanley (1984) for the backbone case, and by Vicsek and Family (1984) for the

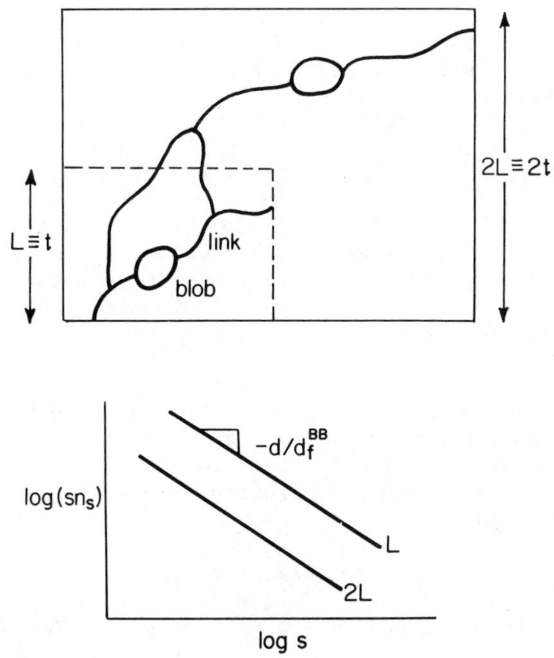

FIG. 5: (a) Schematic illustration of how small blobs become part of larger blobs when the box size L is doubled. (b) Schematic illustration of how the cluster size distribution $n_s(L)$ decreases when L increases.

is changed to d_w^{BB}, which is defined as the exponent connecting the number of steps on the backbone to the range of the walker:

$$N_w^{BB} \sim L^{d_w^{BB}} . \qquad (1.8)$$

Stanley and Coniglio (1984) have demonstrated that the relation

$$d_R = d_w - d_f \qquad (1.9a)$$

obtained in Lecture 2 becomes

$$d_R = d_w^{BB} - d_{BB}, \qquad (1.9b)$$

where d_w^{BB} governs the fashion in which N_w^{BB} (the number of steps in a random walk on the backbone) scales with the range L of the walk. Combining these two equations, we have the "invariance relation"

$$d_w - d_f = d_w^{BB} - d_f^{BB}, \qquad (1.9c)$$

which we sometimes name after my spouse, since she was mildly amused that this simple-looking equation was the only tangible result of a week's research in Naples with Antonio Coniglio.

PART II. FRACTAL VISCOUS FINGERS

Now we turn our attention to another fractal problem of some relevance to oil recovery. Initially oil may rise to the surface under its own pressure. Eventually all oil wells run dry. However there remains a substantial fraction of the original oil in the ground. The puzzle is how to get it out. One trick involves drilling wells nearby and forcing water down these wells. The water being immiscible with oil, pushes the oil ahead of it.

The problem of a fluid (like water) pushing a more viscous fluid (like oil) can be studied profitably using a classic benchtop apparatus called a Hele Shaw cell. This is basically a sandwich, where the role of the bread is played by two transparent pieces of glass separated by a distance b that is fixed at a value between 0.1 mm and 1 mm. Initially, the space between the plates is filled with "oil" but at time $t=0$ water is forced in one end under pressure. The low viscous fluid penetrates the high viscous fluid, producign thereby an instability pattern that resembles, loosely speaking, the finger of a glove. The flow in the Hele Shaw cell is similar to flow in porous media. For both systems, a relation exists between pressure p and flow velocity u, and

$$u = -(k/\mu) \text{ grad } p, \qquad (2.1)$$

where μ is the viscosity and $k = b^2/12$ is the permeability.

Recently, Daccord has made detailed measurements for a somewhat different system than oil/water. The interfacial tension introduces a critical wavelength (Chuoke et al 1959), and only perturbations of larger wavelength will grow. Hence Daccord's experiments use two fluids, water and a high molecular weight polysaccharide aqueous solution. Polymer solutions are non-Newtonian, with a viscosity that decreases under high shear, by as much as an order of magnitude. This has an important practical consequence: it becomes possible to pump the fluid easily in a thin cell without causing the Hele Shaw plates to bend.

Since a zero critical wavelength would imply that perturbations of <u>all</u> wavelengths can grow, we anticipate the possibility of fractal behavior. Indeed, the first experimental fingers were quite ramified, so we analyzed them for fractal behavior. Our methods were so simple that they merit some discussion here. The first step was to trace the figure on a piece of ruled paper. The second step is to digitize the figure--i.e., to replce the continuum finger by a discrete finger composed of, typically, 300 cells. The third step is to place a tiny L x L sandbox around each cell and count the occupied cells ("mass") within the sandbox (Fig. 7a). Step 4 is to average

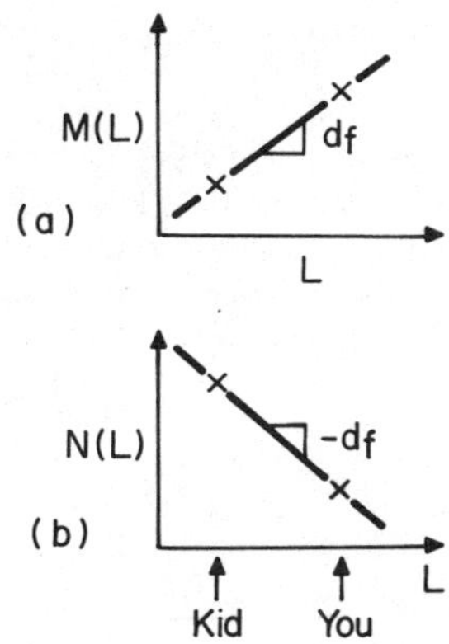

FIG. 7: Schematic illustration of two complementary measurements of the fractal dimension, (a) the sandbox method and (b) the coastline method. See explanation in text.

over all possible cells, and so obtain a mean mass M(L). The fifth and last step is to repeat the entire procedure for several value of L, and to make a log-log plot of M(L) against L. To our immense surprise, this graph was linear! The slope, which is the fractal dimension d_f of the finger was about 1.4.

We confirmed this number by a complementary calculation (Fig. 7b). We calculated the fractal dimension of the hull of the fractal by walking around it with different step lengths L. The idea here is as follows. Suppose you walk around Long Island with your kid, whose step length is half of yours: the kid will take <u>more</u> than twice as many steps because the kid can enter the tiny invaginations of the coastline that you with your long legs will simply step over.

The fractal dimensions of the bulk and of the coastline will agree only if there are no loops. If there are loops, as in percolation clusters, these two will be quite different; e.g., in percolation, we noted in lecture 2 that $d_f = 91/48 = 1.895$, while $d_{hull} = 1.75$. We find that both the bulk and coastline fractal dimensions are the same, to within the accuracy of the measurements--which, incidentally, have been considerably improved in that the "hand digitizer" used in the July 1984 measurements has been replaced by a machine digitizer so that instead of representing the fractal by roughly 300 cells there are almost two orders of magnitude more. The measured fractal dimension barely changes.

INTERPRETATION

From Eq. (2.1), and the fact that the divergence of the velocity field is zero, it follows that the pressure obeys the Laplace equation

$$\nabla^2 P = 0, \tag{2.2}$$

so long as we can neglect any spatial dependence of the viscosity. A random walk model immediately comes to mind. The pressure has one value at the water inlet and another elsewhere. Accordingly, the random walkers can be said to be absorbed if they hit the growing finger, and are absorbed if they hit any of the four walls of the cell. For this reason, the ultimate fractal structure will become one-dimensional in the limit of an infinitely long Hele Shaw cell. Conversely, for a cell whose width is much longer than its length, the structure should resemble DLA since the walls should play no role (cf. Fig. 8). Actually, the structure is a sort of semi-plane of DLA since all the incoming walkers are confined to a half plane (Fig. 9).

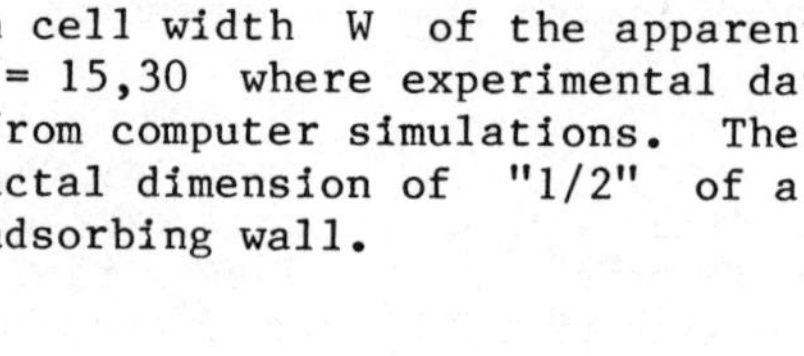

FIG. 8: Dependence on cell width W of the apparent fractal dimension, showing the values W = 15,30 where experimental data were obtained and the solid curve obtained from computer simulations. The horizontal line at 1.71 corresponds to the fractal dimension of "1/2" of a DLA cluster, obtained by placing a seed on an adsorbing wall.

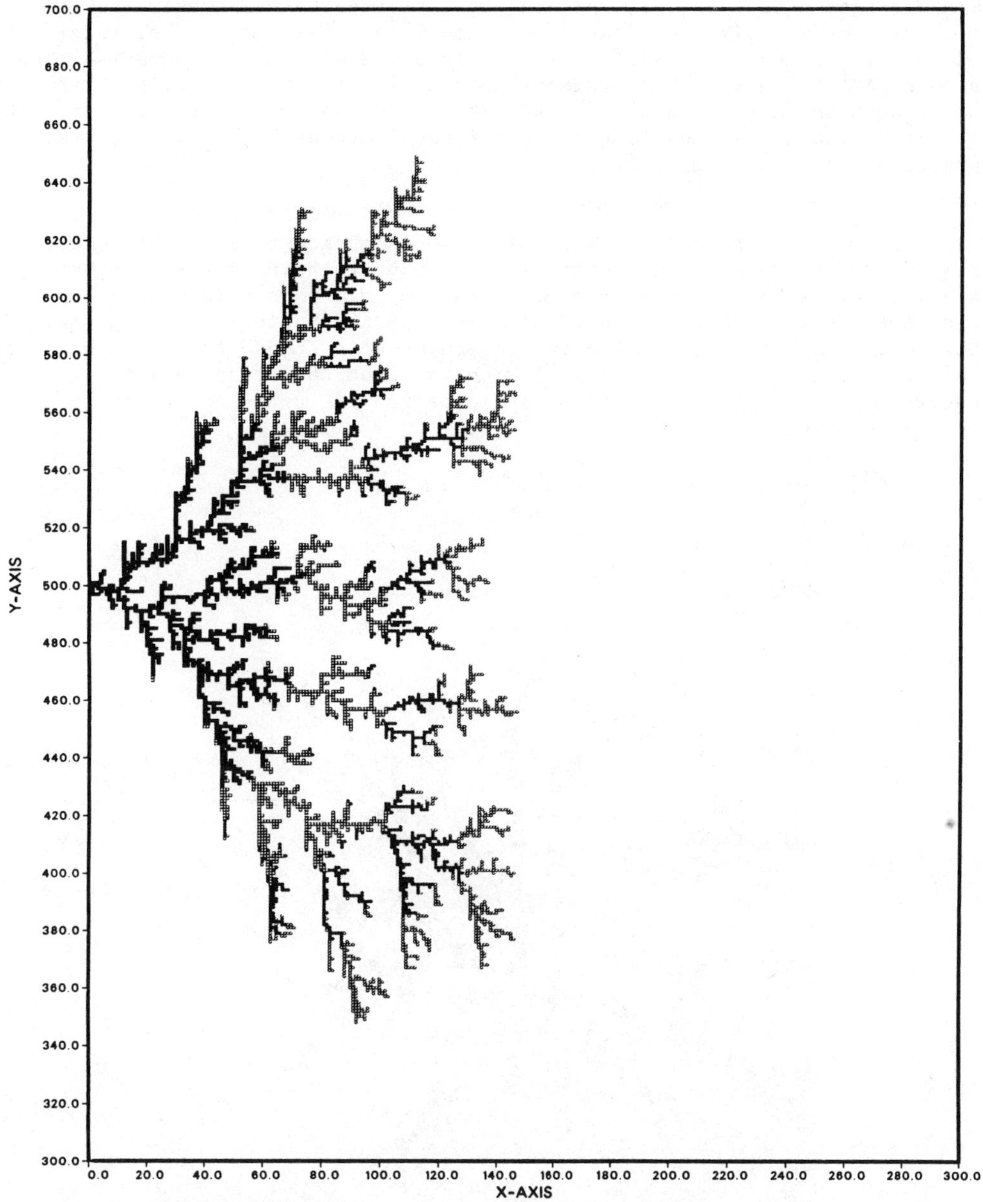

FIG. 9: Simulation of a Hele Shaw cell in the limit that the cell width is much larger than the finger thickness. This object, which appears to be a bona fide fractal object, will eventually cross over to a one-dimensional non-fractal object.

In order to avoid ambiguities with boundary conditions, some of which are somewhat controversial, Daccord et al (1985a) have recently reported observing radial viscous fingers in using a geometry that is completely free of walls. The water is simply brought into the middle of a sandwich through a hose in the center. The fluids cannot know that the system is finite until the finger becomes comparable in size to the diameter of the sandwich. The resulting radial viscous fingers have fractal dimension $d_f = 1.7$, as predicted for DLA (Fig. 10).

Clearly this subject is one that demands much further study before any statements can be made with confidence. However the potential for using statistical mechanics in general (and random walk theory in particular) to shed new light on basic instabilities in fluid mechanics is intriguing. With this optimistic outlook on the future, perhaps we should bring these three lectures to a close. Once again I wish to thank the organizers for their generosity in inviting me and the audience for its patience in listening to me. Thank you.

FIG. 10: A radial viscous finger, with no boundary conditions and measured fractal dimension 1.7. After Daccord et al (1985a).

REFERENCES

Chuoke RL, Van Meurs P and Van der Poel CJ 1959 J Petrol Tech 11 64

Coniglio A 1981 Phys Rev Lett 46 250

Coniglio A 1982 J Phys A 15 3829

Daccord G, Nittmann J and Stanley HE 1985a Phys Rev Lett (submitted)

Daccord G, Nittmann J and Stanley HE 1985b Proc Les Houches Conference on Finely Divided Matter (to be published by Springer, M Daoud and N Boccara eds)

De Arcangelis L, Redner S and Coniglio A 1985 Phys Rev B 31 4757

Feng S and Sen PN 1984 Phys Rev Lett 52 216

Herrmann HJ, Hong D and Stanley HE 1984 J Phys A 17 L261

Herrmann HJ and Stanley HE 1984 Phys Rev Lett 53 1121

Kantor Y and Webman I 1984 Phys Rev Lett 52 1891

Kirkpatrick S 1978 AIP Conf Proc 40 99

Nittmann J, Daccord G and Stanley HE 1985 Nature 314 141

Pike R and Stanley HE 1981 J Phys A 14 L169

Shlifer G, Klein W, Reynolds PJ and Stanley HE 1979 J Phys A 12 L169

Stanley HE 1977 J Phys A 10 L211

Stanley HE and Coniglio A 1984 Phys Rev B 30 522

Vicsek T and Family F 1984 Phys Rev Lett 52 1669

Zallen R 1983 The Physics of Amorphous Solids (Wiley, New York)

EDITORS NOTE

Throughout this three contributions of this volume Professor Stanley refers to lectures one, two and three. For pedagogical reasons the order of presentation in these proceedings is not the same as that at the meeting. Lecture one begins on page 49; lecture two on page 307 and lecture three on page 85.

A REVERSIBLE REACTION LIMITING STEP IN IRREVERSIBLE IMMUNOGLOBULIN AGGREGATION

Jens Feder and Torstein Jøssang

Institute of Physics, University of Oslo
Blindern, 0316 Oslo 3, Norway

1. Introduction

The aggregation of monomers and the clumping of clusters with clusters is of central interest in biology, immunology, polymer – and colloid chemistry, metallurgy and in all kinds of nucleation phenomena at phase transitions.

The studies by Forrest and Witten of smoke-particle aggregates[1], and the subsequent computer simulations by Witten and Sander,[2,3] and extended by several others[4–17] literally gave a new dimension to the understanding of irreversible aggregation processes.

Using electron microscopy Forrest and Witten[1] found that vaporized Fe, Zn, and SiO_2, formed aggregates characterized by a Hausdorff or fractal dimension[18] D, by the relation

$$N \sim R^D. \tag{1}$$

Here N is the number of particles inside a square of dimension R, as measured on the electron micrograph. The experimental values of D were found to be in the range 1.5 – 1.6, which is significantly below the spatial dimension $d = 2$, of the pictures. They also determined the density-density correlation function:

$$g(r) = \langle \rho(r)\rho(0) \rangle \sim r^{-A}. \tag{2}$$

The power-law decay is controlled by the exponent A, related to the fractal dimension of the aggregates by $A = d - D$. Analyzing the pictures using Eq. 2, they obtained $D = 1.55 - 1.69$.

They also made the important observation that the number of particles in a cluster i, is related to the 'capture radius' R of the cluster by

$$R_i \sim i^{1/D} = i^{\beta}. \tag{3}$$

The cluster exponent $\beta = 1/D$, is the inverse cluster fractal dimension, and therefore measurements of R_i determine the fractal dimension of aggregating clusters directly.

The quantity R_i is the *'radius of influence'* introduced by Smoluchowski in his discussion of colloidal stability[19–20]. R_i is also proportional to the radius of gyration of the clusters.

Witten and Sander[2–3] simulate diffusion limited aggregation by letting 'particles' do random walks on a lattice. The particles stick irreversibly on contact. Such simulations[2–6,8,11–12,14,16–17] give $D(2) = 1.7$ and $D(3) = 2.5$. Generalizations where monomers are present initially and where the clusters also may diffuse[9–10,13,15] generate ramified clusters with $D(2) = 1.4 - 1.45$, and $D(3) = 1.75 - 1.8$. Early studies and simulations by Vold[21], Hutchinson and Sutherland[22], Medalia[23–25], and by Sutherland and Goodarz-Nia[26], are consistent with recent simulations and may now be interpreted in terms of the cluster fractal dimension. Many examples of numeric and experimental results have been presented at a recent conference[27].

The experimental results for cluster fractal dimensions have been obtained for several systems. Essentially 3-dimensional clusters of metal particles[1] give $D(2) = 1.55 \pm 0.06$, when analyzed as 2-dimensional objects in electron microscope pictures. Two dimensional clusters of small spheres[28] give $D(2) = 1.64 \pm 0.1$, a value close to the 1.70 of diffusion limited Witten-Sanders clusters. Electron microscopic studies of carbon black clusters[25], give $D = 2.3$. Sputtered $NbGe_2$ films have clusters with $D = 1.9$ [29]

Gold colloids aggregated by the addition of pyridine give $D(3) = 1/\beta = 1.7 \pm 0.1$, and $D(3) = d - A = 1.8 \pm 0.1$, from the analysis of electron microscope pictures[30–31]. These aggregates have also been studied by light scattering[31–32], and by small angle neutron scattering[31]. All these results show that gold colloid aggregates have fractal dimension of 1.7 well below the Witten-Sander results and are consistent with the simulations on cluster-cluster aggregation[9–10,13,15].

Colloidal silica particles have been studied by a combination of small angle X-ray scattering and light scattering[33] thus covering an exceptional range of length scales. The silica colloid clusters are found to have a fractal dimension of $D = 2.12 \pm 0.05$.

A small angle neutron study[34] of water-suspended powders of aggregates of of fine silica particles determined a fractal dimension of $D = 2.52 \pm 0.05$.

We have studied the aggregation kinetics of immunoglobulins[35–37] by quasielastic light scattering and find that the kinetics of the aggregation process can be consistently described in terms of the Smoluchowski kinetic equations assuming the clusters formed have a fractal dimension of $D = 2.5$.

Quasielastic light scattering is a very powerful method well suited for the study of aggregation processes. The determination of cluster fractal dimensions is, however, not simple and can be performed reliably only for certain types of aggregation kinetics as discussed in the following sections. We first summarize the relevant expressions for the scattering of light from macromolecular solutions. Then we discuss the Smoluchowski aggregation equations and show that the effective hydrodynamic radius as measured by photon correlation spectroscopy gives the cluster fractal dimension directly. We then investigate some modifications of the Smoluchowski aggregation equations and discuss the effect on the fractal dimension deduced by fitting correlation functions. Our experimental results for the aggregation kinetics of pooled human IgG immuno-globulins are analyzed in terms of the kinetic models presented.

Preliminary results on the H-fraction of the IgG monomers, are presented. We find that a kinetic model where a reaction step precedes irreversible aggregation describes these experimental results.

2. Theory

Macromolecules diffuse in solution and scatter light. If the molecules form aggregates the scattering intensity will increase and the correlation function of the light scattered quasielastically will show a characteristic increase in the correlation time. If the scatterers are identical light scattering results are readily interpreted in terms of the molecular weight and the hydrodynamic radius of the macromolecules. For an aggregating system additional information is contained in the time development of the observed effective molecular weight and hydrodynamic radius.

In the next section we summarize the well known expressions for the scattering intensity and the correlation function and the scattered intensity. We limit ourselves to the case where the aggregates are small as measured by the wavelength of light so that the Rayleigh scattering theory applies.

2.1. Scattering of Light from Macromolecules

The quasielastic scattering of light has been extensively described and reviewed in the literature[38–40] The following discussion relates to scattering from a solution of aggregates consisting of spherical monomers of radius R_o and mass M.

Let N_i be the number density of i-mers with molecular weight $M_i = iM$. If we assume also the i-mers to be spherical with a radius R_i, then the diffusion constant is given by the Stokes- Einstein relation

$$D_i = kT/6\pi\eta R_i \,, \tag{4}$$

where η is the solvent viscosity, k the Boltzmann constant and T the absolute temperature. This expression is valid in the limit of dilute solutions. At finite concentrations corrections due to hydrodynamic interactions and coulomb interactions between the molecules modify the simple expression for D_i.

Consider the scattering from such a solution. With a scattering angle θ, the magnitude of the scattering vector Q, is given by

$$Q = \frac{4\pi n}{\lambda} \sin\frac{\theta}{2} \,. \tag{5}$$

where λ is the laser wavelenght and n is the refractive index of the solution.

The time averaged scattering intensity $\langle I(Q) \rangle$, is given in terms of the monomer scattering intensity I_o, the molecular weights M_i and the cluster size distribution by[40–42]

$$\langle I(Q) \rangle = I_o \sum_i N_i (M_i/M)^2 S_i(Q). \tag{6}$$

Here the static structure factor of the i-mer $S_i(Q)$, is given in terms of the density-density pair correlation function $g_i(r)$ for a single i-mer by [40–42]

$$S_i(Q) = \frac{1}{M_i} \int d^3\mathbf{r}\, g_i(r) \exp(i\mathbf{Q}\cdot\mathbf{r}) \,. \tag{7}$$

For *fractal* clusters the hydrodynamic radius R_i, is a characteristic length scale beyond which the pair correlation function must decrease rapidly and deviate from the form established for infinite fractals and given in Eq. 2. A finite size scaling form for the pair correlation function has been discussed by Martin and Ackerson[42] and may be written

$$g_i(r) = r^{D-3} f(r/R_i) . \tag{8}$$

The crossover function $f(x) \to 1$ for $x \ll 1$, and vanishes faster than a power law for $x \gg 1$.

The angular dependence of the scattering intensity is therefore given by the static structure factor $S_i(Q)$ averaged over the size distribution. There are two limits to consider: In the *Guinier* limit the clusters satisfy $QR_z \ll 1$, and the angular dependence vanishes. The z-averaged cluster radius R_z, is defined by

$$R_z^2 = \sum_i R_i^2 i^2 N_i . \tag{9}$$

The scattering intensity then reduces to the form

$$\langle I \rangle / I_o = \sum_i i^2 N_i (1 - \frac{1}{3} Q^2 R_z^2 + \ldots) . \tag{10}$$

The intensity is therefore proportional to the average cluster size - or the so called 'z-averaged' molecular weight.

$$\langle i \rangle_z = \sum_i i^2 N_i . \tag{11}$$

We conclude that the scattering intensity is insensitive to the fractal structure of the scattering clusters in the $QR_z \ll 1$. However, we see that the time dependence of the scattering intensity is useful in probing the cluster size distribution since it measures the z-average cluster size.

In the *Porod* regime $QR_z \ll 1$, the angular dependence of the scattering intensity contains information both about the cluster size distribution N_i, and about the cluster structure as described by $g_i(r)$. For a complete analysis of experimental results we therefore need models for N_i and for the pair correlation function. Martin and Ackerson[42] have shown that for a system at the gel point $I(Q) \sim 1/Q^{D(3-\tau)}$, where τ is an exponent characteristic for the cluster size distribution at the gel point.

We limit our discussion to the range $QR_z < 1$ and will not consider the complications of the Porod limit further. However, in detailed analysis of the fractal nature of clusters produced in an aggregation process the experimental results for the complete QR_z range are required.

In the Gaussian approximation[39-40] the correlation function of the scattered intensity at time 0 , $I(0)$, and at a later time t , $I(t)$, is given by[40]

$$C(t) = \langle I(0)I(t) \rangle = \langle I \rangle^2 + B\{\sum_i N_i M_i^2 S_i(Q) \exp(-\Gamma_i t)\}^2 . \tag{12}$$

Here B is a constant that depends on instrument geometry, fluid and monomer polarizabilities etc. The photoelectron count correlation function measured by a digital correlator also has the form given in eq. 12, but B is modified by experimental parameters such as the sample time, clipping level and detector coherence area.

The correlation function $C(Q,t)$, consists of a superposition of exponentials with a decay constant Γ_i for each cluster size given by

$$\Gamma_i = D_i Q^2 + A(QR)D_r \sim D_i Q^2 \,. \tag{13}$$

The cluster translational diffusion constant D_i, is given by eqn. 4, and the rotational diffusion constant is D_r is proportional to D_i/R_i. The amplitude $A(QR)$ is small for essentially spherical clusters. In the limit we consider $QR_z < 1$, the rotational term is of order $(QR_i)^4$ and may be neglected[41–42]. Also in this limit $S_i(Q) = 1$, and a simplified expression for $C(t)$ is obtained.

In order to obtain parameters that characterize the scatterers by fitting the experimental results, it is useful to perform a cumulant expansion[40] of eq. 12. Let us introduce the distribution

$$P(t) = \frac{1}{2}\ln\{C(t) - \langle I \rangle^2\} = \frac{1}{2}\ln B + \ln F(0) + \sum_{n=1}^{\infty} \frac{(-1)^n}{n!} F(n)t^n. \tag{14}$$

This series expansion has the following terms easily determined from the experimental data:

$$F(0) = \sum_i N_i M_i^2 \sim \langle I \rangle, \tag{15}$$

$$F(1) = \{\sum_i N_i M_i^2 \Gamma_i\}/F(0) = \langle \Gamma \rangle. \tag{16}$$

For a given cluster size distribution N_i, the expected correlation function may be calculated from eq. 12. However, by fitting experimental results reliable estimates can normally be obtained only for the scattering intensity $\langle I \rangle \sim F(0)$, and for the *effective hydrodynamic radius* $\langle R \rangle$ defined in terms of $\langle \Gamma \rangle$ by

$$\langle R \rangle = kTQ^2/6\pi\eta\langle \Gamma \rangle. \tag{17}$$

We use the equations 5 and 15–17 to find the expressions

$$\langle I \rangle / I_o = \sum_i n_i i^2, \tag{18}$$

$$\langle R \rangle / R_o = \{\sum_i n_i i^2\}/\{\sum_i n_i i^2 (R_o/R_i)\}. \tag{19}$$

Here $n_i = N_i/N_1(t=0)$ is the cluster size distribution normalized to the monomer concentration at the start of the aggregation process.

The scattering intensity $\langle I \rangle$ and the effective hydrodynamic radius $\langle R \rangle$, may be estimated using quasielastic light scattering. The effective hydrodynamic radius depends on the cluster size distribution *and* on the cluster hydrodynamic radii R_i. If the cluster hydrodynamic radius R_i is taken to satisfy the fractal form given in eq. 1 :

$$R_i = R_o i^{1/D}, \tag{20}$$

then to the extent that this relation is a good approximation to the actual hydrodynamic radius of i-mers we find that the averaged effective hydrodynamic radius estimated from a determination of $\langle \Gamma \rangle$ is given by

$$\langle R \rangle / R_o = \{\sum_i n_i i^2\}/\{\sum_i n_i i^{(2-1/D)}\}. \tag{21}$$

Computer simulations[43] for 3-dimensional cluster-cluster aggregation where the hydrodynamic radius is calculated gives the form 20 with a fractal dimension $D_h = 1.83 \pm 0.05$, and a radius of gyration which behaves as 20 with a fractal dimension $D_g = 1.81 \pm 0.13$, so the fractal dimension determined from observations of the hydrodynamic radius should give the cluster fractal dimension.

We see that the theoretical expression for $\langle R \rangle$ contains the cluster fractal dimension D explicitly. In the next section we show that D may be determined from the experimental results for the case where the aggregation process may by described by the Smoluchowski equation.

2.2. Smoluchowski Aggregation

Smoluchowski[19–20] , in his pioneering work on colloidal stability formulated the kinetic equations for irreversible aggregation of monomers into clusters, and the further clustering of clusters with each other. Smoluchowski considered a system of diffusing particles which stick together irreversibly upon collision. More precisely, if the distance between two diffusing particles becomes less than a 'radius of influence' $2R_1$, they are assumed to form a dimer that itself diffuses and has a radius of influence R_2. We identify the radius of influence with the hydrodynamic radius R_i of the cluster. Thus two particles will stick irreversibly if their distance becomes less than $R_{ij} = (R_i + R_j)$.

Smoluchowski showed that the relative diffusion of a cluster with diffusion constant D_i, and another cluster with diffusion constant D_j, can be described by choosing one particle as the reference. In this frame of reference the other particle diffuses with an effective diffusion constant

$$D_{ij} = D_i + D_j \, . \tag{22}$$

By solving the diffusion equation in spherical coordinates with a fixed concentration of diffusing particles at infinite distances and zero concentration at the distance R_ij, Smoluchowski found that the rate at which i-mers at a concentration N_i and j-mers at a concentration N_j form $(i+j)$-mers is given by

$$J_{i+j} = 4\pi D_{ij} R_{ij} N_i N_j \{1 + R_{ij}(\pi D_{ij} t)^{-\frac{1}{2}}\} \, . \tag{23}$$

In order to simplify our calculations we note that for times

$$t \gg \tau_D = R_{ij}^2 / \pi D_{ij} \, , \tag{24}$$

the second term in the bracket in eq. 23 becomes negligible. The characteristic diffusion time τ_D, is a microscopic time approximately $5 \cdot 10^{-7}$ seconds for the *IgG* immunoglobulins we study. Since the concentration of k-mers increases with the rate $J_{i+j=k}$, and decreases by the rate J_{k+j} the kinetic equations describing irreversible aggregation may be written

$$\frac{dn_k}{dt} = \sum_{i+j=k} n_i K_{ij} n_j - 2 \sum_j n_k K_{kj} n_j \, . \tag{25}$$

In this form of the Smoluchowski equation we have used $n_i = N_i / N_\circ$, and $N_\circ = N_1(t=0)$. The dimensionless time t in eq. 23 is expressed in units of the inverse Smoluchowski collision rate γ_S^{-1}. The collision rate γ_S is given by

$$\gamma_S = 8\pi D_\circ R_\circ N_\circ p \, . \tag{26}$$

Here D_o and R_o is the monomer diffusion constant and hydrodynamic radius. The sticking probability p introduced by Smoluchowski may be less than unity without any change in eq. 25. The collision matrix K_{ij} is given by

$$K_{ij} = \frac{1}{4D_oR_o}(D_i + D_j)(R_i + R_j) . \tag{27}$$

Using the Einstein-Stokes expression 4 for the diffusion constant, and the fractal form 20 for the hydrodynamic radius we find that the transition matrix may be written

$$K_{ij} = \frac{1}{4}(i^{-1/D} + j^{-1/D})(i^{1/D} + j^{1/D}) . \tag{28}$$

The transition matrix K_{ij}, depends on the cluster fractal dimension D. However, $K_{ii} = 1$, independent of i and D, and in our numeric solutions of the Smoluchowski equation 25 with the transition matrix 28, we find that the D- dependence in K_{ij} has a small effect on the cluster size distribution.

No general solution of the Smoluchowski equation 25 exists. However, Smoluchowski made the simplification $K_{ij} = 1$, for all i and j, and gave the exact solution when only monomers are present initially:

$$n_i(t) = \frac{t^{i-1}}{(1+t)^{i+1}} . \tag{29}$$

This solution is shown in Fig. 1. The monomer concentration steadily decreases, and the i-mer concentration first increases, has a maximum and then decreases. Let i^* be the cluster size that has its maximum concentration n^* at time $t \gg 1$. We find that

$$\begin{aligned} i^* &= (1+2t) , \\ n^* &= \frac{t^{2t}}{(1+t)^{2+2t}} \rightarrow t^{-2}e^{-2} , \text{ for } t \gg 1 . \end{aligned} \tag{30}$$

In figure 2 we have plotted $n_i(t)/n^*$ as function of i/i^*, for times in the range 2 to 50. We see that all the curves fall on a single universal scaling curve independent of time t. In fact the exact solution 29 may be written[36] in the scaling form proposed by Vicsek and Family[44]

$$n_i(t) = t^{\theta z} f\left(\frac{i}{t^z}\right) , \tag{31}$$

with the exponents $\theta = 2$ and $z = 1$, for the Smoluchowski case, and the scaling function is $f(x) = \exp(-x)$.

The scaling form 31 valid in the limit $t, i \gg 1$, expresses the fact that there is no intrinsic cluster size. It is time that determines the characteristic cluster size. We find for the Smoluchowski solution 27 that the scattering intensity is proportional to the z-averaged cluster size and is given by

$$\langle I \rangle / I_o = \langle i \rangle_z = \sum_i n_i i^2 = (1+2t) \sim t^z . \tag{32}$$

Here the asymptotic form is obtained using the scaling form for the cluster size distribution. As $t \rightarrow \infty$, we find that the characteristic cluster size and the scattering

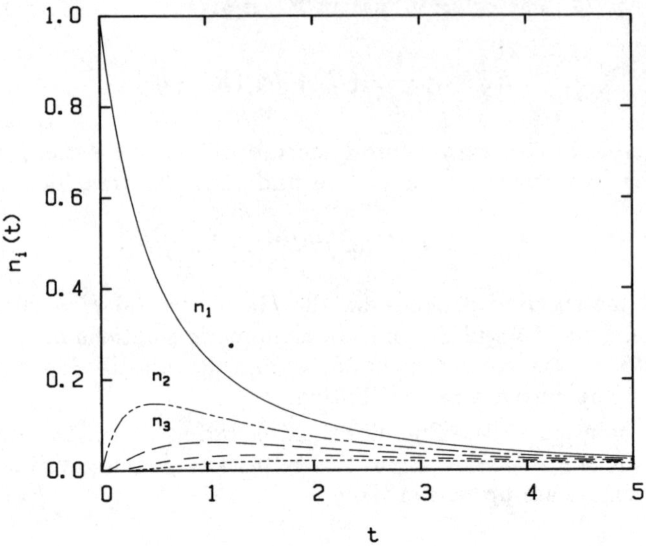

Figure 1: The time dependence of the cluster concentrations for Smoluchowski aggregation.

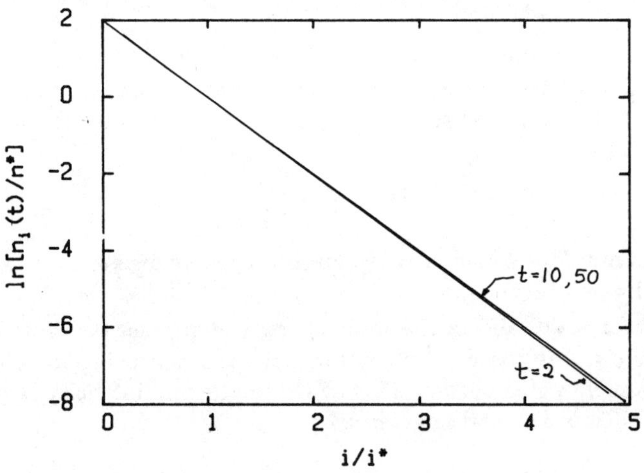

Figure 2: The scaled clustersize distribution $n_i(t)/n^*$ as function of scaled cluster size $i/i^* = 1/(1+2t)$, for various times t, for Smoluchowski aggregation.

intensity diverges. The divergence is controlled by the dynamic exponent z. This is analogous to the divergence of the correlation length as the critical temperature is approached at second order phase transitions, and infinite time corresponds to the critical point.

The asymptotic result for the effective hydrodynamic radius $\langle R \rangle$ is obtained by using the scaling form of the cluster size distribution in the expression 21 for $\langle R \rangle$

$$\langle R \rangle / R_o \sim t^{z/D} . \tag{33}$$

The asymptotic behavior of $\langle R(t) \rangle$ therefore determines the cluster fractal dimension if the dynamic exponent z is known from the asymptotic behavior of the scattering intensity.

Generalizations of the Smoluchowski assumption has been discussed by several authors recently[45–50]. Transition matrices that have no intrinsic size scale satisfy the scaling relation

$$K_{bi,bj} = b^{2\omega} K_{i,j} , \tag{34}$$

and lead in general to size distributions of the form 31, with $\theta = 2$, and $z = 1/(1-2\omega)$, as shown by Ernst et. al.[45–47] and Botet and Jullien[48]. Botet and Jullien also give the general form of the scaling function

$$f(x) = \frac{(1-2\omega)^{1-2\omega}}{\Gamma(1-2\omega)} x^{-2\omega} \exp[-(1-2\omega)x] . \tag{35}$$

This expression is valid for $\omega < \frac{1}{2}$. In the range $\frac{1}{2} < \omega < 1$, the size distribution reverts at a finite time — the gelation time, so that the cluster size distribution is no longer normalized and infinite clusters occur.

Note that both the full transition matrix K_{ij} in eq. 28, and Smoluchowski's simplified K_{ij}, satisfy the scaling relation 34 with a scaling exponent $\omega = 0$. With this value of ω , we find $z = 1$ and the general scaling function $f(x)$ for the cluster size distribution reduces to the Smoluchowski result $f(x) = \exp(-x)$.

Botet and Jullien[48] have generalized the form 28 for the transition matrix to the case where the particles diffuse in a space of dimension d by a random walk along a trajectory that has a fractal dimension d_w, ($d_w = 2$, for ordinary Brownian motion), and where the the velocity along the trajectory is $v_k = k^a$. They find the expression

$$K_{ij} = (i^{2a} + j^{2a})^{\frac{1}{2}} (i^{1/D} + j^{1/D})^{d-d_w} . \tag{36}$$

This expression reduces to an approximation of our expression 28 for the diffusion process $d - d_w = 1$, and the exponent $a = -1/D$. In their model the diffusion constant depends on size by $D_i \sim i^a$. The expression 36 scales with an exponent $2\omega = a + (d - d_w)/D$, and therefore the dynamical exponent z is found to depend on a and D :

$$z = (1 - a - (d - d_w)/D)^{-1} , \tag{37}$$

which gives $z = 1$ for $d - d_w = 1$ and $a = -1/D$. Their own simulations[48] are consistent with these results for $d = d_w = 2$ and $a = 0, -1, -2$.

Numeric simulations of cluster-cluster aggregation in 2- dimensions by Vicsek and Family[44], gave the non classical value $z = 1.4 \pm 0.2$, for the dynamical exponent z. In this simulation they assumed that the clusters moved with a mobility independent of their size, so from eq. 37 we would have expected $z = 1$.

Meakin, Vicsek, and Family[51] simulated 3-dimensional cluster-cluster aggregation and found that the dynamical exponent z depends on the exponent relating the cluster diffusion constant D_i to the cluster size. They also use $D_i = i^a$, which corresponds to our choice ($D_i \sim i^{-1/D}$, for $a = -1/D$). They find z in the range 1.3 to 1.4. for $a = -\frac{1}{2}$. This is consistent with eq. 37. Such clusters have a fractal dimension of $D = 1.75$, and it follows that the expected value of the exponent is $z = 1.33$.

It should be noted that the Smoluchowski equation involves no spatial dependence of the cluster concentrations. Clearly, in a real or simulated system there will be fluctuations in the cluster concentrations. The Smoluchowski equation is a *mean field* equation and should be generalized to include spatial concentration fluctuations. However, there are indications that such a generalization may not be necessary for 3-dimensional systems. Kang and Redner[52] study fluctuation effects in Smoluchowski kinetics for a simplified model where $R_i = R_o$, and $D_i = D_o$, independent of clustersize. Thus $K_{ij} = 1$, and has the scaling exponent $\omega = 0$. They describe spatial fluctuation effects in a phenomenological way by assuming that the kernel K_{ij} is time dependent. They find that fluctuation effects are important and consistent with their simulations for $d < 2$. For $d > 2$, the Smoluchowski molecular field result is recovered and fluctuation effects should not change the Smoluchowski results in $d = 3$.

2.3. Scattering from Smoluchowski Aggregates, $K_{ij} = 1$

In systems where the aggregation process may be described within the Smoluchowski approximation $K_{ij} = 1$, the size distribution n_i is given by Eq. 29. The quantities of interest are then conveniently expressed in terms of sums of the form

$$S_\mu(t) = \sum_i n_i(t) i^\mu = \frac{1}{t(1+t)} \sum_i \left(\frac{t}{1+t}\right)^i i^\mu . \tag{38}$$

For $\mu = 0$, this sum is a simple geometric series, and we find that the total number of clusters is given by

$$n_c = \sum_i n_i = S_0(t) = \frac{1}{1+t} . \tag{39}$$

Other sums with special values of μ may easily be obtained using a relation that results by taking the derivative of Eq. 38.

$$S_{\mu+1}(t) = t(1+t)\frac{dS_\mu}{dt} + (1+2t)S_\mu(t) . \tag{40}$$

Using this result and Eq. 39, we find that the number of molecules is conserved and n_i is normalized:

$$\sum_i n_i i = S_1(t) = 1 . \tag{41}$$

The scattering intensity is given by Eq. 18, and using Eqs. 40 and 41, we find the scattering intensity to be independent of D and given by

$$\langle I \rangle / I_o = \sum_i n_i i_2 = S_2(t) = (1+2t) . \tag{42}$$

This result was first obtained by Troelstra and Kruyt[53] and by Oster [54].

The effective hydrodynamic radius $\langle R \rangle$ is given by Eq. 21, and may be written

$$\langle R \rangle / R_\circ = S_2(t)/S_{2-1/D}(t) \,. \tag{43}$$

We are interested in $\langle R(t) \rangle$ for cluster fractal dimensions D in the range $1 < D < 3$. The sums in Eq. 43 converge also for D well outside this range. By evaluating the expression 43 for $\langle R \rangle$, we find the characteristic power-law behavior shown as the $K_{ij} = 1$, curve in figure 4.

Unfortunately there are no simple expressions for $S_{2-1/D}(t)$ with arbitrary D, and we consider asymptotic and numerical results in the following sub-sections.

2.3.1. The long time limit

In the limit $t \to \infty$ we may obtain an asymptotic form for $S_\mu(t)$. The sum may be written

$$S_{2-1/D}(t) \to t^{1-1/D}\Gamma(3 - 1/D) \,, \text{for } t \to \infty \,. \tag{44}$$

Here $\Gamma(x)$ is the gamma function. The resulting expression for $\langle R \rangle$ is

$$\langle R \rangle / R_\circ \to \frac{2}{\Gamma(3 - 1/D)} t^{1/D} \,, \text{for } t \to \infty \,. \tag{45}$$

We conclude that the effective hydrodynamic radius $\langle R \rangle$ determined from the first cumulant in experiments on a system following Smoluchowski aggregation exhibits a power-law time dependence with the exponent $1/D$.

2.3.2. The short time limit

The short time expansion of the sum $S_\mu(t)$ is easily derived

$$\begin{aligned} S_\mu(t) &= \textstyle\sum_i t^{i-1}\{1 - (i+1)t - \ldots\} i^\mu \\ &= 1 + (2^\mu - 2)t + \ldots \,. \end{aligned} \tag{46}$$

This expansion has already been discussed by Versmold and Härtl[55]. We note that the first two terms in Eq. 46, gives the exact result Eqs. 41 and 42 valid for all times, for $\mu = 0$ and 1, respectively. The hydrodynamic radius $\langle R \rangle$ increases linearly with time in this limit.

The short-time and long-time limits may be combined in the following cluster growth law

$$\langle R \rangle / R_\circ = (1 + \gamma_R t)^{\alpha_R} \,. \tag{47}$$

where the radius exponent $\alpha_R = 1/D$, and the growth rate constant γ_R in the two limits is given by

$$\begin{aligned} \lim_{t \to 0} \gamma_R &= 4D(1 - 2^{-1/D}) \,, \\ \lim_{t \to \infty} \gamma_R &= \left(\frac{2}{\Gamma(3 - 1/D)}\right)^D \,. \end{aligned} \tag{48}$$

The power-law form given in Eq. 47 directly measures the cluster fractal dimension and is well suited for fitting experimental data and is the form we use in later sections.

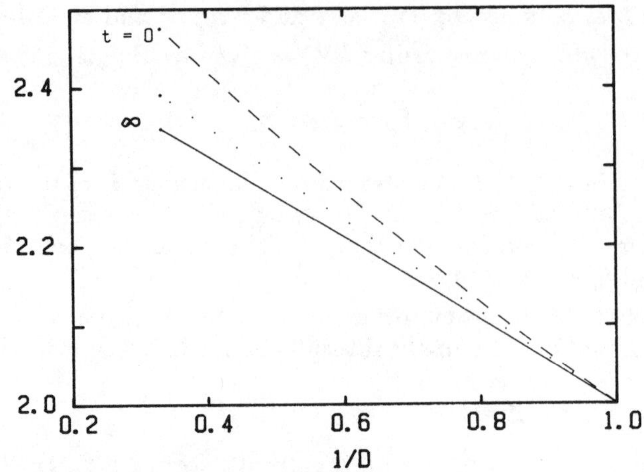

Figure 3: The asymptotic expression for the radius growth rate constant γ_R as function of the cluster fractal dimension. — the $t \to \infty$ limit, – – – the $t \to 0$ limit. ··· - fits to numerical results in the range $0 < t < 5$.

In the range $1 < D < 3$, the two limiting expressions for γ_R, differ by at most 5 %. An impression of the accuracy of the power-law form 47 may be obtained from figure 3.

As seen in Fig. 3, the expression for the growth rate γ_R are surprisingly close. The points represent fits of the power-law form to numerical results obtained by evaluating 43, in the range $0 < t < 5$, which is typical for our experiments.

We conclude that for Smoluchowski aggregation kinetics the scattering intensity increases linearly with time as given by eq. 42. The effective hydrodynamic radius $\langle R \rangle$ has a power-law time dependence given by 47 from which the cluster fractal dimension may be obtained accurately even for results that cover a limited time range of $0 < t < 5$.

2.4. Scattering from Aggregates formed with the Complete transition matrix

With the complete transition matrix K_{ij} given by equation 28, the Smoluchowski equation has to be studied by numerical methods. We have solved the Smoluchowski equation 25, using a fourth order Runge-Kutta method. In the calculations we must assume that there is a maximum clustersize m, and we set $K_{ij} = 0$ for $i+j > m$. This modification ensures that the number of particles in the system is conserved. With $m = 100$, adequate accuracy is obtained for times $t < 10$. In figure 4 the calculated $\langle R \rangle$, is compared with the result obtained by evaluating the expression 43 for the case $K_{ij} = 1$.

The approximate result for $\langle R \rangle$ with $K_{ij} = 1$, is somewhat below the correct result. However, the discrepancy is small for the time range of interest in our experiments, and a qualitatively correct understanding of the time dependence of the

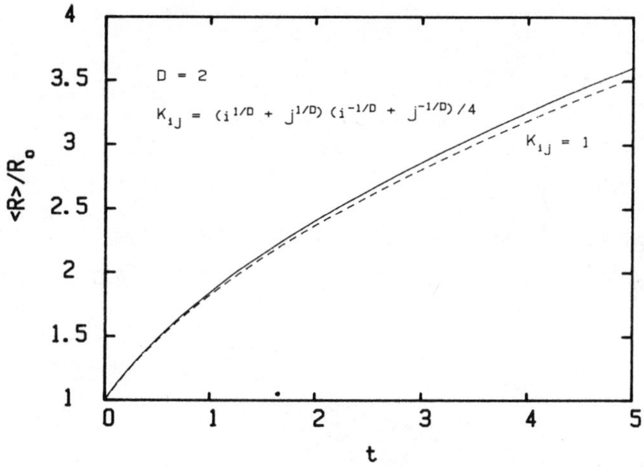

Figure 4: The effective hydrodynamic radius as function of time evaluated with with the complete transition matrix (full curve), and with the simplified form $K_{ij} = 1$ (dotted curve).

effective hydrodynamic radius is obtained from the simplified Smoluchowski theory.

The effective hydrodynamic radius depends on the cluster fractal dimension as shown in figure 5.

By fitting the results in figure 5 with the power-law given in equation 47, we find that the radius exponent is given by

$$\alpha_R = (1.003 \pm 0.006)D^{-1} . \tag{49}$$

We conclude that the effective radius evaluated by the solution of the Smoluchowski equation with the full transition probability 28, has a $1/D$ power-law time dependence. We conjecture that the power-law

$$\langle R \rangle / R_o = (1 + \gamma_R t)^{1/D} , \tag{50}$$

is general for all transition matrices that scale as given by eq. 34, with a scaling exponent $\omega = 0$. In the next section we discuss examples of aggregation kinetics that do not fall into this universality class. The scattering intensity increases almost linearly with time as shown in figure 6.

The scattering intensity increases somewhat with decreasing cluster fractal dimension, but the deviation from linear behavior is small and the cluster fractal dimension cannot be determined from the time dependence of the scattering intensity for Smoluchowski aggregation.

2.5. Monomer Addition Aggregation: $K_{ij} = 0$ for $i \wedge j > 1$

In the original Witten-Sander model monomers diffuse and adsorb on stationary clusters. Thus the clusters can grow only by monomer addition. In the Smoluchowski

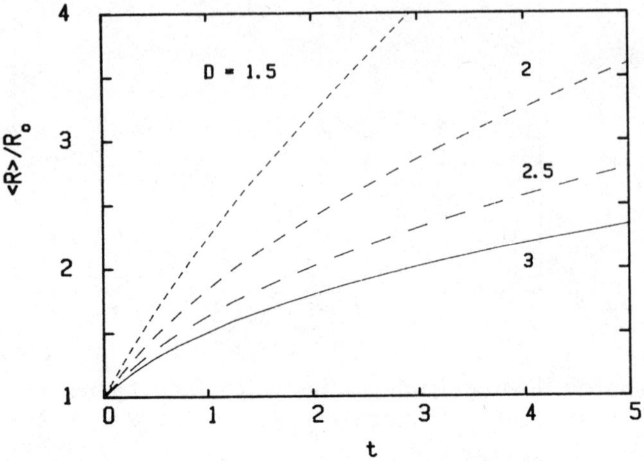

Figure 5: The effective hydrodynamic radius obtained with the full transition matrix, as function of time for several values of the cluster fractal dimension D.

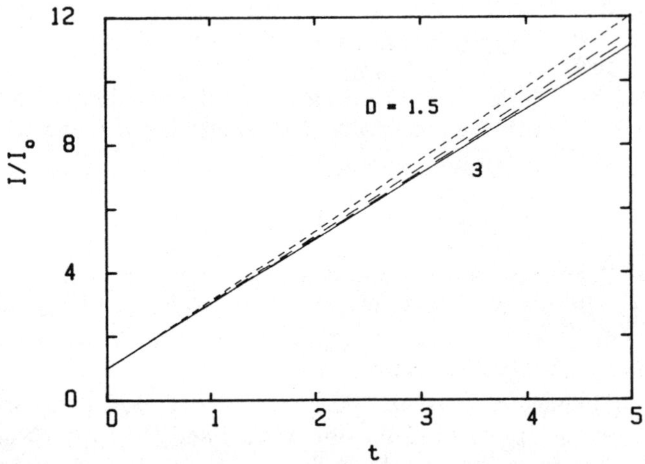

Figure 6: The scattering intensity as function of time calculated from the Smoluchowski equation with the full transition matrix for several values of the cluster fractal dimension.

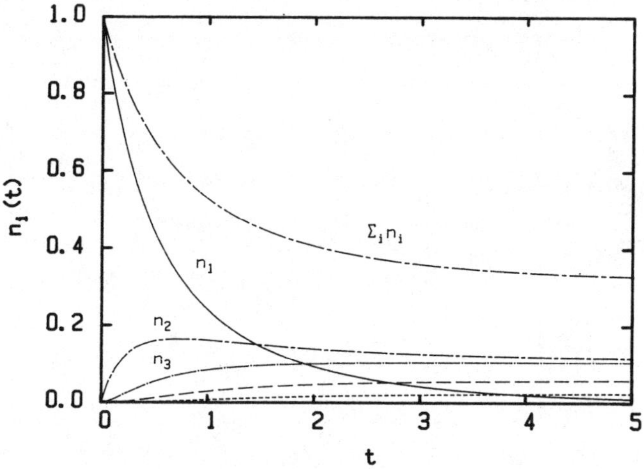

Figure 7: Cluster size as function of time for monomer addition aggregation

equation 25 this model corresponds to a transition matrix of the form

$$\begin{aligned} K_{1j} &= \tfrac{1}{4}(1+j^{1/D})(1+j^{-1/D}) = K_{j1}\,, \\ K_{ij} &= 0\,, \text{ for } i \text{ and } j > 1\,. \end{aligned} \tag{51}$$

This transition matrix has the intrinsic cluster size $i = 1$, and therefore K_{ij} does not satisfy the scaling form 34, and we expect a different behavior. Note that with the transition matrix 51 any monomer may act as a seed for the Witten-Sander growth process and therefore the resulting dynamics will be different from the one where a limited number of nucleation centers are selected beforehand.

We have solved the Smoluchowski equation 25 with this transition matrix and find results qualitatively different from the original Smoluchowski result.The cluster size distribution as function of time is shown in figure 7.

The monomer concentration decays exponentially, whereas the dimers, trimers etc. concentrations saturate at finite concentrations. Also in contrast with the Smoluchowski result $\sum_i n_i = 1/(1+t)$, the total number of clusters in monomer addition aggregation saturates at a finite value of approximately 0.3 for the present case where we use $D = 2$. This behavior results because the supply of monomers is rapidly depleted and the aggregation process effectively quenched. It follows that the estimated hydrodynamic radius saturates at a value < 2, as shown in figure 8.

The scattering intensity also saturates in an exponential way similar to the behavior shown in figure 8.

We conclude that monomer addition aggregation exhibits a characteristic saturation of the scattering intensity and of the effective hydrodynamic radius. For monomer addition aggregation the fractal structure of clusters is unimportant since no large clusters are generated with a finite supply of monomers.

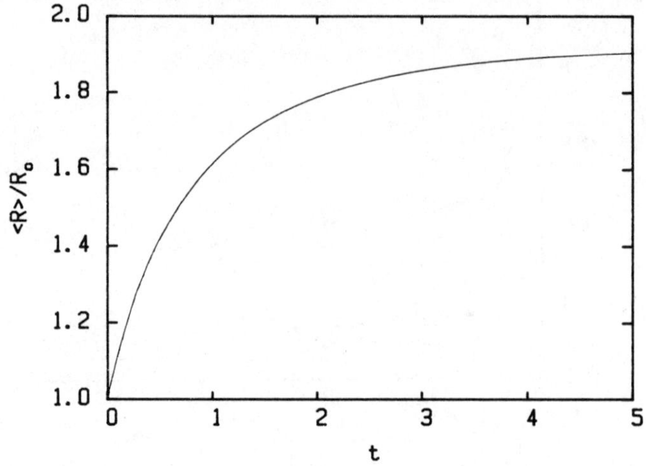

Figure 8: The effective hydrodynamic radius as function of time for monomer addition aggregation, with a cluster fractal dimension $D = 2$.

2.6. Reaction Limited Aggregation

The interpretation of our experimental results on the H-fraction presented in section 4, require an extension of the Smoluchowski aggregation process.

Consider a system where the monomers must be *'activated'* in some way before they can aggregate. For example, for a solution of proteins that may convert from the native N - state to the denatured D - state according to the *reversible* reaction

$$N \rightleftharpoons D \,, \tag{52}$$

with the reaction rates k and k', and ignoring aggregation for the moment, we find the kinetic equations

$$\begin{aligned} \frac{dn_0}{dt} &= -\; k(Kn_0 - n_1)\,, \\ \frac{dn_1}{dt} &= \;\; k(Kn_0 - n_1)\,. \end{aligned} \tag{53}$$

Here, $n_0 = N_0(t)/N_0(t=0)$, and $n_1 = N_1(t)/N_0(t=0)$, are the native and denatured monomer concentrations normalized to the number of monomers (all native) present initially. The equilibrium constant $K = [n_1^\circ]/[n_0^\circ] = k'/k$, relates the reaction rates to the equilibrium concentrations.

We suggest that essentially the same description applies to the process of gold colloid aggregation induced by pyridine and studied by Weitz et. al.[30–32]. The gold particles react with n pyridine molecules at a concentration $[P]$ according to the equation

$$Au + nP \rightleftharpoons D, \tag{54}$$

to produce uncharged gold particles at a concentration $[D]$ that may aggregate. In this case we find again the kinetic equations 53 but K in equation 53 is replaced by $K[P]^n$, where K now is the equilibrium constant for the reaction 54. The kinetics of this aggregation process may therefore be changed by altering the pyridine concentration.

The simplest possible aggregation model for such a system assumes that only the *denatured* monomers participate in the *irreversible* aggregation process and that the aggregating clusters follow the Smoluchowski equation in the form

$$
\begin{aligned}
\frac{dn_0}{dt} &= -\; k(Kn_0 - n_1) \;, \\
\frac{dn_1}{dt} &= k(Kn_0 - n_1) \qquad -2\textstyle\sum_j n_1 K_{1j} n_j \;, \\
\frac{dn_k}{dt} &= \textstyle\sum_{i+j=k} n_i K_{ij} n_j \;\; -2\sum_j n_k K_{kj} n_j
\end{aligned}
\tag{55}
$$

The additional denaturation or reaction step in these equations singles out the monomers even when the transition matrix is given by the expression 28 and does not have an intrinsic cluster size scale. The cluster size distributions obtained by the solution of equation 55 are therefore non- universal and depend on the rate constant k, and the equilibrium constant K.

We have solved the new kinetic equations numerically with the full transition matrix 28. In calculating the scattering intensity and the effective hydrodynamic radius from the size distributions we have used the following modified versions of the equations 18 and 19.

$$\langle I \rangle / I_\circ = n_0 + \sum_i n_i i^2 \;, \tag{56}$$

$$\langle R \rangle / R_\circ = \frac{\langle I \rangle / I_\circ}{n_0 + \sum_i n_i i^{(2-1/D)}} \;. \tag{57}$$

Here we again assume that the cluster hydrodynamic radius is given by $R_i = R_\circ i^{1/D}$, as before, and that the hydrodynamic radii of the native and denatured monomers are the same.

We have found that $k = 4.4$, and $K = 0.13$, with $D = 2$, give results that fit our experimental results on the H-fraction nicely. We therefore use these values as reference values denoted by k_H and K_H in the presentation of our numerical solutions. The cluster size distribution as function of time for the reference values of the constants in the equation is shown in figures 9 and 10.

We see that the cluster size distribution is quite different from the Smoluchowski result shown in figure 1. The concentration of monomers in the native state decreases slowly and since $k > 1$ and $K \ll 1$, a quasi-equilibrium between denatured and native monomer is rapidly established. The native monomers act as a reservoir for the aggregation process. However, the total number of clusters is decreasing and eventually all monomers will be aggregated into a single cluster.

We have calculated the effective hydrodynamic radius $\langle R \rangle$ and the scattering intensity for a set of equilibrium constants. Both the scattering intensity and the effective hydrodynamic radius now show a behavior qualitatively different from ordinary Smoluchowski aggregation. However, as the equilibrium constant K is increased the characteristic power-law behavior of the effective radius and the linear increase

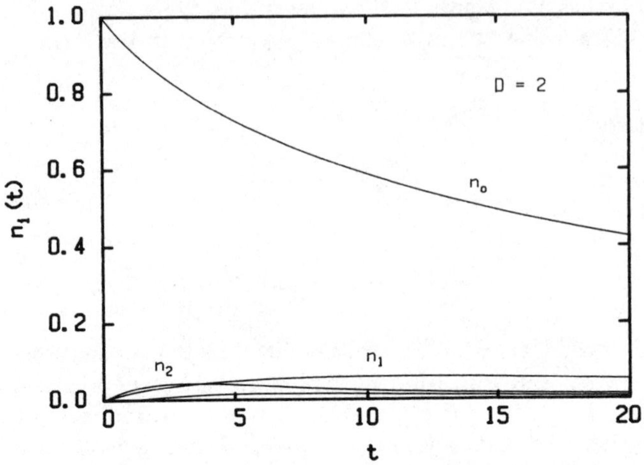

Figure 9: The i-mer concentration $n_i(t)$ as function of time for Smoluchowski aggregation with a reaction step. n_0 is the native (unreacted) monomer concentration. $D = 2$, $k = 4.4$, $K = 0.13$.

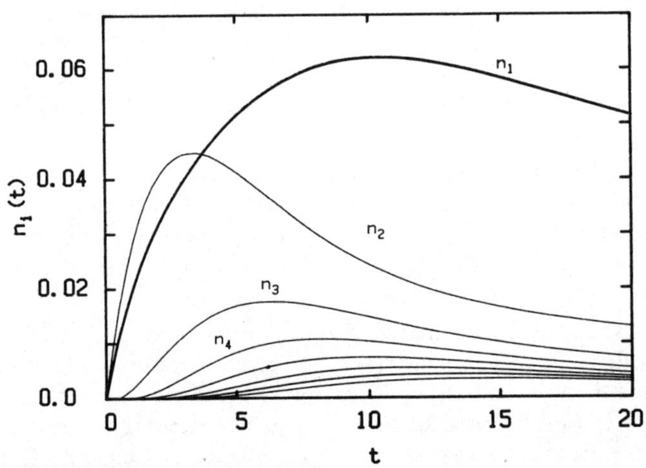

Figure 10: The i-mer concentration $n_i(t)$ as function of time for Smoluchowski aggregation with a reaction step. $D = 2$, $k = 4.4$, $K = 0.13$.

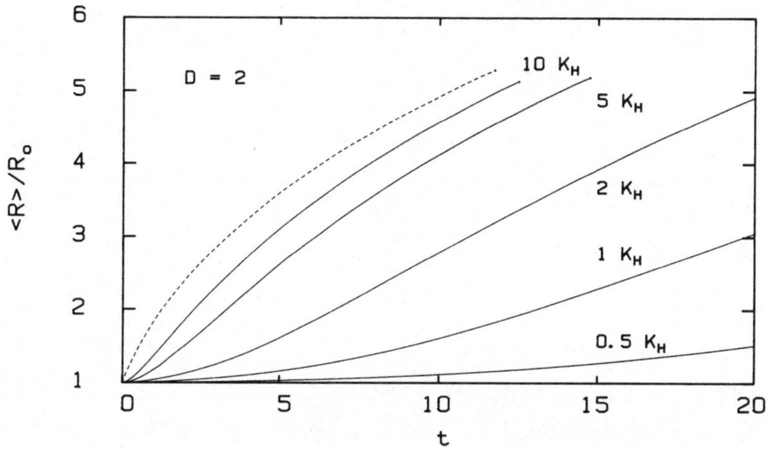

Figure 11: The effective hydrodynamic radius as function of time for aggregation with a denaturation step. $D = 2$, $k = 4.4$, and for equilibrium constants $K = 0.5, 1, 2, 5, 10 \cdot K_H$. The dotted curve is the solution of the Smoluchowski equations without any denaturation step.

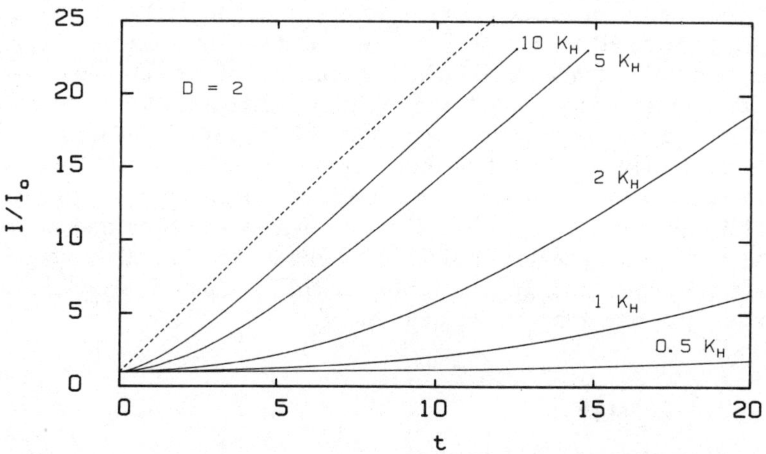

Figure 12: The scattering intensity as function of time for aggregation with a denaturation step. $D = 2$, $k = 4.4$, and for equilibrium constants $K = 0.5, 1, 2, 5, 10 \cdot K_H$. The dotted curve is the solution of the Smoluchowski equations without any denaturation step.

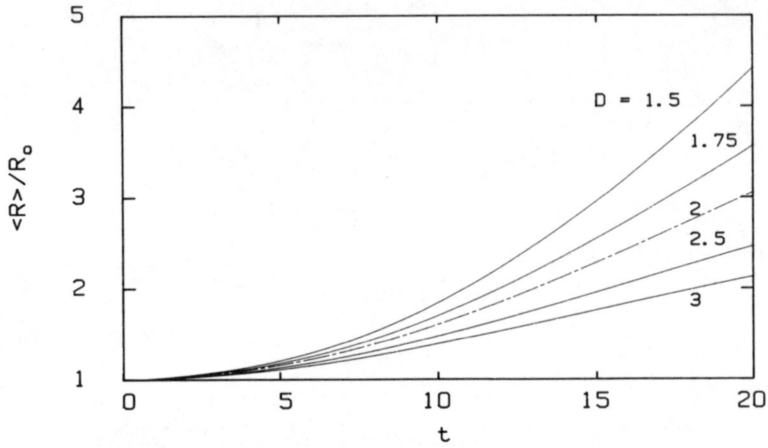

Figure 13: The effective hydrodynamic radius as function of time for various values of the cluster fractal dimension for reaction limited aggregation with $k = k_H$ and $K = K_H$.

of the scattering intensity is recovered. The Smoluchowski aggregation results give limiting curves valid for equilibrium constants $K >> 1$.

For $K < 1$, the time dependence of the effective hydrodynamic radius is dominated by the reaction step, and is insensitive to the cluster fractal dimension D, at least for the initial period where $\langle R \rangle / R_o < 10$, which is the region we study.

In figures 13 and 14 $\langle R \rangle$ and $\langle I \rangle$ for $k = k_H$, and $K = K_H$ are plotted as function of time for various values of the cluster fractal dimension. With low values of D, the clusters get more ramified and this is reflected in an increase in $\langle R \rangle$. However, there is no qualitative change in the time dependence of $\langle R(t) \rangle$, and in fact, a small change in the kinetic parameters k and K, may compensate for the change in $\langle R \rangle$. Again we find that the scattering intensity is only weakly dependent on the cluster fractal dimension. In practice it is therefore not possible to determine the cluster fractal dimension from the kinetic behavior of reaction limited aggregation, unless experimental data for $\langle R(t) \rangle$ up to the time where the time dependence crosses over to the power-law behavior or the kinetic parameters k and K are known from independent experiments. Light scattering for such systems is probably most useful in determining just these kinetic parameters.

3. Experiments

In diagnostic immunology, heating of serum at 56 °C for 30 minutes has become a well established method for inactivating complement[56]. On the other hand, heating of immunoglobulins IgG at 63 °C is a widely used method to produce soluble IgG aggregates. These aggregates possess many biological properties similar to antigen-antibody complexes, and they are used as models for the rheumatoid factor and as

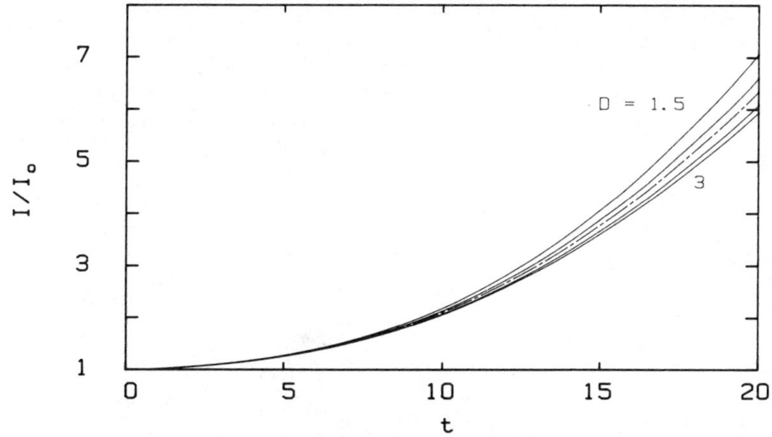

Figure 14: The scattering intensity as function of time for the cluster fractal dimension $D = 1.5$, 1.75, 2, 2.5 and 3; for reaction limited aggregation with $k = k_H$ and $K = K_H$.

standards in various methods for immune-complex assay [57].

Gel filtration profiles of normal human Gammaglobulin are shown in figure 15. Small molecules and clusters are delayed in the flow through the *Biogel A5m* column and larger clusters are eluted first.

The optical density is approximately proportional to the protein concentration. The relation between elution volume and cluster molecular weight is, however, nonlinear and compresses the large molecular weight aggregates into a single peak as shown in figure 15c.

In figure 15a, we see that the *IgG* solution supplied contains a fraction of dimers. The effective hydrodynamic radius measured for a series of fractions exhibits the correlation between elution volume and $\langle R \rangle$. In figure 15b a rather broad distribution of aggregates is seen together with the unreacted monomer peak. Finally heating to 62 °C for 6 hours produces white precipitates. Figure 15c shows that a fraction of the original proteins remain as monomers. Gel filtration and ultracentrifuge studies show that 48 ± 5 % of the monomers belong to the 'heat stable' fraction (H- fraction)[35].The fraction marked H in figure 15c is the H-fraction which we have investigated separately and preliminary results are discussed in a later section.

We prepared monomeric *IgG* from pooled human immunoglobulins (Gammaglobulin Kabi 16 %, AB Kabi, Stockholm, Sweden), by gel filtration. The hydrodynamic radius of *IgG* monomers is $R_o = 5.51 \pm 0.03\,nm$,at 20 °C in water[35], and the molecular weight is $M = 163000$. The dimer radius is $R_2 = 7,5 \pm 0.1\,nm$. Note that if we insist on the relation $R_i = R_o i^{1/D}$ even for $i = 2$, we find $D = 2.2$. Samples with 2, 4, 8 and $16\,mg/ml$ monomers in a $0.2 M NaCl$, $Tris - HCl$ buffer at $pH = 7.6$ were heated to 47, 53, 56, 59 and 62 °C and $\langle I \rangle$ and $\langle R \rangle$ measured as function of time.

Heating causes these proteins to aggregate as seen by the increase in $\langle I \rangle$ and $\langle R \rangle$. The process is irreversible - it may be halted by cooling the sample, and it

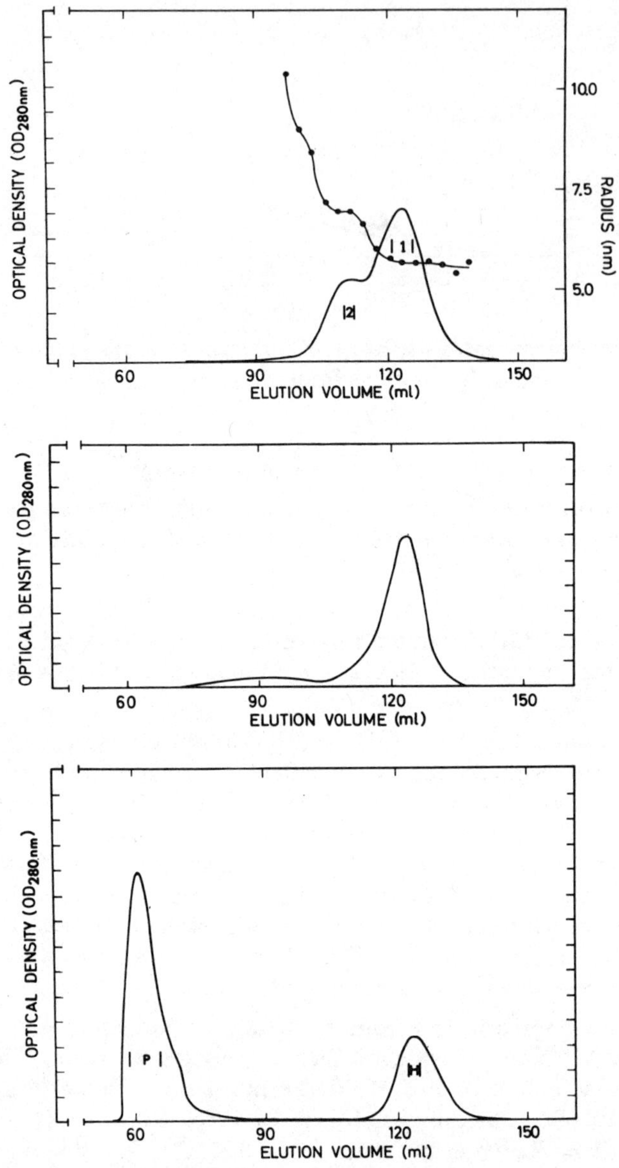

Figure 15: Gel filtration profiles of human *IgG*. (a) Gamma Kabi as received. The points are the effective hydrodynamic radii measured on the elution fractions. (b) Gel filtration profile after 30 minutes at 59 °*C* (c) After 6 hrs at 62 °*C* .

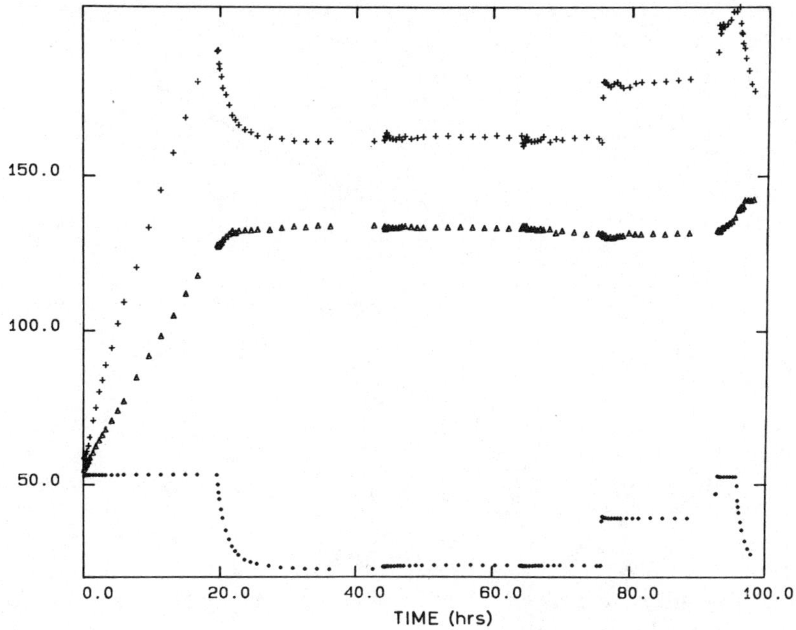

Figure 16: Time dependence of the radius $\langle R \rangle$ +, intensity Δ , and temperature •

resumes when heated again. The general behavior is seen in figure 16. A sample at $8\,mg/ml$ is heated to 53 °C and $\langle R \rangle$ and $\langle I \rangle$ increase with time. Turning off the heater causes the temperature to drop to 23 °C .

As the temperature is decreased the intensity remains constant indicating that the z-averaged cluster size is not changed, and we conclude that the aggregation process is irreversible. The effective hydrodynamic radius $\langle R \rangle$ on the other hand decreases with temperature and follows increases in temperature. This behavior is due to a linear thermal expansion[35] of the clusters . The aggregation process resumes when the temperature is increased to 53 °C again.

The aggregation of IgG is a strongly temperature dependent activated process as seen in figure 17. The experimental result can be accurately fitted with the expressions

$$\langle I \rangle / I_o = (1 + \Gamma_I t) \ , \tag{58}$$

$$\langle R \rangle / R_o = (1 + \Gamma_R t)^{\alpha_R} \ , \tag{59}$$

and obtain the aggregation rates Γ_R and Γ_I which increase rapidly with temperature. The experimental results as function of the reduced time $\tau = \Gamma_R t$ and $\Gamma_I t$ are shown in figure 18. We find a very satisfactory data collapse demonstrating the scaling property of the aggregation process.

The scattering intensity increases linearly with time and we conclude that the dynamical exponent $z = 1$, for this aggregation process.

We find that the aggregation rates may be expressed by

$$\Gamma_R = \Gamma_R^o \exp\{-\Delta H^*(\frac{1}{RT_o} - \frac{1}{RT})\} \ , \tag{60}$$

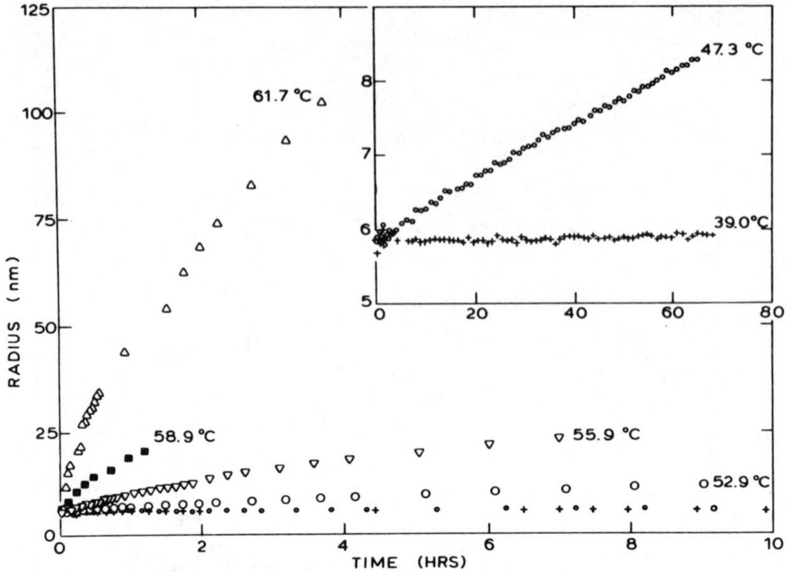

Figure 17: The effective hydrodynamic radius $\langle R \rangle$ as function of time for various temperatures for a 15.4 mg/ml sample of IgG.

with an enthalpy of activation for the aggregation process given by $\Delta H^* = 120 \pm 5\, kcal/mol$.

At the reference temperature $T_o = 329\, K$, the radius growth rate is aproximately proportional to the initial monomer concentration $c(t = 0)$, and given by $\Gamma_R^o = \rho c(t = 0)$, with $\rho = 0.18 \pm 0.01\, ml/mg\, hrs$. This result may be used together with the Smoluchowski collision rate expression 26 to conclude[35] that the sticking coefficient for the IgG collisions is very small and approximately $1.1 \cdot 10^{-9}$. Therefore the heat aggregation process of human IgG is not diffusion limited at the conditions of our experiments. The molecules may diffuse relatively large distances before they successfully add to a cluster by an activated process, and the Smoluchowski equation 25 without a spatial dependence of the cluster concentration is therefore a reasonable model for this process.

The estimate based on all our experimental results for the observed radius exponent is $\alpha_R = 0.48 \pm 0.05$. However, before the cluster fractal dimension may be determined we must correct for the presence of nonaggregating species. We have solved the Smoluchowski equation 25 numerically with the full transition probability 28, and computed $\langle I \rangle$ and $\langle R \rangle$ as function of time with the assumption that 48 % of the monomers present initially do not participate in the aggregation process. This nonaggregating H-fraction does, however, contribute to the scattering intensity and the hydrodynamic radius. We obtain good fits of the form 58 and 59, to the numerical results[58] and find $\alpha_R = 0.0153 + 1.198 \cdot D^{-1}$, for the cluster fractal dimension D in the range $2 < D < 3$. With this result we may use the experimental value of α_R and obtain an estimate

$$D = 2.56 \pm 0.2 \,,$$

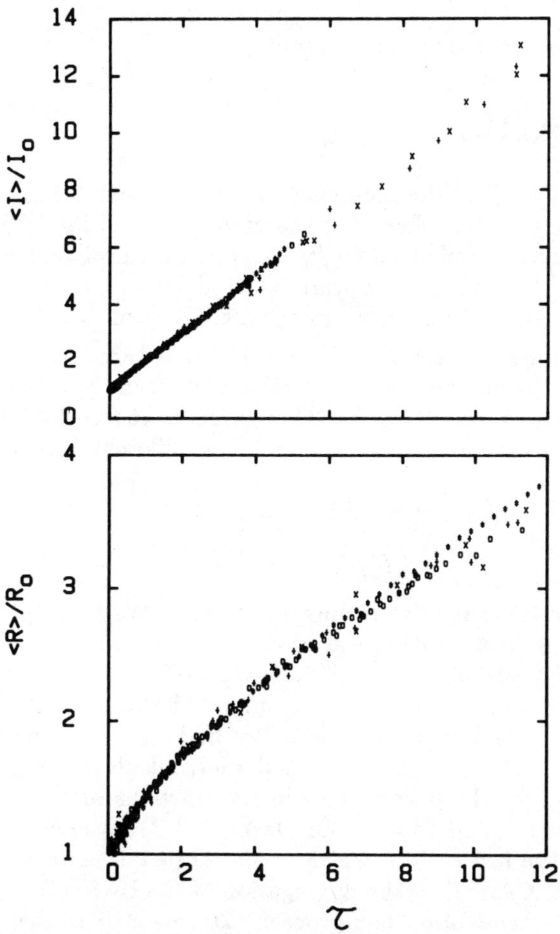

Figure 18: (a) The normalized scattering intensity, and (b) the normalized effective hydrodynamic radius as function of of the reduced time $\tau = \Gamma t$, for heat aggregation of IgG monomers at 2,4,8 and $16\,mg/ml$, and at 47, 53, 56 and 62 $°C$ [36].

for the cluster fractal dimension of normal IgG aggregation.

We conclude that normal IgG follows Smoluchowski aggregation, and that experimental results on the effective hydrodynamic radius $\langle R \rangle$, determined from the first cumulant of the observed intensity correlation function, are consistent with a fractal structure od the IgG aggregates. From the time dependence of the scattering intensity we have found that the dynamic exponent is $z = 1.0$. Clearly, further experiments extending to larger clusters, and by other methods such as neutron small angle scattering are desirable in order to obtain more precise results for the cluster structure. Such experiments are in progress.

4. The H-fraction

We have found that 48 % of the monomers initially present do not participate in the aggregation process just described. We therefore prepared the H-fraction by heating a solution of Gamma- Kabi at $30\,mg/ml$ in our standard buffer at $pH = 7.6$ for 1 hour at 62 $°C$. The monomers were isolated by column chromatography on a $Sephadex\,G - 200$ gel, and concentrated by ultrafiltration to $7.9\,mg/ml$.

By heating samples of the H-Fraction we find that these IgG molecules aggregate slowly as shown by the increase in the effective hydrodynamic radius $\langle R \rangle$, and in the scattering intensity $\langle I \rangle$, see figure 19. Compared to the normal IgG aggregation it is clear that the H-fraction exhibits a completely different kinetics. We have found that the expressions

$$\langle R \rangle / R_\circ = 1 + (\Gamma_R t)^{\delta_R} \,, \tag{61}$$

$$\langle I \rangle / I_\circ = 1 + (\Gamma_I t)^{\delta_I} \,, \tag{62}$$

fit our preliminary experimental results very well. We find $\delta_R(56) = 1.52 \pm 0.01$ and $\delta_R(59) = 1.49 \pm 0.02$, with an average of $\delta_R = 1.50 \pm 0.01$. Similarly we find $\delta_I(56) = 1.78 \pm 0.02$, $\delta_I(59) = 1.68 \pm 0.02$ and $\delta_I(62) = 1.78 \pm 0.03$. We find that $\delta_I = 1.72$ is the best estimate from all the data. The rate constants may be written in the form 60 of an activated process. We find that the enthalpy of activation for the H-fraction is $\Delta H_H^* = 125 \pm 5\ kcal/mol$, which is approximately equal to the values we found for the normal IgG in the previous section. The rate constant $\Gamma_R^0 = 0.032 \pm 0.02\,hrs^{-1}$, and $\Gamma_I^0 = 0.030 \pm 0.01\,hrs^{-1}$. This should be compared to the result obtained in the last section which at the same concentration is 1.42 hrs^{-1}, or a factor of 47 higher. Therefore the aggregation of the H- fraction has approximately the same enthalpy of activation but proceeds at only 2 % of the rate of the normal IgG sample. With the scaled time $\tau = \Gamma_R t$ and $\Gamma_I t$ for $\langle R \rangle$ and $\langle I \rangle$ respectively we find that our experimental results exhibits a nice data-collapse shown in figures 20 and 21

By solving the equations 55 for irreversible Smoluchowski aggregation preceded by a denaturation step, and calculating $\langle R \rangle / R_\circ$ and $\langle I \rangle / I_\circ$ we find the curves shown in figures 20 and 21 by choosing th equilibrium constant $K = 0.13 \pm 0.03$ and a rate constant $k = 4.4 \pm 0.1$, in units of the Smoluchowski aggregation rate constant, and with a cluster fractal dimension taken to be $D = 2$. However, as explained in the section on reaction limited aggregation the results are not sensitive to changes in D. In order to change from the dimension-less time in the theory to the reduced time of the experimental results shown in figures 20 and 21 we use $\Gamma_R t = t_S/10.2$

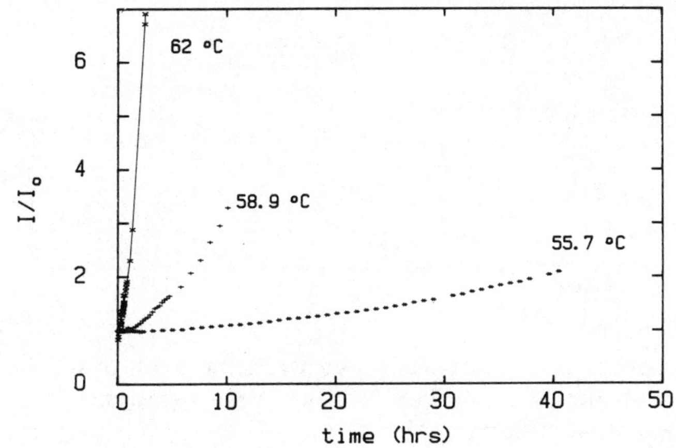

Figure 19: The scattering intensity of the *IgG* H-fraction as function of time at various temperatures.

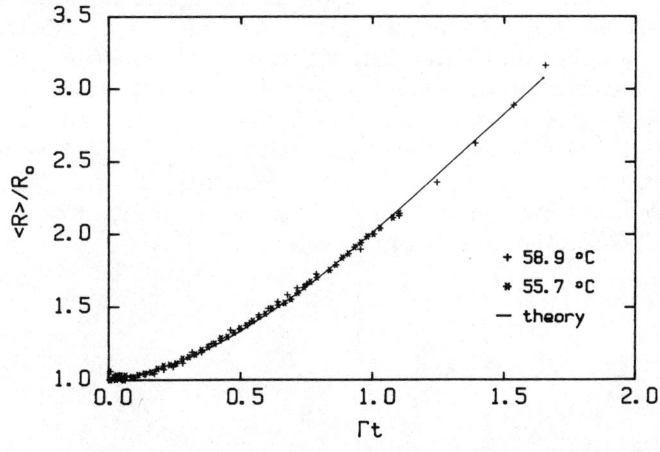

Figure 20: The normalized effective hydrodynamic radius as function of reduced time $\Gamma_R t$ for the H-fraction. at 58.9 °C and 55.7 °C . The curve represents the solution of the Smoluchowski equation with a reaction step.

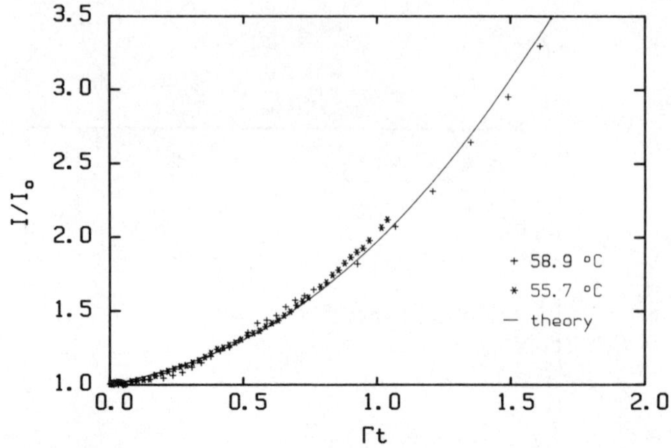

Figure 21: The normalized scattering intensity as function of reduced time $\Gamma_I t$ for the H-fraction. at 58.9 $°C$ and 55.7 $°C$ The curve represents the solution of the Smoluchowski equation with a reaction step.

and $\Gamma_I t = t_S/6.77$, where t_S is the dimensionless time of the equations 55. Note that intensity and radius may have different scale factors since that simply amounts to having different coefficients in front of the time dependent parts of equations 61 and 62 — which are to be considered as approximations to the present theoretical results.

We find the present agreement between the kinetics of aggregation of the H-fraction and our theory for the Smoluchowski aggregation process preceded by a reaction step encouraging. The enthalpy of activation $\Delta H \simeq 123$ $kcal/mol$, is the same for both fractions and in agreement with typical values for protein denaturation. However, for the H-fraction our preliminary results indicate that a denaturation step is required before aggregation occurs - this is different from the aggregation of the complete solution which is able to aggregate without a denaturation step - although also there an activated process is required. Whether the H-fraction represents a biologically significant fraction of the immunoglobulins, or is just a part of the pooled Gamma-Kabi IgG solution remains to be determined [59]

5. Conclusions

In summary, we have shown how the scattering intensity and the average hydrodynamic radius for an aggregating system may be calculated from the time-dependent cluster size distribution and the cluster fractal dimension.

We have discussed the Smoluchowski equation of irreversible aggregation, the scaling properties of the size distribution and our numerical results. In our quasielastic light scattering experiments we find that pooled human IgG aggregates in a way that can be consistently described by a Smoluchowski aggregation process, producing clusters with a fractal dimension $D = 2.56$. However, the process is not diffusion limited as only $5 \cdot 10^{-9}$ of the molecular collisions succeed in passing the activation barrier and become parts of a growing cluster.

We have also introduced kinetic equations that have a *reaction limited reversible reaction step* followed by an *irreversible Smoluchowski aggregation process* of the denatured monomers. The resulting equations have non-universal timedependent size distributions and the kinetics depends on the equilibrium constant K, and the reaction rate constant k. Preliminary experimental results on th aggregation of the H- fraction[59], show that that the H-fraction of pooled human IgG is described by this type of reaction kinetics. We conclude that the aggregation of the H-fraction is limited by a *reversible* denaturation step and proceeds by irreversible aggregation of the denatured molecules.

The interplay between a reversible reaction step to produce monomers that may then proceed to aggregate irreversibly to form clusters is more general than the native-denatured transformation of proteins and we have given a form of the kinetic equations that should apply to the gold colloid aggregation studied by Weitz et. al.[30–32]. Our numeric results show that the kinetic equations we have introduced indeed predicts a crossover from the 'slow' behavior observed at low Pyridine concentrations analogous to our H-fraction results, to the rapid aggregation at high concentrations and analogous to the normal IgG aggregation kinetics.

Acknowledgements

We thank Einar Rosenqvist for a stimulating cooperation on protein aggregation. We gratefully acknowledge the support given by Statoil and by NAVF.

REFERENCES

1 S. R. Forrest, and T. A. Witten, Jr.
J. Phys. A: Math. Gen. **12** L109–L117 (1979)

2 T. A. Witten, Jr., and L. M. Sander
Phys. Rev. Lett. **47** 1400–1403 (1981)

3 T. A. Witten, Jr., and L. M. Sander
Phys. Rev. B **27** 5686–5697 (1983)

4 D. Bensimon, E. Domany, and A. Aharony
Phys. Rev. Lett. **51** 1394 (1983)

5 P. Meakin
Phys. Rev. A **27** 1495–1507 (1983)

6 P. Meakin
Phys. Rev. A **27** 604–607 (1983)

7 P. Meakin
J. Chem. Phys. **79** 2426–2429 (1983)

8 T. A. Witten, Jr., and P. Meakin
Phys. Rev. B **28** 5632–5642 (1983)

9 P. Meakin
Phys. Rev. Lett. **51** 1119–1122 (1983)

10 M. Kolb, R. Botet, and R. Jullien
Phys. Rev. Lett. **51** 1123–1126 (1983)

11 P. Meakin
Phys. Rev. B **28** 6718–6732 (1983)

12 R. Jullien, M. Kolb, and R. Botet
J. Physique Lett **45** L211–L216 (1984)

13 P. Meakin
Bull. Am. Phys. Soc. **29** 353 (1984)

14 R. F. Voss
J. Stat. Phys. **36** 861–872 (1984)

15 R. Jullien, and M. Kolb
J. Phys. A: Math. Gen. **17** L639–L643 (1984)

16 P. Meakin
Phys. Rev. **B29** 3722–3725 (1984)

17 P. Meakin, F. Leyvraz, and H. E. Stanley
Phys. Rev. A **31** 1195–1198 (1985)

18 B. B. Mandelbrot
The Fractal Geometry of Nature,
(Freeman, San Fransisco, Cal. 1982)

19 M. von Smoluchowski
Physik. Zeitschr. **17** 557–571,585–599 (1916)

20 M. von Smoluchowski
Z. phys. Chemie **92** 129–168 (1917)

21 M. J. Vold
J. Colloid Sci. **18** 684–695 (1963)

22 H. P. Hutchison, and D. N. Sutherland
Nature **206** 1036–1037 (1965)

23 A. I. Medalia
in *Surface and Colloid Science* **4**, 1–92 (1971)

24 A. I. Medalia
J. Colloid Interface Sci. **24** 393 (1967)

25 A. I. Medalia, and F. A. Heckman
*J. Colloid Interface Sci.***36** 173–190 (1971)

26 D. N. Sutherland, and I. Goodarz-Nia
Chem. Eng. Sci. **26** 2071–2085 (1971)

27 F. Family, and D. P. Landau, editors
Kinetics of Aggregation and Gelation
(North-Holland, Amsterdam 1984)

28 C. Allain, and B. Jouhier
J. Physique - LETTRES **44** L421–L428 (1983)

29 W. T. Elam, S. A. Wolf, J. Sprague, D. U. Gubser,
D. Van Vechten, G. L. Barz Jr., P. Meakin
Phys. Rev. Lett. **54** 701–703 (1984)

30 D. A. Weitz, and M. Oliveria
Phys. Rev. Lett. **52** 1433–1436 (1983)

31 D. A. Weitz, and J. S. Huang
Kinetics of Aggregation and Gelation
(F. Family and D. Landau,editors ,North-Holland, Amsterdam 1984) 19–28

32 D. Weitz, J. S. Huang, M. Y. Lin, and J. Sung
Phys. Rev. Lett. **53** 1657–1660 (1984)
Phys. Rev. Lett. **54** 1416–1419 (1985)

33 D. W. Schaefer, J. E. Martin, P. Wiltzius, and D. S. Cannell
Phys. Rev. Lett. **52** 2371–2374 (1984)

34 S. K. Sinha, T. Freltoft, and J. Kjems
Kinetics of Aggregation and Gelation
(F. Family and D. Landau,editors ,North-Holland, Amsterdam 1984) 87–90

35 T. Jøssang, J. Feder, and E. Rosenqvist
J. Chem. Phys. **82** 574–590 (1984)

36 J. Feder, T. Jøssang, and E. Rosenqvist
Phys. Rev. Lett. **53** 1403–1406 (1984)

37 T. Jøssang, J. Feder, and E. Rosenqvist
Scattering Techniques Applied to Supramolecular and Nonequilibrium Systems,
edited by S. Chen, B. Chu, and R. Nossal, (Plenum, New York, 1981) 831

38 H. Z. Cummins, and E. R. Pike
Photon Correlation and Light Beating Spectroscopy
(Plenum Press, New York, 1974)

39 B. Chu
Laser Light Scattering (Academic press, 1974)

40 B. I. Berne, and R. Pecora
Dynamic Light Scattering (J. Wiley & sons Inc.,New York, 1976)

41 J. E. Martin, and D. W. Schaefer
Phys. Rev. Lett. **53** 2457–2460 (1984)

42 J. E. Martin, and B. J. Ackerson
Phys. Rev. A **31** 1180–1182 (1985)

43 Zhong-Ying Chen, J. M. Deutch, and P. Meakin
preprint 11 oct 1–13 (1984)

44 T. Vicsek, and F. Family
Phys. Rev. Lett. **52** 1669–1672 (1984)

45 M. H. Ernst
Fundamental Problems in Statistical Mechanics VI .
(E. G. D. Cohen, editor, North-Holland, Amsterdam 1985) 329–364

46 M. H. Ernst, and E. M. Hendriks
Kinetics of Aggregation and Gelation
(F. Family and D. Landau,editors ,North-Holland, Amsterdam 1984) 225–228

47 R. M. Ziff
Kinetics of Aggregation and Gelation
(F. Family and D. Landau,editors ,North-Holland, Amsterdam 1984) 191–199

48 R. Botet, and R. Jullien
J. Phys. A: Math. Gen. **17** 2517–2530 (1984)

49 E. M. Hendriks, M. H. Ernst, and R. M. Ziff
J. Stat. Phys. **31** 519–563 (1984)

50 F. Leyvraz
Kinetics of Aggregation and Gelation
(F. Family and D. Landau,editors ,North-Holland, Amsterdam 1984) 201–204

51 P. Meakin, T. Vicsek, F. Family
Phys. Rev. B **31** 564–569 (1985)

52 K. Kang, and S. Redner
Phys. Rev. A **30** 2833–2836 (1984)

53 S. A. Troelstra and H. R. Kruyt
Kolloid. Chem. Beih. **54** 225 (1943)

54 G. Oster
J. Colloid. Sci. **2** 291 (1947)

55 H. Versmold, and W. Härtl
J. Chem. Phys. **79** 4006–4009 (1983)

56 **J. B. G. Kwapinsky**
Methodology of Immunochemical and Immunological Research (Wiley–Interscience, New York, 1972)

57 **H. Nielsen, and S. Svehag**
Acta. Path. Microbiol. Scand. Sect. **C84** 261 (1976)

58 **J. Feder, and T. Jøssang**
Institute of Physics Report Series, University of Oslo Report No 83–5 (1983)

59 **E. Rosenqvist, J. Feder, and T. Jøssang**
to be published (1985)

NEUTRON AND X-RAY SCATTERING FROM AGGREGATES

J.K. Kjems and T. Freltoft

Risø National Laboratory
DK-4000 Roskilde, Denmark

INTRODUCTION

Scattering experiments are indispensable for the study structures and correlation functions in condensed matter science. Typically, the resulting information can be related to characteristic lengths for the system under study such as the correlation lengths near critical points, the radius of gyration of macromolecules, the mean square displacements for thermal vibrations etc. Also for scale invariant systems without a single characteristic length, such like the fluctuations at the critical point in a continuous phase transition, can scattering experiments be used to determine the characteristic power-law decay of the correlations. Scattering experiments are therefore also well-suited probes for the structures of randomly formed aggregates which appear self-similar over considerable ranges in length and thus can be considered as scale invariant, volume fractals[1-3].

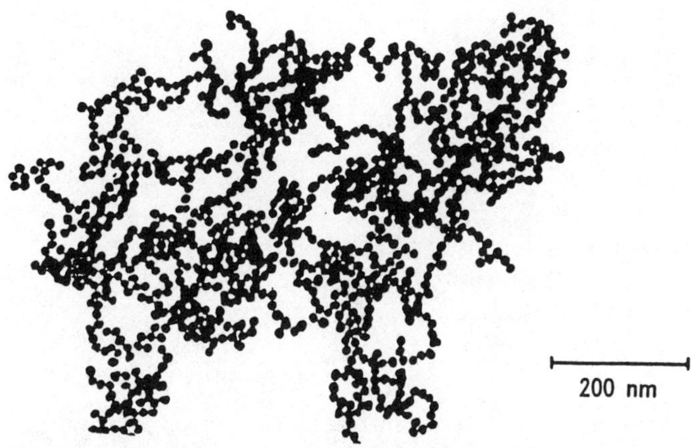

Fig. 1. Example of an aggregates formed by a random diffusion process. Electron micrograph of gold-colloid cluster made by D. Weitz and M. Oliveria[6]. Direct analysis in terms of $M(r) \propto r^D$ gives D=1.75.

In this paper we will discuss the relation between the fractal dimensionality, loosely identified as the Hausdorff dimension, and the scattering law for x-rays and neutrons with special emphasis on the role of the upper and lower length scale cut-offs[4]. This leads to discussions of the effects of polydispersity and of the near neighbour coordination which may obscure/enrich the data depending on the attitudes of the observer.

Finally, we present examples from recent experiments on smoke particles[4] and colloids[5,6].

SCATTERING LAW

A volume fractal is characterized by the fractal dimensionality[3], D, which describes how the mass, M, or number of constituents, N, grows with increasing radius.

$$M(r) = m_o\left(\frac{r}{r_o}\right)^D \qquad (1)$$

It is assumed that there is a particle at the origin and hence the relation (1) represents an integral over the particle-particle correlation function $G(r) = \langle\rho(r)\rho(o)\rangle$, so

$$M(r) = \frac{1}{\langle\rho\rangle}\int_o^r \langle\rho(r')\rho(o)\rangle d^dr' \qquad (2)$$

$M(r) \propto r^D$ implies that $G(r) \propto r^{-\alpha}$ and we can use a simple dimensionality

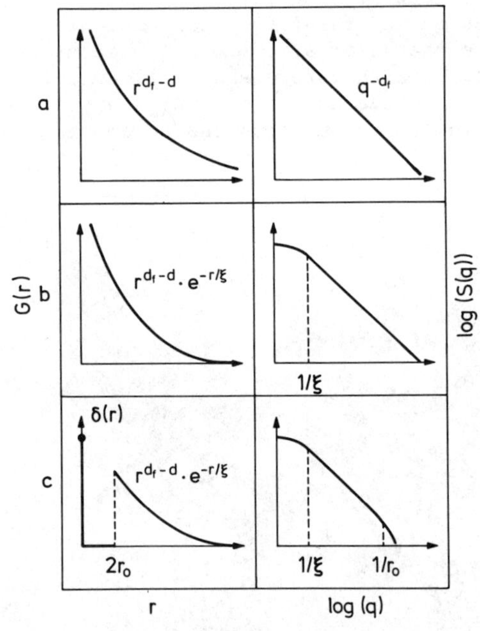

Fig. 2. Schematic illustrations of the correlation function, G(r), and the corresponding scattering function, S(Q). (b) shows the effect of a finite large scale cut-off, ξ, and (c) shows the effect of the "correlation hole" at short distances.

argument to determine the relation between D and α

$$\int G(r)d^dr \propto \int r^{-\alpha}\, r^{d-1}\, dr \propto r^{-\alpha+d} = r^D$$

$$D = d-\alpha \qquad \text{or} \qquad \alpha = d-D \tag{3}$$

A scattering experiment measures the Fourier transform of the pair correlation function and, for rotational invariant systems, one gets (d=3)

$$S(Q) \propto \int_0^\infty G(r)\frac{\sin Qr}{Qr}d^3r$$

$$\propto Q^{-D+3-3}\int_0^\infty (Qr)^{D-3}\frac{\sin Qr}{Qr}d(Qr) \tag{4}$$

$$S(Q) \propto Q^{-D}$$

For real systems this ideal result can only be expected to hold over a limited range of length scales. The lower cut-off, r_o, is given by the size of the primary particles (typically 5-10 nm for colloids) and the upper limit is the aggregate size, ξ, which can be as large as 1 μm for colloids.

The way the scattering law is affected by these limits is shown schematically in Fig. 2. The simplest correction for the finite size of the primary particles is to factorize a formfactor, which assumes that the particles are of nearly identical size and shape, hence the observed intensity can be written like

$$I(Q) = |F(Q)|^2 S(Q) \tag{5}$$

where S(Q) now corresponds to the point correlation function between the particle centers. If r_o is much smaller than the aggregate size, it is a good approximation to assume

$$|F(Q)|^2 = e^{-\frac{1}{3}r_o^2Q^2} \quad , \quad Qr_o \leq 2 \tag{6}$$

where r_o is the Guinier radius for the primary particles, $r_o = \sqrt{\frac{3}{5}}r$. At larger Q values Eq. (6) should be replaced with a form that tends towards the asymptotic, Porod's law $\propto Q^{-4}$.

The modifications due to the finite aggregate size are less obvious and they will involve assumptions and approximations. The simplest approach is to introduce an exponential cut-off in the correlation function

$$G(r) \propto \left(\frac{r}{r_o}\right)^{D-d} e^{-r/\xi} \tag{7}$$

This corresponds to an assumption of spherical aggregates with an average radius, ξ, and this approximation has the virtue that the Fourier transform can be worked out analytically[4]. For d=3 one gets

$$S(Q) \propto \xi^D \frac{(D-1)\Gamma(D-1)}{(1+(\xi Q)^2)^{D/2}} \frac{(1+(\xi Q)^2)^{1/2}}{\xi Q} \frac{\sin[(D-1)\mathrm{atn}(\xi Q)]}{D-1} \tag{8}$$

which has the expected asymptotic behaviour for $\xi Q >> 1$

$$S(Q) \propto \xi^D \frac{\Gamma(D-1)}{(\xi Q)^D} \sin\left[(D-1)\frac{\pi}{2}\right] \tag{9}$$

Note that this scattering vanishes when $D \rightarrow 3$.

This formulation of the scattering law has been used to analyse neutron small-angle scattering data[4] from silica particle aggregates form by flame hydrolysis, Cab-O-Sil. The fits to the data are quite good but the analysis can be further refined by including the effects of the "correlation hole" around the central particle. This is illustrated in fig. 2c where G(r) now has a δ-function at the origin and has the value zero up to $2r_o$ which is the shortest distance between the primary particles

$$G(r) = \begin{cases} A_o\ \delta(r) & r<2r_o \\ A_1\left(\frac{r}{r_o}\right)^{D-d} e^{-\left(\frac{r}{\xi}\right)} & r\geqslant 2r_o \end{cases} \tag{10}$$

Again we can find an analytic, albeit quite complicated form for the scattering function, namely

$$S(Q) = B_o + B_1 \frac{(D-1)\xi^D}{(1+(\xi Q)^2)^{D/2}} \frac{[1+(Q\xi)^2]^{1/2}}{Q\xi} \tag{11}$$

$$* \operatorname{Im}\left[\Gamma(D-1, \frac{2r_o}{\xi}(1-iQ\xi))\frac{1}{D-1}(\cos((D-1)\mathrm{atn}(Q\xi))+i\sin((D-1)\mathrm{atn}(Q\xi)))\right]$$

where $\Gamma(\alpha,z)$ is the incomplete gamma function with complex argument defined by the integral

$$\Gamma(\alpha,z) = \int_z^\infty \omega^{\alpha-1} e^{-\omega} d\omega \tag{12}$$

This form gives an excellent fit to the Cab-O-Sil data (Fig. 3) but the ratio between the parameters B_o and B_1 becomes very delicate. In principle, this ratio can be used to extract information about the nearest neighbour coordination but, unfortunately, the present Cab-O-Sil data does not extend quite far enough in Q to obtain a reliable answer.

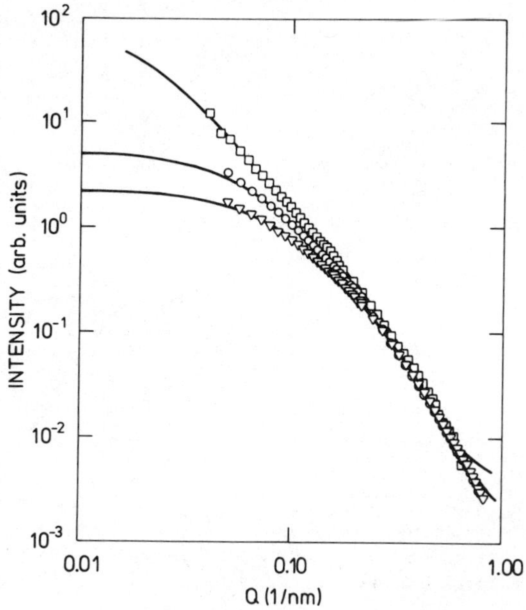

Fig. 3. Log-log plot of neutron scattering intensity versus Q for samples of Cab-O-Sil aggregates. The full lines are based on fits to Eq. 11 as described in the text. The data and curve 1 corresponds to a suspended sample in 8% D_2O/H_2O for which ρ = 0.01 g/cm^3 and D = 2.4±0.1. Data marked 2 and 3 are powders with ρ = 0.13 g/cm^3 and ρ = 0.45 g/cm^3, both give D = 2.55±0.07.

The above formulation of the scattering law takes indirectly the polydispersity of the aggregate size into account through an effective size parameter, ξ. Improvements on this point could be made in cases where additional information about the size distribution is available such as the Smoluchowski analysis of the aggregation process, which has been carried out by Feder, Jøssang and Rosenquist[8] in a light scattering study of the coagulation of blood proteins. However, for very large fractal aggregates the corrections for the upper and lower cut-offs can be neglected over a considerable range in momentum transfer, Q, and the scattering data can be analysed directly to give the fractal dimensionality as shown by the work of Schaefer et al.[5] to be discussed below.

EXPERIMENTAL EXAMPLES

The first in-situ experiment on growing colloid aggregates in a suspension has been reported by Schaefer, Martin, Wiltzius and Cannell[5]. They studied suspensions of silica colloids (Ludox) with primary particle size r_o=2.7 nm, which were made to form aggregates through a change in ph and salinity. The clusters grow to very large sizes > 1 μm over periods of hours and eventually form sediments. The process was followed by light and by x-ray scattering and a combined results at a late stage in the clustering process is shown in Fig. 5. From these data they conclude that D=2.12±0.05.

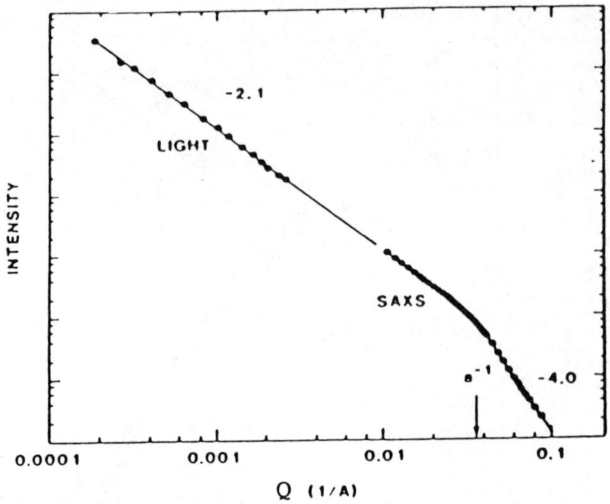

Fig. 4. Combined small-angle x-ray and light scattering results for aggregates formed from Ludox silica colloids. After Schaefer et al.[5], who find the fractal dimensionality, D=2.12±0.05.

A similar process for gold colloids have been studied by Weitz and Oliveria[6] who initially used electron microscopy to analyse the structure of the resulting aggregates and they find D=1.75. Subsequently they have confirmed this value by both light scattering and x-ray scattering[9] and they have also studied the time evolution in detail and established a correlation between the growth rate and the resulting D. Slow growth gives larger D-values.

Together with S.K. Sinha we have made a small-angle neutron scattering study[4,7] of silica particle aggregates formed by flame hydrolysis (Alfasil and Cab-O-Sil). Measurements were made on samples in the form of compressed powders with different densities in the range 0.056 g/cm^3 to 0.45 g/cm^3 and of dilute suspensions in mixtures of D_2O/H_2O (density ~ 0.008 g/cm^3). Examples of our data are shown in Fig. 3 together with the fitted curves corresponding to the theoretical scattering laws Eq. 11.

Figure 5a includes the data for the suspended samples at different contrast levels and Fig. 5b gives the intensity variation with contrast which shows that complete matching can be achieved and hence that all voids are accessible to the suspension fluid. From the fits we derive a value for D=2.55±0.05 for the powders, independent of the density and slightly lower values near

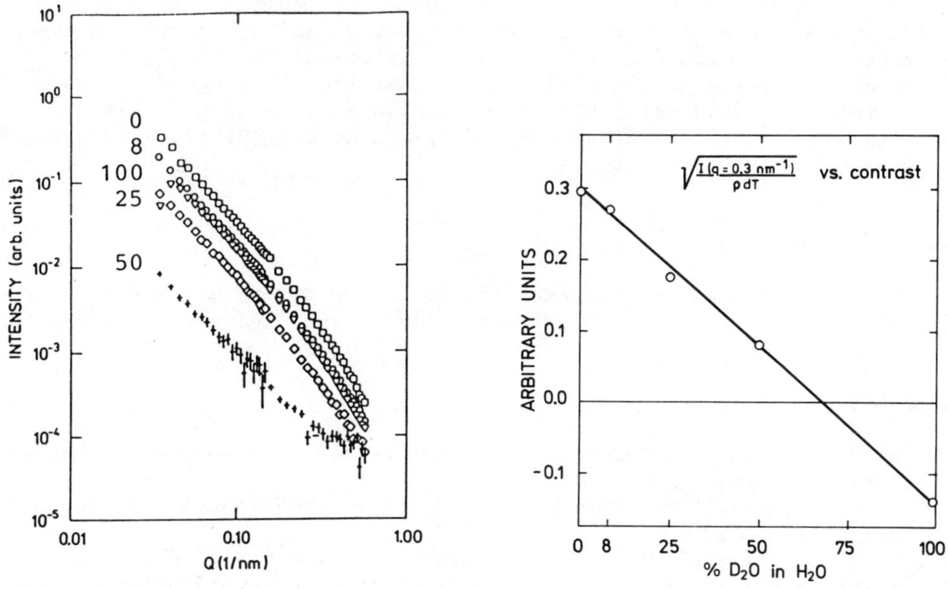

Fig. 5 (a). Small-angle neutron scattering data for Cab-O-Sil aggregates in suspensions with different D_2O/H_2O contrast levels. The data has not been corrected for transmission factors.
(b). Variation of the square root corrected intensities versus contrast at Q = 0.3 nm^{-1} for Cab-O-Sil aggregates suspended in D_2O/H_2O. The linear relation and the matching point value shows that all voids in the aggregates are accessible to the suspension fluid.

2.4 for the suspensions. In all cases we find $r_o = 2.0 \pm 0.1$ nm and, as expected, different values for ξ in the range 10 nm to 50 nm depending on the density.

DISCUSSION AND SUMMARY

All the aggregates studied so far show a scattering law which can be interpreted with a fractal dimensionality, D, well below the spatial dimensionality, d. The D values range from 1.75 to 2.55, which corresponds to the range spanned by the computer simulations for the cluster-cluster models[11] to the Witten-Sander diffusion limited aggregation (DLA) model[1,2]. Clearly, the D-values depend on the aggregation process and the work by Weitz et al.[9] shows that, for the same system, it also depends on the aggregation rate. In some sense D just measures the mass or density of an aggregate and the scattering curves does not give direct information about other structural characteristics like the "chemical" distance between particles as discussed by Havlin and Nossal[10]. Scattering data could possible by analysed in more detail to give information about the near-neighbour coordination and it is also conceivable that an analysis of the energy distribution of the scattering could give information about the dynamics of such clusters. The first attempts in this direction have been made by Martin and Schaefer[12] who have studied the dynamic light scattering from their silica-colloids and they find that the decay rate for the intensity autocorrelation function scales like $Q^{2.7}$. This result is not yet explained satisfactorily and more experiments in this area would clearly be of interest.

ACKNOWLEDGEMENT

Our work on silica-aggregates was initiated during a visit by S.K. Sinha who also has contributed essential elements of theoretical analysis.

REFERENCES

1. T.A. Witten and L.M. Sander, Phys. Rev. Letters 47:1400-1403 (1981).
2. P. Meakin, Phys. Rev. A. 27:1495-1507 (1983).
3. B.B. Mandelbrot, "The Fractal Geometry of Nature" (North Holland, San Fransisco (1983).
4. S.K. Sinha, T. Freltoft and J.K. Kjems, in "Kinetics of Aggregation and Gelation", F. Family and D.P. Landau, editors (North Holland, New York (1984).
5. D.W. Schaefer, J.E. Martin, P. Wiltzius and D.S. Cannell, Phys. Rev. Letters, 53:2371-2374 (1984).
6. D.A. Weitz and M. Oliveria, Phys. Rev. Letters, 52:1433-1436 (1984).
7. T. Freltoft, J.K. Kjems and S.K. Sinha, to be published.
8. J. Feder, T. Jøssang and E. Rosenquist, Phys. Rev. Letters, 53:1403-1406 (1984).
9. D. Weitz, J.S. Huang, M.Y. Lin and J. Sung, Phys. Rev. Letters 53:1657-1660 (1984).
10. S. Havlin and R. Nossal, J. Phys. A: Math. Gen., 17:L427-L432 (1984).
11. R. Jullien, M. Kolb, R. Botet, Journ. de Physique Lettr. 45:211 (1984).
12. J.E. Martin, D.W. Schaefer, Phys. Rev. Lett. 53:2457-2460 (1984).

NEUTRON AND X-RAY STUDIES OF INTERFACES

J.K. Kjems[+] and P. Schofield*

[+]Risø National Laboratory
DK-4000 Roskilde, Denmark
*Materials Physics and Metallurgy Divisions
AERE Harwell, Oxfordshire, United Kingdom

INTRODUCTION

In this lecture we will discuss how information about the interface morphology in complex heterogeneous media can be derived from scattering experiments. The practical systems we have in mind are high surface area aluminas which are used as bearers for catalysts and as components in LiI/ Al_2O_3 composites. The latter show enhanced Li-mobility, probably due to an interface reaction[1]. Other examples are minerals[2], coals[3] and carbon blacks[4] for which evidence of fractal character of the interfaces has been presented.

For the formulation of the scattering law we will assume that the interfaces are infinitely sharp but irregular in the sense that the scattering length density (or electron density) is strictly a two-valued function. This is done mainly for pedagogical reasons since the modifications due to other profiles at the interfaces also can be worked out in the same framework. With the assumption of sharp interfaces it is natural to seek a surface-formulation of the scattering law. This is done in the next section followed by a discussion of how the asymptotic Porod's law follows naturally from this derivation. We then discuss the implications of a fractal interface and illustrate this with a simple one-dimensional example. Finally, we present experimental data which show some of the features we have discussed theoretically.

SURFACE FORMULATION OF SCATTERING LAW

The typical experiment to study interface scattering with either x-rays or neutrons is carried out at small scattering angles and in this region the scattering laws, based on the Born approximation, are identical except that the neutron contrast is given by the nuclear scattering length density $b(\vec{r})$ and the x-ray contrast is given by the electron density $\rho(\vec{r})$. At small angles the variation due to the electron form-factors can be neglected. In the scattering experiment one measures the intensity distribution as function of wavevector transfer $\vec{Q} = \vec{k}_i - \vec{k}_j$; $Q = 2k\sin\theta$

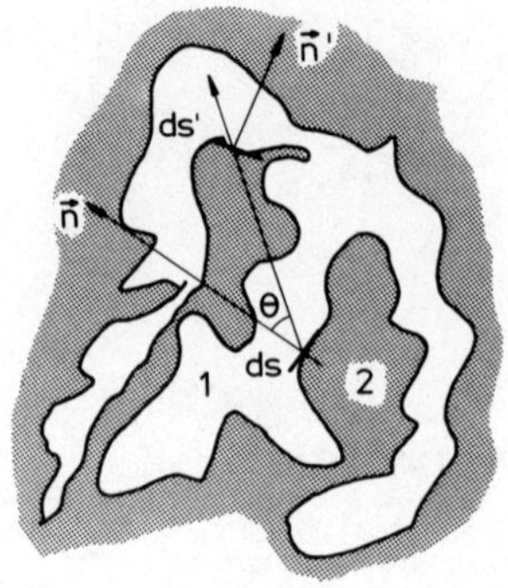

Fig. 1. Schematic representation of a two-component medium.

$$I(\vec{Q}) = \frac{1}{V} \langle |b(\vec{Q})|^2 \rangle$$

with (1)

$$b(\vec{Q}) = \int_V d\vec{r} b(\vec{r}) e^{i\vec{Q}\cdot\vec{r}}$$

Using the mathematical identity

$$\vec{\nabla}\{b(\vec{r})\vec{Q}e^{i\vec{Q}\cdot\vec{r}}\} = iQ^2 b(\vec{r})e^{i\vec{Q}\cdot\vec{r}} + \vec{Q}\cdot\vec{\nabla}(b(\vec{r}))e^{i\vec{Q}\cdot\vec{r}} \tag{2}$$

we can write

$$b(\vec{Q}) = -\frac{i}{Q^2}\sum_j \int dS_j \vec{Q}\cdot\vec{n} b(\vec{r}_j) e^{i\vec{Q}\cdot\vec{r}_j} + \frac{i}{Q^2}\sum_j \int d\vec{r}_j \vec{Q}\cdot\vec{\nabla} b(\vec{r}_j) e^{i\vec{Q}\cdot\vec{r}_j} \tag{3}$$

which, for sharp interfaces becomes

$$b(\vec{Q}) = -\frac{i}{Q^2}\sum_{i,j}(b_i - b_j)\int dS_{ij}\vec{Q}\cdot\vec{n}_{ij}(\vec{r}_s)e^{i\vec{Q}\cdot\vec{r}_s} \tag{4}$$

Finally, with assumption of isotropy in a two-component system one gets

$$I(Q) = \frac{\Delta b^2}{Q^2}\int ds \int ds' \vec{n}(\vec{r})\vec{n}(\vec{r}')\frac{\sin Q|\vec{r}-\vec{r}'|}{Q|\vec{r}-\vec{r}'|} \tag{5}$$

where the integrals are confined to the interfaces and $\vec{n}(\vec{r})$ are their normals.

As shown in the appendix, one can derive a more specific general expression for a two-component medium, namely

$$I(Q) = \frac{2\pi A(\Delta b)^2}{Q^4}\left[1 - \frac{1}{2}Q\frac{\partial}{\partial Q}\right] \mathrm{Re} \frac{(1-P_1(Q))(1-P_2(Q))}{1-P_1(Q)P_2(Q)} \tag{6}$$

where A is the total interface area and $P_i(Q)$ is the Fourier transform of the average length distribution, $P_i(\ell)$, for domains along a line. A random model, $P_i(\ell) = \ell_i^{-1}\exp(-\ell/\ell_i)$, leads to Debye's formula

$$I(Q) = 2\pi A \frac{\ell^4}{(1+Q^2\ell^2)^2} \quad , \quad \ell = \frac{\ell_1\ell_2}{\ell_1+\ell_2} \tag{7}$$

Porod's law

Equations (6) and (7) show that the asymptotic limit for large Q is $\propto Q^{-4}$ and one can use the result given in Eq. 5 to generalize this to effective dimensions less than d=3. For d=1 the integral at large Q-value simply enumerates the number of interface points, N_1, and hence

$$I(Q) \propto \frac{N_1(\Delta b)^2}{Q^2} \qquad d = 1 \tag{8}$$

For d=2 one can show similarly that the large Q contributions to the integral come from the specular scattering from straight sections of the interface curve, integrated to give each L_i/Q hence

$$I(Q) \propto \frac{L(\Delta b)^2}{Q^3} \qquad d = 2 \tag{9}$$

These results have also been shown to be relevant for the discussion of neutron scattering from magnetic domain structures induced by random field effects[6]. Here we shall focus on the effects of interface roughness. Clearly, the asymptotic Porod's law relies on the assumption that the interface can be regarded as smooth below a certain length scale. This assumption may not be valid for systems with very rough and irregular interfaces and in the following we shall discuss the implications of a surface fractal character of such interfaces.

SURFACE FRACTALS

The concept of a surface fractal[7] can be illustrated by the Koch island (Fig. 2) which is a mathematical construction of an interface with infinite length surrounding finite size particle. In any real system there will always be a lower, e.g. the atomic, length scale limit and hence true fractal behaviour cannot be expected. At best, one should find pseudo-fractal character over a limited range of length scales. Nevertheless, as indicated by the celebrated coast line observations[7], it may be a valid concept to be used for structural characterization of complex interface morphologies. It could also be useful for the study of transport in such media. Let us therefore attempt to determine the scattering law in a hypothetical experiment in which we scatter from an object like a Koch island. The divergence of the circumference of the Koch island is given by

Fig. 2. The so-called Koch island illustrates a particle with a fractal boundary.

$$L(\varepsilon) = L(\varepsilon_o)\left(\frac{\varepsilon_o}{\varepsilon}\right)^{D-1} \qquad D-1 = \ell n\left(\frac{4}{3}\right) \tag{10}$$

where ε is the length scale used for the measurement. If we consider objects like that which has a finite lower cut-off scale, ε_1, then the asymptotic scattering for $Q \gg 1/\varepsilon_1$ will follow Porod's law with

$$S(Q) = 4\pi(\Delta b)^2 L(\varepsilon_1)/Q^3 \quad , \quad d = 2 \tag{11}$$

The argument is now that for $Q \ll 1/\varepsilon_1$ the scattering cannot depend on the actual value of ε_1 but, nevertheless should extrapolate to the value given by Eq. 11 for $Q \sim 1/\varepsilon_1$. A form which does just that is

$$S(Q) = 4\pi(\Delta b)^2 L(\varepsilon_o)(Q\varepsilon_o)^{D-1}/Q^3 \quad , \quad d = 2 \tag{12}$$

In general, for d-dimensions it becomes

$$S(Q) = 4\pi(\Delta b)^2 N_o \varepsilon_o^D / Q^{2d-D} \tag{13}$$

where $N_o \varepsilon_o^{d-1}$ is the surface area at scale ε_o. A similar result has been obtained by Bale and Schmidt[3] and by Witten[8], but its relevance has been challenged by a recent analysis by Wong[9].

d=1 "millefeuille"-model

We can improve the plausability of Eq. 13 by considering a one-dimensional analogue of the Koch island. The model starts with a single step which acquires structure under sufficient magnification (see figure 3). Using the result given in Eq. 5 we can write

$$S(Q) \propto \frac{1}{Q^2} |F(Q)|^2$$

with (14)

$$F(Q) = F_o(Q) \cdot F_1(Q) \cdot F_2(Q) \cdots\cdots F_r(Q) \cdots\cdots$$

$$F_r(Q) = e^{-iQd/n^r} - 1 + e^{iQd/n^r} = (2\cos(Qd/n^r) - 1)$$

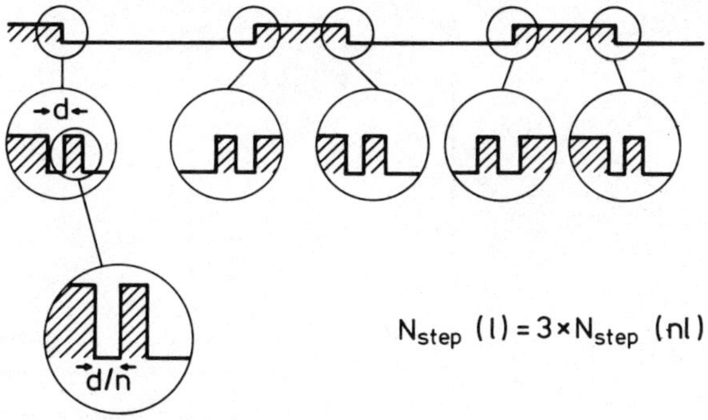

Fig. 3. One-dimensional "millefeuille"-model for a two-component media with fractal interfaces.

This leads to

$$\ln|F(Q)|^2 = \sum_{r=o}^{\infty} \ln(4\cos^2(Qd/n^r)-4\cos(Qd/n^r)+1)$$

$$= \sum_{r=o}^{\infty} \ln A_r(Q) \tag{15}$$

If we neglect the oscillating part, which can be shown to be dampened by randomness, we can regard $A_r(Q)$ as a broad stepfunction that changes from 1 to 3 at $Q = \lambda \frac{n}{d}$. Thus, for a given Q only A_r's up to a maximum $r_m = \ln(Qd/\lambda)/\ln n$ contribute. The result is then

$$|F(Q)|^2 = \exp\{\ln(Qd/\lambda)\ln 3/\ln n\}$$

and (16)

$$S(Q) = (Qd)^{-(2-\ln 3/\ln n)} = (Qd)^{-(2d-D)}$$

$$D = \ln 3/\ln n \quad ; \quad n>3 \text{ for fractal character.}$$

EXPERIMENTAL EXAMPLES

Evidence for fractal character of the interfaces in porous minerals have been suggested by Avnir, Farin and Pfeifer[2] on the basis of an analysis of surface area measurements using gas adsorbtion with molecules of different size. They find power-law relations over limited ranges in length scales. Bale and Schmidt have reanalysed x-ray scattering data from Beulah lignite coal and find deviations from Porod's law with 2d-D = 3.44 i.e. D = 2.56 over two decades in Q. Their data is shown in Fig. 4a. We have made similar x-ray scattering studies on a series of Carbon Blacks which is used as electrode material in batteries and examples of the data is shown in Fig. 4b. These data demonstrate that a range of slopes can be found but that there is no correlation between absolute value of the total surface area and this slope. Differential surface area measurements have not yet been made on these samples but they would be useful for comparison. Interface impedance measurements are in progress.

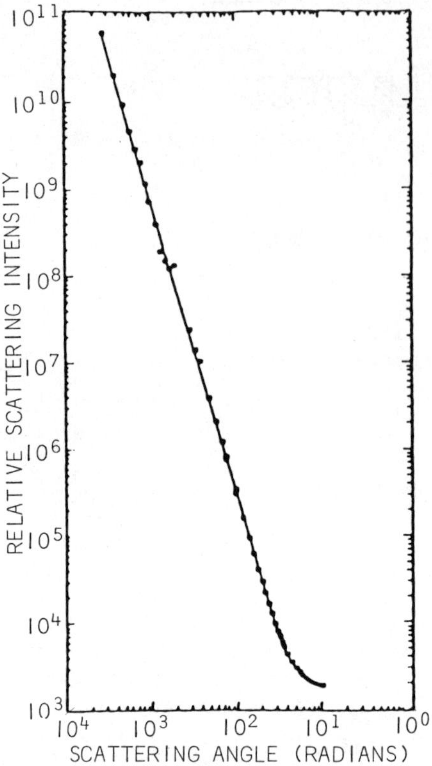

(a)

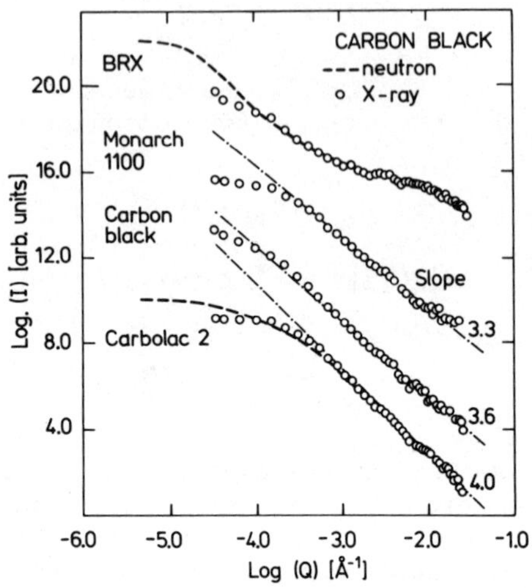

(b)

Fig. 4. (a) X-ray scattering data for Beulah lignite coal by Bale and Schmidt[3] showing a power-law decay with slope 2d-D = 3.44. (b) X-ray and neutron scattering data for a range of highly porous carbon blacks showing different asymptotic trends[4]. The surface areas measured by the BDT-technique range from 200 mg^2/g to 1100 m^2/g. Note that the curves are displaced for clarity so the ordinate scale is arbitrary.

Another interesting system is the LiI/Al_2O_3 composites[1] which show considerable enhancement of the Li-mobility compared to the pure LiI, almost enough for practical applications. This is probably due a the formation of an Li-conducting region at the LiI-alumina interface. The conductivity data (Fig. 5a) show a broad maximum as function of composition around the values which correrspond to the filling of the porous alumina matrix. Powder diffraction data show only the presence of the two original phases in these mixtures and the alumina phase is very non-crystalline with broad diffraction lines. Small-angle neutron scattering data is shown in Fig. 5b for a series of alumina which has been subjected to annealing at 1000°C for different lengths of time. The scattering laws for the fully annealed sample is close to Q^{-4} which is consistent with the platelet morphology seen in the electron-microscope for this sample. The other samples show clear deviations from Porod's law and they appear almost like aggregate clusters in the microscope. Conductivity measurements of composites based on these samples have not yet been carried out but should be instructive for the separation of the geometrical and the chemical contributions to the conductivity enhancement effect.

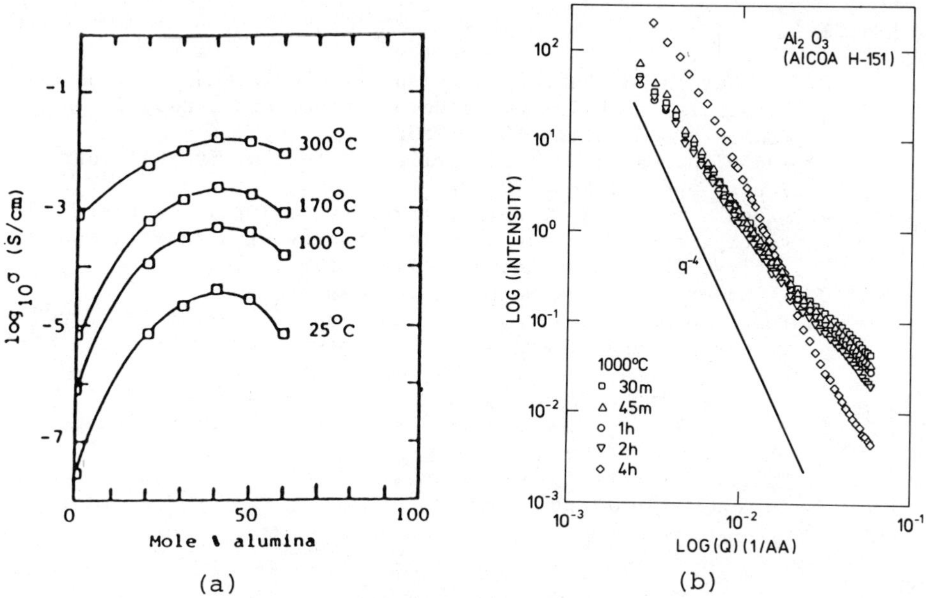

Fig. 5. (a) Ionic ac-conductivity for composites of LiI/Al_2O_3 with varying mole fractions showing the unexpected enhancement called the composite effects. (b) Neutron scattering data for the Al_2O_3 used for the LiI/Al_2O_3-composite as function of annealing for different lengths of time. For the fully annealed material the scattering approaches Porod's law whereas for the intrinsic material the scattering at large Q's is considerably enhanced.

SUMMARY

The notion of surface fractals has not yet been firmly established as a useful concept for the description and understanding of the physical properties associated with complex interface morphologies. The scattering law can be worked out for simplified mathematical models and leads to well-defined deviations from the asymptotic Porod's law but, unfortunately, the interpretation is not unique and more theoretical and experimental work is needed which can establish links between different properties e.g. relating structure and transport properties.

ACKNOWLEDGEMENT

Part of this work was carried out during an exchange visit (JK) to CEN-Saclay, partially supported by NATO. The experimental work on composites has been carried out in collaboration with F.W. Poulsen and T. Freltoft and the x-ray work with A. Tardieu and A. Le Méhauté. Support from an EEC-scientific cooperation contract is acknowledged.

REFERENCES

1. F. W. Poulsen, N. Hessel Andersen, B. Kindl, and J. Schoonmann, Properties of LiI-Alumina Composite Electrolytes, Solid State Ionics 9 and 10:119-122 (1983.
2. D. Avnir, D. Farin, and P. Pfeifer, Molecular Fractal Surfaces, Nature 308:261-263 (1984).
3. H. D. Bale and P. W. Schmidt, Small-Angle X-Ray-Scattering Investigation of Submicroscopic Porosity with Fractal Properties, Phys. Rev. Letters 53:596-599 (1984).
4. J. K. Kjems and A. Le Méhauté, to be published.
5. G. Porod, General Theory, in: "Small Angle X-Ray Scattering", O. Glatter and O. Kratky, ed., Academic Press, London (1982).
6. P.-Z. Wong, J. W. Cable and P. Dimon, , J. Appl. Phys.:55,2377-2382 (1984).
7. B. B. Mandelbrot, "The Fractal Geometry of Nature", Freeman, San Francisco (1983).
8. T. Witten, private communication.
9. P.-Z. Wong, to be published.

APPENDIX

Here we show some of the steps leading from Eq. 5 to Eq. 6. We use polar coordinates with the normal to the interface surface element ds as the polar axis (see Fig. 1). In any direction (θ,ϕ) there will be a sequence of values $|\vec{r}-\vec{r}'|$, $R_n(\theta,\phi)$ where the line from ds cuts the interface. Knowing $R_n(\theta,\phi)$, we can calculate the contribution to $n(\vec{r})\cdot n(\vec{r}')ds_p'$ by calculating the surface normal at R_p. The contributions will alternate in sign as one crossed from medium 1 to 2 and from 2 to 1. The increment in R_p in the polar and azimuthal directions are

$$\delta R_\theta = R_p(\cos\theta\cos\phi,\sin\theta\sin\phi,-\sin\theta)\delta\theta + \frac{\partial R}{\partial\theta}(\sin\theta\cos\phi,\sin\theta\sin\phi,\cos\theta)\delta\theta \tag{A1}$$

and

$$\partial R_\phi = R_p(-\sin\theta\sin\phi,\sin\theta\cos\phi,0)\delta\phi + \frac{\partial R}{\partial\theta}(\sin\theta\cos\phi,\sin\theta\sin\phi,\cos\theta)\delta\phi \tag{A2}$$

which gives for the contribution at R_p

$$\bar{n}\cdot\vec{n}'dS_p' = \left[R_p^2\cos\theta\sin\theta + R_p\frac{\partial R_p}{\partial\theta}\sin^2\theta\right]d\theta d\phi \tag{A3}$$

Therefore, the intensity contribution becomes

$$I_p(Q) = \frac{1}{Q^3} \int dS \int d\phi \int \sin\theta d\theta [R_p\cos\theta + \frac{\partial R_p}{d\theta}\sin\theta]\sin QR_p$$

$$= \frac{2\pi A}{Q^4}\frac{1}{2\pi}\int_0^{2\pi} d\phi \int_0^{\pi} \sin\theta\cos\theta d\theta [2-Q\frac{\partial}{\partial Q}]\cos QR_p \qquad \text{(A4)}$$

Summed over all the crossing points, R_p one finally gets

$$I(Q) = \frac{2\pi A}{Q^4}(2-Q\frac{\partial}{\partial Q}) \sum_p (-1)^p \int_0^{\pi} \sin\theta\cos\theta d\theta \int dR_p P(R_p,\theta)\cos QR_p \qquad \text{(A5)}$$

From a random porosity model where the probability function P only depends on R one gets the result quoted in Eq. 6.

LIGHT SCATTERING EXPERIMENTS IN A GEL SATURATED WITH A LIQUID MIXTURE

Walter I. Goldburg

Department of Physics and Astronomy
University of Pittsburgh
Pittsburgh, Pennsylvania 15260

I. INTRODUCTION

When a binary mixture is quenched through its critical point, domains of differing composition grow in composition contrast and in spatial size. As the process proceeds, droplets form and coalesce. Finally gravity separates the two phases and a sharp interface appears between them. Our concern here is how this scenario is modified when the liquid mixture is saturating a gel. Light scattering results are discussed in this paper[†]; in the following one, S. Satija describes neutron scattering experiments which probe phase separation in the gel at smaller length scales. Both sets of experiments suggest that the gel in effect imposes a static random field, or chemical potential, on the mixture.

In all this work[1] the gel matrix was polyacrylamide (PAA) of pore size, $L \simeq 25Å$. In most of the light scattering measurements the gel was saturated with an isobutyric acid (IBA)-water mixture. In the neutron scattering work a 2,6-lutidine-heavy water (LW) mixture was used. This latter system has an inverted coexistence curve, i.e., it separates into two phases on increasing the temperature. What effect might a gel have on phase separation? At a minimum it should retard the process by impeding fluid flow. In addition capillary effects might altogether block phase separation on a macroscopic scale. In addition the gel strands might prefer to have one component next to them over the other, and we observe this; the PAA polymer swells up manyfold in water but not in either of the other pure components.

To account for the active coupling of the gel with the mixture, DeGennes[2] has proposed that the gel imposes a frozen, spatially random field

[†]My collaborators in the light scattering work are the authors in Ref. 1.

on the mixture. This random field should broaden the transition, produce a metastable glass-like state[3], and modify the scattering. The structure factor now becomes the sum of two terms, one of them a Lorentzian, as in the pure binary mixture, and the other a squared Lorentzian function. The first term has associated with it the normal temporal fluctuations of a pure binary mixture, while in the second, the fluctuation rate $\Gamma = 0$.

II. EXPERIMENTAL

Standard light scattering techniques were used to measure the angular distribution of the scattered light I_q and the fluctuation rate Γ_q, where $q = (4\pi/\lambda)\sin(\theta/2)$ is the photon momentum transfer. The scattering angle is θ and λ is the wavelength of light in the fluid. The incident radiation was the 633-nm line of a He-Ne laser. The intensity I_q is proportional to the structure factor S(q) if multiple scattering is not excessive. In <u>pure</u> binary mixtures near the critical point $S(q) = A/(q^2+\xi^{-2})$ where $\xi = \xi_o\varepsilon^{-\nu}$, with $\varepsilon = |1-T/T_c|$. The critical exponent $\nu \simeq 0.63$. In IBA-water ξ_o = 3.57 Å and in lutidine-heavy water, $\xi_o = 2$ Å. The critical temperatures of these mixtures are $T_c^{\,o} = 26.1^oC$ and 28.6^oC respectively.

In a pure binary mixture $\Gamma_q = Dq^2$. The diffusion constant $D = k_BT/6\pi\eta\xi$, where η is the viscosity. Taking $\eta = 0.02$ poise and $T \simeq 300K$, as in our experiments, gives $\Gamma \simeq 10^4$ sec^{-1} at $\theta = 90^o$ and $|T-T_c| = 10^oC$. At small angles the fluctuations slow down sufficiently to be seen by eye near the critical point.

Contrast these parameters with those of a pure acrylamide gel[4]. Here also, $\Gamma \propto q^2$, but now the constant of proportionality is only weakly temperature dependent. At $\theta = 90^o$ Γ is again approximately 10^4 sec^{-1}. Provided one stays well above the gel collapse temperature, the scattering is isotropic. In PAA this temperature is below 0^oC. A PAA-Water gel scatters light only weakly, except in the forward direction, where spatial irregularities of large size become important. This background contribution was measured far from the critical point and subtracted out.

The samples were generally prepared by placing the water-swollen gel in a large reservoir of the mixture. The reservoir and sample were allowed to equilibrate for two weeks or more before scattering studies were commenced. In the optical studies the samples were ~ 5 mℓ in volume and were usually covered with the supernatant mixture. The neutron scattering samples, being ~ 0.2 mℓ in volume, equilibrated much faster.

To identify a sample of critical composition <u>within</u> the gel, a series of samples were made, of varying concentration (c) of IBA and lutidine. The scattering experiments were performed on the IBA sample with the highest cloud point temperature. The sample selected in this way has an IBA concentration of 45 vol %. This is only 7% higher than the critical value for pure IBA-water. As for the critical temperature of the gel-mixtures, this parameter is ill defined because a sharp phase transition is not observed. This is one of our most notable findings. In pure IBA-water critical opalescence is confined to an interval of ~ $0.1^{o}C$ around T_c. In the gel samples this interval is roughly twenty times larger.

The LW samples have proved more difficult to prepare reproducibly, in part because 2,6-lutidine will slightly dissolve the gel. As a result the cloud point curve has yet to be reliably traced out. All scattering experiments to date have been performed on a sample immersed in a mixture of 33 vol % lutidine concentration; in pure LW, c_c=30 vol% lutidine and in lutidine - D_2O it is the same.

III. RESULTS

The measurements were confined to a temperature range where the scattering was well above the gel background, yet not so strong that multiple scattering was excessive. I summarize our findings in the IBA gel: (a) As already noted, the phase transition is strongly broadened. Though an identifying meniscus is never seen inside the gel, the large-angle scattering ($\theta = 140^{o}$) rises by 10^3 as the temperature is lowered from $24.6^{o}C$ to $20^{o}C$ (recall that in pure IBA-water $T_c = 26^{o}C$). (b) Temporal fluctuations in the gel are virtually absent. At $\theta = 90^{o}$ one does detect an inverse correlation time, $\Gamma \simeq 10^4$ sec^{-1}, but its power spectrum is no more than 1% of the value expected from a pure binary mixture. In addition one does observe fluctuations of large amplitude with $\Gamma \simeq 1/1$ hour. These fluctuations are also observed in the pure gel itself and in gel samplesss made with water containing small polystryne spheres (radius ~ 0.1μ), placed there to increase the scattering. (c) The quenched gel quickly becomes uniformly cloudy; the characteristic time for the process is apparently no longer than the thermal relaxation time. (d) In the interval $21^{o}C \leq T \leq 23.5^{o}C$, the scattering curves can be fitted equally with a Lorentzian or a squared Lorentzian function (see below). The discrimination between these two functions is difficult because the inverse range of the fluctuations, K, is comparable to $q_{max} = 4\pi/\lambda = 2.6 \times 10^5$ cm^{-1}. Temperatures below $21^{o}C$ are inaccessible because of excessive multiple scattering. A fit to a Lorentzian square, $S(q) = const./(q^2 + K)^2$, yields a temperature-dependent K^{-1} which increases from 200 Å to 500 Å as T is reduced further and further below T_c^{o}, i.e., $23^{o}C < T < 21^{o}C$. Near T_c^{o} ($24.7^{o}C < T < 26.6^{o}C$) the gel correlation length K^{-1} starts increasing again. (e) As already noted true phase separation is never seen by eye, nor with a 50-power microscope. If deeply cooled, some samples have remained cloudy for months. If less severely cooled, cloudy samples will eventually become clear. Typical relaxation times in both the lutidine and IBA gels is ~ 10 hours for the small samples and hundreds of hours for the large ones. Cloudy samples can be recycled by returning them to the one phase region for days.

IV. INTERPRETATION

The quenching of the temporal fluctuations, the absence of droplets, the breadth of the transition, and the hysterisis, all suggest that the gel is imposing a random field on the critical mixture. The increase in K^{-1} with decreasing T ($T < T_c^o$) is also consistent with this idea. The gel is certainly doing more than offering a steric hindrance to the growth of droplets; if this were its only effect, the transition would presumably remain sharp and the temporal fluctuations would be present. The transition would also be sharp if the gel were acting like a fully "annealed" third component. Impurities in critical mixtures strongly shift the critical temperature and renormalize critical exponents, but they do not broaden the phase transition[5].

DeGennes[2] has formulated a mean field theory for this problem, utilizing ideas developed for random magnetic systems. This theory yields a dynamic structure factor of the form

$$S(q,t) \propto \frac{e^{-\Gamma\tau}}{q^2 + K^2} + \frac{\ell^{-2}}{(q^2 + K^2)^2} \qquad (1)$$

The theory requires that the correlation length K^{-1} be much larger than the gel mesh size L. The length ℓ which appears here is inversely proportional to the random field strength. This temperature-independent length is related to K through the equation $K^2 = \pm\xi^{-2}(T) + K^{-2}\ell^{-4}$ ($T \gtrless T_c$). Far from the critical point ($\xi \ll \ell$), K^{-1} and ξ become equal. On the other hand $K^{-1}(T)$ never exceeds the gel interaction length ℓ, according to this equation.

The fluctuation rate Γ in the first term of (1) has the same form as in pure binary mixtures except that ξ is now replaced by $1/K^2L$. Note that the Lorentzian square term in S corresponds to frozen fluctuations ($\Gamma = 0$). Since temporal fluctuations are not observed in our experiments, even at small q, our results are consistent with Eq. (1) if $K(T)\ell \ll 1$. Our measurements in the range 21°C to 23.5°C imply that $\ell \simeq \sqrt{\xi/K} \simeq 100$ Å. Since $K \simeq 1/400$ Å in this interval, the above inequality is marginally satisfied.

The metastability exhibited by our gel-mixtures is also consistent with the random field model, but another explanation cannot be ruled out. It may be that changing the temperature of the gel, changes its chemical potential relative to that of the supernatant liquid mixture which surrounds it. In that case this chemical potential difference may act like a dc applied field, adding to the zero-average random field within the gel[6]. The relaxation time we observe (hours to days) may then be a characteristic coupling time between the two systems.

V. ACKNOWLEDGEMENTS

I gratefully acknowledge many discussions with Y. Goldschmidt, correspondence with P.G. deGennes, and the collaboration of L. Rendon. This work was sponsored by the National Science Foundation under Grant No. DMR-8312374.

REFERENCES

1. J.V. Maher, W.I. Goldburg, D.W. Pohl, and M. Lanz, Phys. Rev. Lett. 53, 60 (1984).
2. P.G. deGennes, J. Phys. Chem. 88, 6469. (1985).
3. J. Villain, Phys. Rev. Lett. 52, 1543 (1984).
4. T. Tanaka, L.O. Hocker, and G.B Benedek, J. Chem. Phys. 59, 5151 (1973).
5. M.E. Fisher, Phys. Rev. 176, 257 (1968).
6. David Andelman and J-F Joanny (preprint).

BINARY FLUID PHASE SEPARATION IN GELS: A NEUTRON SCATTERING STUDY

S. K. Sinha and J. Huang

Corporate Research Science Laboratories
Exxon Research and Engineering
Annandale, NJ 08801

and

S. K. Satija*
Physics Department
University of Delaware
Newark, DE 19716

* Supported in part by the University of Delaware research fund grant # LTR 84110

INTRODUCTION

Over the past ten years there has been a considerable amount of interest, both theoretical as well as experimental, in the effects of random fields on phase transitions, in particular, the Random Field Ising model (RFIM). Initially, the crucial issue seems to have been the lower critical dimensionality d_c of RFIM[1]. In the literature arguments have been given in support of d_c being equal to either 2 or 3 (see references 1, 2 and 3 for recent reviews of RFIM). However, recently it has been realized that, experimentally, RFIM systems get trapped in a nonequilibrium state which relaxes very slowly, if at all, towards equilibrium[4]. In recent theoretical papers, Villain[5] and Bruinsma and Aeppli[6] have come to the conclusion that for $d<5$, the random field prevents the growth of an ordered state when the system is quenched to low temperatures from the paramagnetic regime.

In magnetic systems the RFIM is realized by applying a uniform field to a randomly diluted antiferromagnet[7]. The equivalence between these two cases was first pointed out by Fishman and Aharony[8]. In this paper we report the results of neutron scattering experiments on a different system where the ideas of random fields may apply. The phase transition investigated in this study is the critical phase separation of a binary fluid mixture. It has been suggested by light scattering experiments[9] that if a critical binary liquid mixture is incorporated in a gel, the random polymer strands of the gel impose a random field. These ideas have been further developed in a recent theoretical paper by de Gennes[10]. If indeed these ideas of a random field produced by the random network of gel are correct, then it is worth pointing out that in this case we have a proper analog of the RFIM contrary to the case of magnetic systems where, as mentioned above, the RFIM is realized by placing diluted Ising antiferromagnets in a uniform field. Once the gel is made, however, the value of the random field cannot be altered.

We have studied the phase separation of mixtures of 2,6 lutidine and heavy water (LW) in permanently cross-linked gel matrices of polyacrylamide (PAC) 7 wt %. Phase separation in pure mixtures of LW has been well studied before[11]. In order to study the build up of correlations, we have made small angle neutron scattering measurements on the PAC/LW system. This work was done in collaboration with J. V. Maher and W. Goldburg who have also done extensive light scattering measurement on this system which have been reported earlier[9,12]. The results of an analysis of our data show many similarities between PAC/LW system and RFIM studies of diluted antiferromagnets. In particular, in analogy with magnetic systems[7,13] we find that beyond a well defined temperature region, PAC/LW system gets into a metastable state. In this metastable state, the scattering can be described as coming from a system consisting of finite size domains, the size of which, however, is too large to be measured by neutron scattering measurements.

EXPERIMENTAL

The samples of gels were grown at the University of Pittsburgh using standard techniques[9]. The gels were immersed in the appropriate critical mixture of LW for at least few weeks before carrying out the neutron scattering experiments. In order to eliminate multiple scattering, the thickness of the gel samples was limited to 2 mm. Throughout the duration of the experiment, the gel samples were surrounded by a critical mixture of LW, thus keeping the chemical potential constant over the range of temperatures over which the experiment was done. The experiments were done on the small angle neutron scattering facility at Brookhaven National Laboratory. The neutron wavelength used was 7.5Å and Q resolution of the instrument was about $.006Å^{-1}$. We also carried out concurrent scattering measurements on another sample cell which only contained a critical mixture of binary fluid LW.

RESULTS AND DISCUSSION

In contrast to most binary fluid systems, LW has an inverted phase diagram; the two liquids are mixed at low temperatures and phase separate as the temperature is increased. Earlier light scattering experiments showed that PAC/LW has no well defined phase separation temperature as opposed to a pure critical mixture of LW[9]. In PAC/LW, the gel develops a certain degree of opalescence upon increasing the temperature and at yet higher temperatures the gel samples become completely milky in appearance. The gel samples, however, remain milky in appearance upon further increase in temperature, whereas the pure binary fluid LW mixture shows critical opalescence only in a very narrow temperature regime. Clearly, the transition gets broadened considerably in PAC/LW, in comparison to the pure binary fluid LW.

We first briefly report the results of the scattering measurements on the pure LW system. The scattering profiles in this case could be well fitted to a Lorentzian (LZ) line shape at all temperatures

$$I(q) = \frac{A}{q^2 + k^2}$$

where k is the inverse of the correlation length ξ. In Fig. 1, we show a log-log plot of ξ vs t for LW binary mixture which was the supernatant mixture in which the gel sample was immersed. A straight line fit to the data shows

that $\xi = \xi_o\left(\frac{T_c - T}{T_c}\right)^{-\nu}$, with $\nu = 0.61$, $T_c = 28.87$ C and $\xi_o = 2.05$Å. These values are in good agreement with earlier light scattering measurements and with the accepted values for the critical exponents of the 3d Ising ferromagnet. A log-log plot of the amplitude for the supernatant mixture also gives a susceptibility exponent $\gamma = 1.24$, close to the value expected for 3d Ising model.

The scattering line shape in PAC/LW samples is qualitatively different from that in the supernatant mixture. A single LZ lineshape could not be fit at any temperature. It has been shown by Pytte et al.[14,15] that the scattering in RFIM would be given by a sum of LZ and Lorentzian-squared (LSQ) terms,

$$I(Q) = \frac{A}{k^2 + q^2} + \frac{B}{(k^2 + q^2)^2}$$

where again k is the inverse correlation length, the value of which is the same in both the LZ and LSQ terms. Assuming the possibility that strands of gel matrix act like a random field on the binary fluid and carrying out a mean field calculation, de Gennes[10] has also shown that the structure factor should be a sum of two terms, namely LZ and LSQ. In de Gennes' theory the inverse correlation length k is given by

$$k^2 = \xi^{-2} + (k^2 l^4)^{-1}$$

where ξ is the correlation length for the supernatant mixture and the length l characterizes the strength of interaction between the liquid mixture and the gel matrix. The length l is, therefore, related to the strength of the random field in the gel. de Gennes' theory is applicable to the limit k^{-1} much larger than the than the mesh size L of the gel, which in our case of PAC/LW is ~ 25Å.

The neutron scattering data at all temperatures were fit to a form containing a sum of LZ and LSQ terms plus a constant background. This form gives a good description of the experimental results provided the value of k in the LZ and LSQ terms is allowed to be different. For example, at the lowest temperature studied, 3.1 C, the data could be fitted to a LZ (with

$\xi_1 = \frac{1}{k} = 53$Å) plus a LSQ term (with $\xi_2 = 8.5$Å). The amplitude B is less than 10% of the amplitude A. With increasing temperature the correlation length, ξ_1, given by LSQ term varies between 8-12Å. It seems, therefore, that the broad LSQ contribution to the scattering, the width of which is temperature independent, is from the underlying gel matrix[16]. At 21.9 C the scattering profile is quite sharp giving a $\xi_1 = 4200$Å and $\xi_2 = 12$Å. Above ~ 24.0 C, the scattering profiles can be fit by a sharp LSQ component ($\xi_2 >$ 2000Å) and a broad LZ component with $\xi_2 = 20$Å. (We again presume that this broad component is scattering from the underlying gel matrix.)

Thus at a temperature of ~ 24.0 C, there is a qualitative change in the scattering from LZ shape to LSQ shape. At this point it is worth remarking that our PAC/LW sample first showed some degree of opalescence (barely visible to eye) at T ≃ 24 C. Further increasing the sample temperature, increased the opalescence and at the highest temperature studied, 31.5 C, the sample was opaque to the eye. The fact that around 24 C, the scattering profile changes from LZ to LSQ can be demonstrated in a more illustrative manner as follows. Since above 21 C, the correlation length has become very long, the scattering may be written as

$$I(Q) = \frac{A}{q^2} + \frac{B}{q^4}$$

$$\text{or } IQ^4 = Aq^2 + B .$$

Therefore, a plot of Iq^4 vs q^2 should give a straight line, with the slope giving the amplitude of A and intercept being equal to the amplitude B. A crossover from LZ to LSQ form can then be easily seen when the slope goes to zero. Fig. 2 shows such plots for five temperatures in the vicinity of T = 24 C. At all five temperatures, a contribution due to a broad Lorentzian background (ξ = 20Å) of the gel matrix was first subtracted from the data, subsequently obtaining Iq^4 vs q^2 plots using the corrected intensity values. The crossover from LZ to LSQ form is quite apparent in Fig. 2. We interpret the B/q^4 scattering as scattering from the surfaces of frozen-in domains which have formed (Porod scattering).

In light of the above discussion, we loosely define the transition temperature of the PAC/LW sample to be T_c = 24 C. Using this value of T_c, we have plotted in Fig. 1, on a log-log plot, ξ vs t for PAC/LW in the temperature range where the scattering is Lorentzian. Except for the point where the correlation length is ~ 700Å (far outside our resolution range), the data can be fitted to a straight line giving υ = 0.94. As seen from Fig. 1, there is a large amount of scatter in the data. Thus, more work needs to be done before attaching much significance to this exponent. The value of 0.94, however, is very close to υ = 1 for 2d Ising mode.

In Fig. 3, assuming a T_c of 24 C, we show a log-log plot of the amplitude B vs. T (for temperatures greater than 24.4 C). Except for the first point, the data can be fitted to a straight line giving an exponent of 0.96. The plot is reminiscent of an order parameter growth with temperature.

We would like to point out several features of PAC/LW which may be similar to what is observed in RFIM magnetic systems. Neutron scattering data on all the 3d magnetic systems in a random field[4,7], show only LZ profiles at temperatures above or near T_c. At temperatures below T_c, where metastability effects are observed, the LSQ term dominates the scattering. It has been suggested before[4] that the origin of LSQ term is due to the finite-size frozen domains which form in the system below T_c. Both Villain[5] and Bruinsma and Aeppli[6] have shown what the expected size of these domains should be. The data in PAC/LW also point to a picture of frozen-in domains once the correlation length becomes of the order of wavelength of light (domains visible to the eye). In this temperature range, light scattering results also show that all the temporal fluctuations are frozen in. At lower temperatures in PAC/LW where the scattering is predominantly LZ in shape, temporal fluctuations should be observed in light scattering experiments. Careful light scattering measurements, therefore, need to be done in PAC/LW at low temperatures.

Finally, we remark on two unusual features of the PAC/LW system. First the random field theories show that the T_c for PAC/LW should be higher than for the pure binary mixture of LW. This is contrary to what is seen in the experiment. The other unusual feature is that the value of ξ in PAC/LW at low temperatures is higher than seen in pure LW mixture. This is contrary to the picture given by de Gennes where, as mentioned before, the inverse correlation length in gel system is given by $k^2 = \xi^{-2} + (k^2 l^4)^{-1}$). According to this formulae, at a given tempeature the correlation length in PAC/LW should always be less than ξ of pure LW. Both these facts suggest that the gel matrix may not be acting as a rigid, passive random field.

In summary, the neutron scattering measurements in PAC/LW show many of the features observed in RFIM studies. The study of time dependent features of the gel plus binary fluid systems, mentioned in an earlier paper by W. Goldburg, should, therefore, prove to be very interesting.

We wish to thank S. Alexander and P. Pincus for many helpful discussions.

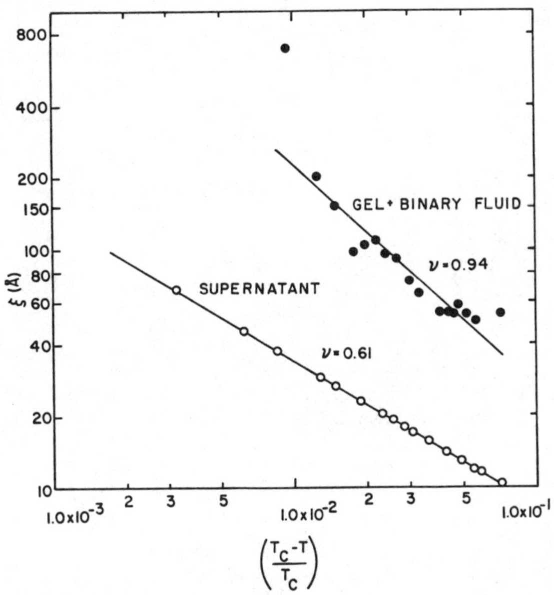

Fig. 1. A log-log plot of ξ vs t in both pure binary mixture of LW (supernatant) and PAC/LW (Gel + Binary Fluid). T_c = 28.87 C (supernatant) and T_c = 24.0 C (Gel + Binary Fluid).

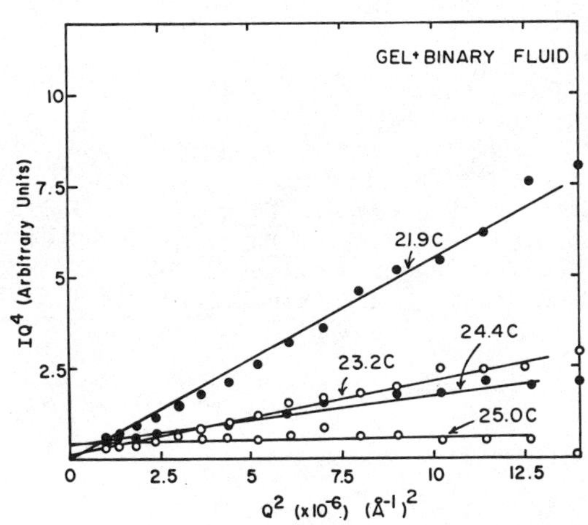

Fig. 2. IQ^4 vs Q^2 for PAC/LW system at five temperatures in the vicinity of T = 24 C.

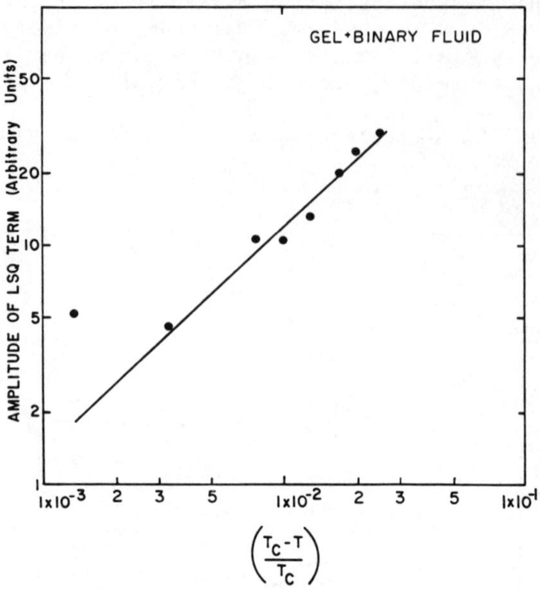

Fig. 3. A log-log plot of the amplitude B of the Lorentzian squared (LSQ) term vs t in PAC/LW assuming T_c = 24 C.

REFERENCES

1. G. Grinstein, J. Appl. Phys. 55:2371 (1984).
2. J. Villain, B. Semeria, F. Lancon and L. Billard, J. Phys. C16:6153 (1983).
3. Y. Imry, J. Stat. Phys. 34:849 (1984).
4. P-Z. Wong, J. W. Cable and P. Dimon, J. Appl. Phys. 55:2377 (1984).
5. J. Villain, Phys. Rev. Lett. 52:1543 (1984).
6. R. Bruinsma and G. Aeppli, Phys. Rev. Lett. 52:1547 (1984).
7. M. Hagen, R. A. Cowley, S. K. Satija, H. Yoshizawa, G. Shirane, R. J. Birgeneau and H. J. Guggenheim, Phys. Rev. B28:2602 (1983).
8. S. Fishman and A. Aharony, J. Phys. C12:L729 (1979).
9. J. V. Maher, W. I. Goldburg, D. W. Pohl and M. Lanz, Phys. Rev. Lett. 53:60 (1984).
10. P. G. de Gennes, J. Phys. Chem. (1985).
11. A. Stein et al., J. Chem. Phys. 56:6164 (1972).
12. W. Goldburg (previous paper).
13. D. P. Belanger, A. R. King and V. Jaccarino, Phys. Rev. Lett. 54:577 (1985).
14. E. Pytte, Y. Imry and D. Mukamel, Phys. Rev. Lett. 43:808 (1979).
15. H. S. Kogon and D. J. Wallace, J. Phys. A14:L527 (1981).
16. T. Tanaka, S. Ishivata and C. Ishimoto, Phys. Rev. Lett 38:771 (1977).
17. P. Debye, H. R. Anderson and H. Brumberger, J. Appl. Phys. 28:679 (1957).

METASTABILITY AND LANDAU THEORY FOR RANDOM FIELDS AND DEMIXING IN POROUS MEDIA

David Andelman[1] and Jean-François Joanny

Physique de la Matière Condensée[2], Collège de France
11 place Marcelin-Berthelot
F-75231, Paris Cedex 05, France

I. INTRODUCTION

In recent years there was a big increase in experiments on physical systems that are realizations of random fields. Just to name a few these include diluted antiferromagnets in a magnetic field, charge density waves pinned by impurities, hydrogen in binary metallic alloys and quite recently also binary liquid mixtures in gels. In all these systems there are annealed degrees of freedom (spin like) and a source of quenched disorder (impurities, random structure, etc) that effectively creates a random field which is coupled to the order parameter. In this article we will concentrate on random field systems where the order parameter is a scalar and the random field is coupled linearly to it (i.e. the random field Ising model - RFIM).

Theoretically, much effort has been made to study critical behavior and the lower critical dimension (d_ℓ) of this model.[1] (d_ℓ is the dimension below which there is no long-range order). Using a simple domain argument Imry and Ma predicted[2] $d_\ell = 2$ whereas other works claimed $d_\ell = 3$.[1] The current belief that indeed $d_\ell = 2$ is based on a refined domain argument[1,3] that takes into account domain wall roughening, some numerical investigations,[4] as well as a rigorous proof[5] on the existence of long-range order at T = 0 for the d = 3 RFIM. At first glance this theoretical prediction of $d_\ell = 2$ seems to be contradictory with neutron scattering experiments[6] on diluted antiferromagnets. Domains (thus lack of long-range order) are seen as the sample is cooled down in presence of a constant magnetic field (field cooling). To resolve this discrepancy, Villain[7] considered the domains as metastable states pinned by the random field. Hence although the long-range order state has the lowest energy, the system gets stuck (or has very long relaxation times) in one of these domain states since it cannot jump over all the energy barriers during the field cooling procedure. Other theories using similar ideas about non equilibrium states also exist.[8]

The purpose of the present work is two-fold : (i) we would like to emphasize the connection between the RFIM investigated so far mostly by its

1 Also : Corporate Science Research Laboratory, Exxon Research and Engineering Co. Annandale, N.J. 08801, U.S.A.

2 Laboratoire rattaché au C.N.R.S. : U.A. 792

magnetic realization and between recent experiments on binary liquid mixtures[9] in presence of random structures such as gels or porous solids. (ii) We propose a Landau theory that takes into account the possibility of metastable domain states and which leads to an explanation of the opalescence seen in the A/B + gel systems close to the demixing curve of the pure A/B mixture.[9] In addition the theory can provide a *possible* explanation of recent neutron scattering experiments[10] on diluted antiferromagnets where an abrupt transition from history-independent to history-dependent behavior occurs. Our theory is also in agreement with recent numerical investigations of the RFIM on finite lattices.[11]

II . BINARY MIXTURES IN GELS AND POROUS MEDIA : RANDOM FIELD SYSTEMS

Recent experiments[9] showed that phase separation and critical behavior of binary liquid mixtures with a well defined demixing curve changes drastically as a gel is immersed in the liquid mixture. One of the most striking phenomena seen in the A/B + gel system (water-lutidine or water-isobutyric acid are chosen as the binary mixtures and agorose or polyacrylamide as the gels) is the appearance of opalescence which follows closely the demixing curve of the pure A/B mixture. The onset of opalescence is not sharp but rather gradual on a range of a few degrees and even well inside the pure two-phase region it persists for long times. This opalescence should be distinguished from critical opalescence usually seen in pure A/B mixtures but only in a range of a few mK° around the consolute point. Nevertheless the fact that the onset of opalescence follows the pure demixing curve is a strong indication that it has to do mainly with the liquid-liquid fluctuations rather than with the gel fluctuations.

Pure A/B liquid mixtures are usually described by lattice gas models ; we would like to extend these models to binary liquid mixtures in a gel. The gel always has an heterogeneous chemical structure. We focus here on rigid gels which thus act on the liquid mixture as a source of quenched disorder (the same assumption is correct for porous solids). In the opposite limit of flexible gels, thermal fluctuations become important and one should consider the binary mixture in the gel as a ternary system. The main effect of the gel on the A/B mixture is a preferential adsorption[12] of one of the liquid components (in our case the water-component A). The gel + A/B system has two characteristic lengths : one is the mesh size of the gel (L) and the other is the bare fluid-fluid correlation length (ξ). The behavior is determined by a comparison between these two lengths. We will describe the system in the regime $L < \xi$ (tight gels or temperatures close to the consolute point). Since ξ is large, the system can be divided into blocks of size L (coarse graining) ; in each block there is an effective random field h(r) proportional to the average gel concentration in the block. The average of h(r), $< h(r) >$, expresses the total preference of the gel plus a contribution from the concentration difference between the liquid inside the gel in the A/B reservoir into which the gel is immersed. The order parameter of the system (analogous to the magnetization in magnetic systems) may be defined as $M = (C_A - C_B)/(C_A + C_B)$ where C_A (C_B) is the number density of A (B). The strength of the random field is characterized by the r.m.s. deviation H.

III. LANDAU THEORY FOR RANDOM FIELDS

Making the analogy between the A/B + gel system and random fields, we proceed by obtaining a description of the former via a Landau theory for the latter. We make a Landau expansion of the free energy (always using the magnetic terminology : magnetization, random fields, etc) in terms of the local order parameter M(r). The expansion includes even terms in M(r) up to fourth order, a gradient term related to the short-range character of the interactions and a linear term which describes the coupling to the random field h(r):

$$F = \int\left[\frac{1}{2}(\nabla M)^2 + \frac{1}{2}\varepsilon M^2 + \frac{1}{4}uM^4 - h(r)\,M(r)\right]d\vec{r}, \tag{1}$$

$\varepsilon = a\,(T - T_c^o)$ is proportional to the distance from the pure critical temperature T_c^o and $u > 0$ is a constant. We restrict the following discussion to the symmetric case $\langle h(r)\rangle = 0$, eventhough for the A/B + gel, $\langle h(r)\rangle = h \neq 0$ can be controlled by changing the concentration in the reservoir around the gel.[12-14] For the diluted antiferromagnets, $\langle h(r)\rangle = h = 0$; a non zero h would correspond to a staggered field (which is not realizable).

Minimizing the free energy functional (1) with respect to M(r) we get an equation for the order parameter M(r)

$$\frac{\delta F}{\delta M(r)} = -\nabla^2 M + \varepsilon M + uM^3 - h(r) = 0 \tag{2}$$

In order to solve (2) which is non linear and has a random source h(r), we split M(r) into two terms : $M(r) = M_0 + \delta M(r)$, where M_0 is the average magnetization of the system, $M_0 = \langle M(r)\rangle = V^{-1}\int M(r)d\vec{r}$ and $\delta M(r)$ is the local fluctuation, $\langle\delta M(r)\rangle = 0$. The equation for the average magnetization M_0 is obtained by averaging (2), thus yielding

$$M_0\,(\varepsilon + 3u\,\langle\delta M^2\rangle) + uM_0^3 = 0 \tag{3}$$

where $\langle\delta M^2\rangle$ is the r.m.s. magnetization fluctuation (the average of odd powers of $\delta M(r)$ vanishes). Subtracting (3) from (2) we get an equation for $\delta M(r)$

$$\nabla^2\delta M + h(r) = (\varepsilon + 3uM_0^2)\delta M + 3M_0u\left[\delta M^2 - \langle\delta M^2\rangle\right] + u\delta M^3. \tag{4}$$

Linearizing (4), we neglect the term $\delta M^2 - \langle\delta M^2\rangle$ and using a decoupling scheme the last term δM^3 is written as $3\langle\delta M^2\rangle\delta M$.* Within this decoupling approximation we get in Fourier space a linear relation between $\delta M(q)$ and $h(q)$

$$\delta M(q) = \frac{h(q)}{q^2 + K^2} \quad \text{where} \tag{5}$$

$$K^2 = \varepsilon + 3uM_0^2 + 3u\langle\delta M^2\rangle \tag{6}$$

and K^{-1} is the correlation length of the magnetization $\delta M(r)$. Hence the contribution to the structure factor from the field fluctuations is a Lorentzian square :

$$S(q) = \langle\delta M^2(q)\rangle \simeq \frac{H^2}{(q^2 + K^2)^2} \tag{7}$$

In position space (5)-(6) implies

$$\delta M(r) = \text{const.}\int d\vec{r}\,'\left[\exp(-K\,|\vec{r} - \vec{r}\,'|)/|\vec{r} - \vec{r}\,'|^{d-2}\right]h(r'). \tag{8}$$

The three physical quantities M_0, K and $\langle\delta M^2\rangle$ which we want to determine as functions of the reduced temperature ε and the strength of the random field $H = \langle h^2\rangle^{1/2}$, are linked by the three equations (3), (6) and (8).

---

* The numerical factor is chosen as the number of ways to do the decoupling. Here it is three.

Squaring (8) and taking the average we get a simplified connection between K and $<\delta M^2>$

$$<\delta M^2> \simeq H^2 \int \frac{d\vec{q}}{(q^2 + K^2)^2} = K^{d-4} \ell^{-2} \qquad \text{for } d < 4$$

$$= L^{-d+4} \ell^{-2} \qquad \text{for } d > 4 \qquad (9)$$

where ℓ is proportional to H^{-1} and expresses the strength of the random field for convenience in units of length. For $d > 4$ lower cut-off L for the field correlation is introduced and is the lattice size (mesh size) for the magnetic case (gel + A/B).

From (3) it is seen that the two solutions for M_0 are: $M_0 = 0$ and $M_0 \neq 0$. For these two cases we get from (6) and (9) :

$$K^2 = \varepsilon + 3u <\delta M^2> \qquad \text{for } M_0 = 0 \qquad (10)$$

$$K^2 = -2\varepsilon - 6u<\delta M^2> = 2uM_0^2 \qquad \text{for } M_0 \neq 0 \qquad (11)$$

Since the expected behavior (10)-(11) depends on whether $d > 4$ or $d < 4$, we will discuss the results obtained from the solution of (10)-(11) separately for the two cases.

III.1 Critical Behavior and Phase Transitions for $d > 4$

For $d > 4$, $<\delta M^2> \rightarrow \ell^{-2} L^{-d+4}$ = const as the correlation length $K^{-1} \rightarrow \infty$. The lower cutoff L expresses the fact that the random fields are correlated up to size L hence $<h(0)\, h(r)> \simeq <h^2>\, e^{-r/L}$. From (11) and (9) we get that the new critical point T_c is smaller than T_c^o in agreement[15] with the infinite range RFIM

$$\varepsilon_c = -\, 3u<\delta M^2> < 0 \,. \qquad (12)$$

For small randomness, $\varepsilon_c < 1$ thus

$$T_c = T_c^o - \frac{3u}{a} <\delta M^2> < T_c^o \,. \qquad (13)$$

The correlation length K^{-1} has a singularity at $T = T_c$, (9)-(11)

$$K^{-1} = (\varepsilon - \varepsilon_c)^{-1/2}, \qquad T > T_c$$

$$= \sqrt{\tfrac{1}{2}}\, (\varepsilon_c - \varepsilon)^{-1/2}, \qquad T < T_c \,. \qquad (14)$$

Notice that we got the usual mean-field value for the critical exponent $\nu = 1/2$, and the usual amplitude ratio. In addition, the average magnetization M_0 vanishes at $T = T_c$ as

$$uM_0 = (\varepsilon_c - \varepsilon)^{1/2} \qquad T < T_c \qquad (15)$$

which gives $\beta = 1/2$ and nothing else except the same shift from T_c^o to T_c.

To summarize, for $d > 4$ we find that the only effect that the term $<\delta M^2>$ has, is to reduce the critical temperature. Its existence, (12)-(15), follows directly from the fact that $<\delta M^2>$ is finite for $K \to 0$ and $d > 4$. For small enough randomness we are getting a second order transition with mean field exponents and amplitude ratios. This should be contrasted with the behavior found for $d < 4$ which is described in the following section.

III.2 Metastability and Equilibrium for d < 4

Since (10) and (11) do not depend on the dimensionality, the main difference between d > 4 and d < 4 comes from (9) where $K \rightarrow 0$ implies $\langle \delta M^2 \rangle \rightarrow \infty$ and cannot be achieved self consistently from (9) and (10)-(11). Figure 1 shows the inverse correlation length K for d = 3, obtained by substituting $\langle \delta M^2 \rangle = \ell^{-2} K^{d-4}$ in (10) and (11). For high temperatures only the solution for which $M_0 = 0$ exists

$$K^2 = \varepsilon + 3u/(\ell^2 K). \tag{16}$$

This solution was already found by de Gennes.[12] For high temperatures, $K^2 \simeq \varepsilon$ as in the pure case whereas for $\varepsilon = 0$, $K^3 = 3u/\ell^2$. Moreover for low temperature this solution ($M_0 = 0$) still exists and $K \simeq -(3u/\ell^2)\varepsilon^{-1} \rightarrow 0$ only for $T \rightarrow 0$.

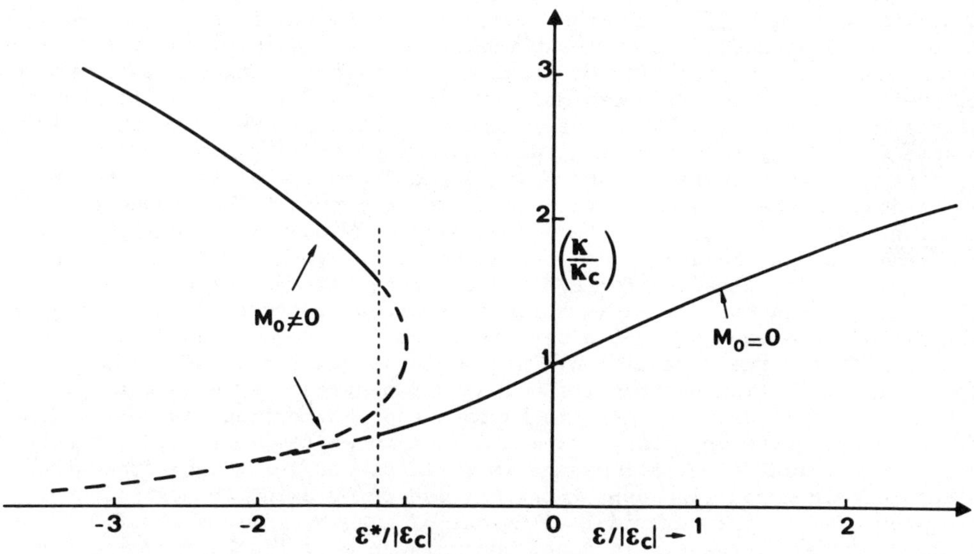

Fig. 1. Inverse correlation length, K, in units of $K_c = (3u/\ell^2)^{1/3}$ as function of reduced temperature, ε, in units of $\varepsilon_c = -3/2\, K_c^2$ for d = 3. One branch exists for $M_0 = 0$ and two for $M_0 \neq 0$. The continuous line denotes the most stable solution whereas the dashed ones are the less stable. Change of stability occurs at $\varepsilon = \varepsilon^*$. M_0 is proportional to K^{-1} for the 2 branches with $M_0 \neq 0$ and $\langle \delta M^2 \rangle \sim K^{-1}$ on all branches.

The other solution $M_0 \neq 0$ exists only for low temperatures such that $\varepsilon < \varepsilon_c$

$$K^2 = 2uM_0^2 = -2\varepsilon - 6u/(\ell^2 K) \tag{17}$$

and it has two branches (Fig.1). As will be discussed later the lower branch has always a higher free energy than the upper one which behaves at low temperatures as in the pure case : $K \simeq M_0 \simeq (-\varepsilon)^{1/2}$, thus has an increasing magnetization and a decreasing correlation length as the temperature is reduced.

Having obtained the two solutions $M_0 = 0$ and $M_0 \neq 0$, the next step is to compare the free energy of the two solutions in order to see which one is more stable. Due to lack of space we discuss here only the results and the full calculation will be presented elsewhere.[16] Substituting back into the free energy functional (1), the various solutions (9), (16)-(17) and using the decoupling scheme to estimate $\langle\delta M^4\rangle \simeq 3\langle\delta M^2\rangle^2$, the evaluation of (1) for the different solutions is straight forward but tedious. It leads us to two conclusions : (i) The upper branch of the $M_0 \neq 0$ solution is always more stable than the lower one in agreement with what is expected physically. (ii) In the interval $\varepsilon^* < \varepsilon < \varepsilon_c$, the $M_0 = 0$ solution is more stable than the $M_0 \neq 0$ solution, whereas for lower temperatures, $\varepsilon < \varepsilon^*$, the latter is more stable. The value of ε^* was found to be $1.085\ \varepsilon_c$.

IV. DISCUSSION AND COMPARISON WITH EXPERIMENT AND SIMULATIONS

The Landau theory discussed in the previous sections is relevant to magnetic realizations of the RFIM as well as to binary liquid mixtures in random structure such as gels and porous media under the conditions discussed in section II ; namely that the random structures are rigid and $\xi > L$. The advantage of the present theory is that it is giving the correlation length K^{-1} as function of T and H even in the critical region whereas all previous theories[7-8,13] assumed a rapid quench from high to low temperatures thus not following the experimental procedure. We also would like to note the strong resemblance between our Fig.1 and Fig.3 of ref 10. Starting at high temperature and doing a field cooling we expect the system to stay on the $M_0 = 0$ solution which we interpret to be the domain state at low temperature. Similar dependence of K on T was seen in ref 10 as the system was field cooled. The explanation why the system get stuck in the domain state, although for $\varepsilon < \varepsilon^*$ the long-range order has lower energy, is similar to the explanation of spinodal decomposition using a mean-field approach. As for the zero field cooled procedure, our theory predict a first-order transition. This could be related to an abrupt, apparently discontinuous transition that was observed experimentally.[10] Although the similarity of our predictions and the neutron scattering data of ref 10 are quite appealing, we would like to emphasize that our theory neglected critical fluctuations and we still do not fully understand the effects of our decoupling approximation. One encouraging support of our theory comes from the numerical solutions of mean-field equation for finite lattices.[11] From these simulations, the existence of (i) history independent region in the T-H plane (only a paramagnetic solution) (ii) history dependent region where the domain state has lower energy than the ferromagnetic state, (iii) history dependent region where the ferromagnetic state has the lowest energy although the domain state also exists was found. These finding are in agreement with our three regions : $\varepsilon > \varepsilon_c$, $\varepsilon^* < \varepsilon < \varepsilon_c$, and $\varepsilon < \varepsilon^*$.

As for the A/B + gel systems, since the opalescence in those systems persists for very long times we believe that it is caused by "freezing" of the system in a domain state, thus K^{-1} is the domains size and is related to the solution with $M_0 = 0$ (Fig.1) even for low temperatures ($\varepsilon < \varepsilon^*$).

As for the other solution with $M_0 \neq 0$, unfortunately a procedure analogue to the zero-field cooling is not possible for the gel (this would mean changing in a controlled way the concentration flucturations of the gel). However an additional external parameter which is available here is the analogue of a constant magnetic field, and phase separation (the equivalent of long-range order) can be induced by changing the relative concentration of the A/B mixture in the reservoir around the gel. Some initial predictions of the influence of such non zero average field are available[13] and could be tested experimentally.

ACKNOWLEDGEMENTS

We would like to thank S. Alexander, P.G. de Gennes and R. Klemm for useful comments and suggestions. One of us (D.A.) acknowledges the support of a Joliot Curie Fellowship.

REFERENCES

1. For a review see : G. Grinstein and S.-k. Ma, Phys. Rev. B 28, 2588 (1983).
2. Y. Imry and S.-K. Ma, Phys. Rev. Lett. 35, 1399 (1975).
3. G. Grinstein and S.-k. Ma, Phys. Rev. Lett. 49, 685 (1982); J. Villain, J. Phys. (Paris), Lett. 43, L-551 (1982); K. Binder, Z.Phys. 50, 343 (1983) ; D.S. Fisher, J. Fröhlich, and T. Spencer, J. Stat. Phys. 34, 863 (1984).
4. J.F. Fernandez, G. Grinstein, Y. Imry, and S. Kirkpatrick, Phys.Rev. Lett. 51, 203 (1983) ; D. Andelman, H. Orland, and L.C.R. Wijewardhana, Phys. Rev. Lett. 52, 145 (1984); D. Stauffer, C. Hartzstein, K. Binder, and A. Aharony, Z. Phys. B 55, 352 (1984).
5. J.Z. Imbrie, Phys. Rev. Lett. 53, 1747 (1984).
6. M. Hagen, R.A. Cowley, S.K. Satija, H. Yoshizawa, G. Shirane, R.J. Birgeneau, and H.J. Guggenheim, Phys. Rev. B 28, 2602 (1983); D. Belanger, A.R. King, and V. Jaccarino, Phys. Rev. Lett. 48, 1050 (1982).
7. J. Villain, Phys. Rev. Lett. 52, 1543 (1984).
8. G. Grinstein and J.F. Fernandez, Phys. Rev. B 29, 6389 (1984) ; R. Bruinsma and G. Aeppli, Phys. Rev. Lett. 52, 1547 (1984) .
9. J.V. Maher, W.I. Goldburg, D.W. Pohl, and M. Lanz, Phys. Rev. Lett. 53, 60 (1984).
10. R.J. Birgeneau, R.A. Cowley, G. Shirane, and H. Yoshizawa, preprint.
11. H. Yoshizawa and D.P. Belanger, Phys. Rev. B 30, 5220 (1984) ; C. Ro, G.S. Grest, C.M. Soukoulis, and K. Levin, Phys. Rev. B 31, 1682 (1985).
12. P.G. de Gennes, J. Phys. Chem. 88, 6469 (1985).
13. D. Andelman and J.F. Joanny, Collège de France, preprint, 1985.
14. D. Andelman and J.F. Joanny, "Proceeding of Les Houches Conference on Physics of Finely Divided Matter, 1985".
15. T. Schneider and E. Pytte, Phys. Rev. B 15, 1519 (1977) ; A. Aharony, Phys. Rev. B 18, 3318 (1978).
16. D. Andelman and J.F. Joanny, to be published.

SCALING IN COLLOID AGGREGATION

D. A. Weitz, M. Y. Lin, J. S. Huang, T. A. Witten,
S. K. Sinha, J. S. Gethner and R. C. Ball*

Exxon Research and Engineering Co.
Route 22 East, Clinton Township
Annandale, New Jersey 08801 U.S.A.

* Cavendish Labs
Department of Physics
Cambridge University
Cambridge, ENGLAND

ABSTRACT

We study the aggregation of aqueous gold colloids and apply modern scaling methods to interpret our results. We find that there are two regimes of aggregation, each with different rate-limiting physics, depending on the sticking probability of the individual colloidal particles. Each regime is distinguished by the fractal dimension of the resultant clusters, the aggregation dynamics and the cluster-mass distribution. The two regimes are diffusion-limited cluster aggregation and reaction-limited aggregation; and the two represent the two limiting universality classes for kinetic cluster-cluster aggregation.

INTRODUCTION

There has recently been a rebirth of scientific interest in the process of irreversible kinetic aggregation of small particles to form larger clusters. This has been due in large part to the recognition that scaling concepts can be successfully applied to the description of the highly disordered structure of the aggregates that result. This has in turn led both to a much deeper understanding of the physics of the aggregation process itself, as well as to some initial attempts to determine the characteristic modification in the physical properties of the aggregates due to the particular type of disorder inherent in their structure. Much of the initial interest was stimulated by theoretical work and computer simulations which suggested that the structure of the aggregates could be described as a fractal[1,2]. This work probed the effect of various types of kinetics on the fractal dimension, d_f, of the resultant clusters, using a variety of extremely simple models for the physics of the aggregation process. However, the true significance of these new concepts must ultimately be judged by their success in describing the properties of real materials which are formed by any of the kinetic growth processes so prevalent in nature. The purpose of this paper is to examine the application of scaling concepts to the description of the structure and dynamics of one such growth process, the irreversible kinetic aggregation of aqueous colloids.

We study a prototypic colloid system, aqueous gold colloids[3], which has the advantage that the aggregation kinetics can be simply and reproducibly varied. In so doing, we show that there are two regimes of aggregation, which represent two distinct universality classes. Each regime exhibits markedly different characteristic behavior, yet each can be understood in a quantitative fashion using scaling concepts to describe both the fractal structure of the clusters as well as the aggregation dynamics that characterize each regime. The first regime is diffusion-limited cluster aggregation, wherein clusters stick to each other irreversibly immediately upon contact and the kinetics are limited solely by the diffusion of the clusters. The fractal dimension of the resultant clusters is $d_f \sim 1.75 \pm 0.05$, while the dynamics exhibit a characteristic power law scaling, and the cluster mass distribution can be approximately characterized as monodisperse. In contrast, the second regime results from a cluster sticking probability that approaches zero, and therefore is called reaction-limited aggregation (RLA). In this case, the cluster fractal dimension is determined statistically by the number of ways that two clusters can touch each other subject to the excluded volume of each, and since some degree of interpenetation is possible, the fractal dimension is somewhat increased to $d_f \approx 2.05 \pm 0.05$. The aggregation dynamics are much slower, and are characterized by an exponential growth, while the cluster mass distribution is also substantially changed, exhibiting a power-law dependence on mass, $N(M) \sim M^{-\tau}$ with $\tau \approx 1.5 \pm 0.2$. In this paper, we show that the physics of each regime can be simply understood with the aid of scaling concepts. In this sense, colloidal aggregation provides an excellent experimental test of the modern theories for kinetic growth, helping to guide their evolution. In addition, however, these regimes of colloid aggregation are specific examples of what has long been known to colloid scientists as "fast" and "slow" coagulation[4]. The increased understanding developed of these regimes provides a much more complete description of the process of colloid aggregation, and resolves a longstanding paradox within the colloid science literature.

The remainder of this paper is organized as follows: In the next section we discuss the properties of the gold colloids and the analysis of the scaling behavior of the structure of the aggregates. The following two sections deal in detail with each of the limiting regimes of aggregation, diffusion-limited and reaction-limited. The final section contains a discussion and concluding remarks.

STRUCTURE OF AGGREGATES

The key to the successful application of scaling concepts to the description of colloid aggregation is our ability to quanitatively characterize the structure of the aggregates in terms of fractals. In this section, we discuss how the gold colloids are made and how the aggregation is induced and controlled. We then describe the experimental techniques used to determine the structure of the aggregates and measure their fractal dimension. These techniques can be divided into two general classes. Transmission electron microscopy (TEM) is used to directly visualize the structure of the aggregates. In addition, in situ measurements of the mass-mass correlation within the aggregate are made using scattering techniques, employing both light scattering and small angle neutron scattering to probe the required range of length scales.

The gold colloids used in these experiments are simply and readily made by following a recipe of Turkevitch[3]. These particular colloids have the advantage that they have been well characterized and their properties are highly reproducible. The typical colloid particles are spheres with diameters of ~ 150 Å, and are highly uniform in size, with the mean variation about the average diameter being $\lesssim$10%. As prepared, their concentration is ~ 10^{12} particles/cc, which results in a mean separation of the particles of nearly 1 m. The interactions between the individual particles can be well described within the DLVO theory[4] for colloid particles. When formed, the surface of the gold particles is highly negatively charged by adsorbed ions[5]. Positive ions in solution screen the surface charges, with a Debye-Huckel screening length of κ^{-1} ~ 100 for the ion concentrations used in the colloid preparation. Thus, except for hydrodynamic effects, the gold particles as prepared are noninteracting until they approach one another to within ~κ^{-1}, at which point they experience the screened coulombic repulsion interaction due to their charged surfaces. This presents a barrier to aggregation which, initially, has a height of ~ 20 kT, making the gold colloid extremely stable. However, if this barrier is reduced, so that two colloiding particles can overcome it, they then fall into an extremely deep attractive well, caused both by the large Van der Waals attraction between the particles, and by the creation of metallic gold-gold bonds when the surfaces actually touch. The depth of this attractive well is at least several orders of magnitude greater than kT, making the aggregation totally irreversible.

The height of the repulsive barrier which stabilizes the sol against aggregation can be reproducibly varied in two ways. The ionic concentration of the solution can be increased, thereby reducing the screening length and hence the effective repulsive barrier[6]. Alternatively, the surface charge on the colloidal particles can be reduced by displacing the adsorbed ions by a neutral adsorbate[7]. In both cases, the reduction of the repulsive barrier leads to aggregation of the colloid. While the use of a salt to reduce κ^{-1} has been the traditional technique used to cause aggregation, we have chosen to reduce the surface charge, since this involves changing fewer features of the colloidal solution and since the resultant particles are not left with a large surface dipole as would be the case when the charge is highly screened by additional salt. In our case, we reduce the surface charge by adding a small concentration of pyridine, which is strongly adsorbed on the surface as a neutral molecule, displacing the charged ions and thereby reducing the charge and the repulsive barrier. We add only a very small amount of pyridine, to concentrations between 10^{-6} and 10^{-2} M. Concentrations above 10^{-3} M displace virtually all the charge from the surface, with the resultant aggregation then being diffusion-limited. The lowest pyridine concentrations reduce the charge only slightly, making the repulsive barrier still $\gtrsim$kT, and resulting in reaction-limited aggregation. In all cases, the aggregation is a kinetic process, which goes to completion when the aggregates become so large that they precipitate out of the solution. This process typically takes several hours for the diffusion-limited kinetics and several days for the reaction-limited kinetics. However, in all cases, the time scale of the pyridine adsorption and the surface charge reduction is much shorter than that of the aggregation.

Transmission electron microscopy (TEM) is an ideal method for visualizing the aggregates and for performing detailed analysis of their structure[8], and the gold colloids are particularily well suited for this because of their high contrast to the electron beam. A sample TEM picture of a typical cluster prepared by diffusion-limited aggregation is shown in Fig. 1a. Our analysis of the structure of the aggregates is in

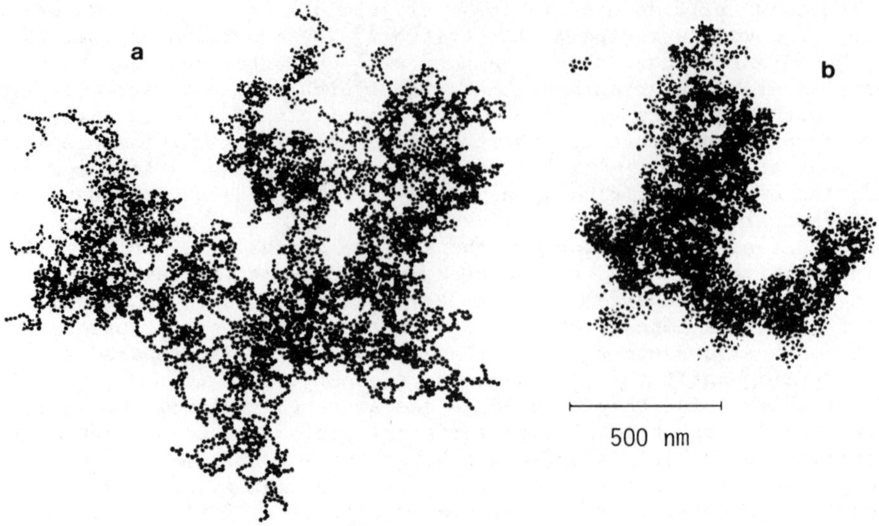

Fig. 1.Transmission electron micrographs of clusters of gold colloids. (a) DLA cluster "4739" containing 4739 colloid particles; (b) RLA cluster of roughly the same mass as in (a).

terms of fractals, for which the mass scales as the size as $M \sim R^{d_f}$, where d_f is the fractal dimension. We have used two separate analysis techniques to show that the structure is indeed characterized as a fractal[8]. In both cases, we take the advantage of the uniformity of the gold particles, so that the mass of a cluster is simply given by the number of particles. In the first method, we measure a series of different clusters from TEM pictures taken of the same colloid sample. For each cluster, we measure the mass by counting the number of gold particles in the cluster and the radius simply from the dimension of the cluster in the picture. A logarithmic plot of the mass vs. the radius for all the clusters results in excellent linear behavior over roughly two decades, and the slope of a least squares fit to the data yields the fractal dimension directly: $d_f = 1.75$. We corroborate this measurement by determining the mass-mass correlations within a single cluster. Thus, for example, we have digitized the positions of the center of each particle in the cluster in Fig. 1a. We then calculate the two-point mass correlation function of the center positions, $C(r)$, accounting for the finite extent of the cluster. We find that $C(r) \sim r^{-\alpha}$ with ≈ 0.25 over about one and a half decades. The parameter α is the co-dimension, $\alpha = d - d_f$ where d is the Euclidean dimension. Since the TEM pictures of the clusters are in fact two dimensional projections of the clusters we use the fact[9] that a 2d projection of 3d fractal remains a fractal with the same fractal dimension, provided $d_f < 2$. Thus, we find $d_f = 1.75$, in excellent agreement with the first method of analysis. However, since the TEM pictures are two dimensional projections, the mass-mass correlation analysis will not be effective for clusters prepared by reaction-limited aggregation, since in this case $d_f > 2$, and the projected image looks essentially solid, as shown by the picture of a typical RLA cluster shown in Fig. 1b. We are, however, still able to obtain an accurate measure of the number of particles in each cluster to determine the fractal dimension through the scaling of mass with the radius. We note that the cluster illustrated in Fig. 1a has become a somewhat ubiquitous example of a fractal object. Since in our analysis we have, quite tediously, determined the exact number of gold particles contained in this cluster, we have christened this picture "4739".

An additional type of analysis that can be done directly from the TEM pictures is the determination of the cluster mass distribution[10], N(M). This is obtained by measuring the mass of a random sample of clusters on a single TEM grid. These can then be compiled into a histogram to determine N(M). This analysis can be simplified substantially by dividing the histogram evenly in the logarithm of the mass. Thus the bin size increases as the mass does, giving meaningful statistics at high mass, even though the number of clusters of large mass is decreasing. Furthermore, the accuracy with which the mass of the large clusters must be determined is decreased.

While the TEM analysis provides direct visual evidence of the fractal structure of the aggregates, it suffers from the fact that the cluster must be dried on a TEM grid. It is desirable to probe the aggregates in solution to corroborate the applicability of our results to the clusters in their natural environment. An ideal technique for this is scattering[11], which yields a fourier transform of the mass-mass correlation function of the clusters[12]. Scattering techniques have the added advantage that they provide direct structural information averaged over a very large number of clusters.

An appropriate analysis of the scattering from fractal objects has been obtained for neutron scattering from cab-o-sil by Sinha et al.[12] The scattering intensity is assumed to be comprised of a form factor for the individual spheres, P(k), and a structure factor, S(k) which reflects the mass correlations of the spheres. Here k is the scattering wave vector, $k = \frac{4\pi}{\lambda} \sin \Theta/2$, with λ the wavelength and Θ the scattering angle. To obtain S(k), they consider a real space correlation function of $C(r) \sim r^{-\alpha} f(\frac{r}{R})$, where the $r^{-\alpha}$ is the fractal portion of the correlations, R is the cluster size, and f(x) is a scaling function which accounts for the finite size of the clusters, and therefore has the properties that $f(x) \to 1$ as $x \to 0$ and $f(x) \to 0$ as $x \to \infty$. For f(x), they chose $f(x) = e^{-x}$, which has the proper form, and has the added advantage that the resulting correlation function can be transformed analytically. The resultant scattering intensity has the form

$$I_R(k) \sim \frac{9[\sin(ka)-ka\cos(ka)]^2}{(ka)^6} \cdot \frac{M^2}{(1+k^2R^2)^{d_f/2}} \cdot \frac{(1+k^2R^2)^{1/2}}{kR} \cdot \frac{\sin[(d_f-1)\arctan(kR)]}{d_f-1} \tag{1}$$

where R is the cluster size and a is the single particle size. Qualitatively, this result has the expected form. At high k, it is dominated by the form factor of the individual spheres, and falls off as k^{-4}, the expected Porod behavior. At intermediate values of k, for $1/R \ll k \ll 1/a$ it exhibits the power law scaling expected for a fractal, $I_R(k) \sim k^{-d_f}$. Finally at very small k, it becomes constant, reflecting the size of the cluster. When we scatter radiation from the solution of aggregates, the measured intensity is integrated over the distribution of clusters:

$$I(k) \approx \int N(M)\, I_M(k)\, dM \tag{2}$$

where $I_M(k)$ is the intensity from a cluster of mass M and size R^{d_f}. However, since the scattering intensity increases rapidly with mass, the total scattering near k=0 will be dominated by the contribution of the largest clusters provided N(M) is not a power law with too sharp a decay. In fact, it can be shown that, provided $\tau < 2$, where $N(M) \sim M^{-\tau}$, the scattering will be dominated by the largest clusters.[10] Since this is the case for both aggregation regimes, we can fit to the functional form in Eq. 1, in which case R characterizes the size of the largest clusters in the distribution.

Since the relevant range of length scales, and hence the range of k, is so large for the fractal aggregates, we use two independent probes to cover the full regime. For the short length scales (large k), we use small angle neutron scattering,[13] while for the long length scales (small k), we use laser light scattering.[14] This leaves an arbitrary scale factor that must be adjusted to fit both data to the expression in Eq. 1. In fact, the absolute cross sections for each probe, the ratio of which is reflected in this scale factor, are quite different. Neutrons are coherently scattered very weakly from the gold colloids, while the water in which they are suspended, contributes a large incoherent background. Thus, for the volume fraction of gold at which the colloids are made, 10^{-6}, the desired signal level is too weak compared to the background from the water. Therefore, the colloid concentration must be increased, which was done by allowing the aggregation to continue until the clusters had become large enough to precipitate. The neutron scattering was then done from the precipitates, which had a substantially larger volume fraction of gold, estimated to be on the order of 1%. In this case, all the clusters are substantially larger than 1/k for the range of k value probed. In contrast, light is scattered very strongly from the gold clusters, which in fact have an optical resonance in the visible giving them their distinctive color. Thus we must ascertain that the scattering intensity is sufficiently weak that there is minimal multiple scattering so the Born approximation, which is implicit in Eq. 1, is indeed valid. We note that if multiple scattering were to occur, it would be within a single cluster, since the concentration of clusters in solution is so low. To test for multiple scattering, we measured the integrated scattering to within 3° of the forward direction. When this was normalized to the number of clusters in the beam, we found that less than 1% of the incident light on each cluster was scattered. Since our light scattering measurements are made only to an angle of 20°, multiple scattering will not be significant, and Eq. 1 can be expected to be valid.

An example of the data obtained is shown in Fig. 2 for clusters prepared with diffusion-limited aggregation. We have plotted the scattered intensity as a function of k on logarithmic scales. The low k data were obtained using a He-Ne laser at 632.8 nm while the high k data were obtained using small angle neutron scattering. The light-scattering data were obtained after the aggregation had proceeded sufficiently that the results were independent of time. The data exhibit the expected behavior. The linear regime on the logarithmic plot, with a slope of -1.79, directly reflects the fractal character of the clusters. At higher k, the data bends over and the slope increases, reflecting the contribution of the form factor of the individual spheres. The bend over occurs at $k^{-1} \sim 75$Å, corresponding to the size of the colloid particles. The solid line is the best fit curve using Eq. 1, obtained with $d_f = 1.79$, a = 74.9 Å, and R = 3.4 μm. The values of d_f and a are in excellent agreement with the values obtained from the TEM analysis, while the value of the characteristic cluster size, R, is sufficiently large compared to all k^{-1} that it contributes very little to the fit. This is consistent both with the fact that the measurement was made after the scattering intensity became constant at all observed values of k, and with the value for the characteristic hydrodynamic radius obtained with

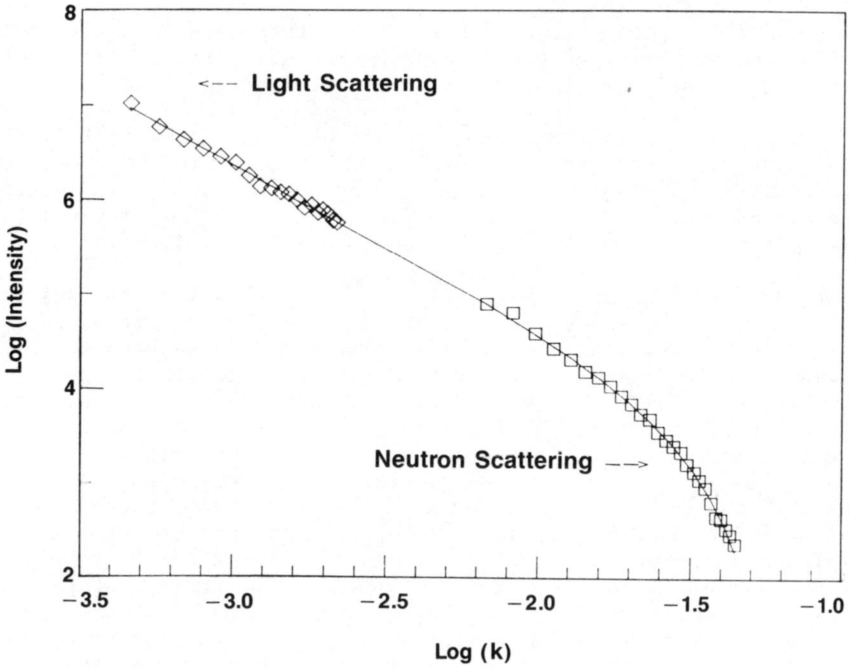

Fig. 2 Light scattering and small angle neutron scattering from DLA clusters plotted on the same graph. Scattering intensities are in arbitrary units. Solid line represents the theoretical fit to Eq. 1. The units of k are $Å^{-1}$.

quasi-elastic light scattering measurements made concurrently with the static light scattering measurement. The values of k spanned by the neutron scattering measurements are most sensitive to the form factor of the individual gold particles, while the correlations between the spheres, which determine d_f, are reflected primarily in the light scattering data. Nevertheless, if the light and neutron scattering data are fit independently by Eq. 1, the same values of the parameters are found, lending credence to their validity.

Both the TEM analysis, as well as the scattering data, show that the structure of the aggregates is well characterized as a fractal. We now use this ability to quantitatively describe the structure of the aggregates to investigate the origin of the fractal structure and the inherent relationship between the aggregation dynamics and the resultant cluster structure.

DIFFUSION-LIMITED AGGREGATION

We first consider the case where the stabilizing charge on the surface of the colloids is entirely removed, so that two clusters stick together immediately upon contact. The aggregation kinetics are then limited solely by the diffusion of the clusters. All of the structural measurements of the clusters discussed in the previous section show that the fractal dimension of the aggregates is $d_f \approx 1.75 \pm 0.05$. The measured fractal dimension is also in excellent agreement with the result obtained from computer simulations for diffusion-limited cluster-cluster aggregation.[15,16] The origin of the fractal structure is the random motion of the diffusion of the clusters, combined with the irreversible bond that is formed when they touch.

An important and illuminating feature of the diffusion-limited cluster aggregation is the dynamics.[17] We have studied this both by using quasielastic light scattering to measure the time-dependent growth of the hydrodynamic radius of the characteristic cluster size and by measuring the cluster mass distribution, and its evolution in time.

Standard techniques were used for the quasielastic light scattering experiments to measure the decay of the intensity autocorrelation function.[18] The decay is highly non-exponential, so we used the first cummulant, or the initial logarithmic derivative, to obtain a mean decay rate. From this, and the Stokes-Einstein relation, we calculated the hydrodynamic radius of the characteristic cluster of the distribution. Each measurement required only a few seconds to complete, so we could follow the time dependence of the cluster growth very closely. While the radius obtained from this measurement reflects an average over the distribution of clusters, we emphasize that it is not the same as that traditionally called the z-average. This is because for most of the time of the measurement, $kR>1$, so that the intensity of scattering from each cluster grows linearly with the mass, rather than as M^2 which is appropriate for scattering from small particles. Thus, we probe a different moment of the distribution of cluster masses than the z-average. Nevertheless, the time dependence of the measured radius accurately reflects the time evolution of the characteristic cluster size of the distribution. We find that the decay rate Γ scales as k^2 even for $kR \gg 1$, as expected for aggregates which move rigidly.[19]

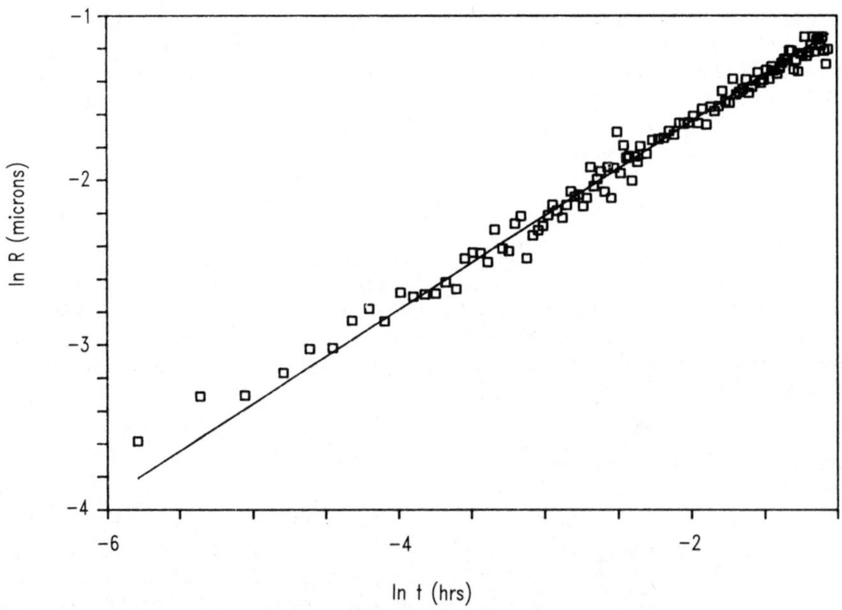

Fig. 3 Scaling of the mean cluster radius with aggregation time. The solid line is a least square fit to data according to $R = 0.60\ t^{0.56}$. The scattering angle $\Theta = 140^o$.

A typical example of the time dependence of the characteristic radius is shown in a logarithmic plot in Fig. 3. The concentration of the colloid was reduced by a factor of three to obtain these data in order to slow down the aggregation and make it easier to follow. Nevertheless, at the earliest times, we made measurements every 10 seconds to follow the rapid initial growth. These measurements were made at a scattering angle of 140^0, so that $kR>1$ is true for the largest extent of time, but we found the expected k^2 dependence in decay rate when other angles were used. The data have a linear behavior in the logarithmic plot in Fig. 3, indicating a power-law growth. From a least squares fit to the data, we find that they are well described as R $t^{0.57}$.

The cluster mass distribution was measured at three different times during a typical diffusion-limited cluster aggregation by taking a small drop of the aggregating solution and preparing a TEM grid. The data are plotted logarithmically in Fig. 4, where we have normalized N(M) by the total number of clusters counted. Each data set exhibits either a peak

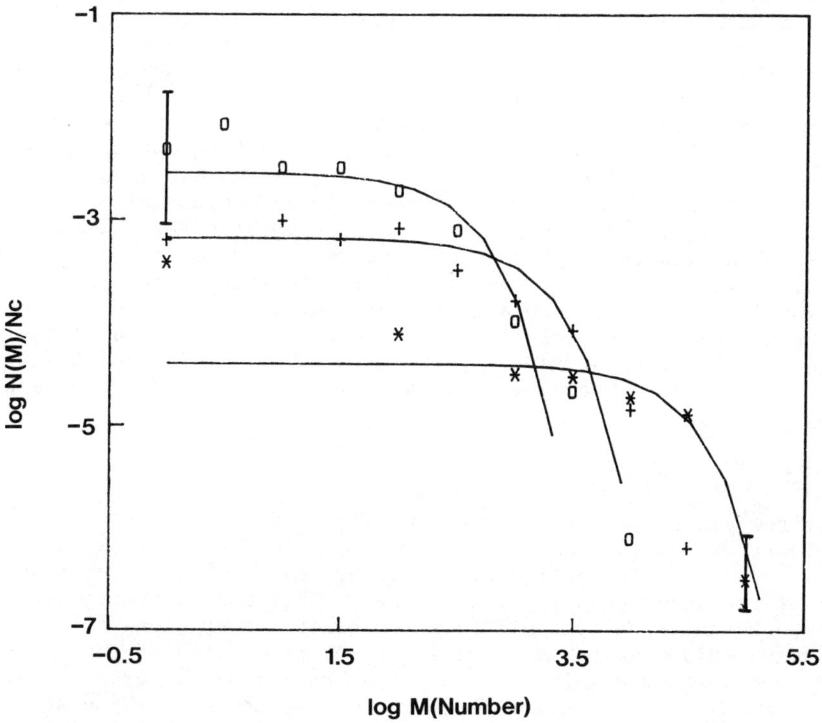

Fig. 4 Logarithmic plot of the cluster mass distribution for DLA at 3 different times of aggregation. N(M), the number of clusters of mass M, is normalized by the total number of clusters N_c. The solid lines represent the theoretical fits according to the Smoluchowski equation (Eq. 3) with a constant kernel. The squares represent measurements taken 1 minute after aggregation, the crosses represent measurements after 10 minutes and the stars represent measurements after 30 minutes.

or a plateau in the mass distribution. The distribution is cut off at some large mass and this cutoff increases with time. The very large error bars shown at low mass reflect the poor statistics obtained sphere making it difficult to determine if there is actually a dip as $M \to 0$, thus we are unable to distinguish between an actual peak in the distribution or simply a plateau in the data.

The traditional way to discuss the dynamics of colloid aggregation has been with the rate or Smoluchowski equations[20] which can be written as

$$\frac{dN(M_i)}{dt} = \frac{1}{2}\sum_{j=1}^{i} K_{j,i-j}\, N(M_j)N(M_{i-j}) - \sum_{j=1}^{\infty} K_{ij}N(M_i)N(M_j) \qquad (3)$$

Here, the kernel, K_{ij} is the rate at which clusters of mass i and j aggregate to form a cluster of mass i+j. The first term on the righthand side of Eq. 3 reflects the gain in the number of clusters of mass i due to all the possible combinations of smaller clusters sticking together to form a cluster of mass i. Similarly, the second term reflects the loss of clusters of mass i due to the formation of larger clusters. In writing Eq. 3, we have assumed that the aggregation is totally irreversible and thus have neglected the possibility that a cluster can break apart to form two smaller clusters. Given the strength of the gold-gold bonds formed, this is a reasonable assumption.

Equations of this sort have been widely studied, and a great deal is known about the behavior of the solutions in a variety of cases. The most important aspect of Eq. 3 is the determination of the kernel, which contains most of the physics. The case of diffusion-limited aggregation has been widely considered, and in fact was the subject of Smoluchowski's own work which was also directed towards the understanding of the aggregation of gold colloids. The kernel is obtained by solving the diffusion equation in the low concentration limit, appropriate for the gold colloids, to determine the flux of particles of radius R_i and diffusion constants D_i, onto a sphere of radius R_j. The boundary conditions used for the solution are that the concentration is the average concentration at infinity and zero at the surface of the sphere. When the j-th cluster is also allowed to diffuse, the kernel becomes

$$K_{ij} \approx (D_i + D_j)(R_i + R_j) \qquad (4)$$

We must now ask how R_i and D_i scale with mass for a cluster with a fractal structure. Because of the hydrodynamic interactions, the diffusion constant for a fractal object scales as it does for a solid object of the same radius, provided $d_f > 1$. This has, in fact, been confirmed by numerical calculations for cluster aggregates.[21] Physically, this means that there is non-draining flow of the fluid around the cluster, despite its very open structure. Therefore, we have $D_i \sim R_i^{-1} \sim M_i^{-1/d_f}$. Similarly, the effective capture radius scales as R_i, since for diffusive motion, which has a trajectory dimension of $d_t = 2$, even the smallest clusters are screened from penetration within a fractal cluster, and are adsorbed on the outer surface. Thus the scaling of the kernel for diffusion-limited aggregation is

$$K_{ij} \approx 2 + \left(\frac{M_i}{M_j}\right)^{1/d_f} + \left(\frac{M_j}{M_i}\right)^{1/d_f} \qquad (5)$$

We note that when $M_i \sim M_j$, the kernel is a constant, and this was the assumption made by Smoluchowski in his original work. Indeed, this kernel deviates substantially from the constant value only far off the diagonal, and thus it is convenient to consider the constant kernel as an analytic solution exists for this case.[22] The behavior of the cluster mass distribution predicted with this solution is shown by the solid lines in Fig. 4, where we have adjusted the cutoff mass in each case to match the data. As can be seen from the figure, the agreement with the data is quite good. The analytic solution for the rate equations also predicts that the characteristic mass should grow linearly with time, and if the values of the cutoff mass used for the fit in Fig. 4 are plotted as a function of time, linear behavior is indeed found. Furthermore, as we shall show below, this is also verified by our measurements of the growth of the hydrodynamic radius with time.

The deviations of the more exact kernel in Eq. 5 from the constant kernel occur primarily for the terms far off diagonal, when $M_i >> M_j$ making the kernel much larger. Physically, this arises from the very small clusters, which diffuse very rapidly, aggregating with the very large clusters, with large capture radius. Thus the rate for these events is much larger than for the diagonal terms, where clusters of equal size aggregate. The effect of this is to decrease the number of small clusters, making the mass distribution more monodisperse,[23] and thus making the plateau in the theoretical curves in Fig. 4 into a peak. If we assume that the cluster distribution is indeed reasonably peaked or monodisperse, we can significantly simplify the rate equations and solve them trivially. We do this by considering the rate of growth of the mass of the characteristic cluster. The growth then arises from the aggregation of clusters of equal mass, so that the change in mass $\Delta M \sim M$. Similarly, the rate at which this growth occurs is given by the solution to the diffusion equation, $(\Delta t)^{-1} \sim C_M RD$. Here C_M is the concentration of clusters of mass M, and is related to the initial concentration, C_o, by $C_M = C_o/M$. Thus we have $dM/dt = \text{const.} \times C_o$, which is identical to what is predicted by the solution of the full rate equation with constant kernel. Since our quasielastic light scattering measures the growth of the radius, we use the fractal scaling of the clusters to obtain

$$R \sim \text{const}\ C_o^{1/d_f}\ t^{1/d_f} \tag{6}$$

This accounts for the power-law dependence of our data in Fig. 3, and the measured slope of 0.57 is in excellent agreement with the predicted value of $1/d_f \sim 0.57$.

This simple model also suggests that the prefactor in the power law dependence of the aggregation rate should scale with the initial concentration in a fashion which again reflects the fractal dimension. Fig. 5, shows a logarithmic plot of the prefactor measured as a function of the initial concentration, normalized to the full concentration of the colloid. We again find the best fit to the data, shown as the solid line, yields a slope of 0.56, as predicted by Eq. 6. Thus the simple model which assumes a monodisperse distribution, and the solution to the full rate equations with the constant kernel, both predict the same sort of behavior, which is in good agreement with our data. Furthermore, the scaling exhibited by the dynamics of the aggregation directly reflects the fractal dimension of the resultant clusters, illustrating the inherent relationship between the dynamics and structure in this kinetic growth process.

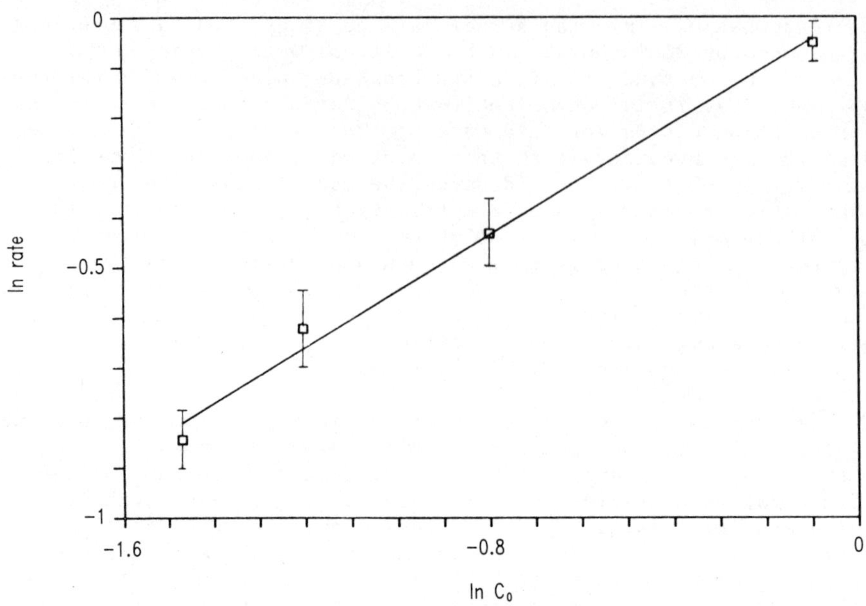

Fig. 5 Logarithmic plot showing the scaling of the aggregation rate with colloid concentration normalized to $1.7 \times 10^{12}/cm^3$. The straight line with a slope of 0.56 is a least-squares fit to the data.

REACTION-LIMITED AGGREGATION

This regime of aggregation occurs when only a small amount of pyridine is added to the solution, so the surface charge is reduced only slightly.[24] The aggregation occurs much more slowly, and exhibits a completely different characteristic behavior. Our goal here is to show that this behavior can be accounted for in a relatively straightforward manner, and that it, in fact, represents a second general class of kinetic cluster aggregation.

The structure of the clusters produced in this regime is quite different from that produced by diffusion-limited cluster aggregation. This is immediately apparent from the TEM micrograph of a typical cluster, shown in Fig. 1b. This cluster has roughly the same mass as "4739" in Fig. 1a, but is obviously much more compact and smaller. The structure is nevertheless still fractal. A logarithmic plot of the mass of a series of RLA clusters as a function of their radii is shown in Fig. 6. It again exhibits a linear behavior, with its slope giving $d_f \approx 2.02 \pm 0.1$. Because the fractal dimension is so high, we are unable to analyze the TEM images to determine the mass-mass correlations. However, we can still do light scattering from the clusters.[14] We again find a linear behavior on a logarithmic plot of I(k) vs k, and the slope gives $d_f \approx 2.05 \pm 0.05$, corroborating the TEM measurements. Thus the clusters formed by reaction-limited aggregation are substantially more dense than those formed by diffusion-limited cluster aggregation.

The dynamics of reaction-limited aggregation are also substantially changed from the power-law dependence observed for the fast

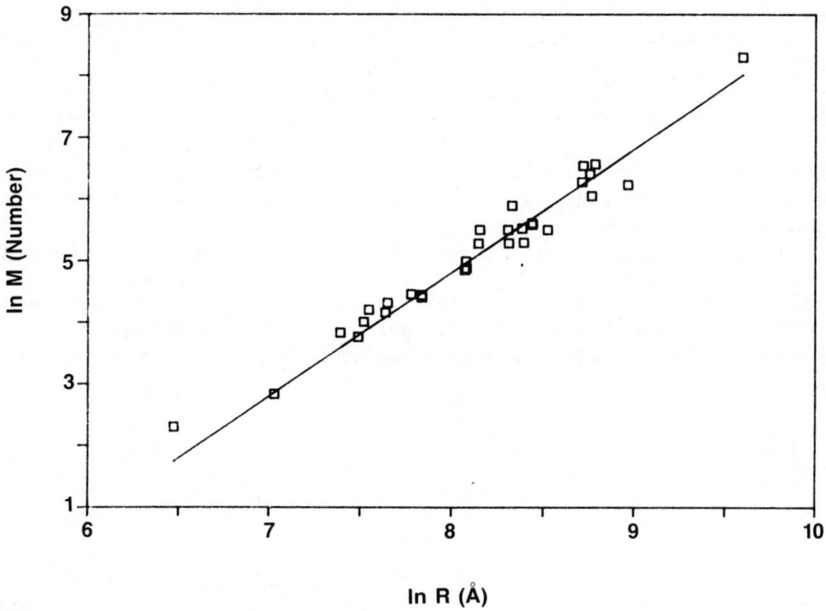

Fig. 6 Logarithmic plot of the mass of the cluster (number of particles in the cluster) vs the size of the cluster for TEM measurements for RLA. Best fit line gives $d_f = 2.02 \pm 0.1$.

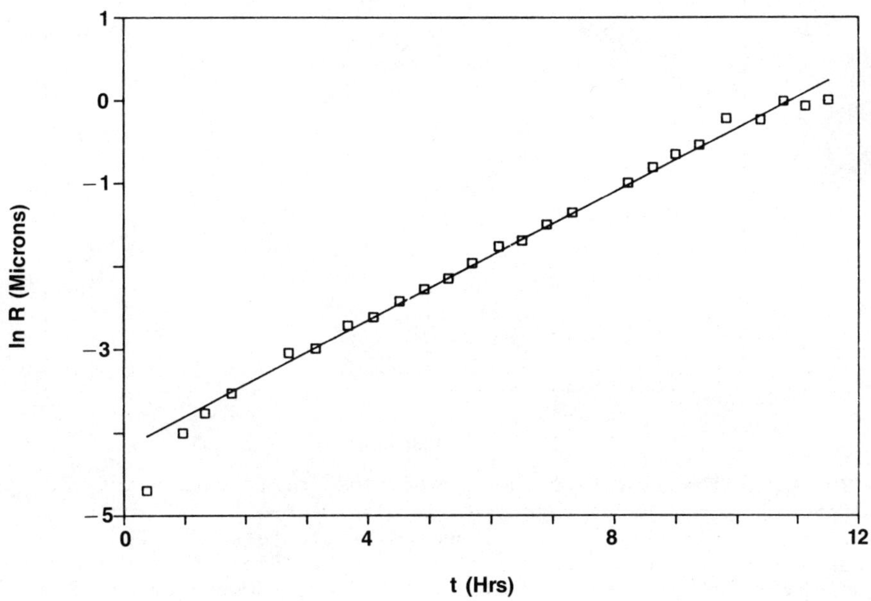

Fig. 7 Exponential growth or the cluster radius R for the RLA process measured at $\Theta = 140^o$.

aggregation. This is shown in Fig. 7, which is a semi-log plot of the growth of the hydrodynamic radius as measured by quasielastic light scattering for a typical reaction-limited aggregation. The solid line shows that the data are well described as exponential growth. These data were obtained at a scattering angle of 140^0, and while there is some sensitivity to angle in the exact values of R measured, we find the same exponential time dependence at all values of k.

Like the structure and the dynamics, the cluster mass distribution for reaction-limited aggregation is also very different than that of the diffusion-limited cluster aggregation. Data obtained for three different times during the aggregation are shown in a logarithmic plot in Fig. 8, where we have again normalized each data set to the number of clusters counted. In this case the data are very well described by a power law, $N(M) \sim M^{-\tau}$, with $\tau = 1.5 \pm 0.2$. In addition, each data set lies on top of the other at the smaller masses, but then has a cutoff at different points.

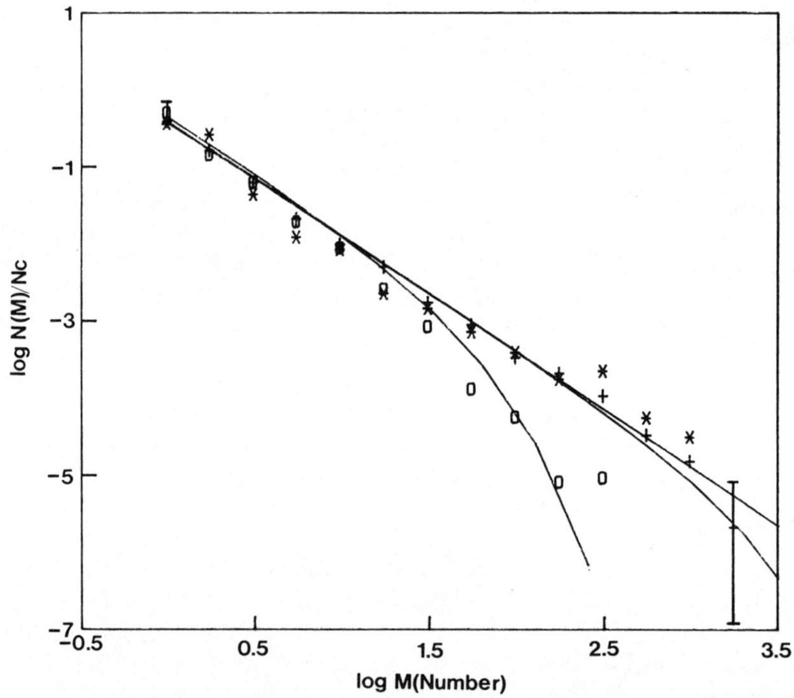

Fig. 8 Logarithmic plot of the cluster mass distribution for RLA at 3 different times of aggregation. N(m), the number of clusters of mass M, is normalized by the total number of clusters N_c. The solid lines represent the theoretical fits according to Smoluchowski equation with sum kernels. Open circles denote measurements taken after 2 hours, the crosses denote measurements taken after 6 hours and the stars denote measurements taken after 8 hours. The limiting slope of power law behavior shows $N(M) \sim M^{-1.5}$.

The physics of reaction-limited aggregation can be understood by first considering the initial rate at which two single gold particles will stick.[25] Since there is still a substantial charge on their surface, they will be able to approach to within κ^{-1}, which is roughly equal to a single particle radius, but will then be prevented from sticking by the repulsive barrier, of height E_b, due to the surface charges. The probability that they will stick on each collision is simply $p = e^{-E_b/kT}$, and the rate is given by $k=\nu p$, where the attempt frequency, ν, is determined by their concentration. Reaction-limited kinetics occur when $E_b > kT$, in the limit where p is finite but approaches zero. In this case the particles have an opportunity to explore, in a statistical sense, all possible mutual configurations except those that would cause them to overlap and occupy the same region of space. Thus, although their motion is diffusive, this very low sticking probability precludes the screening effect that prevents particles from penetrating within a larger cluster as occurs for any diffusion-limited kinetics. This then allows all the possible configurations be sampled, and the bonding configuration can be chosen, without bias, from among these. Thus the trajectory dimension here is effectively $d_t = 0$. Since there is now the possibility of interpenetration of the clusters, the fractal dimensions can be expected to increase. However, since the bonds formed are nevertheless rigid, the clusters will not fill all of space and will still have a fractal structure. Jullien et al[26] have simulated reaction-limited aggregation on a computer using their hierarchical model, which enforces a monodisperse cluster mass distribution, and they find $d_f = 2.0$ in 3d. Brown and Ball[27] have done a more complete simulation where the correct polydispersity is included, and they find $d_f \approx 2.1$, in excellent agreement with our experimental data.

The dynamics of the reaction-limited aggregation can again be treated with the rate equations,[25] provided the correct kernel is found. It is instructive to do this, as much of physics is contained in this treatment. The key here is the fact that $d_t = 0$, so that all possible bonding configurations are explored. In this case, the rate that two clusters will stick is directly proportional to the ratio of the number of bonding configurations to the total number of possible configurations. This is simply obtained by determining the number of ways one cluster can touch the other, without rotating either. We emphasize that the different rotational configurations must be averaged, not summed, since they contribute equally to both the bonding configurations as well as the non-bonding configurations. Thus, for example, for two solid spheres, the probability will scale as the surface area. More generally, two equally sized clusters with fractal structure can be expected to have a somewhat rougher surface of mutual contact, so we allow the surface to scale with their radius as R^{α}. We can immediately set bounds on α by noting that if the clusters were solid, $\alpha=2$, while if they were totally transparent, so that the clusters could interpenetrate each other freely, $\alpha=3$. This free interpenetration is expected[28] whenever $2d_f<d$. Thus the kernel scales as $K_{ij} \sim M^{\lambda}$ for $i \sim j$, where $\lambda = \alpha/d_f$, and is bounded by $2/d_f \leq \lambda \leq 3/d_f$. To solve the rate equations, we also need the scaling of the kernel off diagonal, when $i \gg j$. Since the smaller cluster can penetrate freely into the larger up to a length scale roughly equal to its own, we divide the larger cluster up into units of the same size as the smaller cluster. The number of possible bonding configurations between the smaller cluster and these sub-units then scales the same way as that for two clusters of equal size. Thus, $K_{ij} \sim M_i M_j^{\lambda-1}$ for $i \gg j$.

Knowledge of the scaling of the kernel for both the diagonal and off diagonal terms is sufficient in this case to solve the rate equation.[29,30] For $\lambda < 1$, N(M) exhibits a well defined peak and the characteristic mass of the distribution increases with time as a power-law, $M_c \sim t^{1/(1-\lambda)}$. For $\lambda = 1$, N(M) has a power law dependence, $N(M) \sim M^{-\tau}$, with $\tau = 1.5$, and the characteristic mass grows exponentially with time, $M_c \sim e^{t/\tau_M}$. Finally, for $\lambda > 1$, N(M) is again a power law, but with a quite different power, $\tau = (3 + \lambda)/2$, so that $\tau \geq 2.5$. The characteristic mass in this case has a power law divergence $M_c \sim (t_g - t)^{-2/(\lambda-1)}$, and gelation occurs at a finite time, t_g. Obviously, the case that best fits the observed data is $\lambda = 1$.

The solution for $\lambda = 1$ is in fact rather singular. If λ cnanges slightly to either side of 1, the cluster mass distribution changes dramatically, losing the power law behavior entirely, and becoming monodisperse if $\lambda < 1$, and increasing from $\tau = 3/2$ to $\tau \geq 5/2$ if $\lambda > 1$. This singularity actually enforces the $\lambda = 1$ solution on the reaction-limited kinetics through the adjustment of d_f. This can be understood physically as follows. If λ becomes slightly greater than 1, the cluster mass distribution must rapidly become skewed to produce a larger power law, causing a relative increase in the number of small clusters compared to big clusters. However, small clusters can penetrate and stick well within the larger clusters, and so any new clusters which result from the aggregation will tend to have a larger d_f, thus reducing λ. Similarly, if λ becomes slightly less than 1, the cluster mass distribution must very rapidly become monodisperse by losing many more small clusters than large. However, the aggregation processes that ensue would be primarily between equally sized clusters, which would interpenetrate the least, thereby reducing d_f and increasing λ. Thus λ is driven toward 1 through the adjustment of d_f.

These results can be put on a more rigorous basis. Ball[25] has linearized the rate equations using this kernel with $\lambda = 1$ and has shown that a small number of clusters with $\lambda > 1$ can be tolerated since they are very rapidly consumed through aggregation with clusters having $\lambda = 1$. Thus, the rate equations are stable against deviation toward $\lambda > 1$. Similarly if $\lambda < 1$, the cluster mass distribution becomes monodisperse. However Jullien et al[26] have actually measured λ for perfectly monodispersive RLA, and in 3d find $\lambda > 1$. Thus the kinetics are also stable against deviations toward $\lambda < 1$.

For $\lambda = 1$, the kernel behaves very nearly like the sum kernel, $K_{ij} \sim M_i + M_j$, for which an analytic solution exists.[22] In fact, in Fig. 8 we have plotted as solid lines the analytic calculation for the cluster mass distribution obtained from this solution. We have fit the degree of aggregation, as reflected in the cutoff mass, to the observed data. If this fitted characteristic mass is plotted against time, we find exponential behavior as expected.

The fractal structure, as well as the actual fractal dimension, play key roles in RLA. Indeed, the fractal dimension is actually uniquely determined by the aggregation kinetics. We note that it is the fractal structure of the clusters, not the size of the individual gold spheres, that control the reaction kinetics. Thus we believe that this picture of RLA may resolve a paradox that has existed within the colloid science literature for over 30 years. The regime that we call reaction-limited aggregation corresponds to what the traditional colloid science literature has referred to as slow coagulation. The usual theoretical picture of this has been in terms of a diffusive process with a potential energy barrier, modeled within the traditional DLVO potential picture.[4]

However, since this process involves a balance between attractive van der Waals forces and repulsive screened coulomb forces, this prediction suggests that there should be a marked dependence of the behavior on the size of the primary colloid particles. However, experimentally this is not found. In contrast, in our picture, the kinetics depend solely on the fractal structure of the aggregates, and not on the size of the individual particles, accounting for the experimental observation and resolving the paradox.

CONCLUSIONS

We have studied two regimes of kinetic aggregation in detail. They correspond to the two limiting cases of the sticking probability between two clusters. When the sticking probability is essentially unity, so two clusters stick on contact, diffusion-limited aggregation results. When the sticking probability approaches zero, so a very large number of attempts is required before two clusters stick, reaction-limited kinetics result. In fact, we believe that these may be the only classes of aggregation at least for the gold colloids. If we add an intermediate amount of pyridine, we initially see kinetics that resemble the exponential dynamics of RLA, but eventually the dynamics crossover and behave in a fashion consistent with the power law dependence of diffusion-limited aggregation. This observation of the crossover behavior suggests that the two limiting regimes are in fact the only two universal classes of aggregation. Furthermore, the two values of the trajectory dimension, d_t, are the two limiting values that are physically accessible. This suggests that we can also put limits on the fractal dimension obtainable with cluster-cluster aggregation, 1.75 ± 0.05 d_f 2.05 ± 0.05. We note that strictly speaking, this holds only for the gold colloids. However, the only crucial parameters in the aggregation process would seem to be that κ^{-1} is not considerably larger than a and that the aggregation process be completely irreversible and form rigid, or nearly rigid clusters. These conditions are often met for colloid aggregation, so these results should be more generally applicable.

Our results provide a consistent picture for all other reports in the literature concerning scaling of cluster-cluster colloid aggregation in three dimensions. Schaefer et.al.[11] have measured the fractal dimension of the cluster produced by the irreversible aggregation of silica colloids. They find that the fractal dimension is $d_f \approx 2.1$, while the dynamics are exponential. These results can now be understood as reaction-limited aggregation. Similarly, Von Schultess and co-workers[31] have found a power-law cluster-mass distribution, with $\tau = 1.4 \pm 0.1$ for the aggregation of polystyrene spheres induced by an antigen-antibody reaction. They also inferred[32] that $d_f \approx 2.0$. These results can again be interpreted as reaction-limited aggregation. Thus, we have a coherent picture of all the studies reported to date, and the seemingly different fractal dimensions can be rationalized.

There is one example however, for which our results would not seem to apply. This is the aggregation of human immunoglobin,[33] reported at this conference. In this case, the reaction is much more complicated. It would seem to be not totally reversible, and the clusters formed may not be totally rigid. Thus a different model has been suggested, for the aggregation kinetics, such as would seem to be appropriate.

The aggregation of aqueous colloidal particles has been the subject of considerable scientific effort for over one hundred years. The recent resurgence of interest in this process reflects the great success that modern scaling concepts have had in explaining many aspects of this very complex phenomenon. In this paper we have attempted to show that these concepts are indeed extremely valuable in developing a more complete understanding of the behavior of a colloid system.

We thank J. Sung and J. Dunsmuir for their excellent technical assistance in this work.

REFERENCES

1. T. A. Witten and L. M. Sander, Phys. Rev. Lett. 47, 1400 (1981).
2. "Kinetics of Aggregation and Gelation" ed. by F. Family and D. P. Landau (North Holland, Amsterdam 1984).
3. B. V. Enustin and J. Turkevitch, J. Am. Chem. Soc. 85, 3317 (1963).
4. E. J. W. Verwey and J. Th. G. Overbeek "Theory of the Stability of Lyophobic Colloids," (Elsevier, NY 1948).
5. D. A. Weitz, M. Y. Lin and C. J. Sandroff, Surf. Sci., to appear.
6. J. Turkevitch, A. Garton and P. C. Stevenson, J. Colloid Sci. 9, 26 (1954).
7. J. A. Creighton, C. B. Blatchford and M. B. Albrecht, J. Chem. Soc. Faraday Trans. II 75, 790 (1979).
8. D. A. Weitz and M. Oliveria, Phys. Rev. Lett. 52, 1433 (1984).
9. B. B. Mandelbrot, "The Fractal Geometry of Nature" (Freeman, San Francisco, 1982).
10. M. Y. Lin and D. A. Weitz, to be published.
11. D. W. Schaefer, J. E. Martin, P. Wiltzius and D. S. Cannell, Phys. Rev. Lett. 52, 2371 (1984).
12. S. K. Sinha, T. Freltoft and J. Kjems in Ref. 2, p 87.
13. S. K. Sinha, D. A. Weitz, J. S. Huang, M. Y. Lin, J. S. Gethner and R. Pynn, to be published.
14. D. A. Weitz, J. S. Huang, M. Y. Lin and J. Sung, Phys. Rev. Lett. 54, 1416 (1985).
15. P. Meakin, Phys. Rev. Lett. 51, 1119 (1983).
16. M. Kolb, R. Botet and R. Jullien, Phys. Rev. Lett. 51, 1123 (1983).
17. D. A. Weitz, J. S. Huang, M. Y. Lin and J. Sung, Phys. Rev. Lett. 53, 1657 (1984).
18. B. J. Berne and R. Pecora, "Dynamic Light Scattering" (Wiley, New York, 1976).
19. J. E. Martin and B. J. Ackerson, preprint.
20. M. Von Smoluchowski, Phys. Z. 17, 557, 585 (1916).
21. P. Meakin, Z. Y. Chen and J. M. Deutch, J. Chem. Phys. 82, 3786 (1985).
22. R. J. Cohen and G. B. Benedeck, J. Phys. Chem. 86, 3696 (1982).
23. S. K. Friedlander and C. S. Wang, J. Colloid and Interface Sci., 22, 126 (1966).
24. D. A. Weitz and J. S. Huang, in Ref. 2, p. 19.
25. R. C. Ball, D. A. Weitz, T. A. Witten and F. Leyvraz, to be published.
26. R. Jullien, M. Kolb and R. Botet, J. Phys. Lett. 45, L211 (1984).
27. W. D. Brown and R. C. Ball, to be published.
28. R. C. Bull and T. A. Witten, J. Stat. Phys. 36, 873 (1984).
29. F. Leyvraz, Phys. Rev. A29, 854 (1984).
30. P. G. J. Van Dongen and M. H. Ernst, Phys. Rev. Lett. 54, 1396 (1985).
31. G. K. von Schulthess, G. B. Benedek and R. W. DeBlois, Macromolecules 13, 939 (1980).
32. D. Johnston and G. B. Benedek, Ref. 2, p. 181.
33. J. Feder, T. Jossang and E. Rosenqvist, Phys. Rev. Lett. 53, 1403 (1984) and J. Feder and T. Jossang, this volume.

POSSIBLE FRACTAL STRUCTURE OF CEMENT GELS

A. J. Allen and P. Schofield

AERE Harwell, Didcot, Oxon, U.K.

INTRODUCTION

The understanding of the mechanism by which a newly mixed cement turns into a gel structure which gives it rigidity and compressive strength remains poorly understood. Recently, however, new light has been shed on the hydration process in an extensive series of measurements using small-angle neutron scattering (SANS) in which the evolution of the scattering has been studied as a function of time[1,2,3]. The most detailed work has been carried out on Ordinary Portland Cement (OPC), but experiments on other starting materials show very little difference in the evolution of the scattering (though the time-scales may differ) and in the final product.

OPC contains four major phases, in cement notation there are C_3S, C_2S, C_3A and C_4AF where $C \cong CaO$, $S \cong SiO_2$, $A \cong A\ell_2O_3$ and $F \cong Fe_2O_3$. The ground powder consists of sintered aggregates with particles in the size range 1-200μm with a mean size around 40μm. On mixing with water a complex series of hydration reactions take place of which the main products are an amorphous calcium-silicate hydrate (CSH) gel-like structure and crystalline calcium hydroxide which accounts for 20% of the volume of hydrated material in aged cements. A micrograph of CSH gel (figure 1) indicates a very open porous structure, having the appearance of self-similarity. The pores contain saturated solution of calcium hydroxide. Note, however, that the scale of their micrograph is much larger than the scale of the SANS measurements which is indicated by the .2μm marker.

SANS RESULTS

The results of the SANS experiments are summarised in figure 2, which shows a composite picture of results from the SANS spectrometer 'D11' at ILL at various settings, and for larger scattering vector, Q, from the PLUTO instrument at Harwell. Similar experiments have been carried out by Windsor and Epperson (private communication) on the IPNS at Argonne National Laboratory, giving similar results.

An important feature of these results, not shown in the figure is that at the highest values of Q shown there is a well-defined Guinier region in the scattering where $\Sigma(Q)$ varies as $\exp\{-\frac{1}{3}Q^2R_g^2\}$ corresponding to scattering from objects with radius of gyration about 2nm. The growth

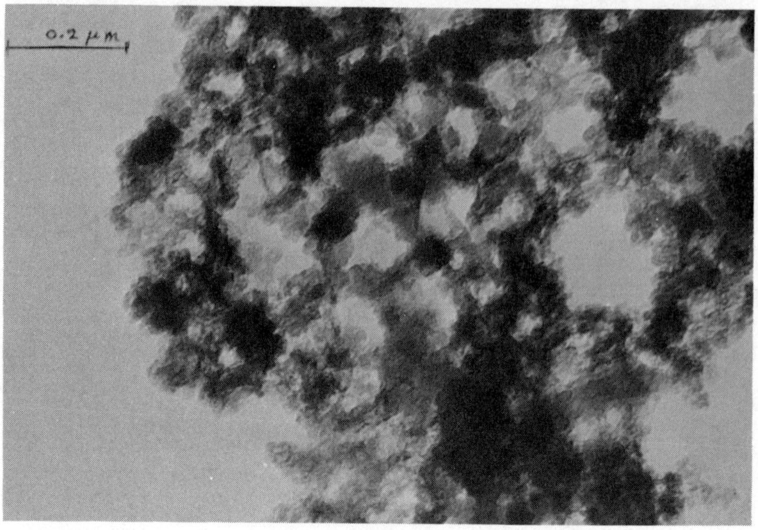

Fig. 1 Wet cell transmission electron micrograph of CSH gel (from reference 1)

of this scattering follows closely the heat output of the cement and it is deduced, therefore, that it is associated with the growth of the CSH gel. 'Contrast variation', substituting D_2O for H_2O led us to conclude that the 2nm 'objects' correspond to pores between 'globules' of gel which form the building blocks of the CSH structure.

Figure 2 is in the form of a Porod plot, showing in $Q^4\Sigma(Q)$ versus lnQ. Scattering from interfaces with curvature >> Q^{-1} would thus give a line of zero gradient on this plot. This is roughly what is observed in the dry powder for $Q < 0.1Å^{-1}$. Up to one hour after adding water there is a general reduction in intensity due to the different contrast of OPC/air and OPC/water, with a dip beginning to develop below $0.01Å^{-1}$ probably arising from a softening of the interface by the formation of a layer of CSH around the particles as the first stage of hydration.

Following the incubation period there is an increase in the scattering at large Q leading to an ultimate $Q^{-2.5}$ behaviour between 0.01 and $0.1Å^{-1}$. At the same time, the scattering below Q = 0.01 exhibits a short range where $\Sigma(Q) \propto Q^{-3.2}$.

DISCUSSION

The formation of CSH gel in cement is a complex interaction of chemical and kinetic process and it is not yet possible to give a definitive interpretation of the SANS results. However, the power law behaviour observed in the 28-day cement can be interpreted as evidence of fractal behaviour, though such an interpretation must be treated with caution in view of the limited length scales over which such behaviour is observed.

$Q > 0.01Å^{-1}$

Here the power law (-2.5) is close to that found for diffusion-limited aggregation[4]. A mechanism by which this could occur is by the diffusion of silicate ions into calcium hydroxide solution. This is consistent with the view that the dissolution of silicate is delayed over that of the calcium by an 'osmotic barrier' at the clinker surface, this is supported by micrograph evidence of 'silica garden' type protruberances[5]. However, the existence of the 2nm Rg porosity shows this is over simplified and indicate that at this scale the gel is constructed of close-packed globules of ∿10nm diameter. On the other hand, cluster-cluster aggregation of such globules is ruled out since this would lead to a much lower dimensionality. One possibility is that thermo-chemical effects stabilise this size of globule, which then form the nuclei for further diffusion controlled aggregation.

An alternative explanation is that the power law reflects a distribution of pore sizes. A pore size radius distribution $R^{-\nu}$ (or pore volume distribution $R^{3-\nu}$) would lead to a $Q^{-(7-\nu)}$ behaviour corresponding to $\nu = 4.5$.

$Q < 0.01Å^{-1}$

Here, again, the appearance of a 3.2 power law is intriguing. A tentative interpretation is that it corresponds to a surface fractal dimension Ds = 2.8 associated with the decoration of the surfaces of the larger pores as is illustrated in figure 1.

In conclusion, it may be said that it is highly plausible that diffusion-controlled aggregation processes play a role in the formation of cement gel, giving fractal characteristics. Work is continuing to give a more quantitative basis to these ideas in terms of the amounts of various products formed in the cementation process.

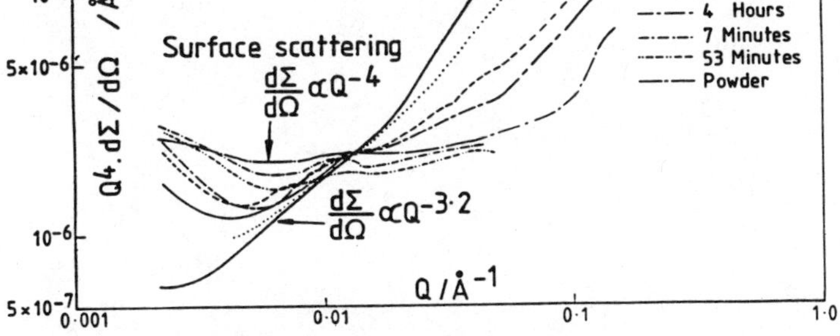

Fig. 2 SANS from Ordinary Portland Cement.

ACKNOWLEDGEMENT

Part of this work is supported by a CEC Scientific Cooperation Contract with Risø National Laboratory. We wish to thank J. Kjems, D. Pearson and C. G. Windsor for helpful discussions.

REFERENCES

1. D. Pearson, A. J. Allen and C. G. Windsor, An investigation on the nature of porosity in hardened cement parts, M. McN. Alford and D. D. Double, J. Mat. Sci., 18:430 (1983)
2. A. J. Allen and D. Pearson, Porosity and gel structure of cement, in "Microstructural characterisation of materials", N.H. Anderson et al, eds. Risø National Laboratory, Roskilde, Denmark, (1984)
3. D. Pearson and A. J. Allen, A study of ultrafine porosity in hydrated cements, J. Mat. Sci. 20:303 (1985)
4. P. Meakin, Diffusion controlled cluster formation in 2-6 dimensional space, Phys. Rev. A., 27: 1495 (1983)
5. D. D. Double, An examination of hydration of Portland cement by electron microscopy, Silicate industrielle 11:233, (1978).

DIFFRACTION ON FRACTALS

C. Allain and M. Cloitre

LHMP UA/CNRS 857 - ESPCI
10 rue Vauquelin, Paris 75005

INTRODUCTION

The fractal dimension of an object may be determined through the relation $M(r)=r^D$ between its mass M and its radius r or the pair correlation function $g(r)=r^{D-d}$ (d is the dimension of the euclidian space embedding the object). Yet, it is possible to have access to the value of D when studying the variation of the intensity I(q) which is scattered by a fractal at a wavevector q: $I(q)=q^{-D}$. This relation has been used to interpret several small angle scattering experiments on silica (1). In this paper, we discuss diffraction experiments on deterministic fractal gratings (F). The intensity I(q) at wavevector q is the Optical Fourier Transform (OFT) of the grating F. This allowed a direct determination of D and led us to take in account other geometrical caracteristics of fractals.

EXPERIMENT

The recursive fractals studied here are calculated on a microcomputer and drawn on a graphics plotter; then, they are photographed using a high resolution 24x35 mm film. Because of the width of the lines drawn by the graphics plotter (0.2 mm), we are limited to objects obtained from a small number of iterations ($\lesssim 7$); consequently, the ratio between the smallest scale l of the fractal to the largest one L is about l/L=1000. Other limitations arise from the finite size of an elementary grain on the film, here lmin is 5 μm, which defines the smallest scale which may be recorded on the film, and from the size L , which determines the resolution of I(q) measurements.

The fractal objects F are then disposed in an optical Fourier transform arrangement. Such a set up is based on the capacity of a simple converging lens to produce a spatial FT in its focal plane (2). The diffraction patterns of F are recorded using a photomultiplier connected to a multichannel analyzer. A small pinhole ensures the selection of the proper wavevector q; its displacement is controled by a high precision motorized micrometer.

Experimental data obtained in the case of Cantor triadic bars (3) are shown on figure 1. Such a fractal is obtained by dividing recursively a bar into three equal segments, then removing its middle third and so on ad infinitum.

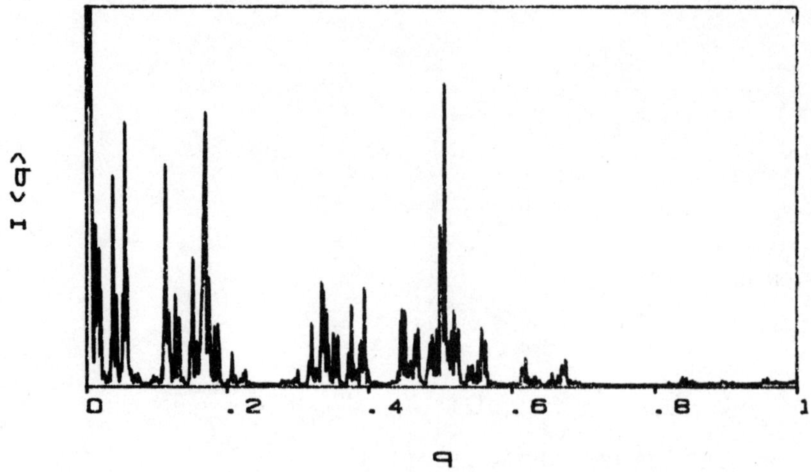

Fig. 1: experimental diffraction curve I(q) of Cantor bars.

The largest spatial frequency 1. on figure 1 corresponds to the frequency associated with the smallest scale l on F, while the smallest frequency which may be resolved (that is the frequency of oscillations on figure 1) is associated with L; l and L may be determined in this way. It is noteworthy that q=1, corresponds to a scattering angle of 2 deg. with an uncertainty of 10^{-2} deg.; moreover, data at very low angles (less than 10^{-2} deg.) are affected by the contribution of the central peak, which is due to the incident beam.

PROPERTIES OF THE STRUCTURE FACTOR OF DETERMINISTIC FRACTALS

Let us show now how data of figure 1 enable us to determine directly the fractal dimension D. The Fourier transform of cantorian triadic bars generated at iteration n can be calculated analytically quite easily :

$$I(q)=I_o\left[\prod_{0}^{n-1}\cos(2\pi 3^i q l)\right]^2 \times \left[\frac{\sin(ql/2)}{ql/2}\right]^2$$

Here, the first term on the right side of this equation is the geometrical structure factor S(q) while the second one is a form factor which will be neglected in the following. In effect, on figure 2, S(q) is computed up to q=0.25; comparing with figure 1, one sees that S(q) is not affected by the form factor at low angles.On figure 2, it appears that the structure factor S(q) of the Cantor's bars is composed of n frequency bands which are scale invariants over a dilatation of factor 3. Now, calculating the mean energy $\langle S(q)\rangle$ which is scattered over each of these frequency bands, we find:

$$\langle S(q)\rangle=q^{-D}$$

where D is the right fractal dimension.
This relation, obtained after an averaging operation in order to break up any regularity due to the deterministic nature of the Cantor's bars, is the same as the one expected for random structures.

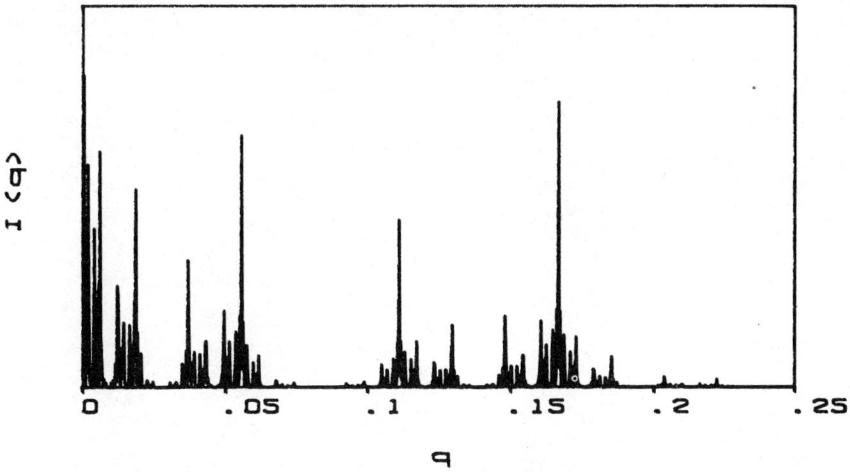

Fig. 2: geometrical structure factor (computed) of Cantor bars

We checked that the experimental data of figure 1 enable a direct determination of D. However, because of the discrete geometry of such deterministic fractals, we are limited by a too small number of iterations; consequently, the precision is poor (10%), but this limitation falls in the case of random fractals where, moreover, the averaging operation on S(q) is not necessary. Such a method is to be particularly well suited to analyze experimentally obtained aggregates, for instance, two dimensional clusters resulting from the aggregation of small particles floating on a liquid (4).

CONCLUSIONS

We have shown that the relation $S(q)=q^{-D}$ holds for deterministic fractals with the proviso that S(q) is in this case a mean average value over well chosen frequency ranges. We have extended this result to the case of some two dimensional fractals, like Viscek aggregates(5).

However, many questions remain unanswered concerning either the effect of other geometrical properties of fractals (ramification,lacunarity, existence of branches) or, in the case of fractal aggregates, the influence of the polydispersity on the values of D which might be measured. For instance, in the case of the Cantor's bars, the averaged structure factor $\langle S(q) \rangle$ gives access to D, but the precise repartition of the intensity scattered over a frequency band depends on the lacunarity. In the same vein, the application of this OFT method to two dimensional infinitely ramified fractals such as Serpinski carpets is now under progress.

REFERENCES

(1) J.E. Martin, D.W. Schaefer. Phys. Rev. Lett.,53, pp 2457-2460, 1984
(2) J.W. Goodman. Introduction à l'optique de Fourier et à l'holographie p 78, Masson et Cie, Paris,1972
(3) B.B. Mandelbrot. The fractal geometry of Nature, chap.8, W.H. Freeman and Company, New York,1982
(4) C. Allain, B. Jouhier. J. Phys. Lett.,44,pp L421-L428, 1983
(5) T. Viscek, J. Phys. A:Math. Gen., 16, pp L647-L652, 1983

ICOSAHEDRAL INCOMMENSURATE CRYSTALS

Per Bak

Physics Department
Brookhaven National Laboratory
Upton, New York 11973 USA

INTRODUCTION

One of the basic facts of condensed matter physics is that the symmetry of 3d crystals can be described in terms of space groups composed of Bravais lattices of 3d translational symmetries and point groups of rotation and reflection symmetries. Only 2-fold, 3-fold, 4-fold and 6-fold rotation symmetries are allowed. In a striking experiment on an Mn-Al alloy, however, Shechtman et al.[1] have observed an electron diffraction pattern with a 5-fold symmetry axis, and an overall icosahedral symmetry (Fig. 1). How can this be possible? The diffraction spots are quite sharp, so there is long range positional ordering, but there can not be translational invariance in view of the considerations above. The spectrum has a "scaling" structure: if the pattern is magnified by a factor G (the golden mean) the positions of peaks remain fixed.

The situation is not entirely new. Incommensurate modulated systems also display a nonperiodic spectrum reflecting the lack of a discrete translational symmetry. However, their diffraction patterns have peaks at positions which can be formed as combinations of the reciprocal lattice vectors $\vec{\tau}_i$ of a basic lattice and the modulation vector(s) g_i of the incommensurate structure[2]. In the icosahedral structures there are no basic lattices, and there is in fact only one length scale so a discussion of symmetries and physical properties must start from scratch.

Here, it will be shown that the symmetries of icosahedral structures can be described in terms of six-dimensional space groups, and that the 5-fold axis is related to a 5-fold axis in the six-dimensional superspace. The actual physical density is the density along a 3d plane in 6d space. In principle the basis to be associated with each unit cell of the 6d Bravais lattice can be determined from an x-ray diffraction experiment. The stability of icosahedral structures can be understood from a simple Landau theory. No attempt will be made to explain the stability from first principles, which would require an enormous electronic structure calculation, well beyond the state of the art today. A simpler two-dimensional case with a 5-fold axis will also be analyzed. This structure has the same symmetry as the two-dimensional space-filling tilings invented by the British mathematician R. Penrose[3].

The icosahedral structures have 6 continuous symmetries compared with 3

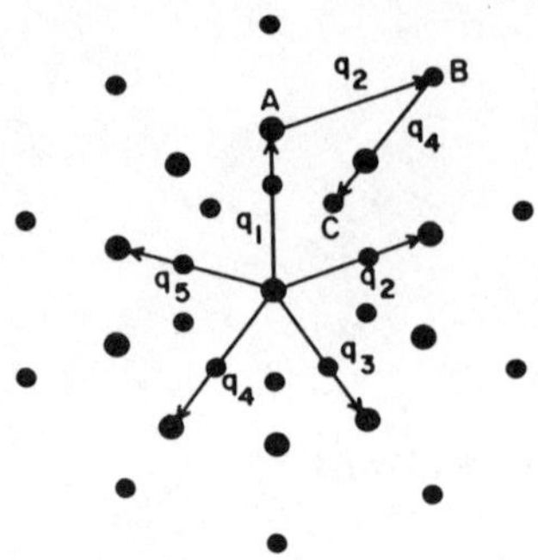

Fig. 1. Diffraction pattern of $Mn_{0.14}Al_{0.86}$ as measured by Shechtman et al., Ref. 1 (schematic). Note the self-similar scaling properties and the lack of translational invariance.

translational symmetries for regular "3d" crystals. The remaining continuous symmetries describe internal rearrangements or phase shifts of the atoms. There are 5 elastic constants associated with displacement strains and phase strains, and there may be 6 soft phonon and phason modes breaking the six continuous symmetries. A more complete account of the work described here is given in Ref. 4.

CRYSTALLOGRAPHY AND ELASTIC PROPERTIES OF ICOSAHEDRAL CRYSTALS AND PENROSE STRUCTURES

The condensation of ordered structures from a three-dimensional rotationally and translationally invariant liquid is conveniently described in terms of spontaneous formation of "mass-density-waves" breaking the translational symmetry[5]. A single density wave

$$\rho(\vec{r}) = \rho_0 + \rho \cos(\vec{q}\cdot\vec{r} + \phi) \tag{1}$$

represents a smectic structure, and a shift of the phase ϕ represents a displacement in the direction of $\vec{q}$. A superposition of three density waves with wave vectors forming an equilateral triangle,

$$\rho(\vec{r}) = \sum_{\vec{q}_i \varepsilon \Delta} \frac{\rho}{\sqrt{3}} \cos(\vec{q}_i\cdot\vec{r} + \phi_i), \tag{2}$$

gives an ordered structure with the symmetry of a triangular lattice (or a lyotropic rodlike structure in 3d). The relative stability can be discussed in terms of a phenomenological Landau theory in which the free energy F is expanded as a power series in the complex amplitudes (or amplitudes and phases) of the density waves. For instance, the triangular structure may be stabilized by a third order term

$$F_3 = -v_3 \frac{\rho^3}{3\sqrt{3}} \cos(\phi_1 + \phi_2 + \phi_3), \tag{3}$$

which locks the 3 phases onto each other. It has been pointed out by Alexander and McTague[6] that it will often be favorable to combine 6 density waves with 6 pairs of wave vectors $\pm q_i$ forming a regular octahedron

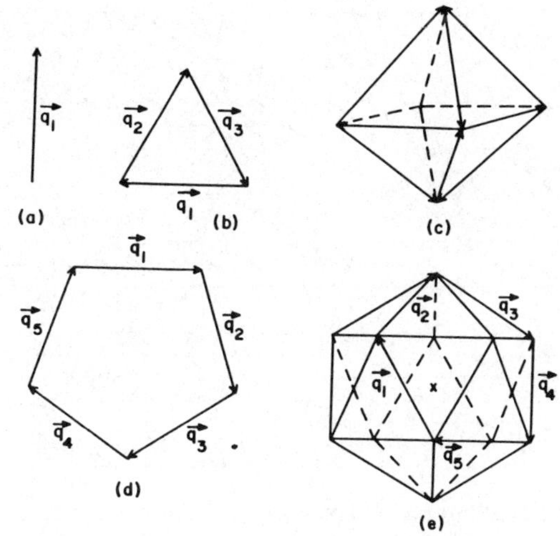

Fig. 2. Wave vector combinations representing a) smectic structure, b) triangular structure, c) bcc structure, d) 2d Penrose structure and e) icosahedral incommensurate structure.

(Fig. 2c), since there will then be contributions to the free energy (3) from each of the 8 triangles constituting the faces of the octahedron. The vectors (110) span the reciprocal lattice of a bcc crystal. In order to form a real bcc crystal with a realistic electronic density to account for the diffraction experiment, higher harmonics must be included. The harmonics do not lower the symmetry of the structure, so the crystallography can conveniently be discussed in terms of the basic density waves.

Extending this line of thinking, it is not difficult to produce an argument for the stability of structures with 5-fold symmetry. A crystal formed by superimposing 5 waves with wave vectors forming a regular pentagon

$$\rho(\vec{r}) = \sum_{q_i \varepsilon \text{ pentagon}} \frac{\rho}{\sqrt{5}} \cos(\vec{q}_i \cdot \vec{r} + \phi_i) \tag{4}$$

gives a contribution to the fifth order free energy

$$F_5 = -\frac{v_5 \rho^5}{25\sqrt{5}} \cos(\phi_1 + \phi_2 + \phi_3 + \phi_4 + \phi_5), \tag{5}$$

which is minimized by locking the 5 phases onto each other, for instance by choosing all phases equal to zero. The resulting structure, illustrated in Fig. 3, has the symmetry of the Penrose tilings. Associating a δ-function atom to each vertex of the original Penrose structure corresponds to a specific choice of higher harmonics. The general structure can not be represented as a 2d space-filling set of tiles.

In analogy with the situation for the triangular case, it may sometimes be advantageous to combine 15 pairs of vectors forming a regular icosahedron (Fig. 2e),

$$\rho(\vec{r}) = \sum_{q_i \varepsilon \text{ icosahedron}} \frac{\rho_i}{\sqrt{15}} \cos(\vec{q}_i \cdot \vec{r} + \phi_i) \tag{6}$$

The icosahedron has 10 pairs of triangular faces each giving a contribution

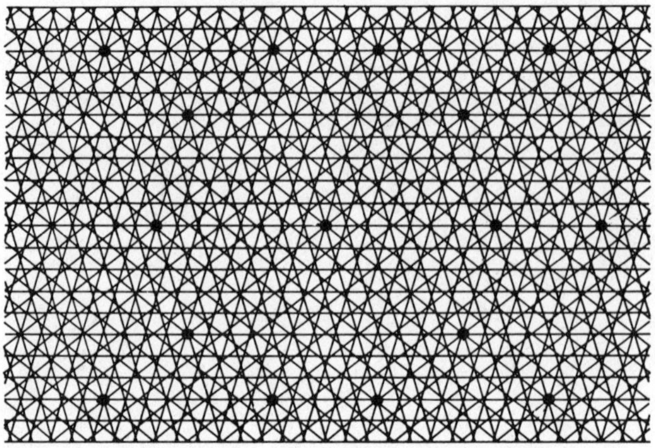

Fig. 3. Ordered structure formed by superposition of 5 waves with wave vectors forming a regular pentagon. The black lines represent the maxima of the individual waves. Atoms may be associated with positions of high density.

to the third order free energy. In addition (Fig. 2e) it is possible to form 6 pairs of regular pentagons out of the 15 pairs of vectors so the icosahedral structure takes advantage of both the third order and the fifth order terms,

$$F_3 = -\frac{\rho^3 v_3}{15\sqrt{15}} \sum_{10\ \text{triangles}} \cos(\phi_i + \phi_j + \phi_k) \tag{7}$$

$$F_5 = -\frac{\rho^5 v_5}{225\sqrt{15}} \sum_{6\ \text{pentagons}} \cos(\phi_i + \phi_j + \phi_k + \phi_\ell + \phi_m).$$

By choosing the parameters v_3 and v_5 suitably, it is possible to stabilize the icosahedral structure relative to the Penrose structures and bcc structures.

Another structure with icosahedral symmetry can be formed by the 6 pairs of vectors $\pm\tau_i$ connecting the center of the icosahedron with the twelve vertices,

$$\rho(\vec{r}) = \sum_{i=1}^{6} \frac{\rho}{\sqrt{6}} \cos(\vec{\tau}_i \cdot \vec{r} + \phi_i) \tag{8}$$

It is not possible to combine subsets of the six τ vectors to form closed polygons. However, there will generally be higher harmonics of the form

$$\rho_{hh}(\vec{r}) = \sum_{\langle ij\rangle} \rho_{hh} \cos((\vec{\tau}_i - \vec{\tau}_j)\cdot\vec{r}) \tag{9}$$

$$= \sum_{i=1}^{15} \rho_{hh} \cos(\vec{q}_i \cdot \vec{r}),$$

where the summation is over nearest neighbor vertices and the q_i's are the 15 edge vectors defined above. The vertex structure thus generally includes the edge-vector waves as higher harmonics[7], and by combining the edge vectors to form triangles and pentagons one can again gain free energy from the third and fifth order terms. These terms will be of sixth and tenth order in the primary order parameter. The length of the 15 q-vectors is 1.05 times the length of the 6 τ vectors; this "almost" degeneracy makes the higher harmonics (9) important since in the absence of higher order terms the two structures are almost equally favorable.

The generalized Penrose structures, defined by (4) with higher harmonics, yield diffraction spots at positions $n_1q_1 + n_2q_2 + n_3q_3 + n_4q_4 + n_5q_5$, so the spots can be indexed by 5 Miller indices $n_1 \ldots n_5$. Since q_5 can be expressed as the sum of the four other vectors, one index is redundant. It is convenient to keep the redundant index to conserve the 5-fold symmetry. A similar situation occurs in conventional crystallographics. A crystal with hexagonal symmetry is usually described in terms of four indices where three indices would be sufficient. The point A in figure 1 may be indexed as (10000). The point B as (11000), and the point C as (11010). The point C could alternatively be indexed as (00-10-1).

The icosahedral vertex structure has diffraction spots at the positions $\sum_i n_i\vec{\tau}_i$, and the diffraction pattern can be labelled by 6 indices. This diffraction pattern coincides with the one measured by Shechtman et al., so we may identify the crystal structure as the vertex type icosahedral (or simple icosahedral, (si)) structure[8].

The icosahedral edge-vector structure has diffraction spots at positions $\sum_{i=1}^{15} n_i\vec{q}_i$ so one might in principle describe the pattern in terms of 15 indices. However, since each edge vectors can be written as sums of two vertex vectors, the diffraction pattern can again be represented by 6 indices, $(n_1 \ldots n_6)$ representing spots at the positions $\sum_i n_i\vec{\tau}_i$. The τ vectors occur in pairs; hence only the peaks for which $\sum_i n_i$ is even are allowed. In anology with cubic crystals one might call such a structure a body centered icosahedral (bci) structure.

How can we characterize the symmetry of the icosahedral structures defined above? There is no translational invariance so they can not be described by any of the usual 230 3d space groups. Consider first a simple one-dimensional incommensurate structure formed by two vectors q_1 and q_2. The ratio between these vectors could be for instance the golden mean $(\sqrt{5}-1)/2$. The most general density can be written

$$\begin{aligned}\rho(x) &= \sum_{i,j} A_{ij}\cos(iq_1x + jq_2x + \phi_{ij}) \\ &= \rho_0(q_1x,\ q_2x).\end{aligned} \qquad (10)$$

Equation (10) defines a two-dimensional function $\rho_0(\theta_1,\theta_2)$,[9] which is periodic in its arguments with periods 2π. Hence, the function represents a two-dimensional periodic lattice. All possible incommensurate structures are characterized by a specific basis to be associated with each unit cell (see Fig. 4), just as for regular crystals. The density in real space is simply the density along the line $\theta_1 = q_1x$, $\theta_2 = q_2x$ whose slope is the golden mean. A change of slope represents a change of the incommensurability. Because of the incommensurability, the line passes infinitely close to all points belonging to the basis. Depending on the choice of basis a variety of situations may occur. Choose first a basis consisting of a line of δ-functions crossing the unit cell diagonally (Fig. 4b). The resulting

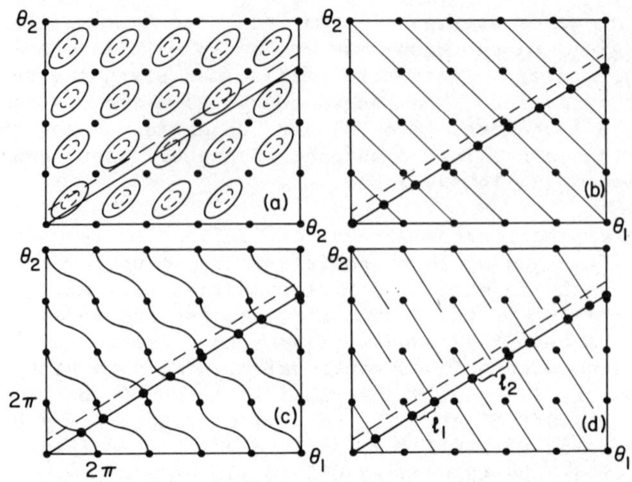

Fig. 4. One-dimensional incommensurate structure as a 2d-crystal. The density along the x direction is the density along the line $(\theta_1,\theta_2) = (q_1x, q_2x)$. The density depends on the basis associated with the Bravais lattice points. a) General case. The basis in b) is a diagonal line of δ functions, in c) a modulated line of δ functions, in d) δ-function lines perpendicular to the real-space line. The broken lines represent the density associated with a phase translation, i.e. a translation of the 2d lattice perpendicular to the line.

structure is an uninteresting equidistant array of δ-function atoms. If the straight line is modulated by a sine function (Fig. 4c), the resulting structure is an array of atoms with a sinusoidally modulated displacement around equidistant positions.

An interesting case arises if the basis is chosen to be a straight line of δ-functions in an off-diagonal direction, for instance perpendicular to the line (q_1x, q_2x) (Fig. 4d). The density function is highly nonanalytic in this case. The resulting structure is a chain of atoms with interatomic distances interchanging in a regular pattern between two different values. This is precisely the "quasicrystal" constructed by Steinhardt and Levine[10], which thus appears as a special (and very unphysical) limit of incommensurate structures. The structure can thus be formed by two "building blocks", or tiles. Note that one can alternatively think of the quasicrystal limit as the projection of 2d lattice points onto the line (q_1x, q_2x). This construction was suggested by Kalugin et al., Ref. 7. However, with this formulation it is difficult to see the connection between the quasicrystal and the more general incommensurate structures. In an experiment, the function $\rho_0(\theta_1,\theta_2)$ can be determined by fitting to the intensities of Bragg peaks at positions $iq_1 + jq_2$, and the actual density can then be found by cutting the 2d space along an appropriate line.

Consider now the generalized Penrose structure (4). The most general density, including all possible higher harmonics, can be written

$$\rho(r) = \rho_0(\vec{q}_1\cdot\vec{r},\vec{q}_2\cdot\vec{r},\vec{q}_3\cdot\vec{r},\vec{q}_4\cdot\vec{r},\vec{q}_5\cdot\vec{r}) \tag{11}$$

$$= \rho_0(\theta_1,\theta_2,\theta_3,\theta_4,\theta_5)$$

The symmetries of the Penrose structure can be expressed in terms of the symmetries of the function $\rho_0(\theta_1,\theta_2,\theta_3,\theta_4,\theta_5)$, defined in the 5d

θ-hyperspace[9]. First, the structure is invariant under the displacements $\theta_i \to \theta_i + 2\pi p_i$, p_i integer. These operations define a 5d Bravais lattice. In addition, the structure is invariant under certain permutation operations forming the point group of the crystal structure. For instance, the 5-fold rotation axis $\theta_i \to \theta_{i+1}$ is a symmetry operation, and the reflections $\theta_1 \to \theta_1$, $\theta_2 \leftrightarrow \theta_5$, $\theta_3 \leftrightarrow \theta_4$ also belongs to the point group. Altogether the symmetry group of the generalized Penrose structures is a 5d space group formed by a 5d Bravais lattice combined with a point group which is isomorphous with C_{5v}, with 24 elements. If the phases ϕ_i in (4) are chosen to be $\pi/2$ the structure has no inversion symmetry and the point group has 12 elements. Equation (11) shows that the actual density in 2d space is the density along a 2d plane in 5d space. In principle, the function ρ_0 can be determined from the intensities of Bragg spots in a diffraction experiment. It is important to note that there is a one-to-one correspondence between symmetry operations of the 5d lattice and the symmetry of the actual quasicrystal.

The si icosahedral structure can be formed by a density $\rho_0(\vec{\tau}_i \cdot \vec{r})$ defined by eq. (8) with higher harmonics. Again this can be viewed as a function of 6 independent variables

$$\begin{aligned} \rho(\vec{r}) &= \rho_0(\vec{\tau}_1 \cdot \vec{r} \ldots \vec{\tau}_6 \cdot \vec{r}) \\ &= \rho_0(\theta_1, \theta_2, \theta_3, \theta_4, \theta_5, \theta_6) \end{aligned} \tag{12}$$

The function $\rho_0(\theta_1 \ldots \theta_6)$, defined in six-dimensional space, has the symmetry of a 6d Bravais lattice since it is invariant under the translations $\theta_i \to \theta_i + 2p_i\pi$. The point group consists of the permutation operations which leaves the function ρ_0 invariant. This group is isomorphous with the icosahedral group and has 120 elements. One of these elements is a 5-fold axis, $\theta_1 \leftrightarrow \theta_1$; $\theta_2 \to \theta_3 \to \theta_4 \to \theta_5 \to \theta_2$. It is this 5-fold axis which is responsible for the 5-fold symmetry observed by Shechtman et al.!

The actual density of the crystal is the density along the 3d hyperplane in 6d space given by Eq. (12). A structure analysis could in principle be done as follows. First, the function ρ_0 can be determined by analyzing the diffraction pattern. The intensities of Bragg peaks should simply be thought of as the Fourier transforms of the 6d function ρ_0. The intensities thus give the necessary information on the basis to be associated with each unit cell of the 6d Bravais lattice. Once ρ_0 has been determined, the density in real 3d space is constructed simply by cutting the 6d space along the 3d plane. Note that since the vectors q_i are incommensurate with the basic 6d lattice, the density along the 3d plane samples every single point within the 6d unit cell, so all the information of the 6d lattice is relevant for the 3d structure.

The choice of basis must be compatible with the space group symmetries defined above. So far we have ignored the fact that the crystal eventually should be formed by an arrangement of Mn and Al atoms. In 6d space the positions of atoms would be given as the intersection of the 3d real space plane with 3d hypersurfaces, just as the positions of atoms in the 1d case (Fig. 2) are given by intersection of a 1d line with 1-dimensional curves or line segments. If the positions of the 3d Al and Mn hypersurfaces are not chosen carefully the resulting arrangement of atoms becomes unphysical. For instance, the surfaces can not intersect each other since this would lead to superposition of two atoms at one point. If the surface is not placed at a symmetry position the point group symmetry requires 120 surfaces in the unit cell, most of them intersecting. However, it is possible to choose a surface which is invariant under all 120 operations. This surface is spanned by three lines perpendicular to the 3d world line in $(11111-\sqrt{5})$ directions. The surface needs not be a 3d plane. A natural arrangement with some Mn atoms surrounded icosahedrally with 12 Al atoms would arise by choosing an Mn hypersurface through the point (000000), and 6 Al surfaces through the points $(\frac{1}{2}00000)$. Thus, the basis of the 6d unit cell consists of one Mn 3d hyper surface and 6 equivalent Al surfaces, and the resulting structure would be stoichiometric, $MnAl_6$. Nonstoichiometric structures can be constructed by "filling" the surfaces partly.

In addition to the operations which leave the structure invariant there are also continuous operations which leave the energy invariant, in analogy with rigid displacements of a 3d crystal. Minimization of the energy F_5 (5) for the Penrose structures leads to one constraint on the phases ϕ_i, so there are four remaining continuous symmetries which do not affect the sum of the phases and thus do not cost energy. Two of these symmetries are rigid displacements of the crystal,

$$u_x = \sum_i \vec{q}_i \cdot \vec{e}_x \phi_i$$

$$u_y = \sum_i \vec{q}_i \cdot \vec{e}_y \phi_i, \tag{13}$$

and the two remaining symmetries represent some internal rearrangement, or phase shifts, of the structure

$$\alpha_x = \sum_i \cos(2V_{ix})\phi_i$$

$$\alpha_y = \sum_i \cos(2V_{iy})\phi_i, \tag{14}$$

where $V_{i\alpha}$ is the angle between $\vec{q}_i$ and the unit vector $\vec{e}_\alpha$. The displacements u_α and the phase shifts α_α transform as two different irreducible representations of the point group, C_{5v}.

Similarly, there are 6 continuous symmetries in the vertex type (si) icosahedral structure, since there are no constraints on the six phases. Three of these symmetries are obviously the usual displacements in x, y, and z directions. The three remaining symmetries describe internal rearrangements. Note that the motions are fundamentally different from the optic displacements in systems with more than one atom per unit cell which cost potential energy. The two types of modes transform as two different three-dimensional irreducible representations of the point group.

The 6 continuous symmetries have a simple interpretation in the 6d super space. The rigid displacements correspond to displacements in 6d space along the 3d hyperplane; the phase displacements correspond to displacements perpendicular to the 3d plane. Consider for simplicity the simple 1d case. Clearly, if the 2d lattice is displaced parallel to the solid line in Fig. 4 representing the actual crystal, the "atoms" or the density is simply shifted along this line. The density along the broken line represents the density associated with a small "phase" shift in the perpendicular direction. In the trivial case (Fig. 4b) nothing happens. In the analytic modulated case (Fig. 4c), a continuous shift of atoms is associated with the phase shift; in the unphysical nonanalytic quasicrystal case (Fig. 4d) only a small number of atoms are affected by the phase shift: the number of atoms which are displaced is proportional to the phase shift.

Once the continuous symmetries have been defined one can construct the elastic energy associated with gradients or strains in these hydrodynamic coordinates[11]. A little group theory reveals that for the Penrose structure one can form two symmetric invariants to second order in the derivatives of the u variables (13). The corresponding two elastic constants are the usual isotropic Lamé coefficients. There are also two elastic coefficients associated with the α strains, and one term which couples the two types of strains, so altogether there are 5 elastic constants. The situation is precisely the same for the icosahedral structures.

In principle, the elementary excitations of the icosahedral structures can be found by diagonalizing the appropriate 6×6 dynamical matrix[11], so there might in principle be 6 hydrodynamic propagating modes with zero gap at q = 0. (There is a possibility that some of the modes may be pinned, in which case there would be a gap at q = 0, or diffusive in which case no well defined excitation will be observable, say in a neutron scattering experiment; see the discussion in Ref. 12). Since there are terms which couple the displacement strains and the phase strains the modes will not be pure phonon or phason modes. Although the u-terms in the elastic energy are isotropic, the coupling to the α-modes causes the phonon modes to have icosahedral, not spherical, symmetry[13].

In conclusion, the icosahedral structures can be described by 6d space group symmetries and their stability can be understood from a simple Landau theory. The task of the crystallographer is to determine the basis associated with the 6d Bravais lattice. The elastic properties and the elementary excitations can be discussed in terms of these symmetries. The ideal icosahedral structures have complete long range order, with sharp Bragg peaks; it is certainly incorrect to consider them to be intermediate between crystals and glasses; they are crystals. It is not difficult to show that the structures are stable with respect to thermal fluctuations, which simply give rise to a reduction of the Bragg peak intensity. The melting transition is first order since the free energy includes third order terms.

In a sense the icosahedral structures are incommensurate structures with only one length scale $2\pi/\tau_i$. The incommensurability comes about because of the relative rotations of the 6 fundamental vectors τ_i: it is not possible to write any of the 6 vectors as sums of rational multiples of the remaining 5 vectors.

REFERENCES

1. D. Shechtman, I. Blech, D. Gratias, and J. W. Cahn, Phys. Rev. Lett. 53:1951 (1984).
2. For a review, see P. Bak, Rep. Prog. Phys. 45:587 (1981).
3. For a popular article on Penrose tilings see M. Gardner, Sci. Am. 236, jan issue, p 110 (1977).
4. P. Bak, Phys. Rev. Lett. 54:1517 (1985); Phys. Rev. B, in press.
5. L. D. Landau and I. E. Lifshitz, "Statistical Physics", 2nd ed., Pergamon, New York, 1968), Chap. XIV.
6. S. Alexander and J. P. McTague, Phys. Rev. Lett. 41:702 (1978).
7. The importance of this fact for the stability of the vertex structure was pointed out by P. Kalugin, A. Kitayev and L. Levitov, ZhETF 41:119 (1985). These authors discovered the 6-dimensional nature of the quasicrystals independently.
8. The fact that the crystal structure of $Mn_{0.14}Al_{0.86}$ is the vertex type (si) structure rather than the edge type (bci) was pointed out to me by D. Nelson.
9. For a description of incommensurate structures in terms of higher dimensional super space groups, see A. Janner and T. Janssen, Phys. Rev. B15:643 (1977).
10. D. Levine and P. J. Steinhardt, Phys. Rev. Lett. 53:2477 (1984).
11. J. D. Axe and P. Bak, Phys. Rev. B26:4963 (1982).
12. H. Brand and P. Bak, Phys. Rev. A27:1062 (1983).
13. P. Kalugin, A. Kitayev and L. Levitov, private communications.

GROWTH OF DOMAINS AND SCALING IN THE LATE STAGES OF PHASE SEPARATION AND DIFFUSION-CONTROLLED ORDERING PHENOMENA

K. Binder and D.W. Heermann

Institut für Physik, Universität Mainz
D-6500 Mainz
Postfach 3980
West Germany

I. INTRODUCTION AND OVERVIEW

These lectures consider the kinetics of phase changes, induced by a sudden change of external thermodynamic parameters. E.g., we treat a system with a second-order transition at a critical temperature T_c (Fig. 1, left part). For $T_0 > T_c$ the system is disordered, while for $T < T_c$ there is an order parameter $\pm\, \psi$ (implying one-component orderings, e.g., an Ising model; later we discuss generalizations). We consider a "quenching experiment" : The system is brought from an initially disordered state at T_0 to a state at T where in equilibrium the system should be ordered[1]. Since no sign of ψ is preferred, the system starts forming locally ordered regions of either sign, separated by domain walls. Due to the unfavorable interface free energy cost, this situation is not thermodynamically stable - there is a driving force to reduce this free energy. Thus the random motion of walls, induced by statistical fluctuations, leads to a growth of a characteristic length L(t) of the ordered regions with the time t after the sudden quench performed at t = 0. Typically one expects $L(t) \propto t^a$ for large t, and the excess internal energy $\Delta E(t) \propto t^{-a'}$: a,a' are the characteristic exponents of interest here. Sometimes even slower growth $[L(t) \propto \ln t]$ might occur, see below.

For this case where space is completely filled with just two kinds of domains actually both phases will form an infinitely connected percolating random structure (for dimensionality d = 3). Hence it is nontrivial to identify from the system configurations what is meant by L(t). We avoid this problem by considering the equal-time structure factor $S(\vec{k},t)$, which is experimentally observable studying the time evolution of scattering intensity under wave vector $\vec{k}$. For systems with perfect long range order we have $[\vec{q} \equiv \vec{k}-\vec{k}_B$, $\vec{k}_B$ describing superstructure Bragg reflections for the considered order]

$$S_T(\vec{k}) \propto \{\psi^2 \delta(\vec{q}) + \chi_\psi/(1+q^2\ \xi_\psi^2)\},\ q < \xi_\psi^{-1}\ ; \qquad (1)$$

here χ_ψ and ξ_ψ denote order-parameter susceptibility and correlation length (for $q \to 0$ the scattering due to short range order has Ornstein-Zernike form). For late stages of a quenching experiment, there comes essentially the same contribution from short range order inside the domains, but the δ-function Bragg peak is smeared out over a finite q-range, which we

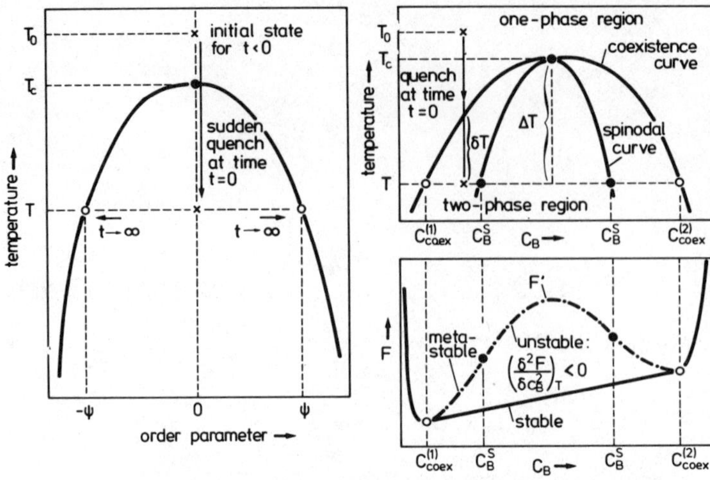

Fig. 1. Order parameter at a 2nd-order transition plotted vs. temperature, indicating changes during a quenching experiment (left part); temperature-concentration phase diagram of a binary mixture (upper right part) and free energy of the mixture plotted vs. concentration (lower right part) at fixed T (schematic). For further explanations cf. text.

take as a measure of $L^{-1}(t)$. Now this part of $S(\vec{k},t)$ is believed to satisfy <u>scaling</u>, i.e.

$$S(\vec{k},t) \propto [L(t)]^d \tilde{S}\{\vec{q}L(t)\}, \quad t \to \infty, \quad q \to 0, \quad qL(t) = \text{finite} \tag{2}$$

This scaling in the late stages of ordering processes is one of the main subjects of these lectures.

Many concepts on this problem have been developed in a somewhat different context : the unmixing kinetics of binary systems AB (Fig. 1, right part). In the phase diagram, the coexistence curve separates the one-phase region (homogeneous mixture stable at all concentrations c_B of B-atoms) from the two-phase region, where equilibrium requires coexistence of macroscopic regions of two phases (with concentrations c^1_{coex}, c^2_{coex} according to the two branches of the coexistence curve). But unlike the ordering problem, the volume fraction $\phi = [c_B - c^1_{coex}]/[c^2_{coex} - c^1_{coex}]$ of the second phase is an important variable: for $\phi \to 0$, we expect well-isolated "islands" of the minority phase (with $c_B = c^2_{coex}$) in a "sea" of the majority phase (with $c = c^1_{coex}$), and similarly $\phi \to 1$ (with the roles of c^1_{coex}, c^2_{coex} interchanged). But for $\phi \approx 1/2$ again interconnected percolating structures of both phases occur (for d = 3).

Sometimes this change of morphology of the phase-separated structures with volume fraction is confused with the spinodal curve c^S_B (Fig. 1), assuming that metastable states (in between the spinodal and the coexistence curve) decay by nucleation and growth of isolated clusters, while unstable states inside the spinodal decaying via spinodal decomposition yield interconnected structures. We emphasize that (i) for systems with short range forces the position of the spinodal is ill-defined, the transition from nucleation to spinodal decomposition is rather gradual[2];

(ii) the late stage morphology is not always related to the early stage morphology : during coarsening an initially interconnected structure may break up[3], but also initially well-isolated droplets may coalesce to an infinite percolating net[4] (which sometimes breaks up again later[4]); (iii) the physical significance of percolation on phase separation is subtle : geometric percolation of clusters of one kind of atoms occurs also in a perfectly random mixture, but does not show up in the structure factor. Thus the right kind of objects to study are not purely "geometrical" clusters but "physical" clusters reflecting order parameter correlations[5-7]. Thus concentrating again on $S(\vec{k},t)$ we find that Eqs. (1,2) still hold $\{\vec{k}_B \equiv 0,\ \psi \equiv (c^{1,2}_{coex} - c^{crit}_B)/c^{crit}_B$, etc.$\}$. But there are also some distinctions : for the mixture, the order parameter is a conserved quantity; for the order-disorder transition it is not. Thus the exponent a in the law $L(t) \propto t^a$ should differ in these two cases, and also the scaling function $\tilde{S}(\vec{x})$ is profoundly different : for non-conserved ψ, $\tilde{S}$ is maximal at $\vec{x} = 0$; for ψ conserved, however, $S(\vec{k} = 0,t)$ is time-independent, it stays at its initial value : the maximum of $\tilde{S}$ hence occurs for finite $\vec{x} \neq 0$. Then $L(t)$ can be defined as the inverse of the wavenumber $k_m(t)$ where $S(\vec{k},t)$ is maximal.

Now questions naturally arise such as : is there a deeper connection between the occurrence of power laws $\{L(t) \propto t^a$ etc.$\}$ and scaling in the present problem, and the occurrence of universal power laws and scaling near a critical point ? For critical phenomena, there is scale invariance of correlations right at T_c : near T_c there is then only one length scale ξ_ψ much larger than all microscopic lengths (lattice spacing, interaction range, etc.); thus, it is plausible that critical exponents etc. do not depend on microscopic details ("universality"). The same argument was conjectured[8-14] for the present case where the length $L(t)$ is so much larger than microscopic lengths, and one may expect the exponents a,a' {and perhaps also $\tilde{S}(x)$} to be universal. Of course, "universality classes" for domain growth are not related to those for critical dynamics[15] - also some features are similarly important for the classification of both problems : to identify which quantities are conserved[9,16-20] and effects due to hydrodynamic modes[8,21,22]; also it matters whether discrete or continuous symmetries are broken[23,24]. But other features relevant for domain growth do not have obvious counterparts in critical phenomena, such as the "sharpness" of the domain wall - systems with "sharp" walls behave differently from those with "soft" walls[25,26]. Moreover, domain growth can be studied for orderings[27] for which the transition at T_c is of first order (e.g., the q-state Potts model[28] for large q), where no critical phenomena occur. For q-fold degenerate orderings q different types of domains compete in their growth - then the above problem that one type of domain may coalesce to a percolating structure becomes less important as q increases. Other reasons why q is important to classify domain growth may also exist[29,30].

In addition, some simulations[16,19,26] indicate that it matters whether in a quench to T = 0 domains still grow unlimitedly, or whether their growth ends in a frozen-in structure[31]. For this case and $T \to 0$ a law $L(t) \propto \ln t$ was proposed[19]. A similarly slow growth has also been suggested when the domain walls get "pinned" due to quenched disorder in the system (e.g. for[32] the random-field Ising model in d = 3. However, preliminary simulations[33] are more consistent with $L(t) \propto t^a$ with $a \approx 1/3$ in this case).

We note another difference between the self-similar structure of critical correlations and those appearing in domain growth or phase separation : critical correlations are isotropic, although there is anisotropy for Ising lattices and short scales. For unmixing, $S(\vec{k},t)$ is isotropic for phase-separating fluids but not for solids (on lattices the interface energy depends on orientation, and the equilibrium shape of a droplet surrounded by the other phase is non-spherical)[34]. Then the scaling function $\tilde{S}\{qL(t)\}$ (which reflects the self-similarity of the droplet size distribution at different times) depends on the direction of $\vec{q}$. This anisotropy is observed experimentally[35] (Fig. 2). These data also show that $\tilde{S}(\vec{x})$ for c_B = const depends on T. But it is possible that this merely reflects a dependence of $\tilde{S}(\vec{x})$ on volume fraction ϕ - such a dependence is expected from several theories[36-38]. What has been discussed so far, rather is a catalogue of open problems than of firmly established results. Firm theoretical evidence only exists for the simplest case, one-component non-conserved ordering (Fig. 2). A variety of theories[11-14,29,39-43], including renormalization group[41], all agree[44] that a = 1/2 independent of temperature : this result is also rather clearly supported by simulations (e.g. Fig. 3b)[45-47]. But we add a caveat : just as for experimental data on unmixing of solids (e.g. Ref. 35) or fluids (e.g. Ref. 48, Fig. 3a), the broad evidence for scaling[16,25-27,45-47,49-53] may be deceptive : often there is much statistical scatter in the data, and only 1 to 2 decades of time range is spanned (and hence one decade or less in L(t)). Thus often scaling is observed, although the exponent a changes with quench depth, ϕ, etc. We suspect that much of these data which "scale" are not yet in the asymptotic regime of late times where scaling actually holds. Theoretical[54] and experimental[55] evidence for this view will be discussed below. On the

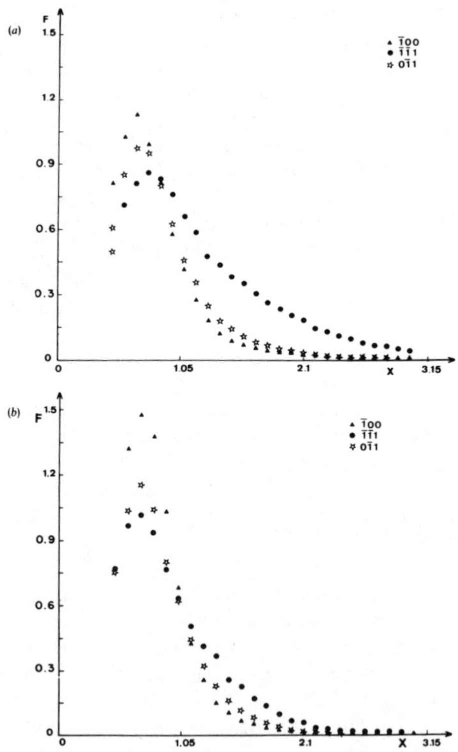

Fig. 2. Scaling function $\tilde{S}(x)$ (denoted by F in the figure) for various directions of $\vec{q}$ in Al-Zn single crystals : (a) Al-12.1 at %Zn aged at 423K (b) Al-12.1 at %Zn aged at 239K. From Ref. 35.

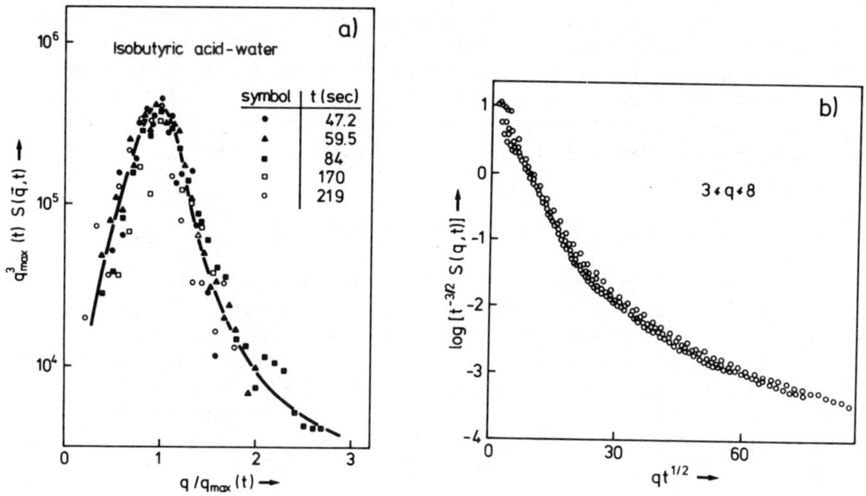

Fig. 3a. Scaling plot for the structure factor of the fluid mixture isobutyric acid-water[48], for a quench through T_c (quench depth 3.4 mK) b) Scaling plot for the structure factor of a nearest neighbour Ising antiferromagnet in zero field, obtained from simulation of a 30 x 30 x 30 sc lattice with periodic boundary conditions (using spin-exchange kinetics). Wave number q is measured in units $(2\pi)/30$, and time in units of attempted exchanges per nearest-neighbour pair. The quench leads from $T = \infty$ to $T = 0.59T_c$. From Ref. 45.

other hand, there is possibly less universality in domain growth than in critical phenomena, and both $\tilde{S}$ and a might depend on parameters such as quench depth, volume fraction (for the case of unmixing), order parameter degeneracy, etc. To explain such a non-universality, Furukawa[56] speculatively suggested a chaotic competition between various growth mechanisms (giving rise to a correction to the exponent similarly as intermittence does in turbulent flow). Indeed, there are many competitive mechanisms [8,9,57]; these lectures give a simplified discussion of some of them. But we still favour the possibility that ultimately one mechanism "wins"; hence crossover between various laws for L(t) may occur, leading to curvature on lnL(t) vs. lnt-plots similar to crossover in critical phenomena. Experimental evidence[58] for this interpretation will be discussed below.

As analytic treatments of these problems are rather incomplete, Monte Carlo simulations have been of crucial importance. In spite of many attempts[49-51,59-62], the exponent a is still unknown for critical quenches ($c_B = c_B^{crit}$) in the Kawasaki[63] spin exchange Ising model of a mixture. But the simulations have been fruitful; while little attention was paid to the original suggestion[8-10] of scaling of the structure factor (Eq. (2)), the observation of this scaling in simulations[49] has stirred up great experimental activity[35,48,51,58,67-71].

These lectures do not intend to fully cover these problems; rather we sketch only some phenomenological ideas on mechanisms leading to scaling in unmixing kinetics. Sec. II reviews the "cluster dynamics" approach[8-10], which generalizes nucleation theory, and permits to treat the Lifshitz-Slyozov[12] evaporation-condensation mechanism and the diffusion and coagulation of clusters[8-10,57] on a common level. Sec. III describes recent simulations of the late stages of phase separation and briefly discusses finite size effects (the latter are very relevant for simulations, and possibly also for experiment !). Sec. IV describes simulations of ordering processes with conserved quantities other than the order parameter[16,17] : the same mechanisms seem to apply there; e.g., the Lifshitz-Slyozov mechanism applies[16] if excess density of the conserved quantity is adsorbed at the walls. Sec. V summarizes our conclusions.

II. SCALING AT LATE STAGES FROM "CLUSTER DYNAMICS"

We consider a binary mixture AB at a lattice, each site being occupied by either A or B. We describe the system configurations in terms of the cluster pattern : a "cluster" labeled by coordinates $(l,\vec{r},s, ..)$ may be defined as a group of l B-atoms, where each atom is nearest neighbour of at least one other atoms of the cluster (Fig. 4a)[9]. Here $\vec{r}$ is the cluster center of gravity, s its surface area, etc. We are mainly interested in a mean cluster concentration $\bar{n}_l(t)$ at time t after the quench (the bar means averaging over all "cluster coordinates" other than l).

In Fig. 4a clusters are defined purely geometrically, as in most numerical work. Since for d = 3 percolation of geometric clusters occurs even at the coexistence curve for $T \gtrsim 0.9\, T_c$ geometric clusters are not physically meaningful in the critical region[5-7] : a modified droplet definition must be used, where (in nearest neighbour Ising models models with exchange J) each bond in a geometrical cluster is "active" with probability[6,7] $p_B = 1-\exp(-2J/K_BT)$ (for all dimensionalities). Only atoms connected by active bonds belong to the same cluster : the

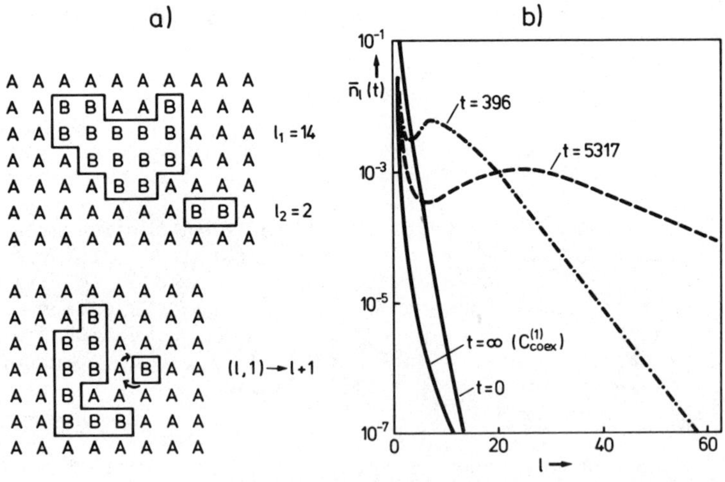

Fig. 4. a) Clusters defined from contours around groups of B-atoms and an example of a "cluster reaction" due to a nearest-neighbour interchange b) Cluster size distribution $\bar{n}_l(t)$ plotted vs. size l at various times after the quench from $T = \infty$ to $T = 0.6T_c$ for $c_B = 0.1$. For further explanations see text. From Ref. 9.

percolating cluster of the minority component is thus destroyed[82]. While this definition of physical clusters allowed work[83] near T_c and deep in the metastable region[8,84], it reduces to the geometrical definition for $T \to 0$. Even for fairly high T droplet size distributions resulting from various definitions are rather similar, Fig. 5[85]. Thus we shall ignore this problem and base our discussion on the geometric cluster picture, although most parts of it are more general.

If atomic exchanges take place, the cluster pattern evolves in time (Fig. 4a). Some exchanges lead to "cluster reactions" (the coordinate l is changing), others lead to "cluster diffusion" only (r is changing). The reactions also involve cluster diffusion (the centre of gravities of the l-cluster and the (l+1)-cluster in Fig. 4a differ a little). Since atomic exchanges occur randomly (governed by some transition probability[63]), the center of gravity displacements are random in direction and thus add up to diffusion (in fluids where hydrodynamic flow of parts of a cluster may occur this picture needs modification[22]).

From the master equation describing the atomic exchanges[63] a master equation describing the time evolution of the clusters follows. We approximate this complicated description, assuming[9] a "closed" kinetic equation for $\bar{n}_l(t)$:

$$\frac{d}{dt}\bar{n}_l(t) = \sum_{l'=1}^{\infty} \frac{\bar{n}_{l+l'}(t)}{n_{l+l'}} W(l,l') - \frac{1}{2}\sum_{l'=1}^{l-1} \frac{\bar{n}_l(t)}{n_l} W(l-l',l')$$

$$+ \frac{1}{2}\sum_{l'=1}^{l-1} \frac{\bar{n}_{l'}(t)}{n_{l'}} \frac{\bar{n}_{l-l'}(t)}{n_{l-l'}} W(l-l',l') - \sum_{l'=1}^{\infty} \frac{\bar{n}_l(t)}{n_l}\frac{\bar{n}_{l'}(t)}{n_{l'}} W(l,l') \quad (3)$$

(for $l = 1$, the second and the third sum on the right hand side must be omitted). The first sum describes the gain of l-clusters due to splitting reactions $l + l' \to (l,l')$: the cluster reaction matrix $W(l,l')$ is the number of such reactions per unit time in thermal equilibrium, and $n_{l+l'}$ the cluster concentration in equilibrium (in the one-phase regions, since the reactions depend on local conditions of atomic arrangement only, and not on the global state of the system). Note $l' = 1$ is simply "evaporation" of single atoms from a cluster. The second sum describes loss of l-clusters due to the splittings $l \to (l-l',l')$. (The factor 1/2 corrects for overcounting of pairs). The third sum describes the gain of l-clusters due to coagulation, $(l', l-l') \to l$, the last sum the loss due to coagulation reactions $(l,l') \to l+l'$. The rate of these processes is related to the inverse ones by detailed balance in equilibrium. Of course, far from equilibrium we may redefine splitting rates ($W_c(l,l') \equiv W(l,l')/n_l n_{l'}$) as the independent parameters of the problem. Eq. (3) is based on many approximations : reactions between 3 or more clusters are omitted; only one cluster coordinate (l) is used : the description is mean field like (Eq. (3) is deterministic, no random force term is included accounting for fluctuations); assumptions such as putting the number of coagulation events $(l,l') \to (l+l')$ proportional to the product $\bar{n}_l\bar{n}_{l'}$ imply the neglect of spatial correlations. Cluster diffusion does not show up in Eq. (3) explicitly : to include it, a similar equation for the local concentration $\bar{n}_l(\vec{r},t)$ is needed,

$$\frac{\partial}{\partial t}\bar{n}_l(\vec{r},t) = D_l \nabla^2 \bar{n}_l(\vec{r},t) + \text{splitting + coagulation terms;} \quad (4)$$

D_l is the cluster diffusivity. Neglecting spatial correlations, integration of Eq. (4) over all space yields Eq. (3), which also contains indirect effects of D_l through the appropriate choice of the coagulation rates $W(l,l')$.

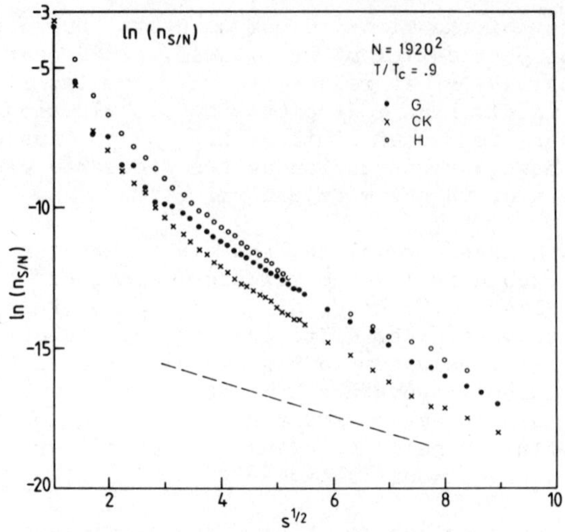

Fig. 5. Droplet size distributions for the nearest-neighbour Ising model at T = 0.9 T_c at the coexistence curve (d = 2), for geometric droplets (G)[7], and the droplet definitions of Coniglio and Klein[6] (CK)x and of Hu (H)o. From Ref. 85.

The approximations of Eq. (3) may seem too drastic : but Eq. (3) generalizes famous approaches to processes far from equilibrium : e.g., neglecting cluster splitting Eq. (3) reduces to the coagulation equation[86-93] which applies for colloid agglomeration[88], aerosol dynamics [87-89], aggregate formation[86-93]; in spite of its "simplicity" and long history[88] it still is an active research area[90-93]. Other limits such as nucleation[1,94,95] and Lifshitz-Slyozov coarsening[72] follow also from[9] Eq. (3), as shown below.

Eq. (3) still satisfies exactly the conservation law for the total concentration c_B. Since each B-atom belongs to some cluster (the B species being minority),

$$c_B = \sum_{l=1}^{\infty} l\, \bar{n}_l(t), \quad \sum_{l=1}^{\infty} l\, \frac{d\bar{n}_l}{dt} = 0 \tag{5}$$

Eq. (3) satisfies Eq. (5), involving the symmetry W(l,l') = W(l',l). We now study large clusters (l >> 1), assuming that $\bar{n}_l(t)$ mainly changes by reactions in small steps (l' << l) so one expands

$$\frac{\bar{n}_{l \pm l'}(t)}{n_{l \pm l'}} = \frac{\bar{n}_l(t)}{n_l} \pm l' \frac{\partial}{\partial l} \frac{\bar{n}_l(t)}{n_l} + \frac{1}{2} l'^2 \frac{\partial^2}{\partial l^2} \frac{\bar{n}_l(t)}{n_l}, \tag{6a}$$

$$W(l-l',l') = W(l,l') - l' \frac{\partial}{\partial l} W; \tag{6b}$$

neglecting deviations from equilibrium for $\bar{n}_{l'}(t)/n_{l'}$ in the nonlinear terms (putting $\bar{n}_{l'}(t)/n_{l'} = 1$ there), one finds the standard nucleation equation[95,9]

$$\frac{\partial \bar{n}_l}{\partial t} = \frac{\partial}{\partial l}\left(R_l\, n_l \frac{\partial}{\partial l} \frac{\bar{n}_l(t)}{n_l}\right),\quad R_l \equiv \frac{1}{n_l} \sum_{l'=1}^{[l/2]} l'^2\, W(l,l') \tag{7}$$

Contributions to the reaction rate R_l with $[l/2] < l' < \infty$ were neglected. Eq. 7 does not satisfy the conservation law, Eq. (5), and is useful for rather early stages only. This failure comes from assuming that the concentrations $\bar{n}_{l'}(t)$ of small clusters, with which the considered l-cluster interacts, and which are those of the supersaturated metastable one-phase state, do not change with time. But actually due to the build-up of large clusters this supersaturation decreases. When the background supersaturation has become so small that further nucleation is suppressed late stage coarsening begins.

As a first approximation for the decrease of supersaturation with time, we assume[9] "local equilibrium" of the small cluster concentrations $\bar{n}_{l'}(t)$ with the time-dependent chemical potential difference $\Delta\mu(t)$ describing the supersaturated background, $\{\bar{n}_{l'}(t)\}\equiv\{n_{l'}(\Delta\mu(t))\}$. Then Eq. (7) is replaced by [$R_l(\Delta\mu)$ is defined below]

$$\partial\bar{n}_l/\partial t = (\partial/\partial l)[R_l(\Delta\mu)\; n_l\;(\Delta\mu)\;(\partial/\partial l)\;\{\bar{n}_l(t)/n_l(\Delta\mu)\}]\,; \tag{8}$$

$\Delta\mu(t)$ is found from imposing the conservation law, Eq. (5). Now the $\Delta\mu$-dependence of $n_l(\Delta\mu)$ should be exponential, $n_l(\Delta\mu) = n_l\exp(-l\Delta\mu)$ {absorbing a factor $(k_BT)^{-1}$ in $\Delta\mu$}. Then detailed belance, $W(l,l')\{n_{l+l'}(\Delta\mu)/n_{l+l'} - (n_l(\Delta\mu)/n_l)\;(n_{l'}(\Delta\mu))/n_{l'})\} = 0$, still holds. Replacing $\bar{n}_{l'}(t)$ by $\bar{n}_{l'}(\Delta\mu)$ in the nonlinear terms of Eq. (3), we get Eq. (8), defining $W_{\Delta\mu}(l,l') \equiv \exp[-(l+l')\Delta\mu]\; W(l,l')$, $R_l(\Delta\mu) = [n_l(\Delta\mu)]^{-1}\sum l'^2 W_{\Delta\mu}(l,l')$. The physical interpretation is, that in the final equilibrium, we have coexistence of infinitely large B-rich domains (concentration c^2_{coex}) with A-rich background (c^1_{coex}), $\Delta\mu\;(t\to\infty) = 0$. For t finite, these domains have still finite sizes {l} and are the l-clusters we consider; they produce strong deviations of $\bar{n}_l(t)$ from the equilibrium distribution n_l at the coexistence curve for sizes l of the order of the typical domain size $[L(t)]^d$. This deviation from equilibrium produces "self-consistently" a supersaturation $\Delta\mu(t)$, with which the small clusters are in equilibrium {for small $\Delta\mu$ one finds $\Delta\mu(t) = -(\sum R_l n_l)^{-1}\sum R_l n_l\;\partial[\bar{n}_l(t)/n_l - 1]/\partial l$}. However there is still an inconsistency : only for small $l' \ll l$ there is quasi-equilibrium $[\bar{n}_{l'}(t) \simeq n_{l'}(\Delta\mu)]$; reaction terms $(l',l) \rightleftarrows l+l'$ with l,l' of the same order in Eq. (3) need be treated differently. Thus we assume l'-clusters in local "equilibrium" only for $l \lesssim l' \lesssim l_c$, l_c being of the order of the "critical cluster size" (in the sense of nucleation theory[1,94,95]) for the supersaturation $\Delta\mu$. For $l' > l_c$, cluster splitting is rather improbable and hence neglected, and sums are replaced by integrals. Instead of Eq. (8) a result combining nucleation-condensation-evaporation-coagulation is found,

$$\frac{\partial \bar{n}_l}{\partial t} = \frac{\partial}{\partial l}\left\{R_l(\Delta\mu)\, n_l(\Delta\mu)\frac{\partial}{\partial l}\frac{\bar{n}_l(t)}{n_l(\Delta\mu)}\right\} + \frac{1}{2}\int_{\ell_c}^{\ell} dl'\,W(l-l',l')\,\frac{\bar{n}_{l'}(t)}{n_{l'}}\,\frac{\bar{n}_{l-l'}(t)}{n_{l-l'}} - \frac{\bar{n}_l(t)}{n_l}\int_{\ell_c}^{\infty} dl'\,W(l,l')\,\frac{\bar{n}_{l'}(t)}{n_{l'}}\,,$$

$$R_l(\Delta\mu) = \frac{1}{n_l(\Delta\mu)}\sum_{\ell'=1}^{\ell_c} l'^2\, W_{\Delta\mu}\;(l,l') \tag{9}$$

In order to check that our central result, Eq. (9), indeed is a valid continuum approximation to Eq. (3), numerical solutions[10] of Eq. (3) were found, with the following assumptions : (i) the initial distribution $\bar{n}_l(0)$ is that of random mixing (= site percolation[96]). (ii) The distribution n_l at coexistence for the d=2 Ising model is well represented[73,74] by the Fisher droplet model[97] for intermediate l. (iii) For the coagulation rate $W_c(l,l')$ a scaling assumption is made,

$$W_c(l,l') = l'^{v}\,\tilde{W}(l/l'),\ v = -3/d,\ \ \tilde{W}(x) = \hat{C}\,[x^{1-2/d} + x^{-(1+1/d)}] \quad (10)$$

Here v and the exponent in $\tilde{W}(x)$ are chosen such that Eq. (3) is consistent [9,10] with both Lifshitz-Slyozov theory[72] and the cluster-diffusion-coagulation mechanism[8,9], see below. Eq. (10) satisfies the symmetry $W_c(l,l') = W_c(l',l)$, but $\tilde{W}(x)$ is rather arbitrary. The constant $\hat{C}^{-1}$ represents the time unit. Fig. 4b then shows the results found from Eq. (3) under these assumptions : $\bar{n}_l(t)$ quickly deviates from the initial monotonous decrease $\{\bar{n}_l(0)\}$, forming a "bimodal" structure : small clusters up to a minimum ($l < l_c$) are soon rather close to the equilibrium distribution $\bar{n}_l(t = \infty)$ of the background; but also a peak appears, corresponding to growing supercritical clusters. As time develops, the peak shifts to larger sizes, so does the minimum, representing the accompanying decrease of background supersaturation. This is precisely the behaviour postulated above.

Returning to Eq. (9), the Lifshitz-Slyozov[72] evaporation-condensation mechanism is found from the approximations : (i) coagulation is neglected. (ii) $R_l(\Delta\mu) \approx R_l(o) = R_l$. (iii) For the limit $l \to \infty$, classical nucleation theory[1,94,95] applies [n_0, f_0 are constants, f_0 being related to the surface tension]

$$n_l = n_0 \exp\,(\Delta\mu l - f_0 l^{1-1/d}) \quad (11)$$

(iv) "Diffusive" terms in cluster size space $[\partial^2 n_l/\partial l^2]$ can be neglected as compared with the "drift" terms of the nucleation equation. This is correct if $(l-l^*)^2 Z^2 \gg 1$, l^* being the critical size resulting from Eq. (11), Z the "Zeldovich factor". Since $Z \propto l^{*-1}$, this condition is valid for the entire distribution $\bar{n}_l$ as $l \to \infty$. Thus

$$\frac{\partial \bar{n}_l}{\partial t} + \frac{\partial}{\partial l}\{\bar{n}_l(t)\,R_l\,[\Delta\mu(t) - f_0\,(1 - \tfrac{1}{d})\,l^{-1/d}]\} = 0 \quad (12)$$

Now we estimate R_l, assuming that the clusters are essentially compact spherical domains, with radius $r_l \propto l^{1/d}$; then the current of B-atoms which impinges on the cluster surface per unit area is $j = D_T(\partial c_B(r)/\partial r)|_{r=r_\ell} \propto r_l^{-1}$ (D_T is a diffusion constant). The total current, and hence R_l, is found by multiplying j with the cluster surface area ($\propto r_l^{d-1}$); so $R_l \propto r_l^{d-2}$, $R_l = R' l^{1-2/d}$. Then we attempt a scaling solution

$$\bar{n}_l(t) = t^y \tilde{n}(l t^{-da}),\ l \to \infty,\ t \to \infty\,. \quad (13)$$

The conservation law, Eq. (5), $(d/dt)\int_0^\infty dl\, l\bar{n}(t) = 0$, implies $y = -2da$. Inserting Eq. (12) into Eq. (5) and integrating by parts, we find $\Delta\mu(t)$ as

$$\Delta\mu(t) = f_0(1-1/d)\int_0^\infty dl\bar{n}_l(t) R_l l^{-1/d} \Big/ \int_0^\infty dl\bar{n}_l(t)\,R_l = c_1 f_0 (1-1/d)\,t^{-a}; \quad (14)$$

{in the last step Eq. (13) was used; c_1 is a constant}. Using Eqs. (13), (14) in Eq. (12) we get $[n' = d\tilde{n}(z)/dz]$

$$\tilde{y n}-da\tilde{n}' = f_0(1-\frac{1}{d})\, t^{1-3a} \frac{d}{dz}[\tilde{n}z^{1-2/d}(z^{-1/d}-c_1)] \tag{15}$$

Hence Eq. (13) is indeed a solution, if $a = 1/3$ {the famous Lifshitz-Slyzov[72] result, see refs. 9,72 for explicit solutions $\tilde{n}(z)$}. The typical linear dimension $L(t)$ follows from

$$[L(t)]^d \equiv \int_0^\infty dl\, l^2 \bar{n}_l(t) \Big/ \int_0^\infty dl\, l \bar{n}_l(t) \propto t^{da} = t^{d/3}. \tag{16}$$

A complementary approximation is to neglect the nucleation-condensation term in Eq. (9) and to try a scaling solution of the resulting coagulation equation. This solution exists if $W_c(l,l') = l'^v \tilde{W}(l/l')$ [Eq. (10)], and one finds [since $l \to \infty$ we put $l_c = 0$][9]

$$da = (1-v)^{-1} \tag{17a}$$

$$y\tilde{n}-daz\tilde{n}' = \int_0^z dz' z'^v \tilde{W}\left(\frac{z-z'}{z'}\right) \tilde{n}(z')\tilde{n}(z-z') - \tilde{n}(z)\int_0^\infty dz' z'^v \tilde{W}\left(\frac{z}{z'}\right)\tilde{n}(z') \tag{17b}$$

Now we estimate the exponent v. <u>If clusters randomly diffuse and coalesce when they touch</u>, the mass Δm which is added per time Δt to a l-cluster is

$$\Delta m/\Delta t = l'(D_{l'}/\Delta^2)\bar{n}_{l'}(t), \tag{18}$$

Δ being the average distance between clusters. Since the volume fraction of the clusters is of order unity, and l and l' are of the same order of magnitude, Δ is of the same order as the cluster linear dimension, $\Delta = Cl'^{1/d}$; the constant C depends on volume fraction. Now we identify

$$\frac{\Delta m}{\Delta t}\bar{n}_l(t) \equiv W(l,l')\frac{\bar{n}_l(t)}{n_l}\frac{\bar{n}_{l'}(t)}{n_{l'}} = l'^v\tilde{W}(l/l')\,\bar{n}_l(t)\,\bar{n}_{l'}(t), \tag{19}$$

and hence

$$l'^v \propto D_{l'} l'^{1-2/d}. \tag{20}$$

Thus to estimate v we need the cluster size dependence of D_l. Originally it was suggested[8] for $T \to 0$ as dominating mechanism for cluster diffusion that atoms evaporate from the surface of a (compact) cluster, diffuse randomly in the matrix, and reimpinge on the surface of the same cluster. Since the average distance travelled by such an atom is finite (independent of l^9), a shift $\Delta \vec{x} \propto 1/l$ of the cluster center of gravity results. Now D_l is found multiplying the number of such processes per unit time (proportional to the surface area $l^{1-1/d}$) with $(\Delta\vec{x})^2$: Thus[8] $D_l \propto l^{-1-1/d}$. From Eq. (20) $v = -3/d$ results [Eq. (10)]. But a more careful study of cluster diffusion reveals at least four competing mechanisms[57] (Fig. 6). For the diffusion of "holes" inside a cluster the rate is proportional to the volume instead of the surface : Hence $D_l \propto l^{-1}$, $v = -2/d$ (Fig. 6c). For the evaporation condensation mechanism of Fig. 6D, the average distance between the two atoms is of the order of the cluster linear dimension, $l^{1/d}$; hence $\Delta\vec{x} \propto l^{1/d}/l$, $D_l \propto l^{1-1/d}(\Delta\vec{x})^2 \propto l^{-1+1/d}$, $v = -1/d$. While near T_c the last mechanism should dominate, the situation is less clear for low T : for a square lattice, mechanism B) involves a temperature dependence e^{-2J/k_BT}, in contrast to e^{-8J/k_BT} for C) and D) (density of holes and isolated atoms). While D always dominates for $l \to \infty$, at intermediate size B) may be more important, and crossover effects might occur. Simulations confirm this picture[57] (Fig. 7). Using these estimates of v in Eq. (17a) yields[8] $a = 1/(d+3)$ (case B), $a = 1/(d+2)$ (case C), $a = 1/(d+1)$ (case D). For $d = 3$, the growth rates resulting from all

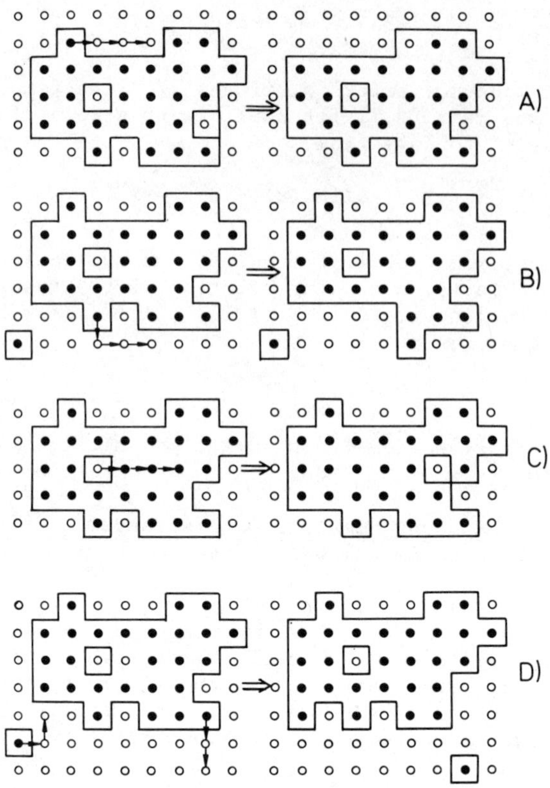

Fig. 6. Schematic illustration of four mechanisms of cluster diffusion in a square lattice gas : A) surface atom moves with no energy cost B) Evaporation from the surface and reimpingement C) Diffusion of holes inside a cluster D) Evaporation of one atom from the surface and condensation of an atom from the surrounding matrix. From Ref. 57.

these mechanisms are smaller than the Lifshitz-Slyozov rate (a = 1/3) which hence should be dominating at late times. For d = 2, however, mechanism D) is competitive with the Lifshitz-Slyozov mechanism. This situation is similar to the case of fluids, where the cluster diffusivity resulting from Stokes law is[8] $D_l \propto r_l^{-1} \propto l^{-1/d}$, $v = 1 - 3/d = 0$ for d = 3, and hence a = 1/d = 1/3 coincides with the Lifshitz-Slyzov result. Note, however, that growth via direct hydrodynamic flow (possible at least for interconnected structures) leads to a much quicker rate[22], a = 1. For solids in d = 3, the situation is complicated since the Lifshitz-Slyozov mechanism similarly to case D) needs single B atoms in the A-matrix; in an Ising model their concentration would decrease $\propto \exp(-12J/k_BT)$, while the rate of case B) still can involve events $\propto \exp(-2J/k_BT)$. Thus one expects[9] the cluster-diffusion-coagulation mechanism to dominate at intermediate times, and then crossover to Lifshitz-Slyozov-laws is observed. Experimental data for Fe-Cr alloys[58] have been interpreted as evidence for this idea, Fig. 8. Even aside from the question whether the data really can be correlated with specific mechanisms, we emphasize that the slope

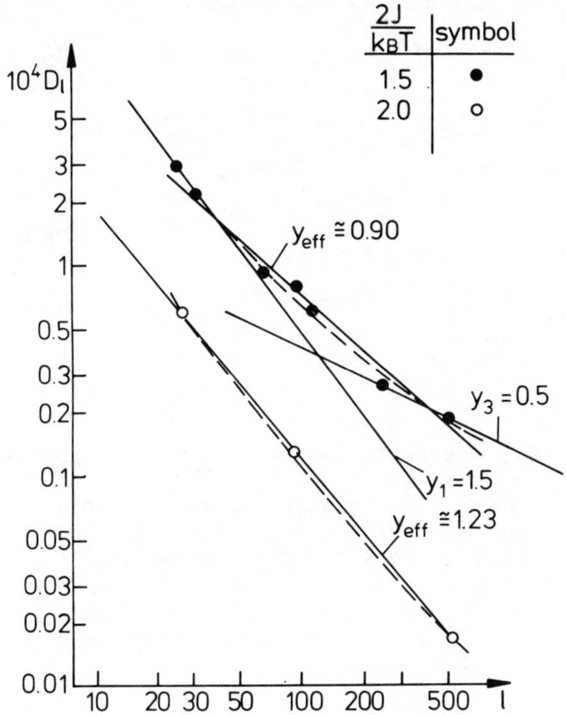

Fig. 7. Log-log plot of cluster diffusivity D_l vs. l (time unit is one attempted exchange/site) for an Ising square lattice. Straight lines indicate various estimates for the exponent of the cluster diffusivity (denoted y_1 for case B) and y_3 for case D)); due to crossover effects the effective exponent y_{eff} lies in between y_1, y_3 and is strongly temperature dependent. From Ref. 57.

of the curve $\ln q_1(t)$ vs. $\ln t$ in Fig. 8 increases with time : if the exponents a observed in such plots were artefacts of a too short observation time interval, as suggested in Ref. 19, claiming that the correct law is $L(t) \propto \ln t$, the slope on the log-lot plot should rather decrease with time.

Of course, it is also a nontrivial matter to relate the scaled droplet size distribution (obtained from either Eq. (15) or Eq. (17b), for instance) to the structure factor $S(\vec{k},t)$). Binder et al[10] suggested to express the concentration correlation function in terms of the conditional probability $g_1(\vec{r})$, that the site $\vec{r}+\vec{r}'$ is occupied by a B-atom if $\vec{r}'$ is within a 1-cluster as

$$\langle c(\vec{r}')c(\vec{r}'+\vec{r})\rangle = \sum_{\ell=1}^{\infty} l\bar{n}_1(t)g_1(\vec{r}) \tag{21}$$

Assuming $g_1(\vec{r})=1$ for $r < r_1(1)$, $g_1(\vec{r})=0$ for $r_1(1) < r < r_2(1)$, and $g_1(\vec{r})= c_B$ for $r > r_2(1)$, where $r_1(1)$ and $r_2(1)$ are fixed such that $S(\vec{k}=0,t)=0$, one obtains a qualitatively reasonable time evolution of the structure factor, Fig. 9 (using for $\bar{n}_1(t)$ the numerical solution of Eq.(3) as

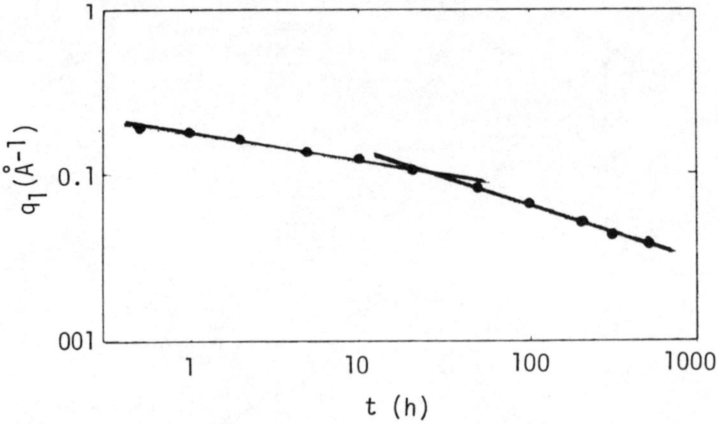

Fig. 8. Log-log plot of $q_1(t) = \sum_q qS(q,t)/\sum_q S(q,t)$ vs. time for Fe Cr 34 at% aged 500°C. The straight lines correspond to the exponent a = 0.17 at early times and a = 0.33 at late ones. From Ref. 58.

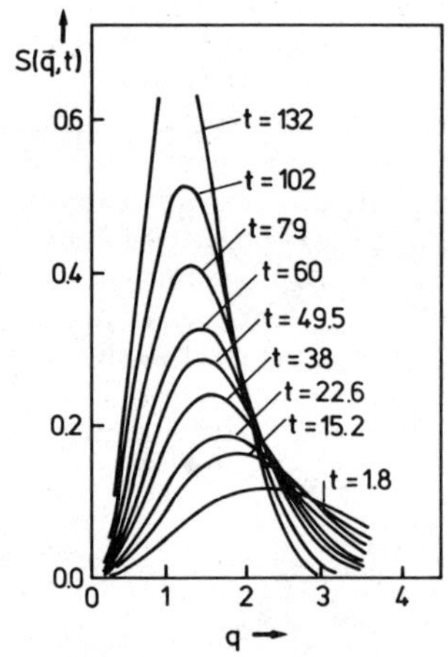

Fig. 9. Structure factor $S(\vec{q},t)$ plotted versus wave-number q for various times after the quench from infinite temperature to $T = 0.6\ T_c$ for a three dimensional Ising model at $c_B = 0.1$, using for $\bar{n}_l(0)$ the site-percolation cluster distribution, for the equilibrium concentration n_l at coexistence the Fisher model[97], Eqs. (3), (10), (21) and Fourier-transforming Eq. (21). Both q and t contain arbitrary scale factors. From Ref. 10.

shown in Fig. 4b, for instance). An improved treatment of this problem along similar lines is found in Ref. 36. In Ref. 10 it was found that over the range of times shown in Fig. 9 one observed for the maximum position $q_m(t) \propto t^{-0.16}$, while the maximum height increases as $S(q_m) \propto t^{0.7}$ (rather than the behaviour $S(q_m) \propto q_m^{-d} \propto t^{0.48}$ which scaling [Eq.(2)] implies). The curvature seen on corresponding log-log plots indicates that the asymptotic regime of times has not been reached (due to numerical difficulties, the solution of Eq. (3) can be found only for rather early times). In fact, numerical solutions of a simplified version of Eq. (8) indicate[54] that one has to follow L(t) over many decades in time, before the asymptotic Lifshitz-Slyozov behaviour is reached.

Fig. 9 resembles both experimental data on spinodal decomposition[35,48,51,58,64-71,98] and results of an approximate non-linear theory[99], where one derives a "closed" kinetic equation directly for the structure factor. In suitable scaled units this equation is[99]

$$\frac{d}{dt} S(\vec{k},t) = -k^2 \{[A(t)+k^2]S(k,t)-1\}. \tag{22}$$

Here the parameter A(t) is determined self-consistently from $S(\vec{k},t)$. Eq. (22) is thought to be a valid description for early times[99]. If we would disregard this caveat and apply Eq. (22) at late times, seeking a scaling solution of the type of Eq. (2), we find[9] for $A(t) \propto t^{a_1}$ that a scaling solution $S(\vec{k},t)-S(\vec{k},\infty) = k_m^{-d}(t)\,\tilde{S}(k/k_m(t))$ indeed exists with $k_m(t) \propto t^{-1/4}$, and $a_1 = -1/2$, but $S(\vec{k},\infty) \propto k^{-2}$. Thus this theory does not yield the Ornstein-Zernike correlation function in Eq. (1), it rather yields an $S(\vec{k},\infty)$ characteristic for a critical state. While the growth law $L(t) \propto t^{1/4}$ is probably incorrect for unmixing, Eq. (22) holds exactly[23] for the n-vector model with conserved order parameter in the limit $n \to \infty$. Here $S(\vec{k},\infty)$ reflects the Goldstone mode singularities at the coexistence curve, and $\tilde{S}$ has been calculated explicitly[23], confirming the scaling analysis of Eq. (22) presented in Refs. 9,10.

We conclude this section by pointing towards two major weaknesses of the "cluster dynamics" treatment sketched in Eqs. (3) - (21) : (i) While it perhaps is correct to treat the rather macroscopic domains of the Lifshitz-Slyozov theory[72] as compact objects, the (smaller) clusters considered in the cluster-diffusion-coagulation mechanism are probably in between compact and ramified (fractal[100]) shapes. This might effect the exponent v and the cluster diffusivity D_1 considerably. Coagulation of fractal clusters has recently been considered in the context of aggreation kinetics[92,93]. (ii) When one quenches to T = 0, for small enough c_B one ends up with a frozen-in cluster size distribution : in the initial state, one has the cluster size distribution of site percolation, with well-separated clusters. Only monomers are freely mobile at T = 0; these monomers diffuse around until they stick at larger clusters, and further evolution stops. For large enough c_B, however, one freezes in a percolating structure[37] in quenches to T = 0; hence there must exist a percolation concentration c_B, where in the final state of quenches to T = 0 a percolating cluster starts to appear. Of course, this percolation is not affected by choosing the other cluster definitions of Refs. 6,7, which at T = 0 coincide with the geometrical one. We expect that this percolation transition should have a pronounced effect on the "cluster dynamics" at low temperatures.

III. MONTE CARLO SIMULATION OF COARSENING

There have been numerous attempts to simulate later stages of phase separation[49-51,59-62,80] and analyse them in terms of laws $L(t) \propto t^a$,

$k_m(t) \propto t^{-a}$. Most of this work was based on studying just one lattice size (30 x 30 x 30 or 50 x 50 x 50 for d = 3). Since this work is reviewed in Refs. 1,50 we shall rather describe only a recent study[101], where effects due to finite lattice of linear dimension $N^{1/3}$ are also considered.

Fig. 10 shows the time evolution of $[L(t)]^d$ obtained in this work[101] (plots of the size of the maximum cluster [not shown] look similar). One observes S-shaped curves : the slope first increases and reaches a value close to unity for times of order 5.10^3 attempted exchanges/site, and then the curves flatten. This flattening is the more pronounced the smaller $N^{1/3}$ and hence a finite-size effect - though the cluster radius $r_1 \propto l_{max}^{1/3}$ is very small compared to $N^{1/3}$ throughout !

A simple possibility to account for finite size in the cluster-dynamic treatment is to replace the upper limit of the sums or integrals in Eq. (3), (5), (9), (14), (16), (17) by a maximal cluster size l_{max} rather than infinity. Instead of Eqs. (13), (14), (16) this leads to

$$\bar{n}_1^{(N)}(t) = t^y \tilde{n}(lt^{-da}, 1/l_{max}), \quad \Delta\mu(t) \underset{t\to\infty}{\rightarrow} \Delta\mu(\infty) = f_o(1-1/d)l_{max}^{-1/d} \tag{23a}$$

$$[L(t)]^d = \int_o^{l_{max}} dl l^2 \, \bar{n}_1^{(L)}(t) / \int_o^{l_{max}} dl l n_1^{(L)}(t) = t^{da}\tilde{f}(l_{max} t^{-da}) \tag{23b}$$

where $\tilde{n}$ rather by the ordinary differential equation, Eq. (15) now is given by a partial differential equation[101]. No attempt to solve it and to obtain the scaling function $\tilde{f}$ has as yet been made. Assuming $l_{max} \propto N$, Eq. (23) suggests to replot the data of Fig. 10 as $[L(t)]^d/N$ versus N/t, since for da = 1 all curves should converge towards a single curve, the scaling function. Fig. 11 shows that only the data for the largest times and largest

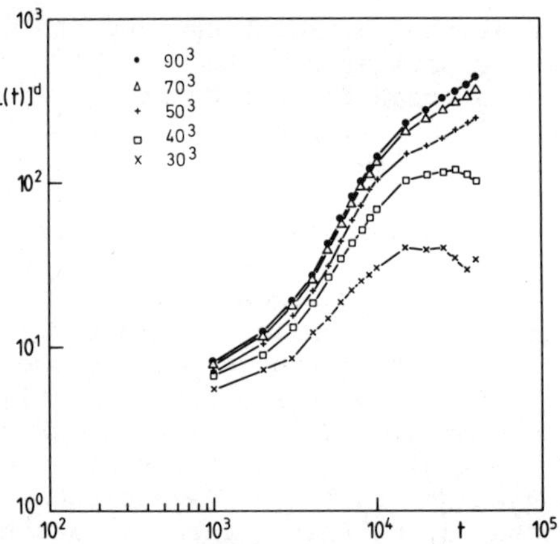

Fig. 10. Log-log plot of the mean cluster size $[L(t)]^d = \Sigma l^2 n_1/\Sigma \, l n_1$, for nearest neighbour Ising models at d = 3, quenched at c_B = 0.048 from infinite temperature to T = 0.59 T_c. Various lattice sizes are shown as indicated. From Ref. 101.

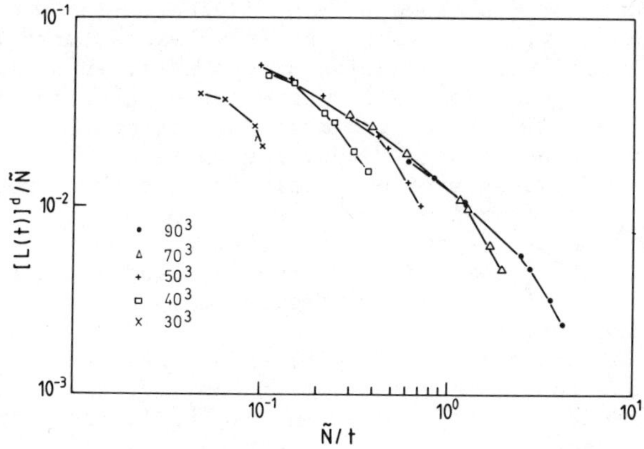

Fig. 11. Log-log plot of average cluster volume $[L(t)]^d/\tilde{N}$ versus $\tilde{N}/t$, using the data of the calculation yielding Fig. 10. $\tilde{N} = [c_B - c^1_{coex}/(c^2_{coex} - c^1_{coex})]N$. From Ref. 101.

sizes are possibly consistent with the scaling hypotheses of Eq. (23b). Of course, the treatment of finite-size effects via a cut-off is only very approximate : another effect is that in the late stages the continuum approximation breaks down - there is only a small finite number of large clusters undergoing coarsening in a finite system, and ultimately just the single largest cluster survives[102]. A somewhat qualitative discussion[102] of this phenomena leads also to a prediction for L(t) consistent with Fig. 10. In any case, we emphasize that the asymptotic Lifshitz-Slyozov[72] behaviour only is reached for extremely late times typically, and also finite size effects are strong even for $L(t) \propto N^{1/d}/10^2$ already - since in real systems grain boundaries, dislocation nets etc. may lead to a similar effective size of sufficiently ideal regions as in Fig. 10, we suggest that similar effect may play a role in experiments, too[103]. As a result, more work is needed to clarify the situation - both in theory, experiment, and simulation.

IV. ORDERING KINETICS WITH CONSERVED DENSITIES

Now, we briefly consider ordering kinetics, as in Fig. 1, where the order parameter is not conserved, while other densities are conserved. For a one-component order parameter[1], such a conservation law has not much effect and still a = 1/2, while for multi-component orderings the situation is different[16,17]. For a q-state Potts model[28] with q = 3 and q = 6 it was found[17] that a $\approx$ 0.2, in simulations of 90 x 90 square lattices where quenches from infinite temperatures to $T = 0.6\ T_c$ (and also $T = 0.8\ T_c$ for q = 3) were performed, using a generalization of the Kawasaki[63] spin-exchange dynamics to the Potts model. By direct observations of the time evolution of the domain pattern, evidence for growth due to domain diffusion and coalescence was obtained[17]. It is possible that also in this case a Lifshitz-Slyzov[72] mechanism might take over at very late times, not accessible in this simulation.

Another study[16] concerned a lattice gas with repulsive interaction between nearest and next nearest neighbours (of equal strength) on the square lattice. For density 1/2, the system orders by forming arrays of alternating full and empty rows at the lattice. This ordering leads to 4 distinct domains; as shown in Fig. 12 there also occur 4 important types of walls : a "heavy" wall with a positive excess mass in the (2 x 2) unit cell, a light wall with a negative excess mass, an antiphase boundary with no excess mass, and a "45°-wall" where the sign of the excess mass alternates along the wall. Since the mass in this model was taken as conserved quantity (spin-exchange[63] dynamics where $S_i = \pm 1$), it is clear that the diffusion of excess mass necessary for growth of domains bounded by the first two types of wall may be a crucial mechanism. Inspection of domains and wall patterns[16] indeed shows, that all four types of walls shown in Fig. 12 persist during the ordering kinetics, and apart from pronounced "roughness" of these walls (or very irregular domain shapes, respectively) the situation seems similar to the T = 0 case considered in Fig. 12. Thus the observation that the domain size L(t) and excess energy ΔE (t) relax with exponents close to 1/3 (Fig. 13) was taken[16] as evidence for the Lifshitz-Slyozov mechanism (Fig. 13). Surprisingly, there is neither a clear-cut lattice size effect in the present case, in contrast to Figs. 10, 11 (Additional data for lattice sizes 120^2 and 800^2 shown in Ref. 16 strengthen this conclusion), nor is there evidence for a domain diffusion and coagulation mechanism as in the other cases described so far. Also, if the same model without conservation law is studied, one simply finds[16] $L(t) \propto 1/\Delta E\ (t) \propto t^{1/2}$, independent of quench temperature. As a further argument, that diffusion of excess mass in this model is crucial, the relaxation of the excess mass $\overline{\Delta m}^2(t)/\overline{\Delta m}^2(0)$ in finite subblocks of sizes ranging from 4 x 4 to 20 x 20 was studied (Fig. 14). It is seen that this excess mass indeed relaxes with a rate consistent with the domain growth law for L(t). At a temperature $T \approx 0.36\ T_c$, however, a slower growth rate $L(t) \propto t^{0.20}$ was observed, this was interpreted as a crossover effect, since for T = 0 the model also exhibited[16] an initially similarly slow growth, which then stops completely at a frozen-in domain pattern. On the other hand, Furukawa[56] takes all these data as evidence for realizations of the chaotic competition between various growth mechanism proposed by him. Thus, more complete studies of the two models of the present section, particularly extending to later times, would be very desirable. In spite of these difficulties, one observes a rather nice scaling behaviour of the structure factor, in accord with Eq. (2), irrespective of these problems about the exponent in the law $L(t) \propto t^a$ (Fig. 15).

V. CONCLUSIONS

In these lectures the basic aspects of some mechanisms which lead to scaling behaviour in the late stages of phase separation and ordering were reviewed, emphasizing particularly the cluster-diffusion-coagulation mechanism and the Lifshitz-Slyozov evaporation-condensation mechanism. Although there is some evidence for the validity of these concepts from both simulation and experiment, there are many challenging open questions : what is the precise relation between geometric structures, expressed in the cluster pattern, and the scattering function ? What is the proper characterization of clusters, and under which circumstances do we have fractal rather than compact objects ? Is the growth law for percolating structures the same as for well-separated objects ? How do we properly model the competition between various mechanisms ? Etc. On the other hand, these mechanisms have broad applications in various branches of physics and chemistry, and putting these very phenomenological concepts on a firm theoretical analytical basis remains an important future task - as well as more systematic and accurate experiments.

Acknowledgements : One of us (K.B.) is indebted to Dr. A. Sadiq for a very fruitful collaboration on research described in Sec. IV (Ref. 16). We are also grateful to Sonderforschungsbereich 41 der Deutschen Forschungsgemeinschaft for partial support.

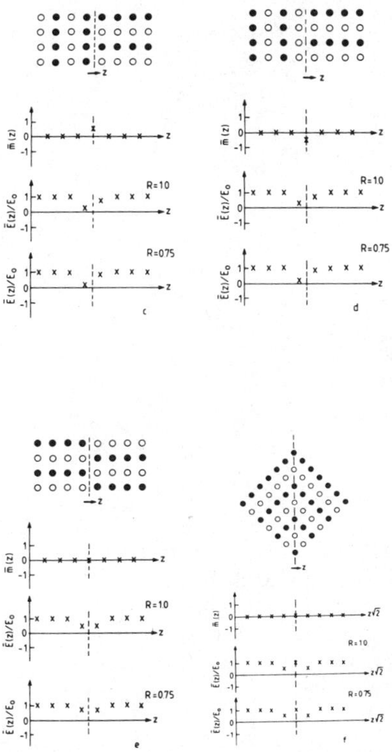

Fig. 12. Different types of walls between the four ordered domains of the (2 x 1) structure at the square lattice (T = 0, density $\rho = (1+\bar{m})/2 = 1/2$). (c) heavy wall, d) light wall, e) antiphase boundary, f) 45°-wall. Profiles $\bar{m}(z)$ and E(z) of the magnetization and energy of the associated Ising magnet, averaged along the wall, are shown in the direction z perpendicular to the wall. The ratio between next-nearest and nearest-neighbour interaction is denoted by R. From Ref.16.

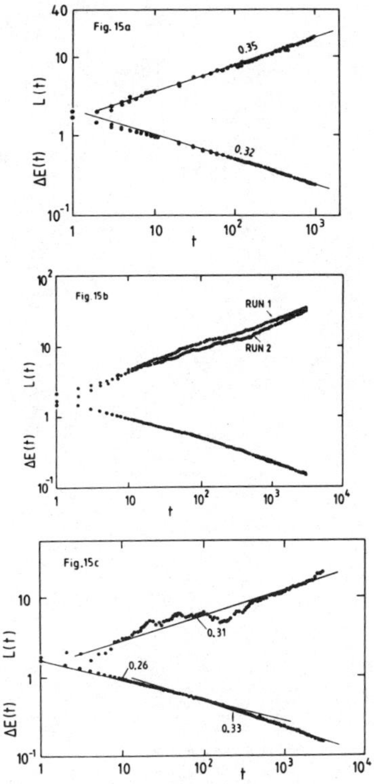

Fig. 13. Log-log plot of $\Delta E(t)$ L(t) vs. time, for the temperature T $\approx$ 0.63 T_c and $\rho = 1/2$, and the lattice sizes 80 x 80 (a) (two averages of 100 runs each), 200 x 200 (two averages of four runs each), (b) and 400 x 400 (c) (one average of four runs). Numbers at the straight lines indicate exponents a, a'. From Ref. 16.

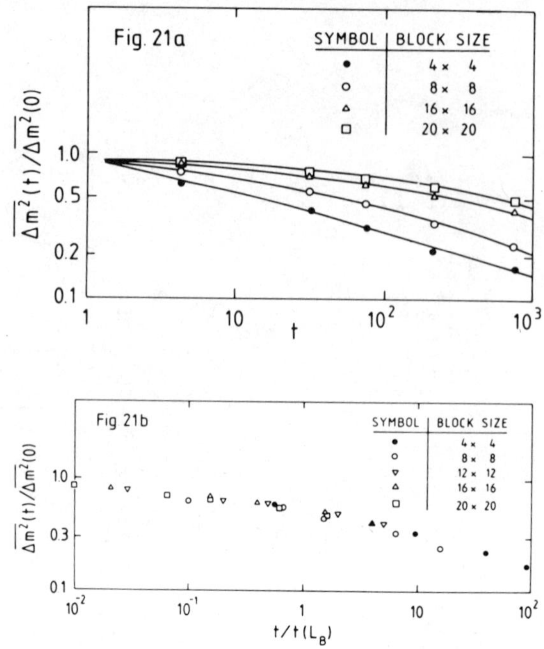

Fig. 14. Time evolution of the mean square density fluctuation (a) and replotted in scaled form (b), for block linear dimensions L_B from L_B = 4 to L_B = 20 as indicated. The characteristic time $t(L_B)$ is defined implicitly with the help of the data of Fig. 13 by $L(t) = L_B$. From Ref. 16.

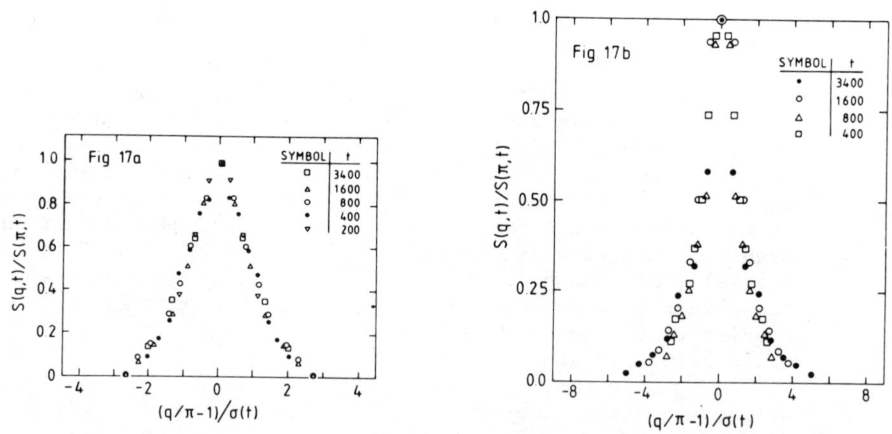

Fig. 15. Structure factor $S(\vec{k},t)$ of the Ising square lattice gas with repulsive interactions between nearest and next-nearest neighbours plotted in scaled form (maximum occurs at Bragg position $k_B = \pi$, lattice spacing = unity, half-width of the peaks is denoted by $\sigma(t)$), for $\rho = 1/2$ and quenches from $T = \infty$ to $T \cong 0.63T_c$ (a) and $T \cong 0.36\ T_c$ (b). Time is measured in attempted exchanges per pair. From Ref. 16.

REFERENCES

1. For an extensive recent review of this area, see J. D. Gunton, M. San Miguel and P. S. Sahni, in "Phase Transitions and Critical Phenomena", Vol. 8, C. Domb and J. L. Lebowitz, eds., Academic Press, New York (1983) p. 267.
2. For more detailed discussions of this point and earlier literature, see e.g. K. Binder, Phys. Rev. A29:341 (1984) and in "Condensed Matter Research Using Neutrons", S. W. Lovesey and R. Scherm, eds., Plenum, New York (1984) p. 1.
3. E.g., this has been seen in Monte-Carlo simulations (K. Binder, Solid State Comm. 34:191 (1980)) and experiments on phase-separating fluid mixtures (W. I. Goldburg, private communication).
4. D. W. Heermann, Z. Phys. B55:309 (1984).
5. K. Binder, Ann. Phys. 98:390 (1976); R. Kretschmer, K. Binder and D. Stauffer, J. Stat. Phys. 15:267 (1976). See also A. D. Bruce and D. J. Wallace, J. Phys. A16:1721 (1983) and A. D. Bruce, preprints.
6. A. Coniglio and W. Klein, J. Phys. A13:2775 (1980).
7. C.-K. Hu, Phys. Rev. B29:5103 (1984).
8. K. Binder and D. Stauffer, Phys. Rev. Lett. 33:1006 (1974).
9. K. Binder, Phys. Rev. B15:4425 (1977).
10. K. Binder, C. Billotet and P. Mirold, Z. Phys. B30:183 (1978); P. Mirold and K. Binder, Acta met. 25:1435 (1977).
11. C. Billotet and K. Binder, Z. Phys. B32: 195 (1979).
12. K. Binder, in "Stochastic Nonlinear Systems in Physics, Chemistry and Biology", L. Arnold and R. Lefever, eds., Springer, Berlin (1981) p. 62.
13. H. Furukawa, Progr. Theor. 59:1072 (1978); Phys. Lett. 66A:60 (1978); 97A:346 (1983).
14. H. Furukawa, Phys. Rev. Lett. 43:136 (1979); Phys. Rev. A23:1535 (1981); A28:1717 (1983).
15. P. C. Hohenberg and B. I. Halperin, Rev. Mod. Phys. 49:435 (1977).
16. A. Sadiq and K. Binder, J. Statis. Phys. 35:617 (1984); Phys. Rev. Lett. 51:674 (1983).
17. G. S. Grest and P. S. Sahni, Phys. Rev. B30:226 (1984).
18. F. G. Mazenko and O. T. Valls, Phys. Rev. Lett. 51:2044 (1983).
19. G. F. Mazenko, O. T. Valls and F. C. Zhang, to be published.
20. J. D. Gunton, J. Stat. Phys. 34:1019 (1984).
21. K. Kawasaki and T. Ohta, Progr. Theor. Phys. 59:361 (1978).
22. E. Siggia, Phys. Rev. A20:595 (1979).
23. G. F. Mazenko and M. Zannetti, Phys. Rev. Lett. 53:2106 (1984).
24. J. K. Bhattacharjee, P. Meakin and D. J. Scalapino, Phys. Rev. A30:1026 (1984).
25. G. S. Grest, D. J. Srolovitz and M. P. Anderson, Phys. Rev. Lett. 52:1321 (1984).
26. O. G. Mouritsen, Phys. Rev. B (1985); B28:3150 (1983) and preprint.
27. P. S. Sahni, G. S. Grest, M. P. Anderson and D. J. Srolovitz, Phys. Rev. Lett. 50:263 (1983); Phys. Rev. B28:2705 (1983); K. Kaski, J. Nierinen and J. D. Gunton, preprint; see also D. J. Srolovitz, M. P. Anderson, P. S. Sahni and G. S. Grest, Acta met. 32:783 (1984).
28. F. Y. Wu, Rev. Mod. Phys. 54:235 (1982).
29. I. M. Lifshitz, Sov. Phys. JETP 15:939 (1962).
30. S. A. Safran, Phys. Rev. Lett. 46:1581 (1981).
31. For the Ising model of a binary mixture at the critical concentration $c_B^{crit} = 1/2$, such frozen-in structures were first studied by P. Meakin and S. Reich, Phys. Lett. A92, 247 (1982) and by A. Levy, S. Reich and P. Meakin, Phys. Lett. 78A (1982).
32. J. Villain, Phys. Rev. Lett. 52:1543 (1984); G. Grinstein and J. F. Fernandex, Phys. Rev. B29:389 (1984).
33. D. Stauffer, C. Hartzstein, K. Binder and A. Aharony, Z. Phys. B55:325 (1984). However, E. Pytte and J. F. Fernandez (Phys. Rev. B31:616 (1985)) do find $L(t) \propto$ Int.

34. C. Rottman and M. Wortis, Phys. Rev. B24:6274 (1981); R. K. P. Zia and J. E. Avron, Phys. Rev. 1352:2042 (1982); J. E. Avron, L. S. Schulman, H. van Beijeren and R. K. P. Zia, J. Phys. A15:L81 (1982); R. K. P. Zia, preprint.
35. J. R. Simon, P. Guyot and A. Fhilarducci de Salva, Phil. Mag. A49:151 (1984).
36. P. A. Rikvold and J. D. Gunton, Phys. Rev. Lett. 49 (1982).
37. K. Tokuyama and K. Kawasaki, Physica 123A:386 (1984); T. Ohta, preprint.
38. P. W. Vvorhees and M. E. Glicksman, Acta met.(1984); J. A. Marqusee and J. Ross, J. Chem. Phys. 80:536 (1984).
39. S. W. Allen and J. W. Cahn, Acta met. 27:1017 (1979); 27:1085 (1979).
40. K. Kawasaki, M. C. Yalabik and J. D. Gunton, Phys. Rev. A17:455 (1978).
41. T. Ohta, D. Gamow and K. Kawasaki, Phys. Rev. Lett. 49:1223 (1982).
42. I. Ohta, ref. 37; M. Grant and J. D. Gunton, Phys. Rev. B38: 5496 (1983); K. Kawasaki and T. Ohta, Progr. Theor. Phys. 67:142 (1982).
43. G. F. Mazenko, Phys. Rev. B26:5103 (1982); G. F. Mazenko and O. T. Valls, Phys. Rev. B27:6811 (1983); G. F. Mazenko and O. T. Valls, Phys. Rev. B30:6732 (1984); F. C. Chang, O. T. Valls and G. F. Mazenko, preprint.
44. Note that this statement refers to domain growth in a system with a second-order transition. If one quenches to a metastable regime underneath a first-order transition, one rather finds $L(t) \propto t$ (S. K. Chan, J. Chem. Phys. 67:5755 (1977)).
45. M. K. Phani, J. L. Lebowitz, M. H. Kalos and O. Penrose, Phys. Rev. Lett. 45:366 (1980).
46. P. S. Sahni, G. Dee, J. D. Gunton, M. K. Phani, J. L. Lebowitz and M. H. Kalos, Phys. Rev. B24:410 (1981).
47. K. Kaski, M. D. Yalabik, J. D. Gunton and P. S. Sahni, Phys. Rev. B28:5263 (1983); E. T. Gawlinski, M. Grant, J. D. Gunton and K. Kaski, Phys. Rev. B31:281 (1985).
48. Y. C. Chou and W. I. Goldburg, Phys. Rev. A23:858 (1981).
49. J. Marro, J. L. Lebowitz and M. H. Kalos, Phys. Rev. Lett. 43:282 (1979).
50. J. L. Lebowitz, J. Marro and M. H. Kalos, Acta met. 30:297 (1982).
51. P. Fratzl, J. L. Lebowitz, J. Marro and M. H. Kalos, Acta met. 31:1849 (1983).
52. F. F. Abraham, S. W. Koch and R. C. Desai, Phys. Rev. Lett. 49:923 (1982); S. W. Koch, R. C. Desai and F. F. Abraham, Phys. Rev. A27:2152 (1983); S. W. Koch and R. Liebmann, J. Statist. Phys. 33:31 (1983).
53. P. S. Sahni and J. D. Gunton, Phys. Rev. Lett. 45:368 (1980); K. Kaski and J. D. Gunton, Phys. Rev. B28:5371 (1983); K. Kaski, M. Grant and J. D. Gunton, preprint; K. Kaski, S. Kumar, J. D. Gunton and P. A. Rikvold, Phys. Rev. B29:4420 (1984); K. Kaski, T. Ala-Nissila and J. D. Gunton, Phys. Rev. B31:310 (1985).
54. J. S. Langer and A. J. Schwartz, Phys. Rev. A21:948 (1980).
55. H. L. Snyder and P. A. Meakin, J. Chem. Phys. 79:5588 (1983).
56. H. Furukawa, Phys. Rev. A30:1052 (1984); A29:2160 (1984) and preprints.
57. K. Binder and M. H. Kalos, J. Stat. Phys. 22:363 (1980).
58. S. Katano and M. Iizumi, Phys. Rev. Lett. 52:835 (1984).
59. A. B. Bortz, M. H. Kalos, J. L. Lebowitz and M. A. Zendejas, Phys. Rev. B10:535 (1974).
60. J. Marro, A. B. Bortz, M. H. Kalos and J. L. Lebowitz, Phys. Rev. B12:2000 (1975).
61. M. Rao, M. H. Kalos, J. L. Lebowitz and J. Marro, Phys. Rev. B13:4328 (1976).
62. A. Sur, J. L. Lebowitz, J. Marro and M. H. Kalos, Phys. Rev. B15:3014 (1978).
63. K. Kawasaki, in "Phase Transitions and Critical Phenomena", Vol. 2, C. Domb and M. S. Green, eds., Academic Press, London (1972) p. 443.
64. C. M. Knobler and N. C. Wong, J. Phys. Chem. 85:1972 (1981).
65. M. Hennion, P. Guyot and D. Ronzaud, Acta met. 30:599 (1982); P. Guyot, preprint.

66. S. Komura, K. Osamura, H. Fujii and T. Takeda, Phys. Rev. B30:2944 (1984).
67. A. Craivich and J. M. Sanchez, Phys. Rev. Lett. 47:1308 (1981).
68. O. Blaschko, G. Ernst, P. Fratzl, M. Bernole and P. Auger, Acta met. 30:547 (1982).
69. P. Guyot and J.P. Simon, in "Solid-solid phase transformations", H. I. Aaronson, D. E. Laughlus, R. F. Sekeka and C. M. Wayman, eds., American Institute of Metals (1982) p. 325.
70. S. Komura, K. Osamura, H. Fujii and T. Takeda, Physica 120B:397 (1983); G. Kosterz, Physica 120B:387 (1983); S. Katano and M. Iizumi, J. Phys. Soc. Japan 51:347 (1982); Physica 120B:302 (1983).
71. D. N. Sinha and J.K. Hoffer, Physica 107B:155 (1981).
72. I. M. Lifshitz and V. V. Slyozov, J. Phys. Chem. Solids 19:35 (1961); see also W. Wagner, Z. Elektro Chem. 65:581 (1961).
73. E. Stoll, K. Binder and T. Schneider, Phys. Rev. B6:2777 (1972).
74. K. Binder and E. Stoll, Phys. Rev. Lett. 31:47 (1973); K. Binder and H. Müller-Krumbhaar, Phys. Rev. B9:3228 (1974).
75. H. Müller-Krumbhaar and E. Stoll, J. Chem. Phys. 65:4294 (1976).
76. D. Stauffer, A. Coniglio and D. W. Heermann, Phys. Rev. Lett. 49:1299 (1982).
77. R. Dickman and W. C. Schieve, Physica 112A:51 (1982).
78. J. Marro and R. Toral, Physica 122A:563 (1983).
79. G. Jacucci, A. Perini and G. Martin, J. Phys. A16:369 (1983).
80. O. Penrose, J. L. Lebowitz, J. Marro, M. H. Kalos and A. Sur, J. Stat. Phys. 19:243 (1978); M. H. Kalos, J. L. Lebowitz, O. Penrose and A. Sur, J. Stat. Phys. 18:39 (1978).
81. H. Müller-Krumbhaar, Phys. Lett. A50:27 (1974).
82. In ref. 7, active and inactive bonds are also introduced in the other component and then there is need to consider clusters of both types, n_1^A and n_1^B. By this approach the symmetry of the disordered phase at $c_B = c_B^{crit}$ above T_c is ensured.
83. D. W. Heermann, A. Coniglio, W. Klein and D. Stauffer, J. Stat. Phys. 36:447 (1984).
84. D. W. Heermann and W. Klein, Phys. Rev. Lett. 50:1962 (1983); Phys. Rev. B27:1732 (1983).
85. D. W. Heermann, to be published.
86. M. von Smoluchowski, Phys. Z. 17:593 (1916); S. Chandrasekhar, Rev. Mod. Phys. 15:1 (1943).
87. R. L. Drake, in "Topics in Current Aerosol Research", G. M. Hidy and G. R. Brock, eds., Vol. 3, Pergamon, New York (1972) p. 201.
88. S. K. Friedlander, J. Meteorology 17:479 (1960); 753 (1961); S. K. Friedlander and C. S. Wang, J. Coll. Interface Sci. 22:126 (1966); S. K. Friedlander, Phys. Fluids 3:693 (1970); J. Pich, S. K. Friedlander and F. S. Lai, Aerosol Sci. 1:115 (1970); C. S. Wang and S. K. Friedlander, J. Coll. Interface Sci. 24:170 (1967); S. K. Friedlander, "Smoke, Dust and Haze", Wiley, New York (1977).
89. G. M. Hidy and J. R. Brock, "The Dynamics of Aerocolloidal Systems", Pergamon Press, New York (1970).
90. F. Leyvraz, Phys. Rev. A29:854 (1984); F. Leyvraz and H. R. Tschudi, J. Phys. A14:3389 (1981); 15:1951 (1982); F. Leyvraz, J. Phys. A16:1861 (1983).
91. R. M. Ziff, E. M. Hendriks and M. H. Ernst, Phys. Rev. Lett. 49:593 (1982); E. M. Hendriks, M. H. Ernst and R. M. Ziff, J. Stat. Phys. 31:519 (1983); R. M. Ziff, J. Stat. Phys. 23:241 (1980).
92. T. Vicsek and F. Family, in "Kinetics of Aggregation and Gelation, F. Family and D.P. Landau, eds., North-Holland, Amsterdam (1984) p. 101; P. Meakin, T. Vicsek and F. Family, Phys. Rev. B31:564 (1985); T. Vicsek and F. Family, Phys. Rev. Lett. 52:1661 (1984); K. Kang and S. Redner, Phys. Rev. A30:2833 (1984).

93. R. Jullien, M. Kolb and R. Botet, in "Kinetics of Aggregation and Gelation", F. Family and D. P. Landau, eds., North-Holland, Amsterdam (1984); M. Kelle, R. Botet and R. Jullien, Phys. Rev. Lett. 51:1123 (1983); R. Botet, R. Jullien and M. Kolb, J. Phys. A17:175 (1984); M. Kolb, preprint.
94. A. C. Zettlemoyer, ed., "Nucleation", Marcel Dekker, New York (1969).
95. K. Binder and D. Stauffer, Adv. Phys. 25:343 (1976).
96. D. Stauffer, Phys. Repts. 54 (1979).
97. M. E. Fisher, Physics 3:255 (1967).
98. For reviews of older experimental data, see ref. 1 and W. I. Goldburg, in "Scattering Techniques Applied to Supramolecular and Nonequilibrium Systems", S. H. Chen, B. Chu and R. Nossal, eds., Plenum, New York (1981) p. 383.
99. J. S. Langer, M. Baron and H. D. Miller, Phys. Rev. A11:1417 (1975).
100. B. B. Mandelbrot, "The Fractal Geometry of Nature", Freeman, San Francisco (1982).
101. D. W. Heermann and K. Binder, to be published.
102. M. Kahlweit, Adv. Coll. Interface Sci. 5:1 (1975).
103 Another effect relevant for late stages is the possible loss of coherence of precipitated clusters with the host lattice, due to building of strong elastic distortions and resulting grain-boundary formation.

FIRST ORDER PHASE TRANSFORMATIONS:

SCALING RELATIONS FOR GRAIN SELF-CORRELATION FUNCTIONS

J. D. Axe and S. M. Shapiro
Brookhaven National Laboratory, Upton, N.Y., 11973

Y. Yamada and N. Hamaya
Faculty of Eng. Sci., Osaka Univ., Toyonaka, Osaka, Japan

INTRODUCTION

At high pressure many alkali halides transform from the NaCl (B1) structure to the CsCl (B2) structure. We have recently studied this transformation in polycrystalline RbI, which transforms at a critical pressure, $P_c = 3.5$ kbar.[1] By observing the time development of the neutron diffraction pattern after sudden increase of hydrostatic pressure from $P<P_c$ to $P>P_c$ we directly deduced X(t), the fraction of the sample converted from metastable to stable phase, as a function of time. We showed that X(t) taken at different P could be approximately scaled onto a universal growth curve by introducing an adjustable characteristic time $\tau(P)$ for each curve, such that

$$X(t,P) = x(t/\tau(P)). \tag{1}$$

This is demonstrated in Fig. 1.

We further noted that the resulting growth curve was very similar to one predicted by a simple model of nucleation and growth in which, when the pressure is instantaneously raised above P_c, infinitesimal grains of stable phase nucleate randomly within the sample at a constant rate, Γ, per unit volume. Although $\Gamma(P)$ is time independent, the

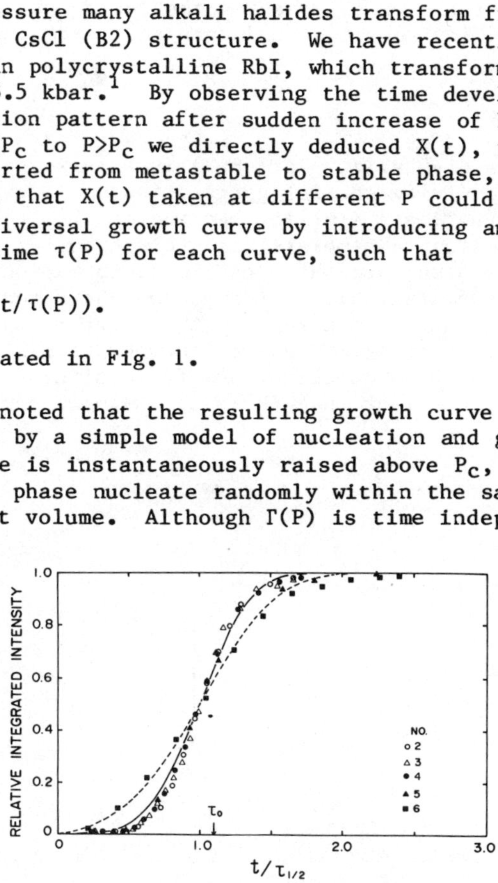

Fig. 1. The scaled growth curves, (1-X(t)), for RbI plotted vs. $t/\tau(P)$. Except for the highest pressure (Curve 6), they are well represented by the Kolmogorov model (solid line).

nucleation rate decreases with time in proportion to the fraction of metastable phase remaining, $(1-X(t))$. Once formed, a grain grows isotropically with constant domain-wall velocity, $v(P)$, until it is impeded by impinging on neighboring grains. An explicit expression for $X(t)$ for this model was first given by Kolmogorov,[2]

$$X(t) = 1-\exp[-\Gamma \int_0^t V(t')dt'] \tag{2}$$

where $V(t')$ is the volume occupied by a hypothetical "free" grain (one that has not yet been impeded by a neighboring grain) at a time t' after nucleation. $X(t)$ given by Eq. 2 for spatial dimension $d = 3$ is shown in Fig. 1 and demonstrates a remarkable similarity to our data.

The success of the Kolmogorov model in fitting $X(t)$ suggests that comparisons of model predictions with other experimental observables be made on the system. For example, by a trivial (in principle) extension of the neutron diffraction techniques described above, one might determine the broadening of the powder diffraction peaks due to finite grain size as a function of time throughout growth. This particle size broadening is related by Fourier transformation to the grain autocorrelation function, $G_s(r,t)$, which measures the ensemble average of the overlap of grains with themselves upon translation of the grain pattern by an amount r. A quantity much studied in connection with heterogeneous reaction kinetics is $G(r,t)$, which measures the inter- as well as intragrain correlations.[3] It is often found that a scaling relation of the form

$$G(r,t) = F(r/R(t)) \tag{3}$$

is at least approximately valid. Here $R(t)$ is a measure of the dominant length scale of the evolving spatial correlations. In early stages of growth $G(r,t)$ and $G_s(r,t)$ are similar, but whereas $G(r,t\to\infty)$ approaches a constant (corresponding to complete transformation to the stable phase), $G_s(r,t\to\infty)$ has structure which carries information about asymptotic grain size distribution. (We assume that each grain retains its unique identity rather than merging with neighbors on contact.) At some intermediate time intergrain separation and grain size are comparable, a situation which poses complications for scaling theories for $G_s(r,t)$, since it is not clear that there is a single dominant length upon which to scale.

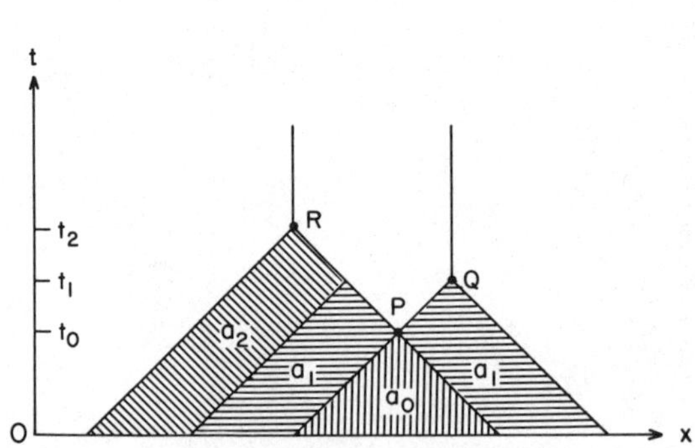

Fig. 2. Diagrammatic representation of domain nucleation and growth for $d = 1$. A grain is formed at P and its ends are progressively pinned at Q and R. The diagonal lines have the slope $r = vt$.

In the following sections we present some results of a study of the scaling properties of $G_s(r,t)$ for the Kolmogorov model for d=1 and d=2. Although the model is highly idealized, it is perhaps the simplest conceivable one which obeys correlation function scaling in early stages of growth and undergoes nontrivial saturation due to volume fraction effects in the late stages. A paper describing the work upon which these results are based is in progress.[4]

AUTOCORRELATION FUNCTION SCALING FOR d=1

Although it is possible that the results have practical applications for linear polymers, d=1 is considered here because it is possible to establish some exact results, in contrast to the computer simulations for d=2 considered in the following Section. d=1 is exceptional in that there are three distinct grain types with pinning index μ=0, 1, or 2, corresponding to the number of boundaries of the grain pinned by impingement on neighboring grains. Each grain grows with velocity $v_\mu = (2-\mu)\,v$. Each newly formed grain progresses from unpinned to fully pinned in a sequence shown diagrammatically in Fig. 2. Through a reasoning process similar to that used by Kolmogorov[2] to derive X(t), it is possible to show[4] that the probability, $P_o(t_o)$, that a grain will nucleate somewhere along a unit length between t_o and $t_o + dt$ is

$$P_o(t_o) = \Gamma \exp-(\Gamma A_o), \tag{4}$$

where A_o is the area shown in Fig. 2. Similarly the probability that a grain will first encounter another grain in dt_1 about t_1 can be shown to be

$$P_1(t_1) = \Gamma \frac{dA_1}{dt} \exp-(\Gamma A_1), \tag{5}$$

and the probability that a partially pinned grain will encounter a second grain in dt_2 about t_2 is

$$P_2(t_2) = \Gamma \frac{dA_2}{dt} \exp-(\Gamma A_2). \tag{6}$$

Eqns. (4) - (6) provide the basis of a complete description of the time evolution of the d = 1 Kolmogorov model.[4] In particular, the grain auto-correlation function is given by

$$G_s(r,t) = \int_r^{2vt} (\ell-r)N(\ell,t)d\ell \qquad r<2vt \tag{7}$$

where $N(\ell,t)$ is the total number of grains per unit length at time t with a length between ℓ and $\ell+d\ell$. It is convenient to introduce a normalization such that

$$\tilde{G}_s(r,t) = G_s(r,t)/G_s(r=0,t),$$

where $G_s(r=0,t)=1-X(t)$, is the volume fraction of the stable phase. In early stages of growth, $t<<(\Gamma v)^{-1/2}$, grains propagate freely and since $N(\ell,t\to0) = \Gamma/2v$, Eq. 7 is easily evaluated.

$$G_s(r,t) = \left(1 - \frac{r}{R(t)}\right)^2 \qquad r<R(t)$$
$$= 0 \qquad r>R(t) \tag{8}$$

showing that $\tilde{G}_s(r,t)$ satisfies scaling at small t, with $R(t) = 2vt$. If scaling were exact at all t, it would be possible to find an R(t) for each t such that eq. (8) was satisfied. Fig. 3 shows the values of R(t) which result from a least square fit to exact values of $G_s(r,t)$ determined numerically by the procedure sketched above. Fig. 4 shows $G_s(r,t)$ for various t plotted vs. the resulting values of r/R(t).

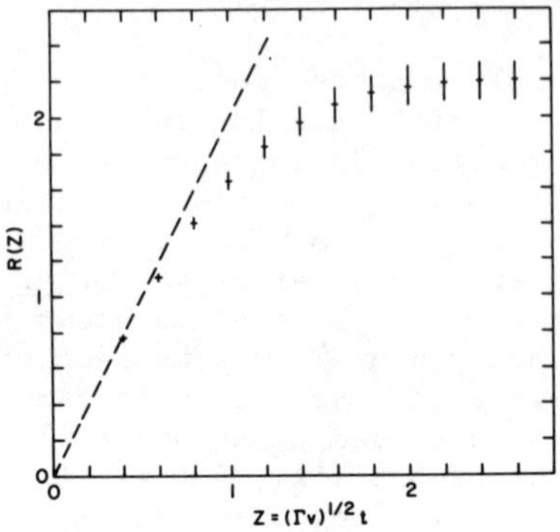

Fig. 3. The scaling length, R(t), for d=1 obtained by least square fitting of $G_s(r,t)$ to eq. 8. The time is in natural units, $z = (\Gamma v)^{1/2}t$.

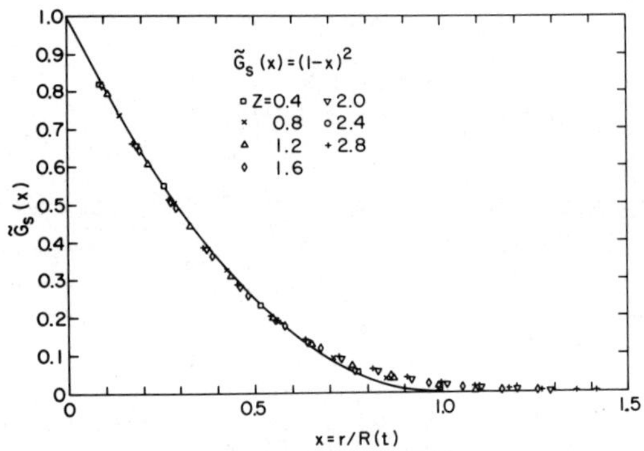

Fig. 4. $G_s(r,t)$ for d=1 and various values of t (in natural units) vs. the scaling parameter $x=r/R(t)$. The line is the scaling prediction (Eq. 8).

Scaling predicts a single universal curve. This is not the case, as can be seen most readily by observing that eq. 8 predicts that $G_s(r,t) = 0$ for $r>R(t)$, whereas in reality it remains small but finite for $R(t)<r<2vt$. Nevertheless eq. 8 does provide a very reasonable approximation over the entire range of t.

AUTOCORRELATION FUNCTION SCALING FOR d=2

Lacking obvious means to generalize the exact d=1 results to higher dimensions, we performed a numerical simulation on a discrete 20x20 lattice with periodic boundary conditions. The procedure, carried out on an IBM 4860 computer with a 4.77 MHz clock cycle, consisted of iterating to completion a two-step cycle. In the nucleation step random sites were chosen as potential nuclei and disgarded only if occupied (i.e. transformed in a previous cycle). In the growth step each grain, including the newly selected point nuclei of the previous cycle, was expanded into any unoccupied nearest neighbor sites. All results were averaged over an ensemble of 16 independent trials.

For d=2 $G_s(r,t)$ is an anisotropic function of the vector r. It is simple to verify that an exact scaling form exists for small t. For the limiting cases of r along an edge or a diagonal of the square array, the explicit forms for $t \ll (\Gamma v^2)^{-1/3}$ and $r < R(t)$ are

$$\tilde{G}_s(r,t) = 1-3/2\left(\frac{r}{R(t)_{edge}}\right) + 1/2\left(\frac{r}{R(t)_{edge}}\right)^3, \tag{9}$$

$$\tilde{G}_s(r,t)_{diag} = \left(1-\frac{r}{R(t)_{diag}}\right)^3. \tag{10}$$

As for d=1, the values of $\tilde{G}_s(r,t)$ derived from the simulation were fitted to the scaling forms suggested above by least squares adjustment of R(t). The values of R(t) so obtained are shown in Fig. 5, and the values of $G_s(r,t)$ plotted vs. the appropriate scale variable r/R(t) are shown in Fig. 6. It is again clear that these simple scaling forms are quantitatively useful approximations, but they cannot be made exact in late stages of growth.

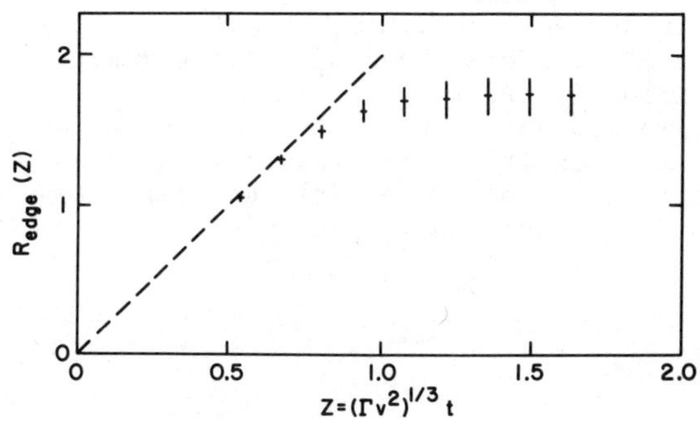

Fig. 5a. Scaling length, $R(t)_{edge}$, obtained by fitting $\tilde{G}_s(r,t)$ for d=2 to Eq. (9). The time is in natural units, $z=(\Gamma v^2)^{1/3}t$.

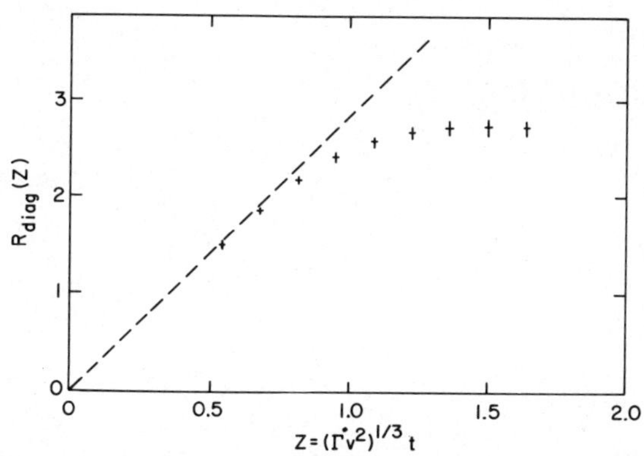

Fig. 5b. Scaling length, $R(t)_{diag}$, obtained by fitting $\tilde{G}_s(r,t)$ for d=2 to Eq. (10).

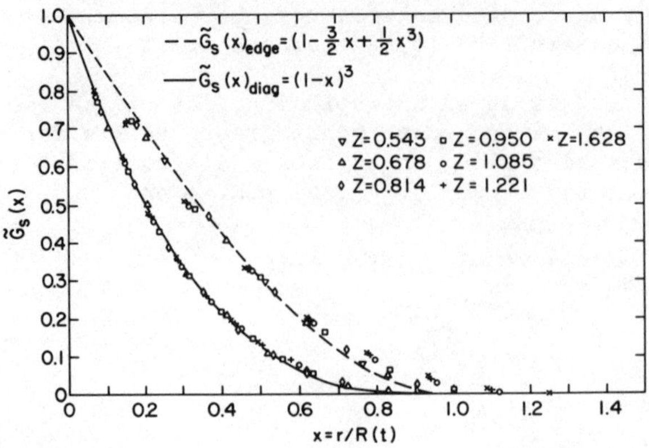

Fig. 6. $\tilde{G}_s(r,t)$ for d=2 and various t (natural units) plotted vs. the scaling parameter x=r/R(t). The lines are the scaling predictions (Eq. (9)-(10)).

REFERENCES

1. Y. Yamada, N. Hamaya, J. D. Axe, and S. Shapiro, Phys. Rev. Lett. 53, 1665 (1984).
2. A. N. Kolmogorov, Bull. Acad. Sci. USSR, Phys. Ser. 3, 335 (1937). This model was also discussed independently by W. A. Johnson and R. F. Mehl, Trans. Am. Inst. Min. Metall. Pet. Eng. 135, 416 (1939) and by M. Avrami, J. Chem. Phys. 7, 1103 (1939); ibid. 8, 212, (1940), and ibid. 9, 177 (1941).
3. For a recent review see J. D. Gunton, M. San Miguel, and P. S. Sahni in "Phase Transitions and Critical Phenomena," C. Domb and J. L. Lebowitz, ed., Vol. 8, p. 267, Academic Press (1983).
4. J. D. Axe and Y. Yamada, to be published.

DYNAMICAL SCALING IN A MODEL STRUCTURAL GLASS

Eric Courtens

IBM Zurich Research Laboratory

8803 Rüschlikon, Switzerland

INTRODUCTION

There is a current active debate on the mechanisms of freezing in model glasses. Experimental freezing temperatures, derived from susceptibilities, are often found to vary approximately linearly with the logarithm of the measuring frequency. The theoretical problem being far from settled, experiments are much in need. To be definitive, they must cover extremely broad frequency ranges. The present paper summarizes the experimental situation for a model glass,[1] mixed crystals of rubidium-ammonium dihydrogen phosphate, which have allowed measurements covering an unprecedented 17 orders of magnitude in frequency.[2,3] The experimental results indicate that dynamics derived from the Vogel-Fulcher (VF) law[4] provide the most appropriate phenomenological description, while dynamics adapted from other current theoretical models of spin-glasses[5,6] seem inadequate for this system.

Susceptibility measurements give information on the pair-correlation function of the frustrated order parameter, in the present case the polarization. For glasses, higher-order correlation functions are also particularly relevant. The present system has allowed a measurement sensitive to fourth-order correlations.[7] Its analysis in terms of the random-phase approximation has led to a surprising and truly remarkable result: A quasi-static Edwards-Anderson (EA) order parameter[8] (pair correlations much slower than those observed in susceptibility measurements) already exists above the VF freezing temperature.[7] It is conceivable that the existence of such a frozen component is precisely what is needed for the VF law to be valid,[9] in which case it may turn out to be a rather universal property.

The frozen component is to be related with an anomalous temperature dependence of the lattice parameters[10] and of the natural birefringence of the tetragonal crystals,[1] which both depart from normal lattice anharmonic behavior below an "onset of freezing". This "onset" occurs at a temperature of the order of ten times the VF freezing temperature T_0. Diffused scattering, observed with both x-rays[10-12] and neutrons,[13] indicates a tendency to form superstructures, in particular for compositions near the ferroelectric limit.[10] This very short-range ordering, which apparently grows on cooling below the "onset" temperature,[10] may actually correspond to a long-range disordered network which should then be related with the quasi-static EA order parameter mentioned above. The dynamics of diffused scattering can now be investigated with neutrons,[13] and will soon allow checking of these conjectures.

Finally, it should also be noted that VF freezing, which was in fashion for spin glasses[14] but recently seems to have lost popularity,[6] does, in fact, account extremely well[15] for some of the most convincing results on spin-glass dynamics.[6]

MIXED CRYSTALS OF RUBIDIUM-AMMONIUM DIHYDROGEN PHOSPHATE

Rubidium dihydrogen phosphate (RDP) and ammonium dihydrogen phosphate (ADP) are isomorphous at room temperature, and have practically the same unit-cell volume. Mixed crystals of $Rb_{1-x}(NH_4)_xH_2PO_4$ (RADP) can be grown over the entire range of x. The room-temperature paraelectric (PE) phase is characterized by acid-proton disorder. At low temperatures, RDP becomes ferroelectric (FE) while ADP becomes antiferroelectric (AFE). The ammonium salts are the only ones with that property in the entire potassium dihydrogen phosphate (KDP) family, the origin of the AFE ordering being related to the hydrogen bonding of the NH_4^+ ions with the nearby PO_4 tetrahedra.

The same phenomenon is responsible for the frustration of the acid-proton ordering which occurs in the mixed crystals.[16] For intermediate concentrations $x_m < x < x_M$, the crystals actually maintain the average PE structure and tetragonal symmetry down to liq.-He temperature ($x_m \simeq 0.22$, $x_M \simeq 0.74$). However, the "onset of freezing" occurs already in the region of 100 K, while the VF temperature is only of the order of 10 K (for $x = 0.35$).

DYNAMICAL SUSCEPTIBILITY MEASUREMENTS AND SCALING

Raman measurements performed on the FE "soft mode",[3,16] indicate that the mode softens rather normally down to ~100 K, at which point the softening becomes incomplete and the mode leaves behind a hard trail. Below ~100 K, the susceptibility is phenomenologically represented by a superposition of Debye relaxators,

$$\chi(\omega,T) = \chi(0,T) \int_0^\infty g(\tau,T)/(1 - i\omega\tau)\, d\ln\tau \,, \tag{1}$$

where $g(\tau.T)$ is a broad distribution of relaxation times normalized to 1 upon integration over $d\ln\tau$. It is rather well approximated by a rectangular distribution

$$g(\tau,T) = \begin{cases} 0, \ \tau < \tau_0 \text{ and } \tau > \tau_c \,, \\ 1/\ln(\tau_c/\tau_0), \ \tau_0 < \tau < \tau_c \,, \end{cases} \tag{2}$$

where $1/2\pi\tau_0$ equals 100 cm^{-1} from the Raman measurements, and τ_c is given by the VF law

$$\tau_c = \tau_0 \exp\left[E_c/(T-T_0)\right] \,, \tag{3}$$

with $E_c = 256$ K and $T_0 = 9$ K. This scaling of $\nu_c = 1/2\pi\tau_c$ has been verified by Raman and Brillouin measurements in the range $10^8 \lesssim \nu_c \lesssim 10^{12}$ Hz, by audio frequency dielectric measurements for $10^2 < \nu_c < 10^5$ Hz, and by polarization saturation measurements in the region of 10^{-4} Hz.[2,3]

At low frequencies, the actual form of the onset of $g(\tau,T)$ becomes important. Using the distribution (2) in (1), one would obtain much too steep an onset for the dielectric losses $\chi''(\omega,T)$. One finds, in fact, that the entire low-frequency data can be scaled on a single curve, assuming that

$g(\tau,T)$ derives from a distribution of energies $f(E)$, where E and τ are related by the VF law,[2,17] with

$$g(\tau,T) = (T - T_0)\ f(E)\ . \tag{4}$$

This suggests precisely that g derives from a pre-existing distribution of barriers, which can be due to the "network" mentioned in the Introduction, while the finite VF temperature is a result of the PE interaction between the remaining free components.

At very low temperatures (below ~15 K), any measurement is by necessity dynamical, as the times to reach equilibrium become of the order of days, and soon are of the order of years (at ~14 K). Applying a voltage step at time $t = 0$ to a thin sample, one can follow its charging which is proportional to $\ln t$ for at least one hour (for $T > 10$ K), beyond which it could not be measured. This indicates that a broad distribution of relaxators continues to exist, and it suggests that the VF law might apply beyond $\nu_c = 10^{-4}$ Hz, although one is unable to determine the cut-off frequency experimentally. However, extrapolating the straight lines of charge vs $\ln t$, one finds that they intercept at some long time t. If the VF law were to apply to those measurements, there should be some divergence of $\chi(0,T)$ as T tends to T_0. Such a strong FE tendency seems highly unlikely.[18] An alternative explanation is that the distribution g that applies at those low temperatures, in fact, is not given by (2) and (3), but is higher and cuts off earlier. This does not imply a breakdown of the VF law in the strict sense, but simply that to obtain the equilibrium g given by (2) one should prepare the sample for times of the order τ_c given by (3), which is of course hardly possible for measurements taken below 14 K. In any case, a main difference with usual critical slowing-down is that the scaling variable is the logarithm of the time rather than the time itself, with the dramatic consequence that the system is unable to reach equilibrium at sufficiently low temperatures.

Finally, it should be emphasized that attempts to scale our data with theories used currently for spin glasses[5,6] do fail. Fitting low-frequency data with those theories, one systematically finds values of τ_0 which are too long to account for the high-temperature, high-frequency data, and are in contradiction to the indications of the Raman spectra.[3,16]

EVIDENCE FOR AN EDWARDS-ANDERSON ORDER PARAMETER

Using Brillouin scattering from sound waves propagating along one of the main crystallographic directions, and in view of the large electrostriction of RADP, one is able to observe strains coupling linearly to $\delta P \delta P$, but whose linear coupling to δP is symmetry forbidden, where δP are the polarization fluctuations.[7] One finds a temperature-dependent dip in hypersound velocity, with an associated peak in the sound attentuation. These effects are proportional to fourth-order correlations in δP. A careful analysis of the signals reveals that their shape can only be understood, within the random-phase approximation, if there is already a quasi-static EA order parameter at temperatures as high as 50 K.[7] It is apparently the first time that the existence of an EA order parameter *above* a freezing temperature is claimed; it may be a very important element in understanding how a VF law can be valid over such an enormous range of temperatures and frequencies.

THE CASE OF SPIN GLASSES

A very careful experimental study of dynamical scaling in $Eu_{0.4}Sr_{0.6}S$ has recently been published.[6] The authors have measured the value of χ''/χ' at small values of χ'', and have used the constancy of that ratio as criterion for freezing. Their results can be interpreted directly in terms of Eqs. (1) to (3). The fit problem involves only the three parameters τ_0, E_c, and T_0, which is the same number of parameters they have in their fits with $T_f \neq 0$. One finds $T_0 = 1.31$ K, $E_c = 6.40$ K, and $\tau_0 = 2.1 \times 10^{-12}$ sec.[15] These values are much more satisfactory than those obtained with the accepted scaling theories, in particular for what concerns τ_0 and T_0. They can, in fact, be used to reproduce approximately the curves of $\chi'(\omega,T)$ and $\chi''(\omega,T)$ as well as the dependence of the maximum of $\chi'(T)$ on ω.[15]

This strongly suggests that the simple phenomenology which applies so well to RADP could also extend successfully to other glasses. Approaches to justify theoretically the VF law are certainly worth pursuing.

ACKNOWLEDGMENTS

Many thanks are expressed to the numerous colleagues who actively participated in this study, and whose names are found in Refs. 3,7,10,11,13,16 and 17. The crystal-growth help of H. J. Scheel and H. Arend, the skillful sample preparation by P. Bayer and M. Margot, as well as discussions with many colleagues whose list would be too long to place here, are also gratefully acknowledged.

REFERENCES

1. E. Courtens, J. Phys. (Paris) Lett. 43:L-199 (1982).
2. Eric Courtens, Phys. Rev. Lett. 52:69 (1984).
3. Eric Courtens and Hans Vogt, to be published.
4. H. Vogel, Z. Phys. 22:645 (1921); G. S. Fulcher, J. Am. Ceram. Soc. 8:339 (1925).
5. K. Binder and A. P. Young, Phys. Rev. B 29:2864 (1984); W. Kinzel and K. Binder, Phys. Rev. B 29:1300 (1984).
6. N. Bontemps, J. Rajchenbach, R. V. Chamberlin, and R. Orbach, Phys. Rev. B 30:6514 (1984).
7. Eric Courtens, François Huard, and René Vacher, to be published.
8. S. F. Edwards and P. W. Anderson, J. Phys. F 5:965 (1975).
9. R. G. Palmer, D. L. Stein, E. Abrahams, and P. W. Anderson, Phys. Rev. Lett. 53:958 (1984).
10. E. Courtens, T. F. Rosenbaum, S. E. Nagler, and P. M. Horn, Phys. Rev. B 29:515 (1984).
11. R. A. Cowley, T. Ryan, and E. Courtens, to be published.
12. S. Hayase, T. Futamura, H. Sakashita, and H. Terauchi, J. Phys. Soc. Japan 54:812 (1985); S. Iida and H. Terauchi, J. Phys. Soc. Japan 52:4044 (1983); H. Terauchi, T. Futamura, Y. Nishihata, and S. Iida, J. Phys. Soc. Japan 53:483 (1984).
13. H. Grimm, K. Parlinski, H. Arend, E. Courtens, and W. Schweika, to be published.
14. J. L. Tholence, Solid State Commun. 35:113 (1980).
15. Eric Courtens, to be published.
16. Eric Courtens and Hans Vogt, J. Chim. Phys. (Paris) to be published.
17. J. Slak, R. Kind, R. Blinc, E. Courtens, and S. Žumer, Phys. Rev. B 30:85 (1984).
18. K. Binder, private communication.

A STOCHASTIC MODEL OF SPIN GLASS DYNAMICS

Paolo Sibani

Nordita, Blegdamsvej 17
DK-2100 Copenhagen Ø
Denmark

ABSTRACT

A theory of spin glass relaxation based on some assumptions about the ground-state morphology and some scaling hypothesis is presented. It yields a closed formula for the AC and time-dependent susceptibilities in close agreement with the experimental results.

INTRODUCTION

Several spin glass models have recently been shown[1,2,3] to have the following ground-state morphology: the spins form connected patches, 'packets of solidary spins', or 'clusters', within which the spins are rigidly correlated at $T = 0$, i.e. they maintain the same relative orientation. We conjecture that a similar picture holds, at least qualitatively, for real spin glasses, and elaborate on the consequences. A fuller account of this work can be found in ref. 4.

LOW TEMPERATURE DYNAMICS

The above mentioned clusters correspond (at least for the models) to exact symmetries of the ground state, i.e. thay can be flipped at no energy cost. Of course, the excited states of the system need not (and in general will not) share the same symmetries as the ground state; i.e. the clusters cannot be considered as non-interacting entities. Nevertheless, on a coarse time scale the system will, for low T, always be in a neighborhood of a ground state and the dynamics can be described in terms of rigid flips of the clusters, without of course implying any coherent microscopic motion. In this coarse-grained picture, the correlation function obeys

$$|\langle\sigma_i(t)\,\sigma_j(t)\rangle| = \begin{cases} \langle\sigma_i(t)\sigma_i(0)\rangle & \text{if i and j belong to the same cluster} \quad (1a) \\ 0 & \text{otherwise} \quad (1b) \end{cases}$$

and simple algebra leads to

$$\frac{1}{N}\langle S(t)S(0)\rangle = \frac{\sum_c (2P_{X_c}(t) - 1)X_c}{\sum_a X_c} , \tag{2}$$

where S is the total spin sum, X_c is the size of cluster c, and $P_{X_c}(t)$ is the probability that it has the same orientation at times zero and t. Since the union of n clusters is again a cluster (it still possesses a reflection symmetry) of size $X_1 + ...X_n$, one should require that the form of the distribution of cluster sizes should be invariant except for a scale factor under n-fold convolutions, for any n. This is the probabilistic version of scale invariance, which is mathematically expressed by $X_1 + ...X_n \overset{d}{=} n^{1/\alpha}X_1$, for all n, where α is a constant and the symbol $\overset{d}{=}$ means that the stochastic variables on both sides of the equation have the same distribution. The above formula means by definition that the X_i's have stable distributions. Referring for further details to Feller[5], we just mention that this entails, among other things, that with high probability the largest of the X_n's is of the same order of magnitude as the sum of all the others. As a consequence, the r.h.s. of eq. (2) will be dominated by the largest term, especially at long times when the term $2P_{X_c}(t) - 1$ has relaxed to zero for the small clusters. In conclusion, we only keep the largest term in eq. (2) and obtain

$$\frac{1}{N}\langle S(t)S(0)\rangle \sim 2P(t) - 1 , \tag{3}$$

where all the subscripts will be omitted from now on. The fluctuation-dissipation theorem now entails

$$\chi_{AC}(\omega) = \frac{2}{T}(1 - i\omega\tilde{P}(i\omega)) , \tag{4}$$

where X_{AC} is the linear response susceptibility at frequency ω, and for any f, $\tilde{f}(s) = \int_0^\infty e^{-st}f(t)dt$. It can be shown that all the quantities of interest can be expressed in terms of the probability density that a cluster flips at time t, $A_1(t)$, namely:

$$\tilde{P}(s) = \frac{1}{s}\,\frac{1}{1 + \tilde{A}_1(s)} \tag{5}$$

and hence

$$\chi_{AC} = \frac{2}{T}\,\frac{\tilde{A}_1(i\omega)}{1 + \tilde{A}_1(i\omega)} . \tag{6}$$

A time-dependent susceptibility can also be defined as

$$\chi(t) = \frac{2}{T}\mathcal{L}^{-1}\left(\frac{1}{s}\,\frac{\tilde{A}_1(s)}{1 + \tilde{A}_1(s)}\right) , \tag{7}$$

where $\mathcal{L}^{-1}$ is the inverse Laplace transform.

The power spectrum $\mathcal{C}(\omega) = \int_0^\infty \langle S(t)S(0)\rangle \cos(\omega t)dt$ is given in terms of A_1 by

$$\mathcal{C}(\omega) = \frac{2}{\omega} \frac{A_1''(\omega)}{(1 + A_1'(\omega))^2 + A_1''(\omega)^2} , \tag{8}$$

where $A_1(i\omega) = A_1'(\omega) - iA_1''(\omega)$.

SCALING LAW FOR HOPPING TIME DISTRIBUTION

In principle $A_1(t)$ should be calculated on the basis of a microscopic description of the system. Since this is not possible, we adopt the view that the form of A_1 does not depend on the details of the dynamics, but rather on some generic property of the spin glass 'phase', which we identify with selfsimilarity. We start with the 'many valley' picture of the energy as a function of the phase space variables, which seems well established in the literature[6,7,8], and which in the present context has the following meaning: in order to go from one ground state to the next, a large energy barrier must be crossed. A closer look at the barrier reveals local minima, or valleys, separated by smaller barriers, and so on and so forth through many orders of magnitude. At a fixed level of resolution, one sees only N local minima, each corresponding to some spin configuration. A path joining the two global minima is defined by the temporal sequence in which (some of) the local minima are visited. Let us first consider the 'easy' path which has the lowest pass height. For each k we assign to the k'th minimum a spin variable σ_n which is 'up' if the minimum has not been passed and 'down' otherwise. The waiting time for going from a ground state to another is

$$S_n = T_1 + \dots T_n , \tag{9}$$

where T_i is the time elapsed between the flipping of spins $i-1$ and i. At this level of resolution, the origin of the T_i's remains unexplored, but if the system is selfsimilar, each of them arises in the same way as S_n, hence the same distribution, except for scales, and

$$S_n \overset{d}{=} n^{1/\alpha} T_1 . \tag{10}$$

The scaling law leads again to stable distributions.

Since the above considerations apply to each path separately, the flipping time associated with each path is stably distributed. At the end one should average over all the paths which require the definition of a measure. It is not clear how this should be done; we just note that this averaging over the paths may be the mechanism which produces the very small order α of the flipping time distribution, which is consistent with the experiments.

NUMERICAL RESULTS

The Laplace transforms of stable distribution are known analytically[5] to have the form

$$\tilde{A}_1(s) = e^{-(sc(T))^{\alpha}} , \tag{11}$$

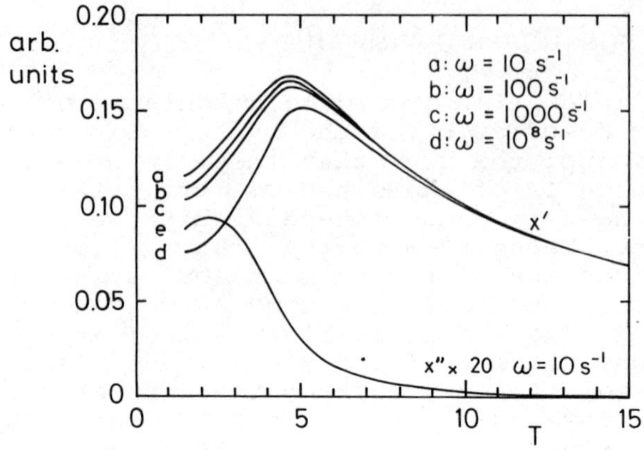

Figure 1. AC susceptibility according to eq. 12 for different frequencies.

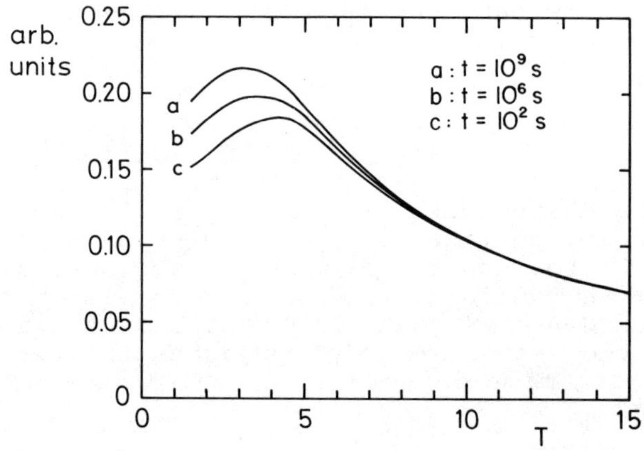

Figure 2. Time dependent susceptibility according to eqs. 7 and 11.

where α, the order of the distribution, lies between one and zero and $c(T)$ is a scale factor. In the present case, both α and $c(T)$ are temperature dependent. Inserting eq.(11) into eqs.(6) and (8) explicit expressions can be obtained. For example

$$\chi_{AC}(\omega) = \frac{2}{T}\,\frac{\exp(-\omega^{\alpha}c(T)^{\alpha}\exp(i\pi\alpha/2))}{1+\exp(-\omega^{\alpha}c(T)^{\alpha}\exp(i\pi\alpha/2))}, \qquad (12)$$

which is our main result. The experimental findings can be reproduced if $\alpha \ll 1$. Writing $f(T) = c(T)^{\alpha(T)}$ and expanding all the other functions of α to first order, we obtain from (12) and (8) the following relations:

$$\chi'' = -\frac{\pi}{2}\frac{d\chi'}{d\ln\omega}, \qquad (13)$$

$$\frac{\chi''}{\chi'} = \frac{f'(T)\,\alpha(T)}{1+\exp(-f(T))} + \mathcal{O}(\alpha^2), \qquad (14)$$

and

$$\mathcal{C}(\omega) = \pi\alpha(T)\exp(-f(T))\frac{1}{\omega}, \qquad (15)$$

the ubiquitous 1/f noise. In eqs.(13) and (14) we use the notation $\chi_{AC} = \chi' - i\chi''$. All the above relations are very well satisfied for a variety of spin glass systems.[9,10,11]

In order to discuss temperature dependences, $\alpha(T)$ and $f(T)$ must be specified. We have assumed that $c(T) = \exp(T_0/T)$ and that $\alpha(T)$ is a smooth function proportional to $1/T^2$ for T large and with a quadratic maximum at $T \approx 3k$. Since $f(T) = \exp(\alpha(T)T_0/T)$ and T_0 need not be small, this leads to an amplification of the shape of $\alpha(T)$, and $f(T)$ must not be expanded. A more detailed discussion of these assumptions can be found in ref.4. Here we just note that we are now in a position to evaluate eq.(12), and, by an approximate Laplace inversion based on the smallness of α, also the time-dependent susceptibility, eq.(7). The results are displayed graphically in figs. 1 and 2 and possess all the characteristic features of experimentally found spin glass susceptibilities.[9,11]

REFERENCES

1. F. Barahona, R. Maynard, R. Rammal and J.P. Uhry, J. Phys. A15:673(1982).
2. J. Vannenimus, J.M. Maillard and L. de Sèze, J. Phys. C12: 4532(1979).
3. P. Sibani and J.A. Hertz, to appear in J. Phys. A (Nordita preprint 84/40).
4. P. Sibani, to appear in J. Phys. A (Nordita preprint 85/8).
5. W. Feller, An Introduction to Probability Theory and its Applications, Vol. II, Wiley, New York (1971).
6. R.G. Palmer, Adv. Phys. 31:669(1983) and in: Springer Lecture Notes in Physics 192(1983).
7. I. Morgenstern and K. Binder, Phys. Rev. Lett. 43:1615 (1979).
8. I. Morgenstern and H. Horner, Phys. Rev. B25:504(1982).

9. L. Lundgren, P. Svedlindh and O. Beckman, J. Magn. Magn. Mat. 15:33(1981).
10. A.J. van Duyneveldt and C.A.M. Mulder, Physica 114B:82 (1982).
11. C. Pappa, J. Hammann, G. Jehanno and C. Jacoboni, Saclay preprint DPhG/SPSRM184-75 (1984).

CELLULAR AUTOMATA AND CONDENSED MATTER PHYSICS*

Stephen Wolfram

The Institute for Advanced Study
Princeton, New Jersey 08540

Cellular automata are mathematical models for systems containing many identical components with local interactions. These notes describe some of their properties, and discuss applications to condensed matter physics.

1. Introduction to cellular automata

A cellular automaton consists of a lattice of sites, with each site carrying one of a finite set of k possible values. The system evolves in discrete time steps, with the new value of each site specified by a fixed rule in terms of the previous values of a neighbourhood of sites at a distance of r or less on each side. Despite the simplicity of their construction, cellular automata are found to be capable of highly complex behaviour, and in fact provide models for a wide variety of complex phenomena in nature. These notes discuss some aspects of cellular automata which potentially relate to condensed matter physics. For discussions of other aspects of cellular automata, see refs. [1-5], and references therein.

As a first example, consider a one-dimensional cellular automaton, in which the value a_x of a site at position x is updated according to the rule $a_x{}' = (a_{x-1} + a_{x+1}) \bmod 2$. Figure 1.1 shows the pattern produced by this rule, growing from a simple "seed" consisting of a single non-zero initial site. The limiting form of this pattern is recognized as a Sierpinski curve. It is a classic example of a regular fractal [6]: portions of the pattern, when magnified, are similar to the whole.

Figure 1.2 now shows patterns produced by the cellular automaton rule of figure 1.1, evolving from "disordered" initial states, in which the value of each site is chosen independently to be zero or one, with probability 1/2. Patterns obtained with three different initial states of this kind are shown: while they differ in detail, their overall qualitative and statistical form is the same. With a given cellular automaton rule, it seems that almost all initial states yield similar patterns.

* Work supported in part by the U.S. Office of Naval Research under contract number N00014-85-K-0045.
Based on lectures given at the NATO Advanced Study Institute on "Scaling Phenomena in Disordered Systems", Geilo, Norway, April 8-19, 1985.

Figure 1.1. Patterns produced by evolution according to the simple one-dimensional cellular automaton rule $a'_x = (a_{x-1} + a_{x+1}) \bmod 2$, starting from a single non-zero initial site. Each site has value zero or one, shown as white or black, and the rule determines the new value of a site in terms of the previous values of its left- and right-hand neighbours on the line above. The first few stages in the growth of the pattern are shown on the left. The result after many time steps is shown on the right. The limiting pattern is seen to be self-similar, and has fractal dimension $\log_2 3 \simeq 1.59$.

Figure 1.2. Patterns produced by evolution according to the cellular automaton rule of figure 1.1, starting from three disordered initial states.

Different rules, however, can give very different kinds of behaviour. Figure 1.3 shows patterns obtained by evolution with the possible $k=2$, $r=1$ rules. Looking at these and many other classes of rules, one can identify four overall classes of behaviour [8].

1. Pattern dies out, leaving a homogeneous final state. This can be considered as a fixed point, to which all initial states evolve.
2. Pattern eventually becomes periodic (but sometimes systematically shifting) with time, and typically consists of a sequence of separated simple structures. Such behaviour is analogous to a limit cycle.
3. Chaotic behaviour is produced, and the pattern never becomes periodic. Such cellular automata are analogous to dynamical systems such as differential equations that exhibit strange attractors (e.g. [7]).
4. Complicated localized structures are produced, and continually-changing behaviour occurs. This seems to happen in $k=2$, $r=1$ cellular automata only when the rules are asymmetrical, or do not leave the zero configuration invariant. When $k>2$ or $r>1$ it seems to happen in a few percent of all cases.

Various quantitative characterizations of these classes of behaviour can be given. One based on dynamical systems theory considers the stability of patterns with respect to small perturbations in the initial state. Figure 1.4 shows the differences between patterns produced according to various rules, starting from initial states that differ by a change in the value of a single site. In class 1 cellular automata, such changes always die out. In class 2 cellular automata, they may

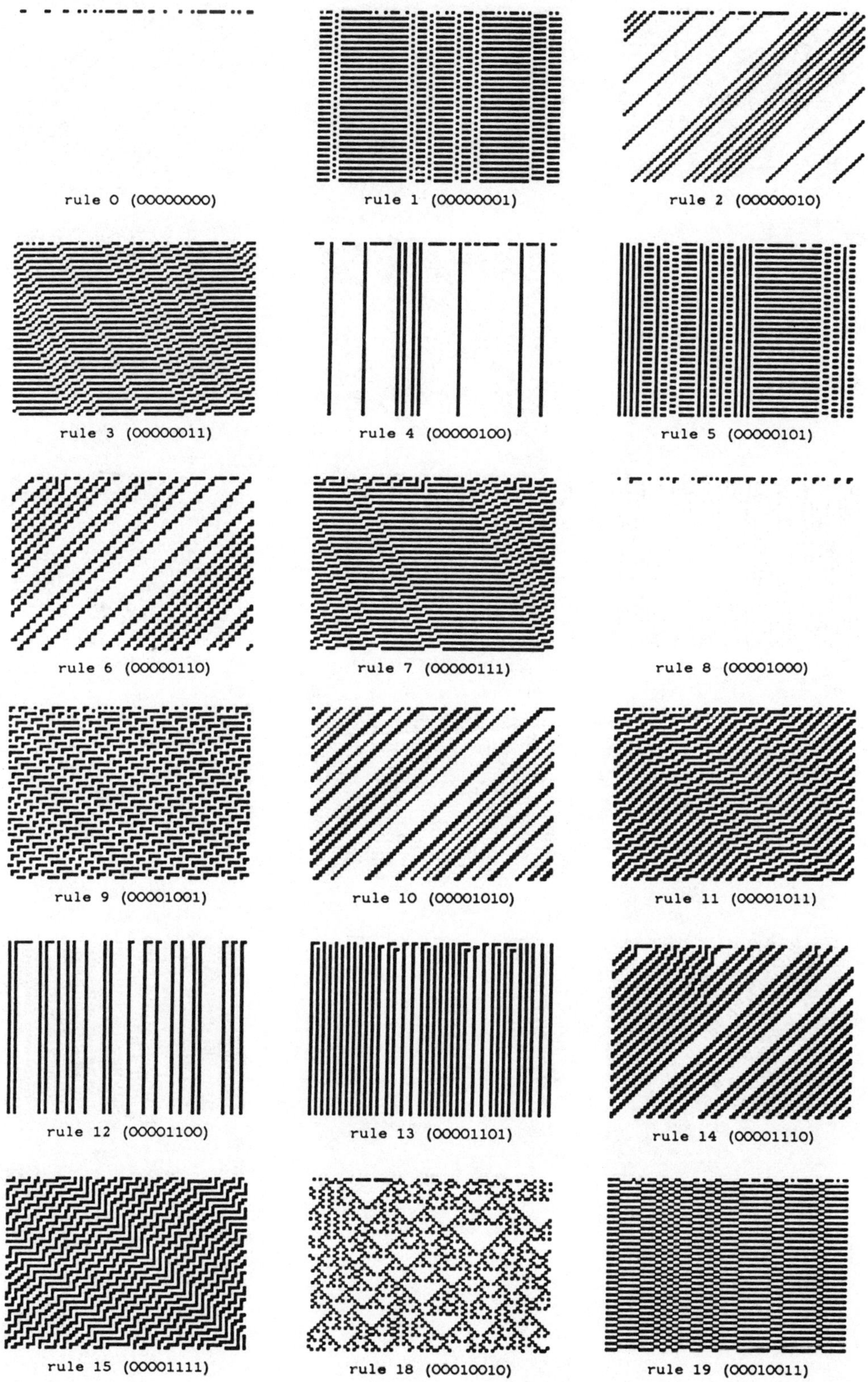

Figure 1.3. (Legend on page 255)

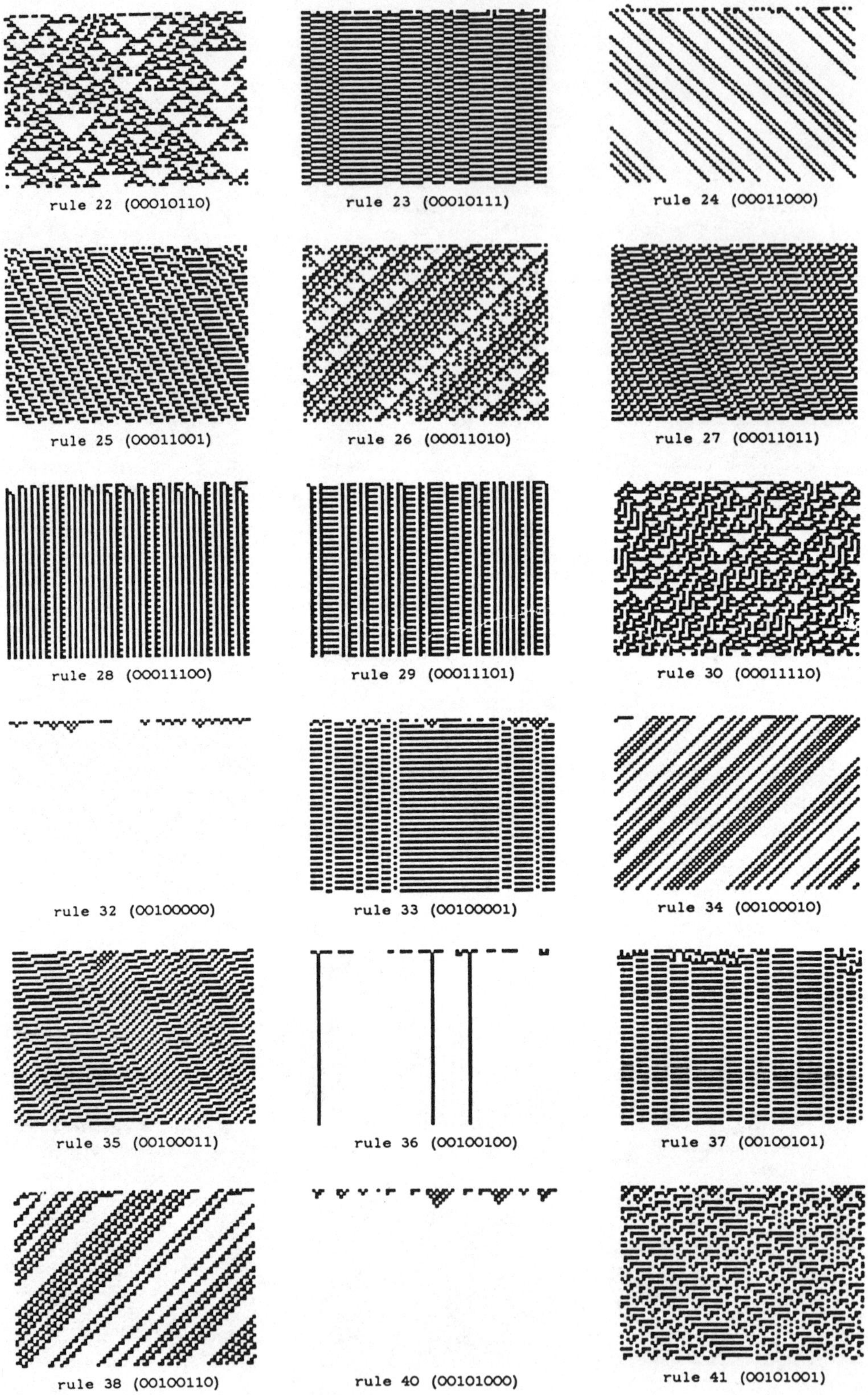

Figure 1.3. (Legend on page 255)

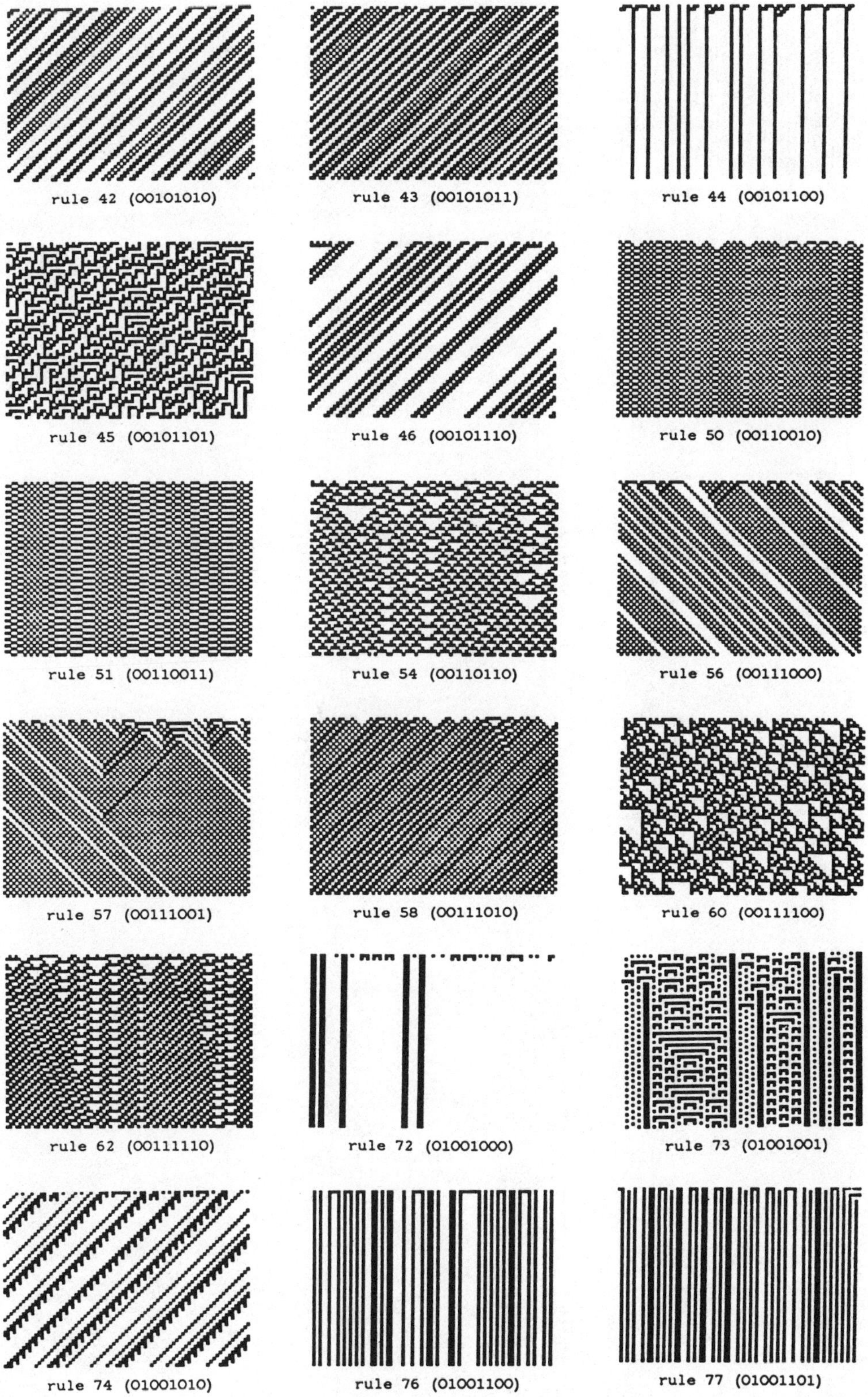

Figure 1.3. (Legend on page 255)

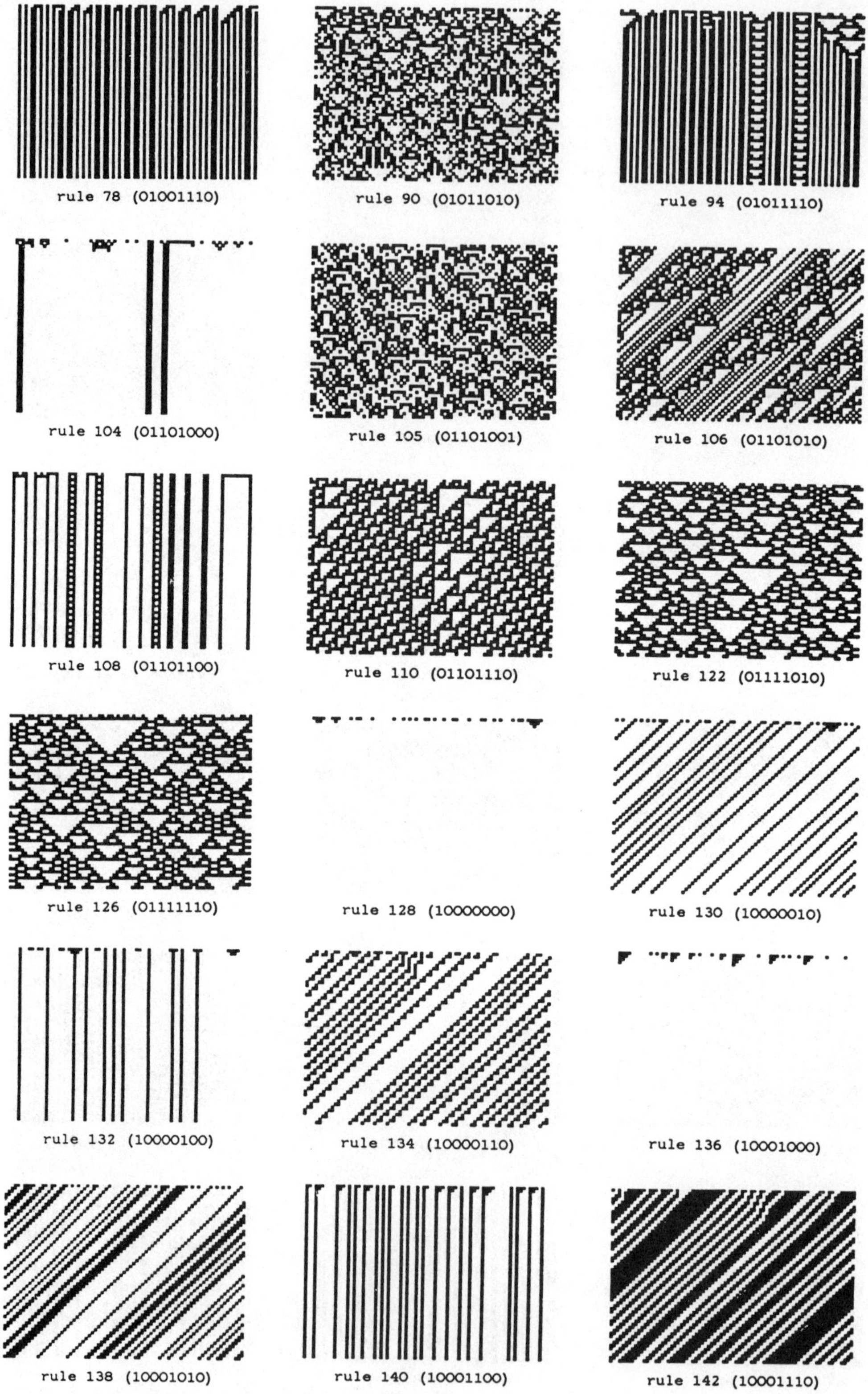

Figure 1.3. (Legend on page 255)

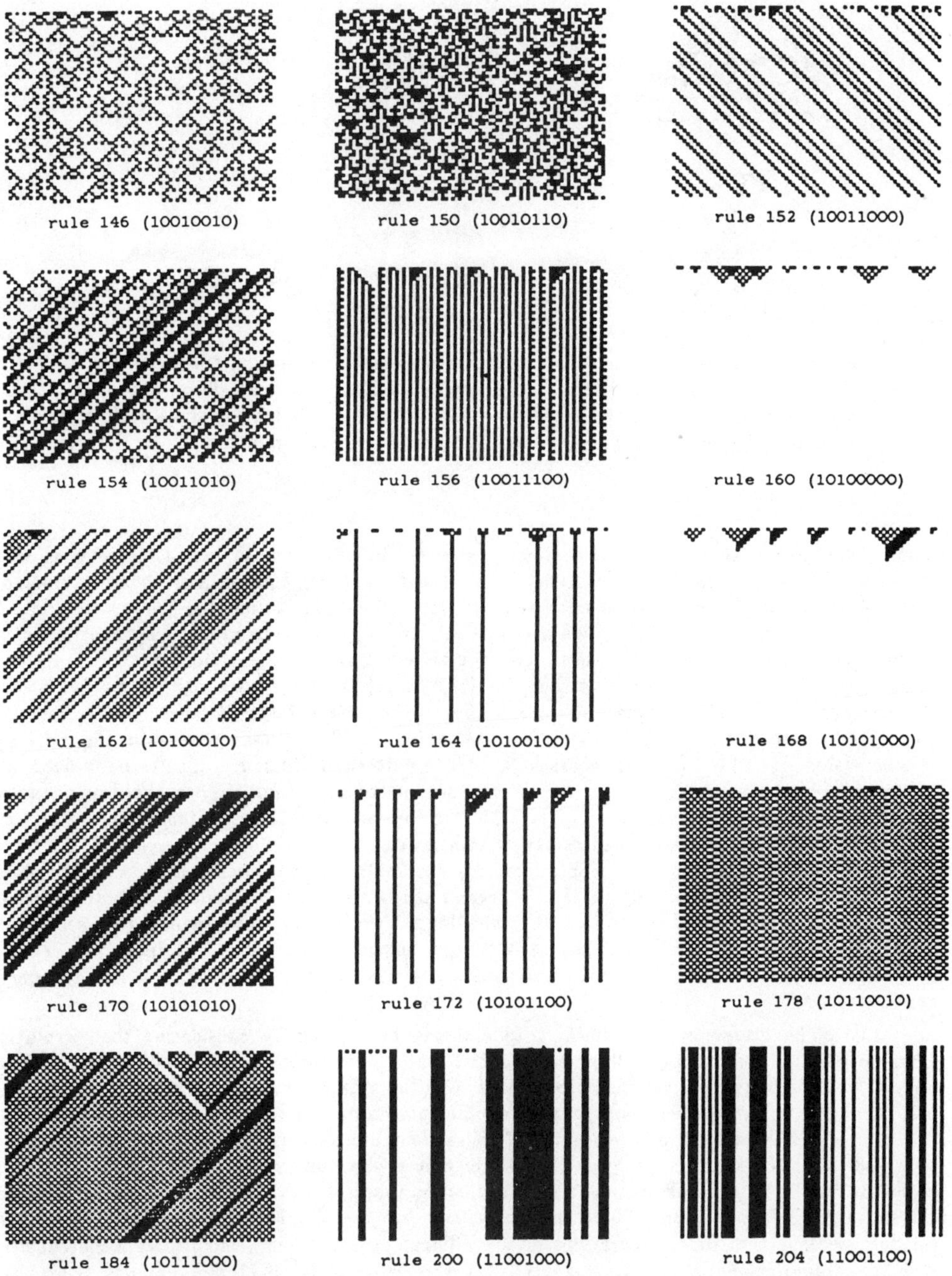

Figure 1.3. Patterns obtained by evolution from disordered initial states according to all the possible distinct one-dimensional cellular automaton rules with $k = 2$ and $r = 1$. The sequences of binary digits in the rule numbers give the images of the $2^3 = 8$ possible three-site neighbourhoods under the cellular automaton rule. The first digit gives the image of 111, the second 011, the third 101, and so on (cf. [3]). Table 3.1 gives various equivalences between rules.

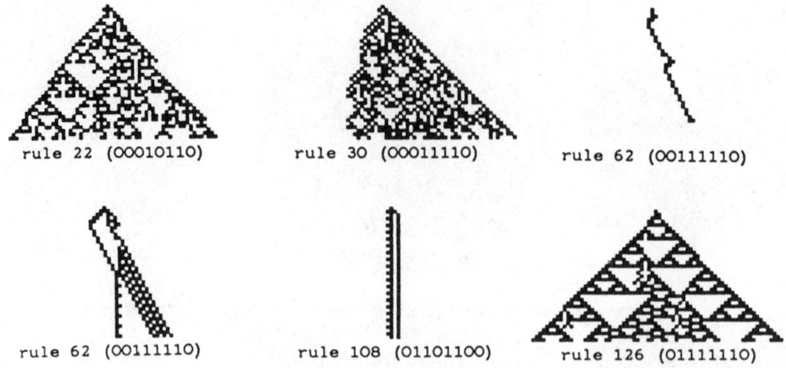

Figure 1.4. Differences in cellular automaton patterns resulting from changes in single site values in disordered initial states. In class 1 and 2 cellular automata, these differences remain localized with time. Class 3 cellular automata yield expanding difference patterns, and have positive Lyapunov exponents.

persist, but always remain localized. Class 1 and 2 cellular automata may thus be considered stable with respect to small perturbations. Class 3 and 4 cellular automata are however unstable to these perturbations. A small initial change in a class 3 cellular automaton grows with time, to affect a linearly-expanding region. The speed of this growth gives a Lyapunov exponent λ for the cellular automaton evolution. (A metric on cellular automaton configurations can be defined by selecting an origin, then evaluating $\sum(|\delta a_x| + |\delta a_{-x}|)\, k^{-x}$ for each pair of configurations. With this metric, configurations are close if they agree over a large distance. The trajectories corresponding to the evolution of configurations in class 3 cellular automata then diverge exponentially with time, with exponents λ [8,9].) The pattern of differences illustrated in figure 1.4 is analogous to a Green function for the cellular automaton evolution: it gives the response to an "impulse" change in the initial state. The region of support for this Green function expands at a speed that represents the rate of information transmission in the cellular automaton. Thus information remains localized in class 1 and 2 cellular automata, but can propagate in class 3 cellular automata. As a consequence, the value of a particular site in a class 3 cellular automaton depends on ever more distant initial site values: the behaviour of the cellular automaton is thus sensitively dependent on its initial conditions. In class 4 cellular automata, the pattern of differences expands irregularly with time, and it is not clear that a definite limiting speed or Lyapunov exponent exists.

The global properties of cellular automata can be studied by considering their evolution from an ensemble containing all possible initial states. In a cellular automaton with a finite number of sites N (say with periodic boundary conditions), there are a total of k^N possible complete configurations. The behaviour of the cellular automaton can then be represented by a finite graph, in which each node represents one of these configurations, and is joined by a directed arc to a node representing the configuration to which it evolves under the cellular automaton rule. Figure 1.5 shows an example of such a state transition graph, for the cellular automaton of figures 1.1 and 1.2. This particular cellular automaton rule has the simplifying feature that it is linear (modulo two), so that its configurations satisfy a linear superposition principle. As a consequence, it is possible to give a complete analysis of this rule using algebraic methods [10]. Each configuration is represented by a characteristic polynomial, and its time evolution corresponds to multiplication by a fixed polynomial. Properties of graphs such as figure 1.5 are then found to be determined by various number theoretical functions.

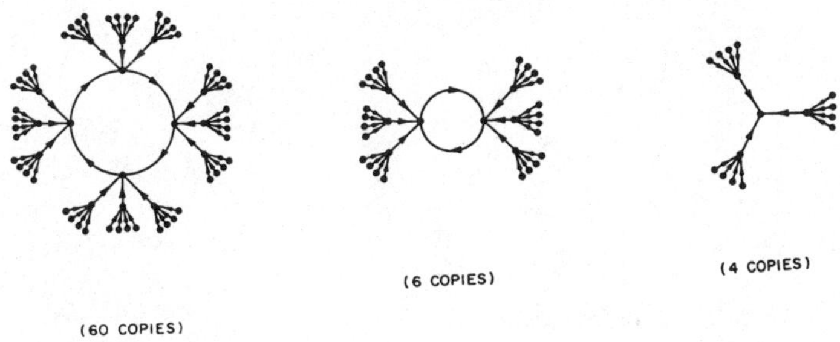

Figure 1.5. Global state transition graph for a cellular automaton with 12 sites (and periodic boundary conditions) evolving according to the rule of figures 1.1 and 1.2. Nodes in this graph represent configurations of the cellular automaton, with particular values for each of the 12 sites.

A significant feature of the graph in figure 1.5 is that, while each node has a unique successor, it may have a varying number of predecessors under the mapping defined by the cellular automaton rule. Paths in the graph, representing trajectories in the phase space of the cellular automaton, can merge with time. The evolution is thus irreversible, and the volumes of phase fluid representing ensembles of cellular automaton configurations can contract with time. The set of configurations that remains at large times can be considered as an attractor for the cellular automaton evolution. Starting from any possible initial configuration, the cellular automaton must ultimately evolve to a configuration on this attractor. The form of this attractor thus determines the large time behaviour of the cellular automaton. So for example, if states on the attractor have some special "organized" form, the cellular automaton will be seen to "organize" itself with time.

Various characterizations can be given of the structure of attractors for cellular automata. One consists in specifying the "volume" of the attractor in the infinite size limit, using entropies or dimensions [8]. These quantities can also be considered to measure the "information content" of cellular automaton patterns. The spatial topological entropy of a one-dimensional cellular automaton configuration can be defined as follows. Consider a block of X sites in the cellular automaton. There are k^X possible sequences of values for the sites in this block. But irreversible cellular automaton evolution may allow only $N(X) \leq k^X$ possible sequences. The topological entropy is then defined as $\lim_{X \to \infty} (\log_k N(X))/X$. This quantity is equal to the fractal dimension of the set of configurations that can occur, represented by a volume in the phase space of the cellular automaton. One can also define a measure entropy, in which blocks of site values appear weighted with their probabilities.

Spatial entropy $d^{(x)}$ describes the set of configurations that can occur at a particular time in a cellular automaton. If the cellular automaton evolution is irreversible, this quantity thus decreases with time. In a class 1 cellular automaton, $d_\mu^{(x)}$ (where the subscript μ signifies measure entropy) tends to zero with time. One can also define a temporal entropy $d^{(t)}$, which describes the sequence of values attained with time by a single site in the cellular automaton. This quantity vanishes for class 2 cellular automata, since they ultimately become periodic with time. In general, the ratio of temporal to spatial information densities in cellular automata is bounded in terms of the rate of information transmission by $d^{(t)}/d^{(x)} \leq 2\lambda$. In class 3 cellular automata, all these quantities are nonzero.

Entropies and dimensions measure the total "volume" of attractors in cellular automata. But at least for one-dimensional cellular automata, it is possible to obtain a much more complete characterization of the structure of attractors, using methods from the theory of computation [11]. The set of configurations produced by evolution from all possible initial states for t time steps

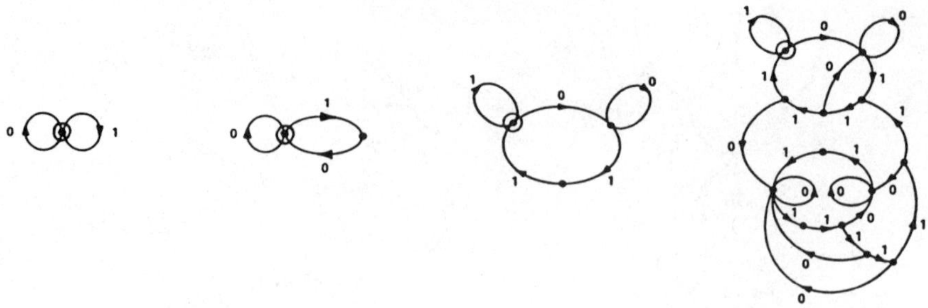

Figure 1.6. Graphs for regular languages representing various sets of cellular automaton configurations. The first graph represents all possible sequences of zeroes and ones, since any sequence can be obtained by following some path. The second graph represents only sequences in which pairs of adjacents ones appear, since on all paths a one must immediately be followed by a zero. The third and fourth graphs give minimal specifications for the sets obtained after one and two time steps in the evolution of a class 3 cellular automaton ($k=2$, $r=1$, rule number 126).

can in fact be shown to form a regular formal language [12]. Possible configurations correspond to possible paths through finite graphs such as those illustrated in figure 1.6. It is always possible to find a graph with the smallest number of nodes that represents a particular set of cellular automaton configurations. This graph then gives a complete, canonical, specification for the set. From it can be obtained quantities such as spatial entropy, which is given by the logarithm of an algebraic integer equal to the largest eigenvalue of the adjacency matrix for the graph. Other quantities can also be calculated from the graphs. One is the "regular language complexity" Ξ of the set of configurations, equal to the number of nodes in the minimal graph. This measures the length of the shortest specification for the set in terms of regular languages. For class 1 and 2 cellular automata, Ξ typically becomes constant with time, or increases at most like a polynomial. The sets obtained in the infinite time limit in these cases also correspond to regular languages. But for class 3 and 4 cellular automata, Ξ is usually found to increase rapidly (probably exponentially) with time, yielding limit sets that are in general not regular languages [13].

While most cellular automata are irreversible, there do exist some reversible ones. Figure 1.7 shows patterns produced by several such systems. Cases with class 2, 3 and 4 behaviour are found, and can be characterized by Lyapunov exponents and temporal entropies.

2. Regular crystal growth

Cellular automata can be used rather directly as models for crystal growth. Values 1 or 0 at sites in the cellular automaton lattice can represent the presence or absence of atoms or blocks of atoms in the crystal lattice. Two-dimensional lattices provide an example [4].

In the simplest model for crystal growth, a site attains value 1 if any of its neighbours have value 1 on the previous time step. Applying this rule starting with a simple initial seed yields the behaviour shown in figure 2.1, corresponding to the growth of a faceted crystal.

But in actual crystal growth, there is often growth inhibition: solidification at one point releases latent heat which prevents growth at the same point until the heat has diffused away. In fact, it appears that a planar solid interface is unstable with respect to corrugation with a

Figure 1.7. Patterns generated by some reversible cellular automata from disordered initial conditions. These particular cellular automata have rules of the form $a_x^{(t)} = \phi(a_{x-1}^{(t-1)}, a_x^{(t-1)}, a_{x+1}^{(t-1)}) \oplus a_x^{(t-2)}$, where $\oplus$ signifies addition modulo two (cf. [14]).

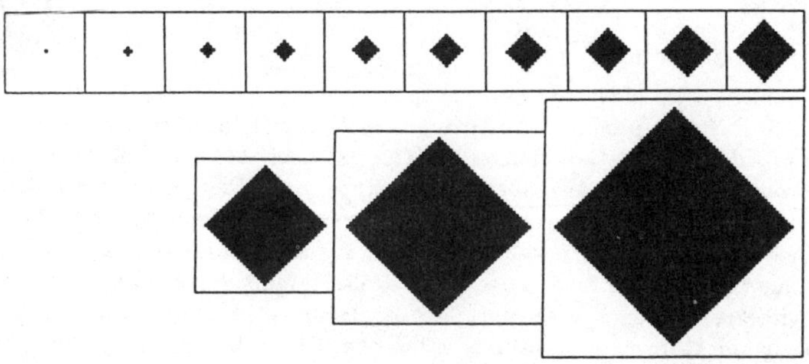

Figure 2.1. Configurations obtained by evolution in a two-dimensional cellular automaton from a simple seed according to a rule that simulates faceted crystal growth. Time steps 0 through 8, 10, 20, 30 and 40 are shown.

particular characteristic length scale determined by the microscopic properties of the system. This phenomenon can easily be incorporated into a cellular automaton model [15]. Each site represents a region whose size is of order the characteristic length. Then the cellular automaton rule specifies that a site becomes solid if exactly one of its neighbours is already solid (and once it has become solid, it remains so forever). The pattern produced in this way is shown in figure 2.2. It has a remarkable resemblance to the patterns obtained in actual dendritic crystal growth experiments.

One notable feature of the pattern in figure 2.2 is that its outline alternates between a faceted and a dendritic form. At times of the form 2^j, the boundary consists of planar facets. Then from each corner a branch grows. This branch sprouts sidebranches, and so on, producing a dendritic pattern. But eventually the sidebranches grow into each other, and at time 2^{j+1} all of them have collided, leaving again a faceted boundary. In the interior, a pattern of holes remain,

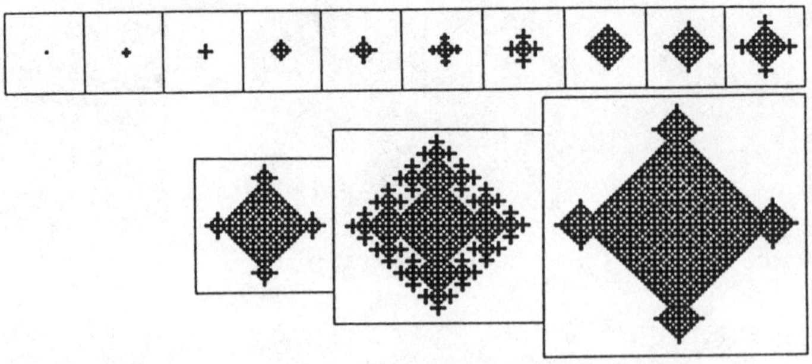

Figure 2.2. Configurations of a two-dimensional cellular automaton that simulates the growth of a dendritic crystal from a simple seed.

giving a "frozen history" of the growth. The detailed pattern of holes depends on the form the cellular automaton rule for neighbourhoods in which several sites are nonzero. Figure 2.3 shows patterns produced with other choices.

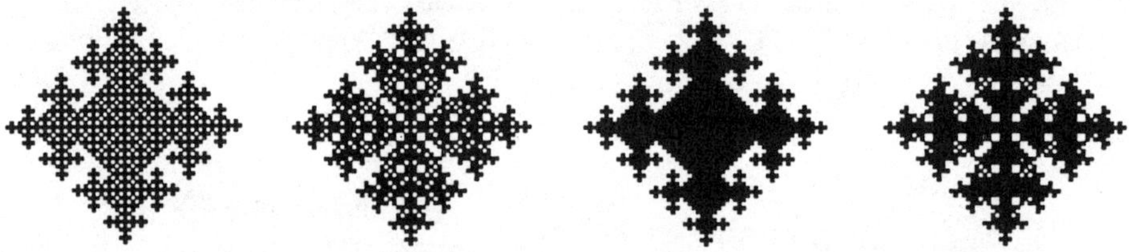

Figure 2.3. Patterns produced by two-dimensional cellular automata with various choices of sidebranch interaction rules. The patterns shown are obtained by evolution from a single site seed for 29 time steps, according to rules which take a site to attain value one if exactly one of its four neighbours already has value one. If two neighbours have value one, then growth is inhibited, and the site retains value zero. In the first case shown, it also retains value zero if three or four neighbours are one. In the second case, it attains value one if three neighbours are one. In the third case, it attains value one if four neighbours are one, but retains value zero if just three are. And in the fourth case, it attains value one if either three or four neighbours are one.

The patterns of figures 2.2 and 2.3 are indeed quite close to those found in actual crystallization processes. Analogous patterns generated on hexagonal lattices produce patterns strongly resembling those found in snowflakes.

Cellular automata seem to provide models that capture the essential phenomena responsible for generating complex global patterns in dendritic crystal growth. They can be viewed as approximations to the reaction-diffusion equations which are ultimately believed to describe the process. In many cases, discrete cellular automaton models indeed seem to represent rather accurate approximations to these partial differential equations (cf. [16]).

3. Scaling behaviour

The spacetime patterns produced by cellular automaton evolution, and shown in figures 1.1 and 2.2 have the property of being scale invariant, or self-similar. Scale invariant patterns are in fact often produced by cellular automata, and one can consider using methods based on the renormalization group (e.g. [17]) to analyse them.

Table 3.1. Simulations of $k=2$, $r=1$ cellular automaton rules (as illustrated in figure 1.3) by other $k=2$, $r=1$ rules under blocking transformations. The notation $R \rightarrow R' : b_0\, b_1\, (T)$ indicates that a configuration containing only blocks b_0 and b_1 will evolve under T steps of rule R just like a configuration with these blocks replaced by single site values 0 and 1 respectively under the rule R'. Blocks up to length 8 are considered. $R = R'$ indicates that rule R simulates R' with blocks b_i of length one, so that the rules are equivalent under interchange of 0 and 1, or left-right reversal. Note that rule 204 is the identity, rule 51 interchanges 0 and 1, and rules 170 and 240 are shifts. Rules 90 and 150 are linear (modulo two). (This table was generated using a computer program by J. Milnor.)

R	R'	$b_0\, b_1\, (T)$
0 →	0:	00 10 (1)
1 →	0:	11 10 (2)
	200:	00 11 (2)
	204:	000 111 (2)
2 →	34:	00 10 (2)
	170:	000 100 (3)
	0:	1000 1100 (4)
3 →	0:	11 10 (2)
	240:	00 11 (4)
4 →	204:	00 10 (1)
	0:	00 11 (1)
5 →	200:	00 10 (2)
	204:	000 100 (2)
	0:	111 110 (2)
	51:	00010 11010 (1)
6 →	184:	00 10 (2)
	34:	00 11 (2)
	170:	0000 1000 (4)
	128:	0100 1100 (4)
	240:	1000 1010 (4)
	85:	10000 11000 (5)
	0:	11000 11100 (10)
7 →	192:	00 10 (2)
	0:	000 100 (2)
	240:	000 111 (6)
8 →	0:	00 10 (1)
9 →	0:	0010 1110 (2)
	170:	1000 0010 (6)
	34:	1000 0011 (6)
	204:	0100000 1100000 (5)
	240:	00000000 11010000 (8)
10 →	34:	00 10 (2)
	170:	000 100 (3)
	0:	010 110 (3)
11 →	240:	00 11 (2)
	0:	010 110 (3)
	15:	000 111 (3)
	128:	1100 1001 (4)
	170:	1100100 1001100 (7)
12 →	204:	00 10 (1)
	0:	100 110 (1)
13 →	192:	00 10 (2)
	0:	100 110 (1)
	204:	10100 10010 (1)
14 →	240:	10 01 (2)
	34:	00 11 (2)
	15:	010 101 (3)
	0:	1100 1000 (4)
	170:	0000 1100 (4)
	128:	1100 0110 (4)
15 →	240:	00 10 (2)
	15:	110 001 (3)
16 =	2	
17 =	3	
18 →	90:	00 10 (2)
	204:	11000 10100 (2)
	0:	10100 11100 (2)
19 →	51:	00 11 (1)
	0:	11 10 (2)
	204:	00 11 (2)
20 =	6	
21 =	7	

(continued)

Table 3.1. (Continued)

State	Key	Entry
22 →		
	146:	00 10 (2)
	90:	0000 1000 (4)
	0:	11011000 11111000 (4)
23 →		
	51:	00 11 (1)
	128:	00 10 (2)
	204:	00 11 (2)
	0:	000 100 (2)
24 →		
	48:	00 10 (2)
	240:	000 100 (3)
	0:	100 011 (3)
25 →		
	0:	1101000 1111000 (7)
	240:	0000000 1101000 (14)
26 →		
	90:	00 10 (2)
	85:	010 110 (3)
	170:	100100 101100 (6)
	0:	10110100 10111100 (8)
	240:	11100100 10011100 (16)
27 →		
	48:	11 10 (4)
	85:	010 110 (3)
	240:	000 100 (6)
	0:	0100 1100 (8)
	170:	100100 101100 (6)
28 →		
	192:	00 10 (2)
	200:	10 01 (2)
	51:	100 110 (1)
	204:	100 110 (2)
	0:	1010 1100 (1)
29 →		
	204:	00 10 (2)
	200:	10 01 (2)
	51:	100 110 (1)
	0:	1010 1100 (1)
30 →		
31 =	7	
32 →		
	0:	00 11 (1)
	128:	00 10 (2)
33 →		
	132:	00 10 (2)
	200:	00 11 (2)
	0:	111 100 (2)
	204:	000 111 (2)
	128:	0000 1010 (4)
34 →		
	170:	00 10 (2)
	0:	100 110 (3)
35 →		
	240:	00 11 (4)
	0:	100 110 (3)
	170:	10100 10010 (5)
36 →		
	0:	00 11 (1)
	4:	00 10 (2)
	204:	000 100 (1)
37 →		
	200:	00 11 (2)
	0:	1111 1100 (2)
	204:	0000 1111 (2)
	170:	100000 110000 (6)
	240:	010000 110000 (6)
	128:	010000 111000 (6)
38 →		
	34:	00 10 (2)
	85:	100 110 (3)
	170:	0000 1000 (4)
	0:	1100 1110 (4)
39 =	27	
40 →		
	128:	00 10 (2)
	0:	00 11 (2)
	170:	11010 10110 (5)
41 →		
	148:	00 10 (2)
	184:	0000 1000 (4)
	176:	0000 1010 (4)
	170:	11010 10110 (5)
	240:	00000000 10000000 (8)
	128:	10000000 10100000 (8)
	0:	11111000 11001000 (8)
	136:	01010000 10101101 (8)
42 →		
	170:	00 10 (2)
	34:	00 11 (2)
	0:	1100 1110 (4)
43 →		
	170:	10 01 (2)
	240:	00 11 (2)
	15:	000 111 (3)
	0:	1001 1000 (4)
	128:	1100 0110 (4)
44 →		
	12:	00 10 (2)
	204:	000 100 (1)
	0:	1000 1100 (1)
45 →		
46 →		
	34:	00 11 (2)
	0:	110 100 (3)
	170:	000 110 (3)
47 =	11	
48 =	34	
49 =	35	
50 →		
	51:	10 01 (1)
	128:	00 10 (2)
	204:	10 01 (2)
	0:	1010 1000 (2)
51 →		
	51:	10 01 (1)
	204:	00 10 (2)
52 =	38	
53 =	39	
54 →		
	50:	00 10 (2)
	51:	1000 0010 (2)
	128:	0000 1000 (4)
	204:	1000 0010 (4)
	170:	1000 1110 (4)
	240:	0010 1110 (4)
	0:	000010 111010 (4)
55 =	19	
56 →		
	240:	00 10 (2)
	128:	10 01 (2)
	184:	010 101 (3)
	0:	1010 0110 (2)
	34:	1010 1101 (4)
	170:	11010 10110 (5)
57 →		
	128:	10 01 (2)
	184:	010 101 (3)
	0:	1010 0110 (2)
	48:	1010 0100 (4)
	34:	1010 1101 (4)
	240:	10100 10010 (5)
	170:	11010 10110 (5)
58 →		
	128:	00 10 (2)
	0:	101 100 (3)
	240:	1100 1110 (8)
	170:	11010 10110 (5)
59 =	35	
60 →		
	60:	00 10 (2)
61 =	25	
62 →		
	240:	1100 1110 (8)
	204:	11000 11010 (3)
	0:	111110 100000 (3)
63 =	3	
64 =	8	
65 =	9	
66 =	24	

67 =	61
68 =	12
69 =	13
70 =	28
71 =	29
72 →	0: 00 10 (1) 4: 00 11 (2) 204: 000 110 (1)
73 →	204: 1100 0110 (2) 51: 11000 11010 (1) 0: 10110 10000 (2)
74 →	34: 00 10 (2) 170: 000 100 (3) 0: 10000 10100 (5) 85: 1110000 1101000 (7)
75 =	45
76 →	204: 00 10 (1) 0: 1010 1110 (1)
77 →	204: 10 01 (1) 128: 00 10 (2) 0: 1010 1000 (1)
78 →	0: 101 100 (1) 204: 11010 10110 (1)
79 =	13
80 =	10
81 =	11
82 =	26
83 =	53
84 =	14
85 =	15
86 =	30
87 =	21
88 =	74
89 =	75
90 →	90: 00 10 (2)
91 =	37
92 =	78
93 =	69
94 →	90: 00 11 (2) 0: 1010 1110 (1) 204: 1010 0101 (2) 51: 10010 11110 (1) 136: 111100 110110 (6) 192: 110110 011110 (6)
95 =	5
96 =	40
97 =	41
98 =	56
99 =	57
100 =	44
101 =	89
102 =	60
103 =	25
104 →	128: 00 10 (2) 4: 00 11 (2) 0: 000 100 (1) 204: 0000 1100 (1)
105 →	150: 00 10 (2)
106 →	170: 00 10 (2)
107 =	41
108 →	76: 00 10 (2) 204: 000 100 (1) 51: 10100 11100 (1) 0: 10010 11110 (2)
109 =	73
110 →	0: 110100 101100 (9) 240: 111000 100110 (9) 170: 10011000 11111000 (16)
111 =	9
112 =	42
113 =	43
114 =	58
115 =	49
116 =	46
117 =	81
118 =	62
119 =	17
120 =	106
121 =	97
122 →	128: 00 10 (2) 90: 00 11 (2) 0: 1010 1000 (2) 204: 11100 11110 (2)
123 =	33
124 =	110
125 =	65
126 →	90: 00 11 (2) 204: 11100 11110 (2) 0: 01110 10001 (2)
127 =	1
128 →	0: 00 10 (1) 128: 00 11 (2)
129 =	126
130 →	34: 00 10 (2) 170: 000 100 (3) 0: 1000 1100 (4) 128: 1000 1111 (4)
131 =	62
132 →	204: 00 10 (1) 128: 00 11 (2) 0: 000 110 (1)
133 =	94
134 →	184: 00 10 (2) 162: 00 11 (2) 170: 0000 1000 (4) 128: 0100 1100 (4) 240: 1000 1010 (4) 85: 10000 11000 (5) 0: 100000 101100 (6)
135 =	30
136 →	0: 00 10 (1) 136: 00 11 (2)
137 =	110
138 →	34: 00 10 (2) 170: 00 11 (2) 0: 010 110 (3)
139 =	46
140 →	204: 00 10 (1) 0: 100 110 (1)
141 =	78
142 →	240: 10 01 (2) 170: 00 11 (2) 15: 010 101 (3) 0: 1100 1000 (4) 128: 1100 0110 (4)
143 =	14
144 =	130
145 =	118
146 →	90: 00 10 (2) 204: 11000 10100 (2) 0: 10010 11110 (2)
147 =	54
148 =	134
149 =	86

(continued)

Table 3.1. (Continued)

150 → 150: 00 10 (2)
151 = 22
152 → 48: 00 10 (2) 240: 000 100 (3) 136: 1000 1111 (4) 0: 01000 11000 (5)
153 = 102
154 → 90: 00 10 (2) 85: 010 110 (3) 170: 1100 0110 (4)
155 = 38
156 → 192: 00 10 (2) 200: 10 01 (2) 136: 10 11 (2) 51: 100 110 (1) 204: 100 110 (2) 0: 1010 1100 (1)
157 = 70
158 = 134
159 = 6
160 → 128: 00 10 (2) 0: 000 100 (1)
161 = 122
162 → 170: 00 10 (2) 128: 10 11 (2) 0: 100 110 (3)
163 = 58
164 → 90: 11 10 (2) 128: 00 11 (2) 204: 000 100 (1) 0: 0000 1100 (1)
165 = 90
166 = 154
167 = 26
168 → 128: 00 10 (2) 136: 00 11 (2) 170: 10 11 (2) 0: 000 100 (1)
169 = 106
170 → 170: 00 10 (2)
171 = 42
172 → 34: 11 10 (2) 204: 000 100 (1) 170: 110 111 (3) 0: 1000 1100 (1)
173 = 74
174 = 138
175 = 10
176 = 162
177 = 114
178 → 51: 10 01 (1) 128: 00 10 (2) 204: 10 01 (2) 0: 1010 1000 (2)
179 = 50
180 = 166
181 = 82
182 = 146
183 = 18
184 → 240: 00 10 (2) 128: 10 01 (2) 170: 10 11 (2) 184: 010 101 (3) 0: 1010 0110 (2)
185 = 98
186 = 162
187 = 34
188 = 152
189 = 66
190 = 130
191 = 2
192 = 136
193 = 124
194 = 188
195 = 60
196 = 140
197 = 92
198 = 156
199 = 28
200 → 0: 00 10 (1) 204: 00 11 (1)
201 = 108
202 = 172
203 = 44
204 → 204: 00 10 (1)
205 = 76
206 = 140
207 = 12
208 = 138
209 = 116
210 = 180
211 = 52
212 = 142
213 = 84
214 = 148
215 = 20
216 = 202
217 = 100
218 = 164
219 = 36
220 = 196
221 = 68
222 = 132
223 = 4
224 = 168
225 = 120
226 = 184
227 = 56
228 = 216
229 = 88
230 = 152
231 = 24
232 → 204: 00 11 (1) 128: 00 10 (2) 0: 000 100 (1)
233 = 104
234 = 168
235 = 40
236 = 200
237 = 72
238 = 136
239 = 8
240 = 170
241 = 112
242 = 176
243 = 48
244 = 208
245 = 80
246 = 144
247 = 16
248 = 224
249 = 96
250 = 160
251 = 32
252 = 192
253 = 64
254 = 128
255 = 0

The basis of these methods is a study of the behaviour of cellular automata under "blocking transformations", which replace sequences of site values by single "blocked" values. The resulting blocked patterns can always be described by some cellular automaton rule, but not necessarily by a rule with the same k and r. But in studying scale invariance, it is of interest to consider cases in which blocking does in fact yield a rule of the same kind.

Since the number of possible length X (>1) blocks is larger than k, such blocking transformations can apply only to configurations that contain specific blocks of site values. So for example one can consider configurations that contain only the two-site blocks 00 and 01, to which one then applies a blocking transformation 00→0, 01→1. If the original configurations were generated according to the rule used in figures 1.1 and 1.2, then it turns out that the resulting blocked configurations, sampled every two time steps, can be reproduced by the same cellular automaton rule. The rule is thus a fixed point of this blocking transformation. This explains the scale invariance of figure 1.1: both the initial configuration and the rule are invariant under the blocking transformation, so the final pattern produced must also exhibit this invariance.

In general, the effective rule for blocked configurations may differ from the original rule, even if it is of the same kind. Table 3.1 specifies how various rules "simulate" others under blocking transformations.

One simple case is when a rule simulates the identity under a particular blocking transformation. In such cases all configurations that contain only the blocks involved must yield patterns that are periodic in time.

In other cases, a rule may simulate under a particular blocking transformation a rule like that of figure 1.1 which is invariant under blocking transformations. Then initial states that contain only the blocks involved produce scale-invariant patterns. There a quite a number of rules which simulate the rule of figure 1.1 when only blocks of the form $00\cdots$ and $10\cdots$ occur. With any of these rules, growth from a "simple seed" consisting of a single nonzero site always yields a self-similar pattern with the same form as figure 1.1. But when other blocks occur in the initial configurations, patterns with different forms can be produced, as shown in figure 3.1.

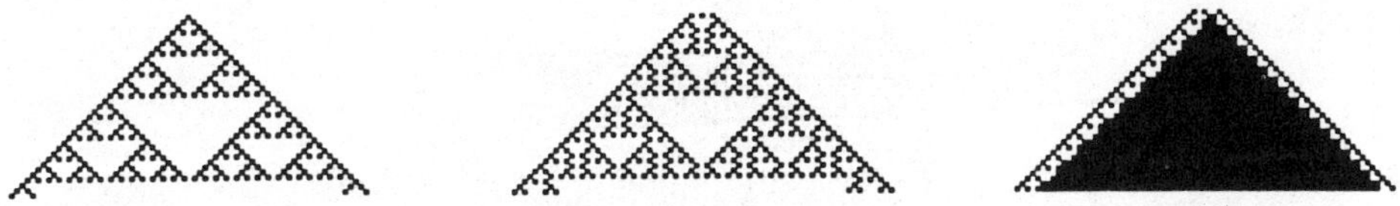

Figure 3.1. Different kinds of patterns produced by the same rule, but with slightly different initial seeds. In the first two cases, the seeds contains only blocks for which the rule used simulates the rule of figure 1.1. But in the third case, the seed contains another block, and a pattern of a different form is obtained. ($k=2$, $r=1$ rule number 218.)

4. Scale non-invariant growth

One might expect that patterns produced by local rules from localized initial seeds could involve no intrinsic scale (other than the size of the local neighbourhood) and so would always be self similar on large scales, like figure 1.1. Figure 4.1 shows however the range of patterns produced by $k=2$, $r=1$ rules. Not all are asymptotically scale invariant. Rule 30 is a clear example. Even though no large length scales were introduced explicitly in its construction, the pattern does not become self-similar even on arbitrarily large scales.

In the simplest cellular automata, a localized seed always gives rise to a simple homogeneous or periodic pattern. More complicated cellular automata yield self-similar patterns, in which sequences of site values are almost periodic. But as figure 4.1 shows, there is a further class of cellular automata that yield patterns which show no self-similarity. Sequences in these patterns are in fact often so complicated as to seem random with respect to statistical tests of randomness [18].

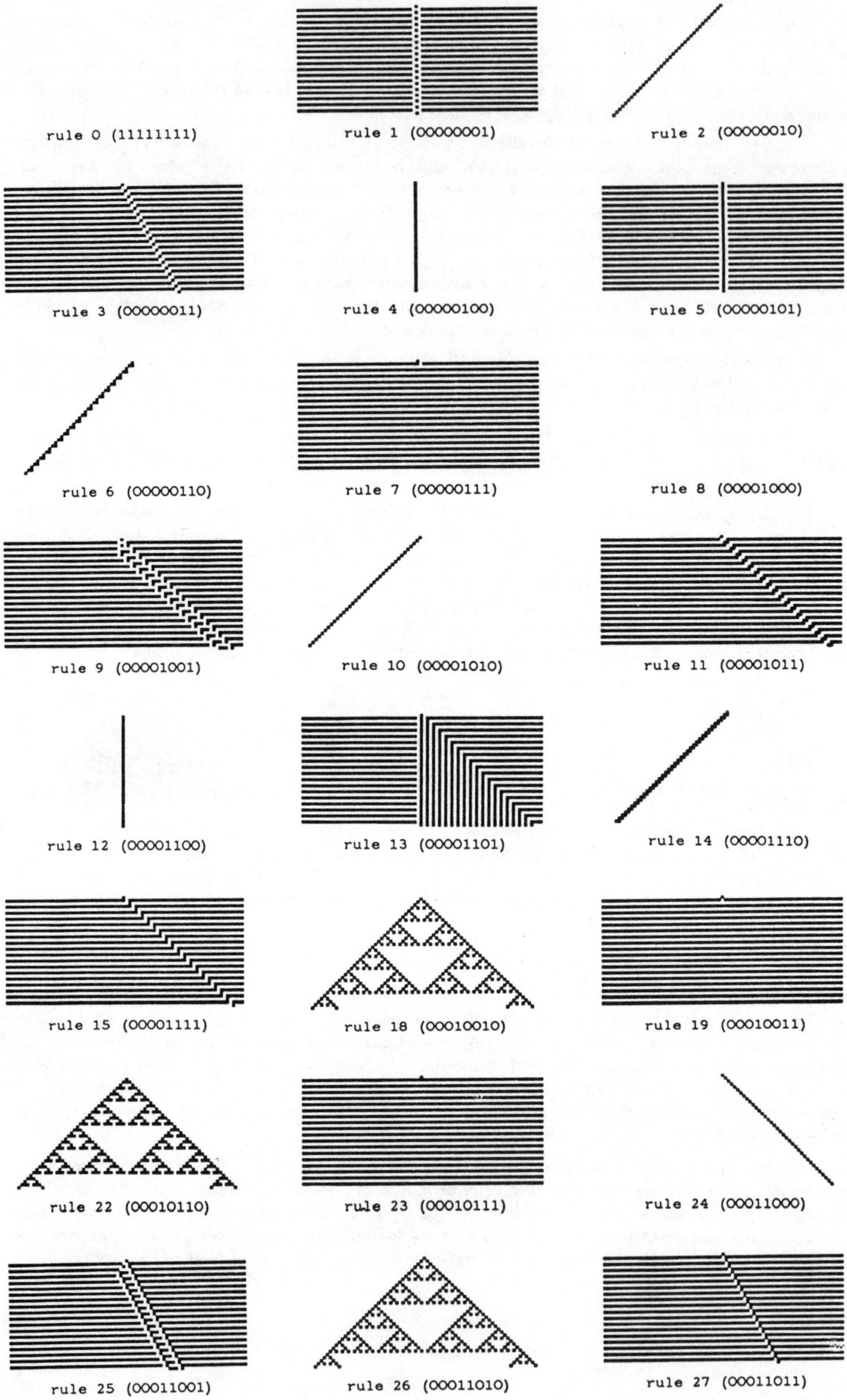

Figure 4.1. (Legend on page 269)

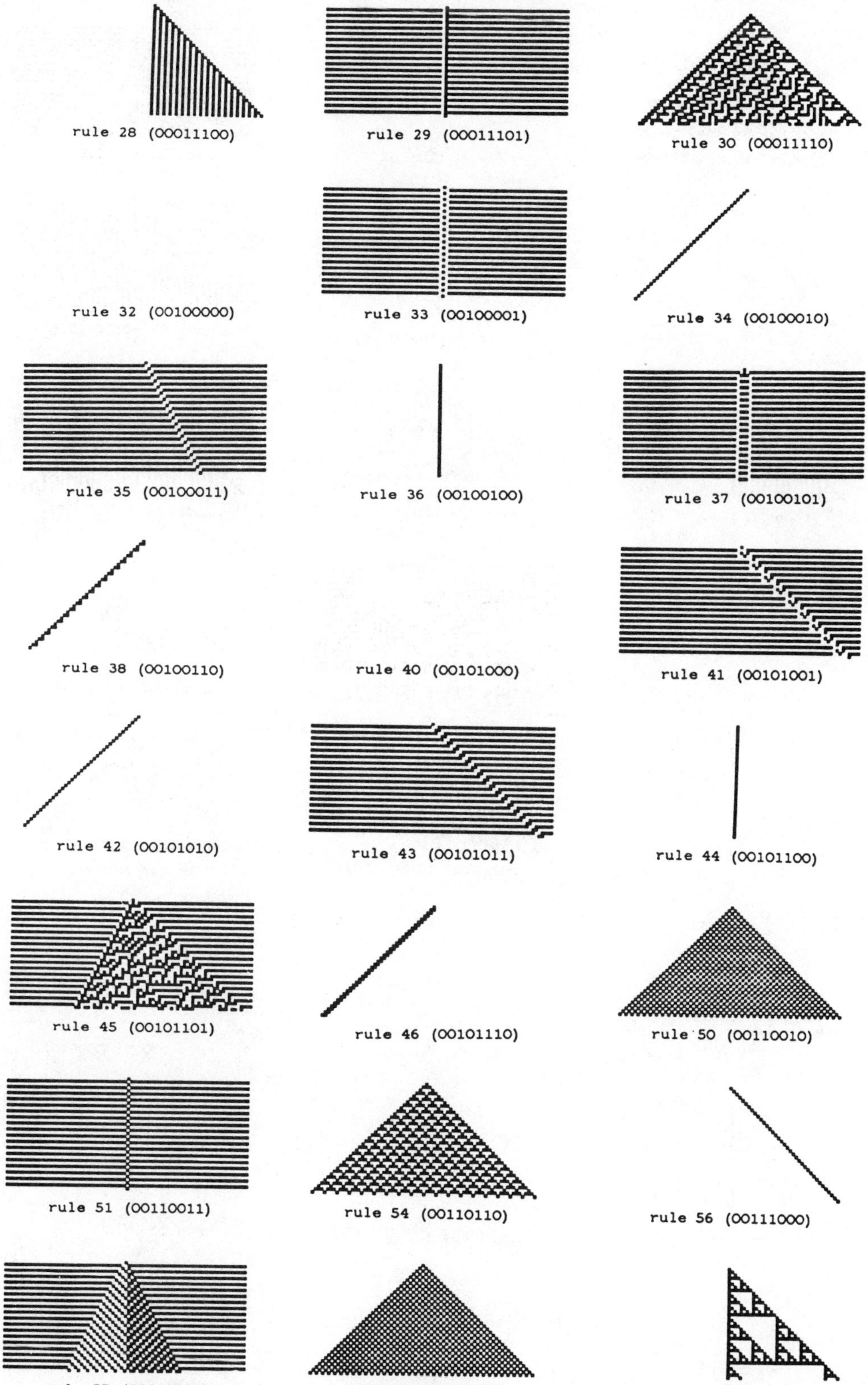

Figure 4.1. (Legend on page 269)

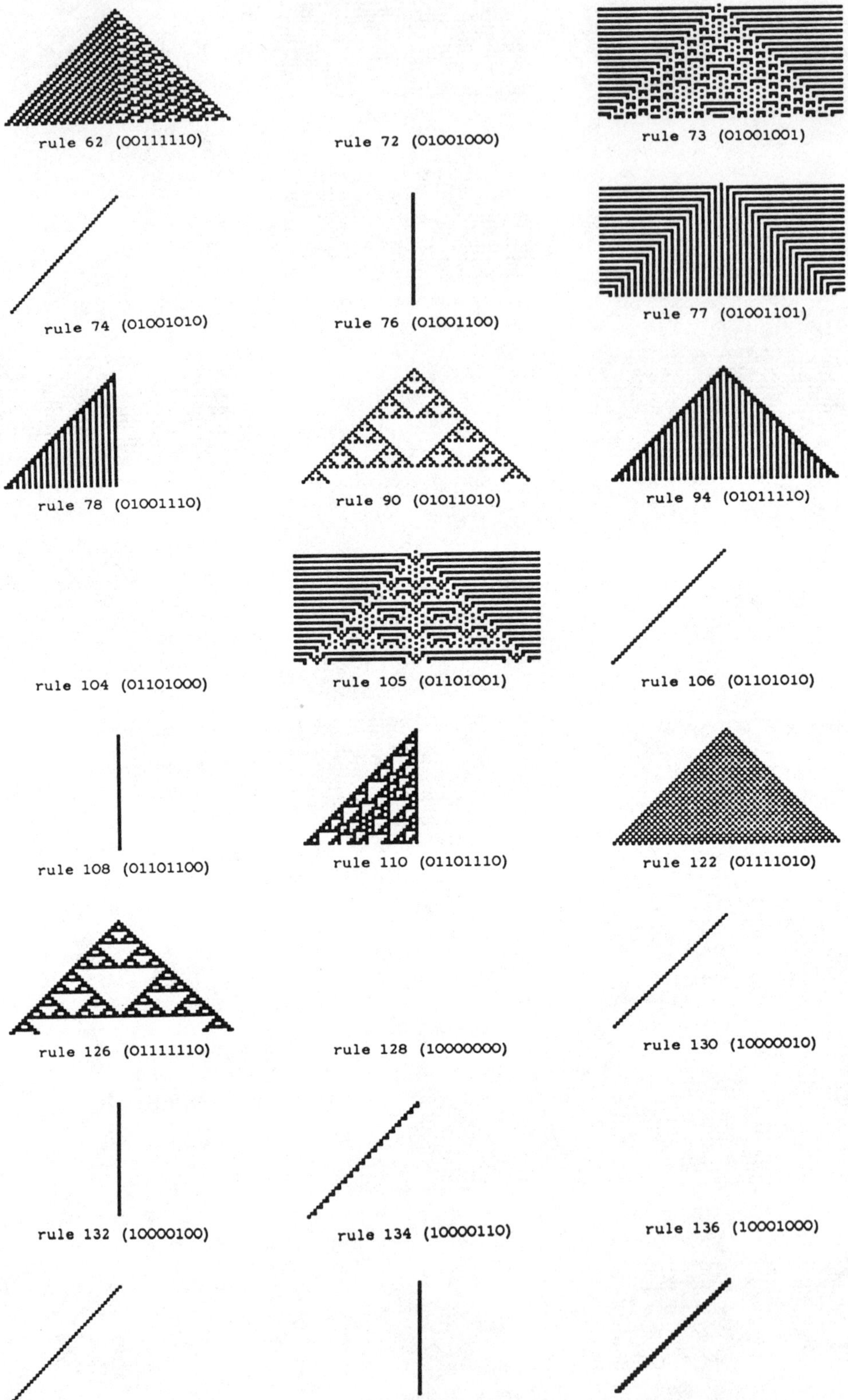

Figure 4.1. (Legend on page 269)

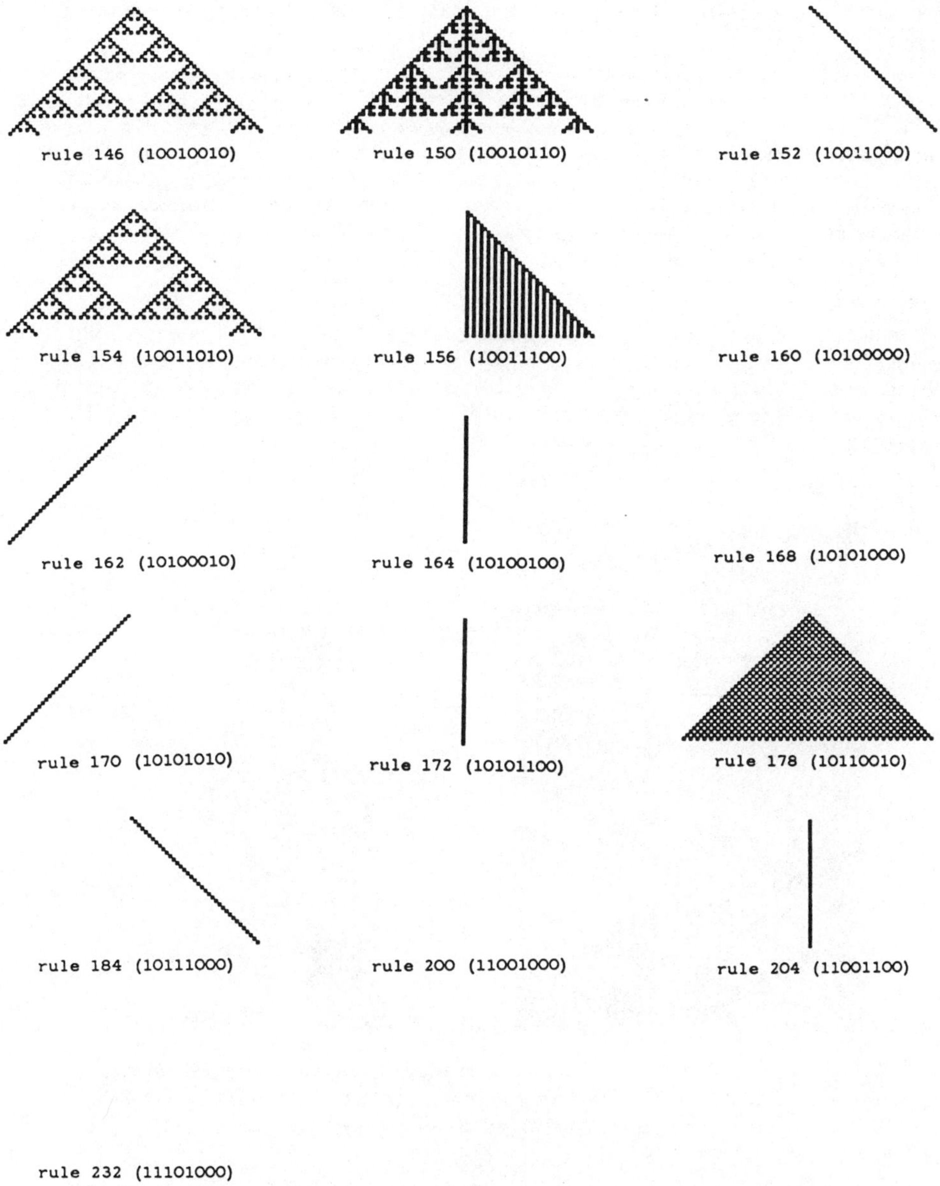

Figure 4.1. Patterns obtained by evolution from a single site seed according to the possible $k=2$, $r=1$ cellular automaton rules. In many cases, asymptotically scale invariant patterns are produced. But in some cases (such as rule 30) no self-similarity is evident, even on large scales.

These cellular automata thus act as pseudorandom sequence generators (e.g. [19]): starting from a simple seed, application of the cellular automaton rule yields sequences that are random for practical purposes.

Locality of microscopic dynamics is sometimes used as an explanation for the occurrence of fractal patterns in a wide variety of systems. But figure 4.1 shows that this not always a sufficient condition. The dynamics must have further simplifying features for fractal patterns to be generated.

Only a few cellular automaton rules have been found to be fixed points of blocking transformations, so that they must generate self-similar patterns. All those that have been found so far have the feature that they are linear (or "additive") modulo k. For such rules there is a superposition principle which gives the pattern produced from any seed as a superposition of patterns produced say from single site seeds. One may suspect in fact that self-similarity is usually associated with such linearity in underlying rules. When completely non-linear processes are present, patterns like figure 4.1, which are not self similar, may generically be produced.

5. Irregular growth

Localized patterns in most cellular automata grow essentially at a fixed rate, typically equal to the quantity r, which is analogous to the "speed of light" in the system. In several dimensions, the boundaries of the pattern can sometimes become dendritic. But even then the tips of the outermost branches still grow at a fixed rate, and if the pattern is sampled only when it takes on a faceted form, it will be seen to grow at a fixed rate.

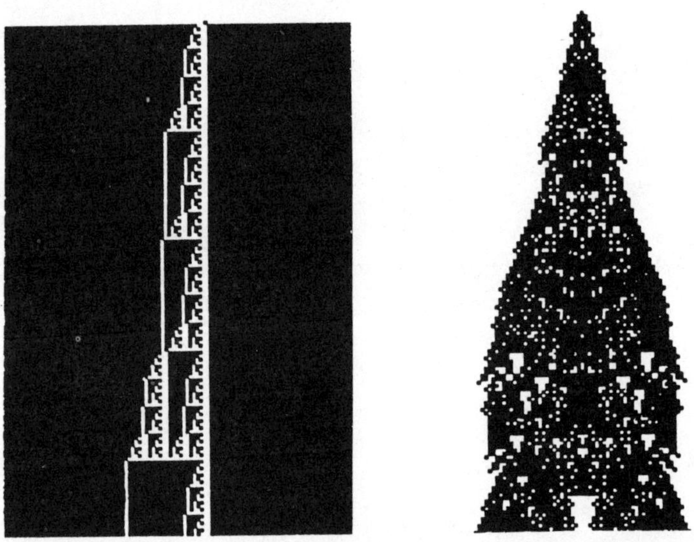

Figure 5.1. Examples of cellular automata in which localized patterns do not grow at the fixed maximal possible rate with time. ($k=2$, $r=1$ rule number 225, shifted one site per time step; $k=5$, $r=1$ totalistic rule code number 985707700, nonzero sites shown black.)

There are however cellular automata in which localized patterns can grow irregularly with time, or slower than linearly. Figure 5.1 shows two examples in one-dimensional cellular automata. In the first case, the pattern grows only with time t only like $\sqrt{t}$ - it expands by 2^j sites only at times of the form 2^{j+1}. In the second case, the pattern expands irregularly with time, at a average rate of about 0.2 sites per time step. In this latter case, expansion seems to occur whenever the outermost site has a particular one of its five possible values. The cellular automaton evolution acts like a pseudorandom generator, producing a sequence of site values that appears quite random. The boundary site value required for expansion occurs essentially as it would if generated just with a particular probability, rather than according to a definite rule.

Figure 5.2. Patterns generated by two-dimensional cellular automata that exhibit irregular grow from a localized seed [4].

Figure 5.2 shows a two-dimensional analogue of the phenomenon in figure 5.1: starting from a localized initial seed, a pattern grows irregularly with time. In the case shown, the pattern seems to tend asymptotically to a circular form. It appears to grow only when a particular feature occurs on its boundary.

The behaviour shown in figure 5.2 is very similar to that found with the Eden model (e.g. [20]), in which a pattern grows by new sites being added to the boundary with a particular fixed probability. This similarity between a stochastic model and an entirely deterministic model is in many respects not surprising. For in a practical computer simulation of the Eden model, one cannot make 'truly random" or "purely probabilistic" choices. Instead, one must make choices according to the output of a pseudorandom sequence generator. Such a generator uses a simple formula (in practice usually of the linear congruential form $x_n = ax_{n-1} + b \mod N$ [19]) to obtain a deterministic sequence of numbers which exhibit at least certain statistical characteristics of randomness (such as equidistribution and lack of correlation). Nevertheless, the results obtained seem to be good approximations to those that would be found by a genuinely stochastic procedure.

As figure 4.1 and 5.1 suggest, cellular automata can generate sequences of site values that seem quite random. In fact, one can use cellular automata as practical pseudorandom sequence generators, and the randomness of the sequences they produce seems often to be much greater than that of sequences produced by standard pseudorandom generators [21]. In standard computer simulations of stochastic processes, pseudorandom numbers are inserted from outside. But when the pseudorandom generator is included, the whole system is entirely deterministic. All that is really different in cases like those in figures 5.1 and 5.2 is that there the pseudorandom generator is included as an intrinsic part of the system, rather than being introduced from outside. One suspects in fact that there should be a deterministic system that provides a close approximation to any stochastic process. The simplicity of the rules necessary to generate apparently random sequences in figure 4.1 suggests that such deterministic systems need not in fact be substantially more complicated in construction than the stochastic systems to which they correspond. Now in fact, it seems likely that in many physical processes there is no intrinsic source of stochasticity. Rather, apparent randomness must arise for example through large numbers of molecular collisions, from fundamentally deterministic processes. What one may suspect is that in fact the apparent randomness associated with thermodynamic behaviour sets in even with a rather simple deterministic system.

6. Domain formation

Figure 6.1. Patterns obtained after 625 steps in the evolution of various two-dimensional cellular automata from disordered initial states [4].

Figure 6.1 shows the outcome of many steps in the evolution of various two-dimensional cellular automata, starting from initial states that consist not of simple seeds, but rather of random or disordered configurations in which the value of each site is chosen with equal probability from the k possibilities. Despite the randomness of the initial state, the irreversibility of the cellular automaton evolution allows various kinds of structures to form.

One common phenomenon is the formation of many "domains", each in a particular "phase". The phases correspond to different sets of states in the cellular automaton. Often they can be characterized as set of configurations in which the cellular automaton simulates some other cellular automaton, as discussed in section 3. Thus for example in a particular phase, only specific blocks of site values might occur.

Rule 33 in figure 1.3 is simple example of a cellular automaton in which long sequences of site with value one are transformed to value zero, and vice-versa. Thus any region with uniform site values behaves periodically with time. The cellular automaton may be considered to support a phase in which it simulates the identity rule (after blocking by a factor two in time). And as illustrated in figure 1.3, a disordered initial state relaxes to produce patterns in which only domains of this phase occur. There are two kinds of domains, corresponding to regions where all sites have value zero on even- or odd-numbered time steps. Once the domains are formed, their boundaries remain fixed for all time. In each domain the system is analogous to a crystal; the boundaries are the analogue of crystal defects. The existence of two distinct kinds of domains is an example of spontaneous symmetry breakdown.

Any set of configurations that admit a blocking transformation, as described in section 3, correspond to a phase in a cellular automaton rule. But only some of these phases are "stable", so that they are commonly produced from disordered initial states. Figure 6.2 shows a cellular automaton that supports two phases, one stable and one unstable. In the stable phase, only the blocks 0101 and 1010 occur, and the cellular automaton simulates the identity rule according to the blocking transformation 1010→0, 0101→1, so that the regions in this phase remain fixed with time. In the unstable phase, the blocks 00 and 01 occur, and the cellular automaton simulates a class 3 rule (with rule number 90) according to the blocking transformation 00→0, 01→1. If the cellular automaton contained only these blocks, it would behave like the class 3 rule. But whenever any of the stable phase occurs, it expands linearly with time, and eventually takes over the

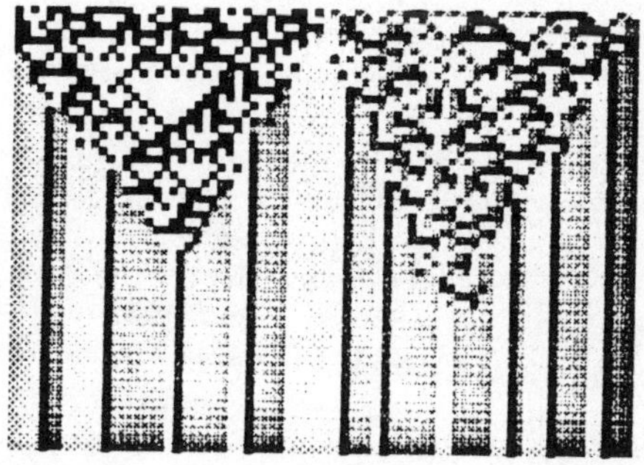

Figure 6.2. A one-dimensional cellular automaton with a stable and an unstable phase ($k=2$, $r=1$ rule number 94). Pairs of sites with values 00, 01, 10, 11 are indicated by progressively darker shades of grey. Alternate time steps are shown. The unstable phase, which shows chaotic, class 3 behaviour, contains two defects, one at the center, and one at the edge (the system is taken to have periodic boundary conditions). Starting from these defects, domains of the stable class 2 phase expand roughly linearly with time.

whole system. It may be possible to define a quantity analogous to free energy which determines the relative stability of different phases.

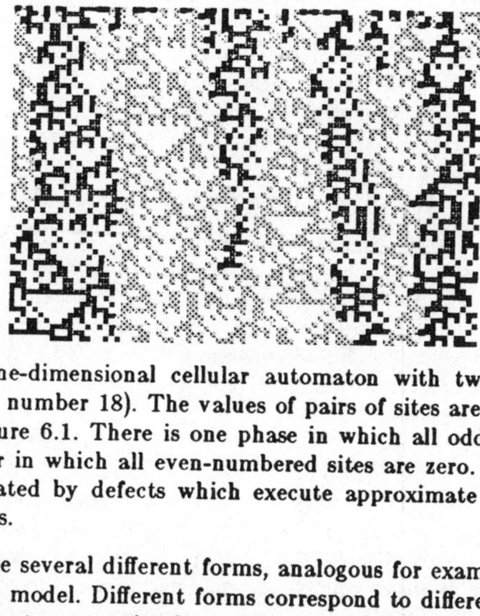

Figure 6.3. A one-dimensional cellular automaton with two kinds of domains ($k=2$, $r=1$ rule number 18). The values of pairs of sites are indicated by shades of grey, as in figure 6.1. There is one phase in which all odd-numbered sites are zero, and another in which all even-numbered sites are zero. Domains of the two phases are separated by defects which execute approximate random walks, and annihilate in pairs.

One phase may have several different forms, analogous for example to different directions of magnetization in an Ising model. Different forms correspond to different blocking transformations under which the cellular automaton simulates the same rule. So for example figure 6.3 shows the behaviour of a cellular automaton which simulates a particular class 3 rule (rule number 90) under both the blocking transformations 00→0, 01→1, and 00→0, 10→1. The rule thus supports two kinds of domains. The domains are separated by walls which execute random walks, and eventually annihilate in pairs. The density of these walls thus decreases like $t^{-1/2}$ [22], and the domains that remain become arbitrarily large.

Notice again that in figure 6.3, seemingly random behaviour is generated even though the evolution of the cellular automaton is entirely deterministic. The lifetime of the domain walls is long enough that each step in their path cannot be determined by a different initial site value. As a consequence, a comparatively short sequence of initial site values must be "processed" by the cellular automaton rule to produce an apparently random sequence of perturbations on the domain wall.

Whenever cellular automaton evolution is irreversible, only a subset of states are produced with time. Often these states contain predominantly a single phase that can be described as simulating some other rule. In such cases, one may consider the simulation of that rule to be a "attractive" one. And when such behaviour occurs, the cellular automaton must behave in the long time, large volume, limit, like a rule that is a fixed point of the blocking transformations that define the simulations. Thus the patterns that are produced must be statistically scale invariant.

There are nevertheless many cellular automata which do not seem to simulate other rules under blocking transformations in the large time limit. It is possible that such rules could be found equivalent to others under more complicated transformations, but then no scale invariance need result.

In one-dimensional cellular automata domain boundaries are single points. But as illustrated in figure 6.1, two-dimensional cellular automata may show much more complicated domain boundaries [4]. In many cases, the boundaries behave as if they had a surface tension. In some cases, the surface tension is positive, and the domains tend to a circular shape. They can presumably be described by an approximate continuous partial differential equation. In other cases, the effective surface tension is negative, and very convoluted, labyrinthine, patterns are produced.

It is comparatively simple to find phases in one-dimensional cellular automata which simulate the identity rule, and are therefore fixed with time. But in two dimensional cellular automata, the identification of such stationary phases can be much more difficult. In fact, even the question of whether there exist configurations which are invariant (or have any particular temporal period) under a given two-dimensional cellular automaton rule, is in general formally undecidable (e.g. [4]). This question is equivalent to the question of whether the plane can be covered by tiles from a given set, with the constraint that all adjacent pairs of edges must have the same colour. And this question is in turn in general equivalent to the problem of whether a computer (such as a Turing machine) can have input on which it will never halt.

If there are any configurations in a one-dimensional cellular automaton which are periodic in time, then there must be some which are also periodic in space, so that a regular "crystalline" structure can occur. But it turns out that there can be two-dimensional cellular automaton in which the only temporally periodic configurations are spatially non-periodic (e.g. [4]). Thus the stationary states of two-dimensional cellular automata may necessarily be spatially non-periodic, and so cannot form a regular crystalline structure (cf. [23]).

7. Correspondence with spin systems

Cellular automata are closely related to spin systems in several ways. The relations are most evident when probabilistic elements are added to the cellular automaton rules. Then the cellular automaton evolution corresponds directly to a generalization of directed percolation, or to a dynamic spin system in the Glauber model. In one parametrization, a cellular automaton can evolve with two possible rules for each site, each chosen with a certain probability. Then as a function of this probability, there can be discontinuous changes in the behaviour of the cellular automaton, directly analogous to phase transitions in spin systems [24]. The transitions seem to fall into a variety of universality classes [24].

Another possibility is to consider the spacetime pattern formed by the evolution of a probabilistic d-dimensional cellular automaton as corresponding to a configuration of a $(d+1)$-dimensional system built up by successive application of transfer matrices [25]. With this approach, the time dependent behaviour of probabilistic cellular automata may be related to the equilibrium behaviour of spin systems.

Entirely deterministic cellular automata can however also correspond to dynamic spin systems. In the Glauber model for dynamic spin systems, the spins flip probabilistically, as if by interaction with a heat bath. This model is appropriate for example when interactions with phonons in a crystal are important. But there are also cases in which such interactions are negligible, and only interactions between the spins themselves should be considered. (An extreme example would be nuclear spin ferromagnetism.) In this case only the spins can appear in the model, and there should be no external source of noise. In addition, if the "spins" indeed represent magnetic dipoles, then the interactions between them must typically conserve energy, so that their dynamics is reversible.

It is possible to make a simple cellular automaton model for the dynamics of spins with a particular (say Ising-like) local energy function [26]. The model takes each spin to have a certain kinetic energy (or momentum), which can be exchanged with potential energy of interaction if the spin flips. The model is entirely deterministic, and conserves energy exactly. Yet at large times it yields configurations whose statistical properties correspond to the equilibrium Ising model.

The model is in fact a direct approximation to the dynamics of actual spins. And the fact that it yields behaviour corresponding to the equilibrium Ising model at large times is a consequence of the fact that such spin systems in fact relax to thermal equilibrium. A noteworthy feature again is that the randomness associated with thermal equilibrium is generated by the deterministic dynamics of the system, and occurs even if the initial state is simple.

This behaviour may well exemplify the general operation of the Second Law of thermodynamics [27]. Even starting from simple or ordered initial conditions, the evolution of systems is often complicated enough to produce behaviour that seems random according to measurements of such statistical quantities as coarse-grained entropy. The widespread validity of the Second Law would then be related to the ubiquity of computational sophistication in physical systems, responsible for example for the production of complex patterns such as some of those in figure 4.1.

The evolution of the cellular automaton just described samples many configurations typical of thermal equilibrium. In fact, the cellular automaton rule can be considered as corresponding to the microcanonical method for sampling equilibrium configurations. In the usual application of this method, configurations are chosen according to a probabilistic rule. But here the dynamics can be considered to provide a source of pseudorandom numbers that effectively emulates this probabilistic rule. Of course, in actual digital computer simulations, pseudorandom numbers are always used. But usually they are obtained by a separate procedure; here they are actually generated in the evolution of the system itself. The result is a method that promises to be somewhat more efficient in implementation, particularly on forthcoming parallel-processing computers [28], and also represents the microscopic physics of spin interactions. The method can be extended to treat systems where "spins" take on many values, corresponding for example to elements of discrete subgroups of continuous groups [29].

A cellular automaton that contains explicit probabilistic elements can be considered as an approximation to a deterministic cellular automaton which generated pseudorandom sequences. The explicit probabilistic elements inserted provide approximations to the actual pseudorandom sequences produced. But the macroscopic measurements which are made are usually insensitive to the degree of randomness of these sequences. As a consequence, pseudorandom sequences which fail even quite simple statistical tests of randomness are usually adequate for Monte Carlo simulations of spin systems and the like. The evolution of these systems does not succeed in "cryptanalysing" the pseudorandom sequences that are used.

Just as there are cellular automata which are closely related to standard homogeneous spin systems, so there can also be cellular automata that correspond to inhomogeneous spin systems, such as spin glasses, or Ising models with random external magnetic fields. A simple cellular automaton analogous to such an inhomogeneous system can be constructed as follows. Take the value of each site to be a two-component vector. The first component corresponds to the value of a spin at that site. But the second component can be used to specify which of a set of rules should be applied to update this spin.

8. Discussion

The main conclusion of these notes must be that many phenomena studied in condensed matter systems already occur in cellular automata whose microscopic construction is very simple. Cellular automata seem to capture the essential mathematical mechanisms that are responsible for many of the complex physical phenomena observed. But the simplicity of their construction potentially allows rather complete analysis to be carried out.

These notes have discussed several approaches to the analysis of cellular automata. One that seems to hold great promise, but which has not been covered here, is based on the theory of computation [30,18].

In this approach, the evolution of a cellular automaton is viewed as a computation, in which the information corresponding to the initial configuration is processed according to the rules of the cellular automaton. The nature of the behaviour and structures produced are then determined by the nature of the computation associated with the evolution. As mentioned in section 1, formal language theory can for example be used to describe the set of configurations produced by a one-dimensional cellular automaton.

Similar methods can be used to describe patterns produced from simple seeds. Periodic and self-similar patterns correspond to computations that can be performed by finite state machines; more complicated and apparently random patterns correspond to more sophisticated computations. Many cellular automata in fact seem to produce such complicated patterns that their behaviour can be found only by irreducible computations that are effectively equivalent to direct simulation [30]. No simplifying features that would allow for more efficient computation or prediction are in general present.

The Second Law of thermodynamics provides a general principle which describes the overall behaviour of many systems studied in condensed matter physics. But the precise operation of this law for systems far from thermal equilibrium remains for the most part a mystery. Cellular automata seem to capture enough of the fundamental mechanisms involved that they can be used to study it. The possibility of generating apparently stochastic behaviour by entirely deterministic processes appears to be an essential element. And through cellular automata such phenomena can be studied in rather direct terms using methods from the theory of computation.

References

1. S. Wolfram, "Cellular automata as models of complexity", Nature 311 (1984) 419.
2. *Cellular automata*, edited by D. Farmer, T. Toffoli and S. Wolfram, Physica 10D (1984) nos. 1 and 2, and North-Holland Publishing Co. (1984).
3. S. Wolfram, "Statistical mechanics of cellular automata", Rev. Mod. Phys. 55 (1983) 601.
4. N. Packard and S. Wolfram, "Two-dimensional cellular automata", J. Stat. Phys. 38 (1985) 901.
5. S. Wolfram, "Twenty problems in the theory of cellular automata", Phys. Scripta T9 (1985) 170.
6. B. Mandelbrot, *The fractal geometry of nature*, Freeman (1982).
7. R. Shaw, "Strange attractors, chaotic behaviour and information flow", Z. Naturforsch. 36a (1981) 80.

8. S. Wolfram, "Universality and complexity in cellular automata", Physica 10D (1984) 1.

9. N. Packard, "Complexity of growing patterns in cellular automata", Institute for Advanced Study preprint (October 1983).

10. O. Martin, A. Odlyzko and S. Wolfram, "Algebraic properties of cellular automata", Commun. Math. Phys. 93 (1984) 219.

11. S. Wolfram, "Computation theory of cellular automata", Commun. Math. Phys. 96 (1984) 15.

12. J. Hopcroft and J. Ullman, *Introduction to automata theory, languages, and computation*, Addison-Wesley (1979).

13. L. Hurd, "Formal language characterizations of cellular automaton limit sets", Princeton University preprint (May 1985).

14. N. Margolus, "Physics-like models of computation", Physica 10D (1984) 81.

15. N. Packard, "Cellular automaton models for dendritic crystal growth", Institute for Advanced Study preprint (May 1985).

16. J. M. Greenberg, B. D. Hassard and S. P. Hastings, "Pattern formation and periodic structures in systems modelled by reaction-diffusion equations", Bull. Amer. Math. Soc. 84 (1975) 1296; B. Madore and W. Freedman, "Computer simulations of the Belousov-Zhabotinsky reaction", Science 222 (1983) 615.

17. R. Stinchcombe, these proceedings.

18. S. Wolfram, "Origins of randomness in physical systems", to be published.

19. D. Knuth, *Seminumerical algorithms*, Addison-Wesley (1981).

20. L. Sander, these proceedings.

21. J. Milnor and S. Wolfram, "Cryptography with cellular automata", in preparation.

22. P. Grassberger, "Chaos and diffusion in deterministic cellular automata", Physica 10D (1984) 52.

23. P. Bak, these proceedings.

24. W. Kinzel, "Phase transitions of cellular automata", Z. Phys. B58 (1985) 229; P. Grassberger, F. Krause and T. von der Twer, "A new type of kinetic critical phenomena", J. Phys. A17 (1984) L105.

25. E. Domany and W. Kinzel, "Equivalence of cellular automata to Ising models and directed percolation", Phys. Rev. Lett. 53 (1984) 311.

26. M. Creutz, "Deterministic Ising dynamics", Ann. Phys., to be published.

27. S. Wolfram, "Computation theory and the Second Law of thermodynamics", in preparation.

28. D. Hillis, "The Connection Machine: A computer architecture based on cellular automata", Physica 10D (1984) 213; S. Wolfram, "Scientific computation with the Connection Machine", unpublished report (1985).

29. N. Packard and R. Shaw, private communication.

30. S. Wolfram, "Undecidability and intractability in theoretical physics", Phys. Rev. Lett. 51 (1985) 735; "Computer software in science and mathematics", Sci. Amer. 251 (Sept. 1984) 188.

FRACTAL GEOMETRY OF PERCOLATION IN THIN GOLD FILMS

Richard F. Voss, Robert B. Laibowitz, and Eileen I. Alessandrini

IBM Thomas J. Watson Research Center
Yorktown Heights, NY 10598 USA

ABSTRACT

Transmission electron micrographs of thin evaporated gold films with thickness varying from 6 to 10 nm were analyzed by computer. The films cover the range from electrically insulating to conducting and thus span the 2D percolation threshold. The film geometry is interpreted in terms of both the scaling theory of percolation and Mandelbrot's fractal geometry. We find that Au-Au and Au-substrate interactions set a small scale correlation length of order 20nm. Small clusters are dominated by these effects and have simple almost-circular shapes. At larger scales, however, the irregular connected clusters are ramified with a perimeter linearly proportional to area. Near the percolation threshold the large scale power-law correlations and area distributions are consistent with the scaling theory of 2nd order phase transitions. The collection of all clusters forms a fractal distribution with D=2 while the largest cluster has $D = 2-\beta/\nu \simeq 1.9$. Many of the usual analytic scaling relations between universal exponents are shown to have fractal geometric basis.

introduction

Thin film configurations of metal-insulator mixtures are technologically important and have been extensively investigated. Changes in the conductivity properties as the material composition is varied have been interpreted in terms of percolation [1]. The conductivity measurements are, however, complicated by the presence of other conductivity mechanisms not related to percolation. Leakage through the insulator or quantum mechanical tunneling between neighboring islands smear the percolation transition and give a temperature dependence to the conductivity. Although detailed structural micrographs of composite materials have been available for many years, there has traditionally been little work directly comparing geometric properties with percolation.

In the idealized percolation problem, a random mixture of conducting (fraction p) and insulating (fraction 1-p) material abruptly exhibits long-range conduction at a critical concentration $p=p_c$. This formulation [2] is mathematically equivalent to a second order phase transition [3], and is ideally suited to computer modeling [4]. The rich variety of universal "scaling" behavior near p_c is associated with the irregular shapes of the connected clusters. Until recently [5-7], however, statistical studies of cluster geometry were limited to computer simulations. In this paper, on the other hand, we use computer analysis of digitized micrographs to directly examine the cluster geometry of thin gold films near p_c.

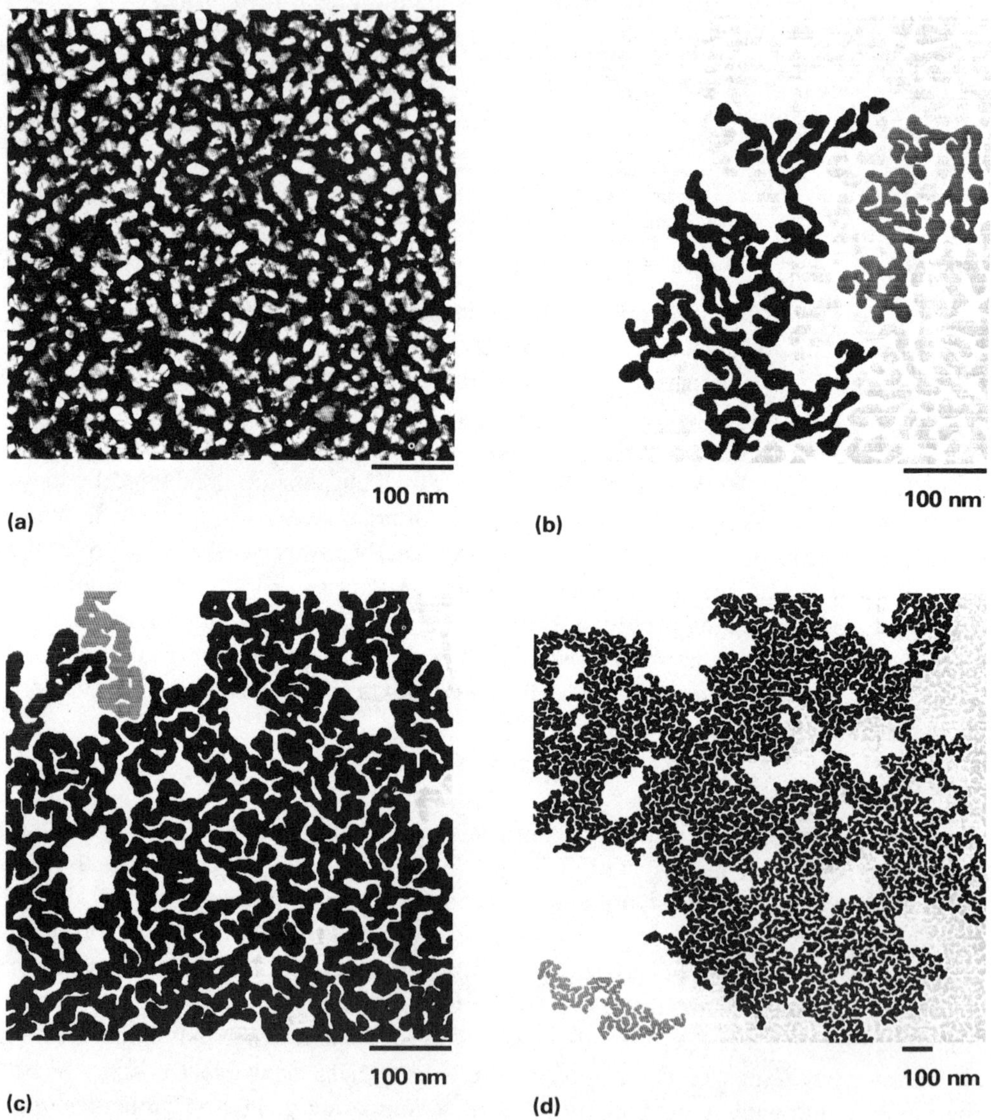

Figure 1. (a) Sample TEM micrograph. (b) Connectivity analysis of Fig. 1 with $p=0.64$ below p_c. The three largest clusters are shown in dark shades. (c) $p=0.76$ just above p_c. (d) Larger scale view with $p=0.71$ just below p_c.

experiment

The thin Au films were made at room temperature by electron beam evaporation onto special insulating substrates developed for transmission electron microscopy (TEM) [8]. The substrates consisted of a layer of amorphous Si_3N_4 deposited on a standard single-crystal Si wafer. The Si backing was then etched away from a portion of the wafer leaving a 30 nm thick Si sub 3 N sub 4 window supported by a Si frame. This window structure allowed the transmission of the electron beam for the TEM studies. A series of substrates was mounted in a vacuum chamber with a base pressure of 2×10^{-7} Torr. An electron beam was used to heat Au metal and cause deposition of Au atoms on the insulating substrate. A moving shutter blocked successive samples from the Au and produced a series with nominal thickness ranging from about 6 to 10 nm that varied from electrically insulating to conducting.

Transmission electron micrographs were taken of each of the samples at varying magnifications. A typical micrograph is shown in Fig. 1(a). The light areas are metal and the dark background is the amorphous substrate. The complicated irregular shapes of the connected clusters are readily visible. The structure within each cluster is due to different crystallographic orientations of the individual Au grains that form a cluster. The micrographs were digitized with a scanning densitometer (typically on a 512 by 512 grid). Figure 2 shows a histogram of the number of pixels (picture elements) having a given intensity of transmitted light for the image in Fig. 1(a). Two peaks are clearly visible. The peak on the left, at low intensity, corresponds to the dark background of the insulating substrate. The one on the right corresponds to the metal. By choosing a threshold intensity at the minimum between the peaks it was possible to automatically distinguish metal from insulator in the digitized images and create a two-state (metal-insulator) 2D array that approximated the actual film.

From this point, the analysis is the same as that used with computer simulations of percolation [2]. An optimal connectivity checking algorithm was used to find all connected clusters in each image. Figure 1(b) shows the result of the threshold separation into metal and insulator and the connectivity checking for the image in Fig. 1(a). The 3 largest clusters are seen in dark shades while the smaller clusters are seen in a uniform light gray. At a fractional Au area coverage $p = 0.64$, Fig. 1(c) is below p_c and the extent of the largest clusters is less than the field of view. Figure 1(c) shows the largest clusters from a film just above p_c with $p = 0.76$ while Fig. 1(d) shows a larger scale image with $p = .71$ just slightly below p_c.

Before discussing the detailed comparison of geometric statistics from such images with 2D expectations, it is important to reiterate the approximations being made in our analysis. Over large scales (>100 nm) the film properties (such as fractional coverage) are uniform. At small scales, however, the Au "congeals" into 8-20 nm wide "sausages" with a 4-8 nm spacing [9]. The transition from insulator to metal in the films as more Au is added is actually a 3 dimensional process complicated by surface tension and substrate wetting effects. Although it is likely that the initial distribution of where Au atoms first hit the substrate is completely random, the atoms can move once on the surface. There is a tendency for the Au to form droplets on the surface. In fact, at elevated temperatures (around 550 C) the Au forms isolated round droplets

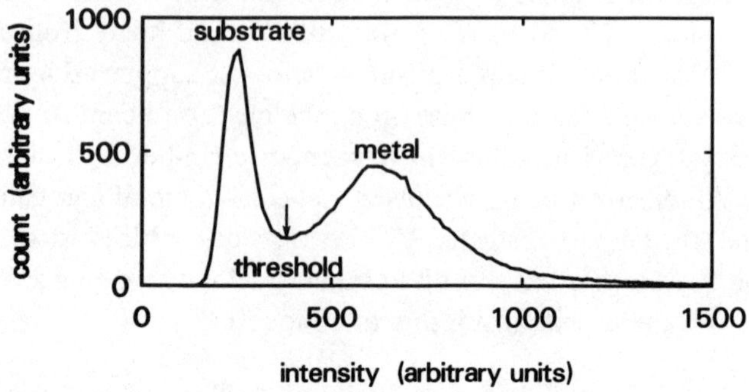

Figure 2. Histogram of the number of pixels having a given intensity from Fig. 1(a).

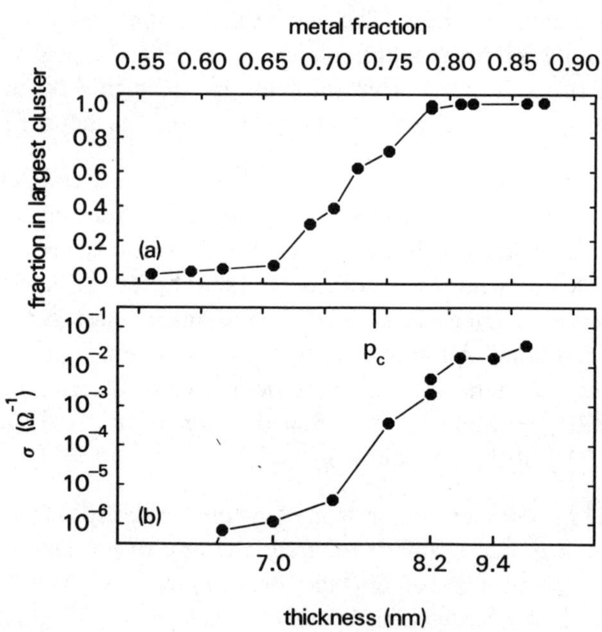

Figure 3. (a) Fraction of image in the largest cluster as p is varied. (b) Measured room temperature conductivity.

rather than the irregular clusters shown in Fig. 1. The small scale roundness (of order 10 nm) is indicative of the beginnings of this process in our room temperature samples. The nominal thickness (here about 20-30 atomic layers) and the 2D metal fraction ($\simeq$0.74) at which long range connectivity develops, thus, strongly reflect the metal and substrate material properties and thermal history. Nevertheless, as we shall see below, the large scale geometric scaling properties of these complicated 3D systems are adequately described in terms of 2D random percolation and fractals.

percolation as a 2nd order phase transition

The percolation problem is one of the simplest realizations of a 2nd order phase transition [2,3]. At the transition, the usual analytic scaling behavior as characterized by various exponents is reflected in the cluster geometry with its fractal dimensions [3,10,11]. P(p), the "percolation probability" (the fraction of the sample in the "infinite" cluster) increases for $p > p_c$ as $P \propto (p - p_c)^{\beta}$. The correlation length ξ diverges both below and above p_c as $\xi \propto 1/|p - p_c|^{\nu}$. The average cluster area below p_c (which corresponds to the average hole size above p_c) diverges as $\langle A \rangle \propto 1/|p - p_c|^{\gamma}$. At p_c the interesting distributions become power-laws. The number of clusters (per unit area) having area A has the form $n(A) \propto 1/A^{\tau}$. The pair correlation function (the probability that two points separated by a distance R are in the same cluster) has the dependence $G(R) \propto 1/R^{\eta}$. Away from p_c the scaling behavior (and power-laws) are valid for length scales λ up to the correlation length ξ . For $\lambda >> \xi$ the properties become homogeneous. Although the exact value of p_c is sensitive to local details (square or triangular lattice, site or bond percolation), the exponents are universal [2,3]. In 2D they have the values $\beta=5/64\simeq0.139$, $\nu=4/3\simeq1.333$, $\gamma=43/18\simeq2.309$, $\tau=187/91\simeq2.055$, and $\eta=10/48\simeq0.208$.

Figure 3 shows a comparison of the geometric and electrical measurements. Figure 3(a) shows the fraction of each image in the largest cluster, P(p), for a series of Au samples as p is varied across p_c. Figure 3(b) shows the measured room temperature conductivity σ for the same samples. Both cases show a broad transition. For the geometric P(p) this is due to the small finite area studied with the micrographs. Although the conductivity measurements sample a much larger area than the micrographs, the presence of other conduction mechanisms masks the sharp threshold in σ at p_c.

fractal dimensions and scaling in the Au clusters

Different physical quantities correspond to different universal exponents in the scaling interpretation and to different geometric objects (with different fractal dimensions) in the fractal interpretation [10,11]. At large scales the film properties as shown in Fig. 1(a) become uniform. All points are close to a metal-insulator boundary and (like a Peano curve) the cluster boundaries "fill" the plane corresponding to a fractal dimension $D = d = 2$. At p_c the largest cluster extends across the sample but does not "fill" the plane. It has a fractal dimension $D_c \simeq 1.9$ that directly reflects the universal exponents. D_c also characterizes the small scale, $\lambda < \xi$, properties of the other clusters.

This fractal behavior is expected to extend from length scales ξ down to the lattice size. For the case of our Au films, however, surface tension and wetting effects give a small scale cutoff much greater than the atomic size. The cutoff is clearly seen in a perimeter,

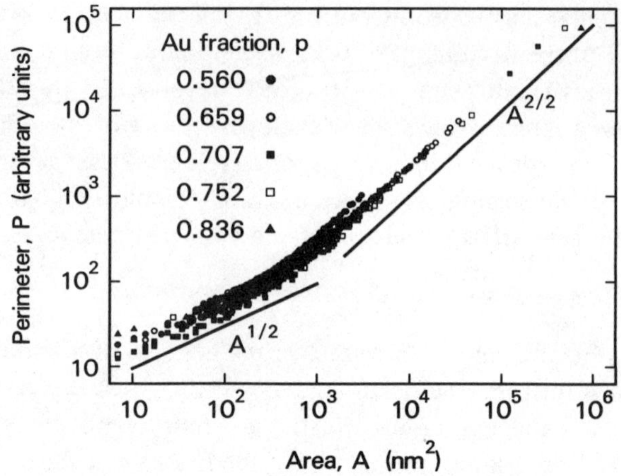

Figure 4. Scatter plot of the perimeter P vs area A for the gold clusters. Each cluster produces one point. The solid line shows the expected $P \propto A$ behavior at large A.

Figure 5. Exact fractal model of percolation geometry.

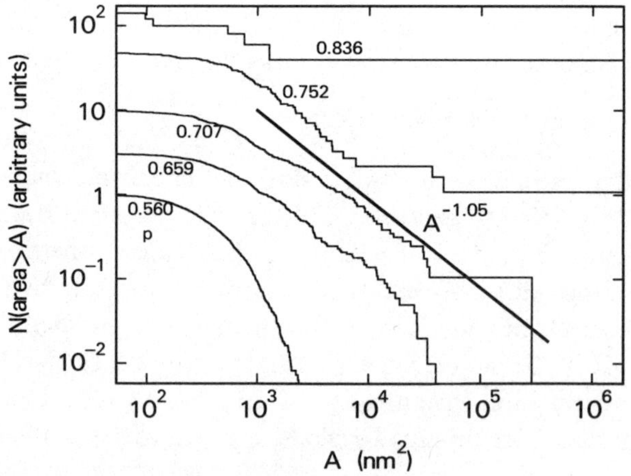

Figure 6. The cumulative distribution N(area>A) vs A at different fractional Au coverages p. The solid line shows the expected $1/A^{1.055}$ dependence at p_c.

P vs area, A plot for all of the clusters. As first suggested by numerical simulation and later proven rigorously [12] for percolation clusters, $P \propto A$ for large A at all p. Figure 4 shows a scatter plot of P (defined as the number of unoccupied sites on the digitized grid adjacent to a given cluster) as a function of A for our films. Each point represents one cluster. The smallest clusters ($A < 250$ nm^2) have simple almost circular shapes and show $P \simeq A^{1/2}$ corresponding to line-like boundaries with $D \simeq 1$. For $A > 600$ nm^2, on the other hand, $P \propto A$ both above and below p_c. The crossover between $P \propto A$ and $P \propto A^{1/2}$ is the same for all p and is related to the Au-substrate interaction.

An important quantity in the scaling theory of percolation [2,3] is n(A), the average number (per lattice site) of clusters of area A as a function of p. As discussed above, at p_c there is no characteristic size scale for the clusters and $n(A) \propto 1/A^\tau$. The fractal interpretation of this law can be illustrated with an exact recursive model for percolation cluster geometry [13] at p_c as shown in Fig. 5. The construction proceeds as a Koch curve [10]. At each stage, a "bond" is replaced by the solid lines in Fig. 5 plus the new (disconnected) cluster shown as the dashed line. Since this construction consists of N=9 parts each of which is scaled by r=1/3 from the original it has a fractal dimension $D = \log(N)/\log(1/r) = 2$ in agreement with percolation. The largest cluster consists of N=8 parts with $D_c = \log 8/\log 3 \simeq 1.89$ in good agreement with the accepted value for the infinite cluster. Other geometric quantities (such as the backbone fractal dimension) are also close to their accepted values in this exact model [13].

For such a scaling model, the number of clusters having a characteristic size scale $\lambda >$ some length L is given by [10]

$$N(\lambda > L) \propto 1/L^D. \tag{1}$$

The area or "mass" of an individual cluster is given in terms of its size λ as

$$\text{area} \propto \lambda^{D_c} \tag{2}$$

and the cumulative area distribution becomes

$$N(\text{area} > A) \propto 1/A^{D/D_c}. \tag{3}$$

Since the cumulative distribution N(area>A) is simply the integral of n(A) above, scaling theory (with $n(A) \propto 1/A^\tau$) corresponds to $N(\text{area}>A) \propto 1/A^{\tau-1}$ and we have a relation between D_c and τ,

$$D_c = D/(\tau - 1) = d/(\tau - 1). \tag{4}$$

Figure 6 shows N(area>A) vs A for some of the Au films studied. The use of the cumulative distribution N(area>A) rather than n(A) eliminates any problems with histogram bin sizes. For p well below p_c, N(area>A) decays rapidly for A beyond the characteristic size ξ. For p well above p_c, N(area>A) becomes independent of A for $A > \xi^2$. Near p_c, however, N(area>A) decays with the expected $1/A^{\tau-1} \simeq 1/A^{1.05}$ dependence for a large range of A.

Many important physical properties of a system are determined by the pair correlation function [3] G(R). For percolation, G(R) is the probability that two points separated by a distance R are in the same cluster. At the critical point G(R) takes the power-law form $G(R) \propto 1/R^{\eta}$ and η can be related to D and D_c. We first consider the conditional correlation function for a single cluster. If the origin is in a specific cluster, then $G_c(R)$ is the probability that a point a distance R away is also in the cluster. $G_c(R)$ is also a power-law with exponent η_c, $G_c(R) \propto 1/R^{\eta_c}$, up to the cluster size λ. For $R >> \lambda$, $G_c(R) \rightarrow 0$. $G_c(R)$ is directly related to the mass distribution given by

$$M(R) \propto R^{D_c} \propto \int_0^R G_c(r)\, r\, dr \propto R^{2-\eta_c}. \tag{5}$$

and $D_c = 2 - \eta_c$. Figure 7 shows $G_c(R)$ for the largest cluster in some of our Au samples. For $p < p_c$, $G_c(R)$ falls off rapidly with increasing R. Above p_c, $G_c(R)$ becomes a constant at large R. Near p_c, however, $G_c(R)$ does show large ranges of power-law behavior. The solid lines in Fig. 6 show least squares fits to these portions where $0.097<\eta_c<0.24$.

G(R), however, involves both the single cluster $G_c(R)$ and the distribution over all clusters sizes λ. G(R) is proportional to $G_c(R)$ times the probability that the origin is in a specific cluster with size $\lambda > R$. Thus,

$$G(R) \propto G_c(R) \times \int_R^{\infty} \text{area}(\lambda)\, n(\lambda)\, d\lambda$$

where $n(\lambda) \propto 1/\lambda^{D-1}$ is the derivative of the cumulative distribution $N(\lambda>L)$ and $\text{area}(\lambda) \propto \lambda^{D_c}$. This leads to

$$G(R) \propto 1/R^{\eta_c} \times R^{D_c} \times 1/R^{D} \propto 1/R^{2\eta_c} \propto 1/R^{\eta}. \tag{6}$$

Eqs. (5) and (6) give another relation between D_c and the universal exponents

$$D_c = d - \eta_c = d - \eta/2. \tag{7}$$

Figure 8 shows the measured G(R) averaged over all clusters at different p. As in Fig. 7, for $p<p_c$, G(R) decays rapidly at large R while for $p>p_c$ G(R) approaches a constant. At p=0.707 G(R) shows a large power-law section with $\eta = 0.44 \pm 0.07$ but decreases more rapidly for R > 500 nm. At p=0.752, on the other hand, just above p_c, G(R) remains very close to a power-law with $\eta = 0.17 \pm 0.02$. The measured G(R) is thus in good agreement with the accepted $\eta \simeq 0.2$ at p_c and consistent with the single cluster $G_c(R)$ in Fig. 7.

conclusion

We have used a geometric analysis of TEM micrographs to show that at large scales actual Au clusters near p_c have the scaling properties one expects from both phase transition arguments and lattice percolation simulations and, hence, belong to the same

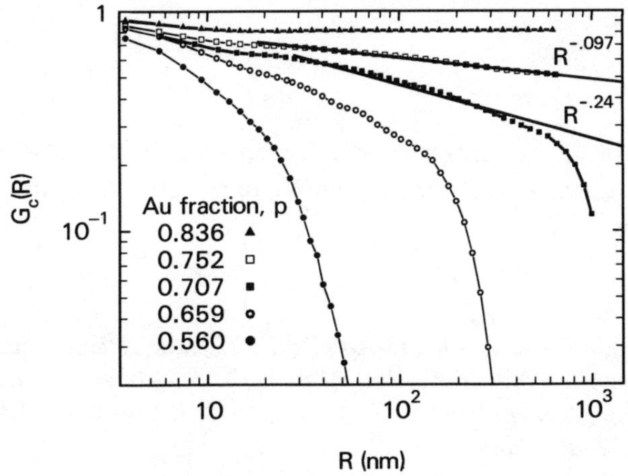

Figure 7. The single cluster correlation function $G_c(R)$ for the largest cluster in each the Au samples. The solid lines show power-law fits near p_c.

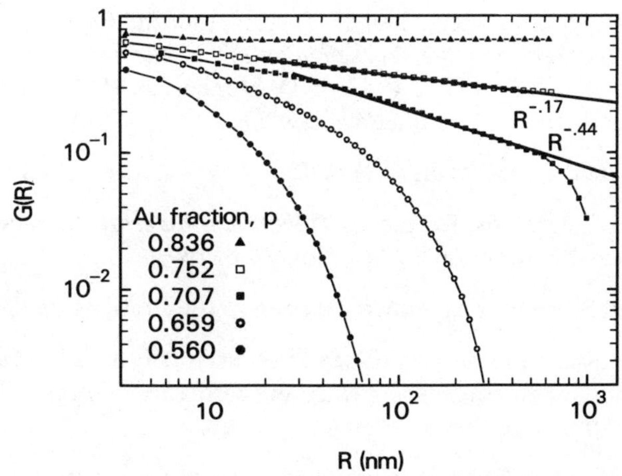

Figure 8. The pair correlation function G(R) vs R for the same samples. G(R) is the probability that two points separated by a distance R are both in the same cluster, averaged over all clusters. The solid lines show power-law fits near p_c to give the exponent η .

universality class. Local Au-Au and Au-substrate correlations, however, set a lower cutoff to the scaling behavior of about 20 nm and raise p_c to about 0.74. The collection of all all clusters at large scales are characterized by D=d=2. At p_c (and for length scales up to ξ away from p_c) individual clusters are characterized by $D_c = d/(\tau\text{-}1) = d\text{-}\eta_c = d\text{-}\eta/2 \simeq 1.9$.

acknowledgement

We are grateful to C. R. Guarnieri for expert sample preparation and to B. B. Mandelbrot, Y. Gefen, S. Kirkpatrick and Y. Imry for illuminating discussions and helpful suggestions.

references

1. See for example B. A. Abeles in *Applied Solid State Science,* edited by R. Wolfe (Academic, New York, 1976) Vol. 6, p. 1; B. A. Abeles, H. L. Pinsh, and J. I. Gittleman, *Phys. Rev. Lett. 35,*247 (1976); or C. J. Lobb, M. Tinkham, and W. J. Skocpol, *Solid State Comm. 27,* 1253 (1978).

2. See the excellent review by D. Stauffer, *Phys. Reports 54,* 1, (1979) and references therein.

3. R. B. Stinchcombe, this volume and H. E. Stanley, this volume.

4. S. Kirkpatrick, *A.I.P Conf. Proc. 40,* 99 (1977) and *A.I.P Conf. Proc. 58,* 79 (1979).

5. R. B. Laibowitz, E. I. Allessandrini, and G. Deutscher, *Phys. Rev. B25,* 2965 (1982).

6. R. F. Voss, R. B. Laibowitz, and E. I. Alessandrini, *Phys. Rev. Lett. 49,* 1441 (1982) and *Proceedings of the Workshop on the Mathematics and Physics of Disordered Media,* Univ. Minnesota 1983, B.D. Hughes, B.W. Ninhan, eds. (Springer-Verlag Lecture Notes in Mathematics, 1983).

7. A. Kapiltunik and G. Deutscher, *Phys. Rev. Lett. 49,* 1444 (1982).

8. R. B. Laibowitz and A. N. Broers, in *Treatise on Materials Science and Technology,* (Academic Press, New York, 1982), Vol. 24 p. 237.

9. The same metal-insulator asymmetry is seen in the Pb films on Ge substrates [7].

10. For a general discussion of fractals see B. B. Mandelbrot, *The Fractal Geometry of Nature* (Freeman, San Francisco 1982) and references therein. Chapter 13 deals specifically with percolation.

11. R.F. Voss, "Random Fractals: characterization and measurement", this volume.

12. H. Kunz and B. Souillard, *J. Stat. Phys. 19,* 77 (1978) and A. Coniglio and L. Russo, *J. Phys. A 12,* 545 (1979).

13. B. B. Mandelbrot and J. A. Given, *Phys. Rev. Lett. 52,* 1853 (1984).

ANOMALOUS DIFFUSION ON PERCOLATING CLUSTERS

Amnon Aharony

School of Physics and Astronomy
Tel-Aviv University
Ramat Aviv 69978, Israel

I. INTRODUCTION

Much of the recent renewed interest in *percolation theory* is related to the realization that percolation clusters are *self-similar*,[1] and may thus be modeled by *fractal* structures.[2] On a fractal structure, all the physical properties behave as *powers* of the relevant length scale, L. This behavior crosses over to a *homogeneous* one (i.e. independent of L, for appropriately defined quantities), on length scales larger than the percolation connectedness (or correlation) length, $\xi \propto |p-p_c|^{-\nu}$. Assuming that ξ is the only important length in the problem, all other lengths should be measured in units of ξ, and thus depend on L only via the ratio L/ξ. This implies *scaling*. For example, above the percolation threshold ($p \geq p_c$) one has[1]

$$M(L) = L^D \, m(L/\xi) \tag{1}$$

for the number of sites on the infinite incipient cluster within a volume of linear size L. The exponent D is the *fractal dimensionality*[2] of the cluster in the self-similar regime, and the scaling function m(x) behaves as a constant for $x \to 0$ and as $m(x) \sim x^{\beta/\nu}$ for $x >> 1$, so that $M(L) \sim L^d P_\infty$. Here, d is the Euclidean dimensionality of space, and $P_\infty \sim \xi^{-\beta/\nu} \sim (p-p_c)^\beta$ is the probability per site to belong to the infinite cluster. Thus, one identifies $D = d - \beta/\nu$.

Similarly, the *average conductivity* of such a volume behaves as

$$\sigma(L) = L^{-\mu/\nu} S(L/\xi), \tag{2}$$

with the homogeneous limit $\sigma \sim \xi^{-\mu/\nu} \sim (p-p_c)^{\mu}$ for $L \gg \xi$ and the anomalous one $\sigma \sim L^{-\mu/\nu}$ for $L \ll \xi$.

Such *scaling* ideas apply to *all* physical quantities. In the present lectures we shall use similar scaling ideas, and various heuristic arguments, to review the time (or frequency) dependent effects related to *diffusion* (or random walks) and *conductivity* on fractals.

The field of anomalous diffusion is currently very active, and new ideas appear at a high rate. Although I attempt to review here some of the established results, with the unavoidable personal emphasis, I chose to devote a large fraction of these notes to less established, preliminary heuristic results. At the risk of being wrong on some of these ideas, I shall try to transfer the flavor of the current discussions in the field and to stimulate further arguments on the many open questions.

II. ANOMALOUS DIFFUSION ON DILUTE CONDUCTORS

The *diffusion coefficient* of a random walker (e.g. on a lattice) may be defined via

$$\mathcal{D} = \frac{d}{dt} \langle r^2 \rangle, \tag{3}$$

where $\langle r^2 \rangle$ is the mean square displacement after t time-steps. Usually, $\mathcal{D}$ is a constant, and one has

$$t \sim \langle r^2 \rangle^{d_w/2}, \tag{4}$$

with the *fractal dimension of the random walk* $d_w=2$.

Following de Gennes,[3] we now turn to the notion of a *random walker* on a dilute network (the "*ant in the labyrinth*"). We start in the homogeneous regime, $L \gg \xi$. In this regime one may expect the conductivity to be related to the density of charge carriers, n, and the diffusion coefficient $\mathcal{D}$ via the Einstein relation,[4]

$$\sigma \propto n\mathcal{D}_{\infty}. \tag{5}$$

The subscript ∞ on $\mathcal{D}$ reflects the fact that on these scales, the d.c. conductivity occurs only through the *infinite* cluster (other clusters are typically smaller than ξ). Therefore, we also have $n \propto P_\infty \propto \xi^{-\beta/\nu}$. Using $\sigma \propto \xi^{-\mu/\nu}$, we find that $\mathcal{D}_\infty \propto \xi^{-\theta}$, with $\theta=(\mu-\beta)/\nu$. Following the same logic as before we thus have[4]

$$\mathcal{D}_\infty(L) = L^{-\theta}\mathfrak{D}(L/\xi), \tag{6}$$

with $\mathfrak{D}(x)\to x^{\theta}$ for $x>>1$. Eq. (6) implies that for $r<<\xi$ one has $\mathcal{D}_\infty(r)\sim r^{-\theta}$. Substitution in (3) therefore yields the *anomalous diffusion* law[4]

$$t \sim r^{2+\theta}, \qquad r\equiv\langle r^2\rangle^{\frac{1}{2}}, \tag{7}$$

i.e. $d_w=2+\theta$. We note that Eq. (7) holds only for motion on the infinite cluster. The same result is expected for all finite clusters whose linear size r_s is larger than $t^{1/(2+\theta)}$. For $r_s<t^{1/(2+\theta)}$ we expect $r\sim r_s$. Averaging over all sizes yields[4]

$$\langle r^2\rangle = t\xi^{-\mu/\nu} f(t/\xi^{2+\theta}), \tag{8}$$

with the limits $\langle r^2\rangle \sim t^{(2-\beta/\nu)/(2+\theta)}$ (for $t << \xi^{2+\theta}$), $\langle r^2\rangle \sim t\xi^{-\mu/\nu}$ (for $p > p_c$, $t >> \xi^{2+\theta}$) and $\langle r^2\rangle \sim \langle r_s^2\rangle \sim \xi^{2-\beta/\nu}$ for $p < p_c$, $t >> \xi^{2+\theta}$).

One standard way to derive the diffusion law (4) starts with the Langevin equation,[5]

$$m\ddot{\vec{r}} + \alpha\dot{\vec{r}} = \vec{F}, \tag{9}$$

with a random force $\vec{F}$ which has zero average and time correlations $\langle\vec{F}(0)\cdot\vec{F}(t)\rangle = K(t)$, and with a friction coefficient $\alpha(t)$ which is related to the effective diffusion coefficient, $\mathcal{D}(t) \propto 1/\alpha(t)$. Since $\alpha(t) \propto t^{-1}\int_0^t ds\, s\, K(s)$, the result $\mathcal{D}(t) \sim r(t)^{-\theta} \sim t^{-\theta/(2+\theta)}$ (with $\theta \neq 0$) can be reproduced[6] if one assumes that $K(t)$ has long time correlations, $K(t) \propto t^{-2/(2+\theta)} \propto r(t)^{-2}$. Anomalous diffusion is thus a result of *long memory*, or of force-force correlations of the form

$$\langle\vec{F}(0)\cdot\vec{F}(\vec{r})\rangle \sim 1/r^2. \tag{10}$$

The exponent 2 on the RHS of (10) is *superuniversal*, independent of d, D or θ. We have recently[6] performed simulations on both percolation and diffusion limited aggregates, and (apart from oscillations in the signs) confirmed Eq. (10). It is intriguing to find a general explanation for this superuniversal behavior.

III. AC CONDUCTIVITY

The arguments presented above yield a special time scale, i.e. $\tau = \xi^{2+\theta}$, at which the crossover from anomalous to normal diffusion on clusters occurs. The assumption that τ is the *only* relevant time scale led to the scaling form in Eq. (8). Consider now the frequency dependent ac conductivity, $\sigma(\omega)$. The above assumption implies that σ depends on ω only via the product $\omega\tau$,

$$\sigma(\omega) = \xi^{-\mu/\nu}\, S(\omega\tau). \tag{11}$$

with this assumption, the high frequency limit $\omega\tau >> 1$ yields

$$\tau(\omega) \propto \omega^{x}, \tag{12}$$

with $x = \mu/[\nu(2+\theta)]$. A Kramers - Kronig transformation on (11) then yields the dielectric constant,[4]

$$\varepsilon(\omega)-1 \propto \xi^{2-\beta/\nu}\, E(\omega\tau), \tag{13}$$

with $\varepsilon(\omega) \propto \omega^{-(1-x)}$ for $\omega\tau >> 1$ and $\varepsilon \propto \xi^{2-\beta/\nu}$ for $\omega\tau << 1$.

An alternative approach to the problem of ac conductivity starts from a *mixture of two conductors*, with basic conductances σ_1 and σ_2. The average conductance is then written in the scaling form[7,8]

$$\sigma = \sigma_1 |p-p_c|^{\mu}\, \bar{S}\, [(p-p_c)(\sigma_2/\sigma_1)^{-a}]. \tag{14}$$

For a metal ($\sigma_1>0$) - insulator ($\sigma_2=0$) alloy, $\bar{S}(x)$ approaches a constant for $x \to 0^{+}$, and we reproduce the usual limit. For a superconductor ($\sigma_1 \to \infty$) - metal ($\sigma_2>0$) we expect that the conductivity divurges as $\sigma=\sigma_2|p-p_c|^{-s}$ when $p \to p_c^{-}$ (this defines the exponent s), and thus $\bar{S}(x) \to (-x)^{-(\mu+s)}$ for $x \to 0^{-}$ and $a=1/(s+\mu)$. We shall return to the general mixture and to Eq. (14) below, in Sec. IV.

If the second component is a dielectric insulator, then one may replace σ_2 by $\sigma_2=i\omega\varepsilon_2/4\pi$. Eq. (14) may then be rewritten as

$$\bar{\sigma} = |p-p_c|^{\mu}\, \bar{\bar{S}}\, (\omega\tau_2), \tag{15}$$

with $\tau_2 \propto |p-p_c|^{-(\mu+s)} \propto \xi^{(\mu+s)/\nu}$. This is an *additional* basic time scale, which is associated with ε_2/σ_1, i.e. with the typical RC relaxation time of the "capacitor" constructed from pairs of clusters and the polarizable medium between them. If one assumes that τ_2 is the only relevant time scale,[8] then Eq. (15) yields Eq. (12) with $x=\mu/(s+\mu)$, and Eq. (13) is replaced by $\varepsilon \sim \xi^{s/\nu} \sim |p-p_c|^{-s}$ for $\omega\to 0$. Eq. (13) represented *the polarizability of finite isolated clusters*, while the result $\varepsilon \sim \xi^{s/\nu}$ contains only the polarizability of the *medium* between the clusters, which is connected with the long range Coulomb *interactions* between charges moving on *different* clusters.

In general, there is no reason *a priori* to ignore either of the time scales τ and τ_2. Thus, one should write

$$\sigma(\omega) = \xi^{-\mu/\nu}\, \tilde{S}(\omega\tau,\omega\tau_2). \tag{16}$$

The results in the various limits will now depend on the order by which these limits are approached, and one may expect different properties in the regions $\omega<1/\tau$, $1/\tau<\omega<1/\tau_2$ or $1/\tau_2<\omega$. These should be checked.

In the limit of high frequencies one might expect the general behavior $\sigma(\omega) \sim \xi^{-\mu/\nu}\,(\omega\tau)^u(\omega\tau_2)^v$, and ξ will cancel provided $u(2+\theta)+v(\mu+s)/\nu-\mu/\nu=0$. This reproduces Eq. (12), but now $x=u+v$, and additional information is needed to determine x.

As we discuss below, the exponent s turns out to be very close to $(\nu-\beta/2)$. If this value is used, then $\tau_2/\tau_1 \propto \xi^{-s/\nu}$. A simple ratio with $v=2$ and $u=-1$ then yields $\sigma \sim \omega^1$, in agreement with some recent experiments.[9] Although it is not clear if the above arguments apply to these experiments (which were done on two dimensional layers, but with three dimensional Coulomb interactions), such double scaling forms may turn out to be important for future developments on this problem. Direct ways to determine u and v are yet to be found.

IV. TERMITES

The scaling form (14) leads to the interesting generalization of the "ant" to a random walker which moves quickly on the "good" conductor (σ_1)

and slowly on the "bad" one ($\sigma_2 << \sigma_1$). Such a walker was called a "*termite*" by de Gennes,[10] and many of its properties are described in Stanley's lectures in these proceedings. This "Boston termite" moves into a bond whose conductance is σ_i with a (normalized) probability which is proportional to σ_i, and thus moves (σ_1/σ_2) steps on "good" bonds for each step on a "bad" one.[11] One should thus measure the time t in units of σ_1 or σ_2. A natural combination of Eqs. (8) and (14) is thus[11,12]

$$<r^2> = t\sigma_1 \xi^{-\mu/\nu} F[t\sigma_1/\xi^{2+\theta}, (p-p_c)(\sigma_2/\sigma_1)^{-1/(s+\mu)}]. \quad (17)$$

An alternative way to write (17) is

$$<r^2> = t\sigma_2 \xi^{s/\nu} \bar{F}[t\sigma_1/\xi^{2+\theta}, t\sigma_2/\xi^{2-\bar{\theta}}], \quad (18)$$

where $\bar{\theta} = (s+\beta)/\nu$. This identifies a *third basic time*, $\tau_3 \propto \xi^{2-\bar{\theta}} \propto \tau/\tau_2$.

The "Tel Aviv termite"[12] was constructed to correspond to the *superconducting* limit, $\sigma_1 \to \infty$. Thus, it has an equal probability to move from one site to any other site on a "superconducting" cluster. The time is kept constant until the termite moves to a "normal" bond, on which it acts like the "ant." Taking the limit $\sigma_1 \to \infty$ in Eq. (18), the first argument in $\bar{F}$ should disappear, and

$$<r^2> = t\sigma_2 \xi^{s/\nu} \bar{F}(\infty, t\sigma_2/\xi^{2-\bar{\theta}}), \quad (19)$$

in complete symmetry with (8) (provided -s replaces μ). This yields the homogeneous diffusion coefficient $<r^2>/t \propto \xi^{s/\nu}$ for $t >> \tau_3$, as expected, and the novel anomalous behavior

$$<r^2> \propto t^{(2-\beta/\nu)/(2-\bar{\theta})} \quad (20)$$

for $t << \tau_3$. The former was indeed confirmed in two dimensional simulations of both the Boston and the Tel Aviv termites.[11,12] The latter seems more problematic, for reasons similar to those detailed following Eq. (16); the result might depend on the order of limits in the double scaling form (18). The behavior of the termite at short time scales remains the subject of current research.

Before concluding this section we mention that there have been several recent efforts to relate the exponent s to "static" exponents, like β and ν. In his ε-expansion, in $d=6-\varepsilon$ dimensions, Stephen[13] obtained the result $s=2\nu-\beta$, consistent with the limit $\omega\to 0$ of (13). This is probably the correct result for the polarizability of *non-interacting* clusters. On the other hand, Keretsz[14] noted that numerical values of s for the superconductor - metal alloy are very close to one half of that value, $s=\nu-\beta/2$. Recently, Coniglio and Stanley[15] gave heuristic arguments which reproduce the Kertesz result above a lower dimensionality d_ℓ, which they identify as $d_\ell \simeq 2.1$. We present here a modified version of their arguments, which gives a different result. When the Tel Aviv termite "lands" on a large superconducting cluster of linear size r_s, it will jump back and forth on the cluster. Very often it will get off the cluster, onto a perimeter "normal" site, perform a few "normal" steps, and then move back on the cluster. It may also move onto another small superconducting cluster (which may be completely surrounded by the large one), and then back to the large cluster. The probability that the termite leaves the region of the large cluster is thus proportional to the surface to volume ratio, of a sphere of radius r_s. Thus, the typical time spent in this region is of order $\tau_s \propto r_s^d / r_s^{d-1} \propto r_s$, and the typical distance traveled is also of order r_s.

Consider now the motion of the termite as a sequence of m such motions within large clusters. The total displacement and time are thus

$$\vec{R} = \sum_{i=1}^{m} \Delta\vec{R}_i, \quad t = \sum_{i=1}^{m} \Delta t_i, \tag{21}$$

with the averages

$$<|R|^2> = m<|\Delta\vec{R}_i|^2> = m<r_s^2>, \qquad <t> = m<\tau_s> \propto m<r_s>. \tag{22}$$

Using the usual cluster size distribution,[16] one always has $<r_s^2> \propto \xi^{2-\beta/\nu}$. However, the average of r_s is of order $\xi^{1-\beta/\nu}$ only if $\beta/\nu<1$, i.e. for $d \lesssim 4$. For higher dimensions, $\beta/\nu>1$ and $<r_s>$ remains finite for $p\to p_c$.

For large m, the diffusion coefficient becomes $\mathcal{D}=<|\vec{R}|^2>/<t> \propto <r_s^2>/<r_s>$, and we find $\mathcal{D} \propto \xi^{s/\nu}$ with

$$s = \begin{cases} 2\nu-\beta, & d \gtrsim 4, \\ \nu, & d \lesssim 4. \end{cases} \tag{23}$$

The result for $d \gtrsim 4$ agrees with Stephen[13] and with Eq. (13).[4] The result for $d \lesssim 4$ was also obtained by Sahimi,[17] using different arguments, and agrees at two dimensions with Aharony and Stauffer's[18] proposal for the exponent μ (which should equal s at d=2 by duality). However, the value ν seems to be somewhat too large compared to available data at d=3 ($s \simeq 0.7 - 0.8$, $\nu \simeq 0.88$).

For very short times, the termite may spend all its time around a *single* large cluster. In this case, one should probably write $\mathcal{D} \propto \langle|\Delta\vec{R}|^2/\Delta t\rangle \propto \langle r_s^2/r_s\rangle \propto \langle r_s\rangle$, yielding a constant for $d \gtrsim 4$ and $D \propto \xi^{1-\beta/\nu}$, i.e. $s=\nu-\beta$ for $d \lesssim 4$. Note that the Kertesz conjecture is the average between our two values, ν and $(\nu-\beta)$. Could there be a delicate crossover between the two scales?

An alternative procedure,[15] in which one replaces $\langle r_s\rangle$ by $\langle r_s^2\rangle^{1/2} \propto \xi^{1-\beta/2\nu}$, would reproduce the Kertesz value, but such a replacement is not clearly justified. Ref. 15 contains additional arguments in favor of $s=\nu-\beta/2$ for $d \gtrsim 2.1$.

V. FRACTONS

Given Eq. (7), the linear size of the region of sites visited by the "ant" after t time steps is $r \sim t^{1/(2+\theta)}$, and therefore their number is of order $S \propto r^D \propto t^{d_f/2}$, with

$$d_f = 2D/(2+\theta) = 2D/d_w. \tag{24}$$

Since each of these sites must have been visited many times, the probability of return to the origin is proportional to $t^{-d_f/2}$, and thus the related density of states (on the single cluster) behaves as ω^{d_f-1}. In analogy to the usual ω^{d-1}, Alexander and Orbach[19] called the related (localized) excitations "fractons," and d_f the "fracton dimensionality." It is also called the "spectral" dimensionality.[20]

The density of states must crossover from ω^{d-1} at low frequencies to ω^{d_f-1} at high frequencies. Assuming that $\tau \propto \xi^{2+\theta}$ is the only time scale, one may write[21]

$$N(\omega) = \omega^{d_f-1}\mathcal{N}(\omega\tau), \tag{25}$$

and the normalization condition on N yields an interesting step-like increase in N around $\omega \sim 1/\tau$.[21,22]

Much of the theoretical interest in the fracton dimensionality arose from Alexander and Orbach's observation, that $d_f \simeq 4/3$ for percolation clusters at $d \simeq 2$. Following Rammal and Toulouse,[20] Leyvraz and Stanley[23] presented plausibility arguments that $d_f = 4/3$ should hold exactly, at least at high dimensionalities. These arguments concentrate on the number of visited sites, S(t), and are specific for the percolation infinite cluster at p_c. One constructs the cluster simultaneously with the motion of the "ant:" At each time step, the ant may either move to one of the already visited sites, or try to move into one of G growth sites never investigated before. In the latter case, this site may either be unoccupied, whence G decreases by $\Delta G = -1$ and the number of blocked sites, B, increases by unity, or it is occupied, whence $\Delta G > 0$ and the ant moves there. If there are no correlations then G is dominated by random fluctuations: $G \propto (S+B)^{\frac{1}{2}} \propto S^{\frac{1}{2}}$. The probability to increase S per unit time is[20] $dS/dt \propto G/S \propto S^{-\frac{1}{2}}$, and hence $S \propto t^{2/3}$.

The above argument clearly breaks down at d=1: Then S(t) consists of a straight segment, $G \equiv 2$, $dS/dt \propto 1/S$ and $S \propto t^{\frac{1}{2}}$, i.e. $d_f = d = 1$. A generalization of such a boundary effect might say[18,24] that G is always on the *compact narrow surface* of S. Since $S \propto r^D$, this yields $G \propto r^{D-1} \propto S^{(D-1)/D}$ and

$$d_f = 2D/(D+1), \tag{26}$$

yielding in particular $\theta = D-1$ and $\mu = (d-1)\nu$. Generally, one might expect a competition between the available compact surface, of $S^{(D-1)/D}$ sites, and the number of randomly added sites, of order $S^{\frac{1}{2}}$. This led Ref. 18 to suggest that Eq. (26) may hold only for $D \lesssim 2$, i.e. $d \lesssim 2.1$.

Recent accurate numerical studies indicate that neither $d_f = 4/3$ nor Eq. (26) holds at d=2, where one observes[25] $G \propto S^{0.49}$ (instead of $S^{\frac{1}{2}}$ or $S^{0.47}$) and the growth sites are not located on a narrow boundary of the visited sites. It has recently also become clear that, although the Alexander-Orbach "rule" $d_f = 4/3$ holds exactly for d>6, it breaks down[26] in $d = 6-\varepsilon$ dimensions. At the moment there exists neither a satisfactory explanation for the excellent approximation $d_f \simeq 4/3$ nor an alternative exact relation between μ (or θ) and "static" exponents (β and ν).

One possible source for the breakdown of $d_f=4/3$ concerns the neglect of *correlations* in Leyvraz and Stanley's argument. If one assumes that the steps in G are correlated, $\langle \Delta G_i \Delta G_j \rangle = A\delta_{ij} + B|i-j|^{-x}$, then $\langle G^2 \rangle = \sum_{i,j=1}^{S+B} \langle \Delta G_i \Delta G_j \rangle \simeq AS + BS^{2-x}$. For $x>1$ this corresponds to a factor $(1+S^{1-x})$ in G(S), and thus to

$$S(t) \propto t^{2/3} \, (1 + Ct^{-2(x-1)/3}). \tag{27}$$

So far, fits to the form (27) have not been very accurate. However, small values of (x-1) and a negative C may explain the observation of an effective power which is smaller than 2/3.

To obtain a value of d_f which is *smaller* than 4/3 one must have oscillating correlations, such that the (otherwise leading) term AS cancels, and one ends up with $\langle G^2 \rangle \sim S^{2-x}$, where $x=2(2-d_f)/d_f>1$. The ε-expansion yields[26] $x=1+\varepsilon/7$, and it would be interesting to identify x in terms of other exponents.

Before concluding this section we emphasize again that all the above arguments apply only to percolation clusters. Neither the Alexander-Orbach nor the Aharony-Stauffer relations apply to ordered structures, like the Sierpinski gaskets,[28] to the "parasite" problem (ants on lattice animals)[29] or to diffusion on diffusion limited aggregates.[30]

VI. CONCLUDING REMARKS

Some of the open problems in the field were listed above:

(a) What are the force-force correlations for a random walk on a fractal? Why is the exponent describing them superuniversal?

(b) What are the different time scales relevant to anomalous diffusion? How should one treat the related multivariable scaling functions?

(c) How do termites behave on short time scales?

(d) Is there an exact relation between s and "static" exponents?

(e) Is there an exact relation between μ and "static" exponents? What are the corrections to the Alexander-Orbach rule? How is the fracton dimensionality affected by correlations in the motion?

There exist many problems which we had no space to discuss, but are also quite important. Of these, we mention only one:

(f) The behavior of *elastic* networks seems to differ from that of resistor networks.[31] A new exponent T replaces the exponent μ, and much of our scaling arguments must be repeated.

ACKNOWLEDGEMENTS

These lectures would not be possible without the important contributions from and collaborations with J. Adler, S. Alexander, Y. Gefen, B.B. Mandelbrot, H. Nakanishi and D. Stauffer. Discussions with A. Coniglio, R. Orbach and H.E. Stanley are also appreciated. This work was supported in part by the U.S. - Israel Binational Science Foundation and by the Israel Academy of Sciences and Humanities.

REFERENCES

1. e.g. A. Kapitulnik, A. Aharony, G. Deutscher and D. Stauffer, J. Phys. A16:L269 (1983).
2. B.B. Mandelbrot, *The Fractal Geometry of Nature*, (Freeman, San Francisco, 1982).
3. P.G. de Gennes, La Recherche 7:919 (1976).
4. Y. Gefen, A. Aharony and S. Alexander, Phys. Rev. Lett. 50:77 (1983).
5. e.g. F. Reif, *Fundamentals of Statistical and Thermal Physics* (McGraw Hill Kogakusha, Tokyo 1965), Chap. 15.5.
6. H. Nakanishi, Y. Gefen and A. Aharony, unpublished.
7. J.P. Straley, Phys. Rev. B15:5733 (1977).
8. A.L. Efros and B.I. Shklovskii, Phys. Status Solidi (b) 76:475 (1976); D.J. Bergman and Y. Imry, Phys. Rev. Lett. 39:1222 (1977).
9. R.B. Laibowitz and Y. Gefen, Phys. Rev. Lett. 53:380 (1984).
10. P.G. de Gennes, J. de Phys. Lett. 200:2197 (1979).
11. A. Bunde, D.C. Hong, I. Majid and H.E. Stanley, J. Phys. A18:L137 (1985).
12. J. Adler, A. Aharony and D. Stauffer, J. Phys. A18:L129 (1985).
13. M.J. Stephen, Phys. Rev. B17:4444 (1978).
14. J. Kertesz, J. Phys. A16:L471 (1983).
15. A. Coniglio and H.E. Stanley, Phys. Rev. Lett. 52:1068 (1984).

16. D. Stauffer, Phys. Rept. 54:3 (1979); *Introduction to Percolation Theory* (Taylor and Frazer, London, 1985).
17. M. Sahimi, J. Phys. A17:L601 (1984).
18. A. Aharony and D. Stauffer, Phys. Rev. Lett. 52:2368 (1984).
19. S. Alexander and R. Orbach, J. de Phys. (Paris) Lett. 43:L625 (1982).
20. R. Rammal and G. Toulouse, J. de Phys. (Paris) Lett. 44:L13 (1983).
21. A. Aharony, S. Alexander, O. Entin-Wohlman and R. Orbach, Phys. Rev. B31:2565 (1985).
22. See also lectures by R. Orbach in these proceedings.
23. F. Leyvraz and H.E. Stanley, Phys. Rev. Lett. 51:2048 (1983).
24. S. Alexander, Ann. Israel Phys. Soc. 5:149 (1983).
25. H.E. Stanley, I. Majid, A. Margolina and A. Bunde, Phys. Rev. Lett. 53:1706 (1984).
26. A.B. Harris and T.C. Lubensky, J. Phys. A17:L609 (1984).
27. A. Aharony, H.E. Stanley and A. Margolina, unpublished.
28. Y. Gefen, A. Aharony, B.B. Mandelbrot and S. Kirkpatrick, Phys. Rev. Lett. 47:1771 (1981).
29. S. Wilke, Y. Gefen, V. Ilkovic, A. Aharony and D. Stauffer, J. Phys. A17:647 (1984); S. Havlin, Z. Djordjevic, I. Majid, H.E. Stanley and G.H. Weiss, Phys. Rev. Lett. 53:178 (1984).
30. S. Havlin, Phys. Rev. Lett. 53:1705 (1984).
31. D.J. Bergman and Y. Kantor, Phys. Rev. Lett. 53:511 (1984); Y. Kantor and I. Webman, Phys. Rev. Lett. 52:1891 (1984); L. Benguigui, Phys. Rev. Lett. 53:2028 (1984).

MAGNETIC PROPERTIES NEAR PERCOLATION

Amnon Aharony

School of Physics and Astronomy, Raymond and Beverly Sackler
Faculty of Exact Sciences, Tel-Aviv University
Tel-Aviv 69978, Israel

Yuval Gefen

Institute of Theoretical Physics, University of California
Santa Barbara, California 93106

Yacov Kantor

Exxon Corporate Research Laboratory
Clinton, New Jersey 08801

INTRODUCTION

At the percolation threshold, clusters of connected sites exhibit self-similar fractal geometry, in which many geometrical properties scale as powers of the relevant length scales. This motivated both the construction of fractal models, on which physical problems may be solved exactly,[1,2] and the detailed study of various geometrical characteristics of percolation clusters.[3,4] In this lecture we shall review the effects of this fractal geometry on the magnetic correlations of spins at and near the percolation threshold. The results reflect on neutron scattering experiments from percolation clusters and similar fractals.

Already the first study of magnetic models on fractals[5] showed that their universal exponents depend on various geometrical characteristic parameters, in addition to the fractal dimensionality. This was followed by detailed studies on various fractals,[6-8] which will not be repeated here. More recently, we undertook[10] to investigate the structure factor which is expected when neutrons are scattered from dilute magnets near percolation. There exist many related experimental studies,[11] and it turns out that the theoretical predictions which result from the fractal structure imply significant changes in the functional forms to which these data should be compared.

Since some of the material covered in this lecture was included in Ref. 10, we proceed with a short summary, which is followed by some more recent results.

The magnetic correlations between two spins propagates only through the *backbone* of the cluster connecting them. It is currently believed[4] that on length scales which are short compared to the percolation correlation length ξ_p (i.e. all length scales at the threshold $p=p_c$) this backbone has the shape shown schematically in Fig. 1. There are "blobs" of multiply connected spins, between which there exist chains of "singly connected" bonds. At the threshold there exists no typical length scale. If the distance between the two end points is r then all the dependences on r should be via power laws. In particular, the average number of singly connected bonds scales as

$$L_1(r) = A_1 r^{\tilde{\zeta}_1}, \tag{1}$$

and Coniglio[4] showed that $\tilde{\zeta}_1 = 1/\nu_p$, where ν_p describes the divergence of the correlation length, $\xi_p \propto |p-p_c|^{-\nu_p}$.

Another quantity of interest is the electrical resistance between the two end points, if each bond has unit resistance. The resistance is believed to scale as

$$L_R(r) = A_R r^{\tilde{\zeta}_R}. \tag{2}$$

At dimensions d=2,3 one has $\tilde{\zeta}_1 = 3/4$, 1.14 and $\tilde{\zeta}_R \simeq 0.97$, 1.25.[10]

At and below p_c all the spins belong to finite clusters. The probability that our two end spins belong to the same cluster, G(r), behaves as

$$G(r) \propto r^{-(d-2+\eta_p)} \tag{3}$$

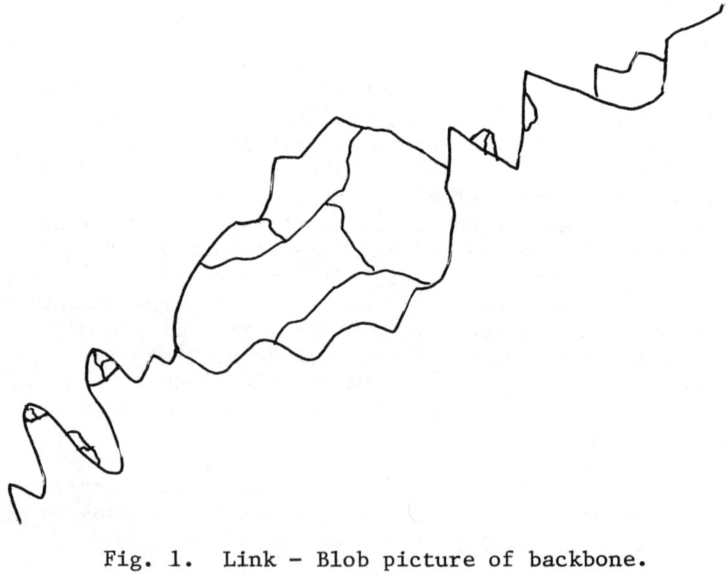

Fig. 1. Link - Blob picture of backbone.

at $p=p_c$ (or $r<<\xi_p$), and as $r^{-(d-1)/2}\exp(-r/\xi_p)$ for $p<p_c$. The exponent η_p has values $\eta_p \simeq 0.2$, -0.01 for d=2, 3.

MAGNETIC CORRELATIONS

Consider first the Ising case. For a linear chain of $L_1(r)$ bonds in series it is easy to see that the spin-spin correlation function is[10]

$$<S(0)S(r)> = (\tan h\, K)^{L_1(r)} = e^{-L_1(r)/\xi_1}, \tag{4}$$

with $K=J/k_BT$, where J is the nearest neighbor exchange, T is the temperature and k_B is the Boltzmann constant. The one-dimensional thermal correlation length is given by

$$\xi_1 = (-\ell n \tan h\, K)^{-1} \simeq \tfrac{1}{2}\, e^{2K}. \tag{5}$$

If two spins are connected via two chains, we can identify pairs of bonds such that disconnecting the two bonds in a pair disconnects the two spins. If the number of these pairs is $L_2(r)$ then to leading order in e^{-2K} one has[10]

$$<S(0)S(r)> \simeq 1 - 2L_2 e^{-4K} \simeq e^{-\frac{1}{2}L_2/\xi_1^2} \tag{6}$$

Thus, to leading order in e^{-2K} one may say that the multiply connected bonds create rigid correlations; the spin correlation function decays only on the singly connected spins, and[4]

$$<S(0)S(r)> \simeq \exp\,[-L_1(r)/\xi_1]. \tag{7}$$

However, this is true only if $L_1/\xi_1 << L_2/\xi_1^2$. On the percolation clusters, it turns out[10] that $L_2(r) \propto L_1(r)^2$, and therefore Eq. (7) holds only for $L_1(r)<\xi_1$. At larger distances one encounters a faster decay. For practical calculations we shall use a sharp cutoff, i.e. assume that (7) holds for $L_1(r)<u\xi_1$, while $<S(0)S(r)> = 0$ for larger distances.

For Heisengerg spins on a linear chain one still has $<\vec{S}(0)\cdot\vec{S}(r)> \simeq \exp(-L_1/\xi_1)$, but now (for K>>1)

$$\xi_1 \propto K = J/k_BT. \tag{8}$$

If two chains, of lengths L_1 and L_2, connect the same two spins, then $<\vec{S}(0)\vec{S}(r)> \simeq [1-a/(K_1+K_2)]$, with $(1-a/K_i) = (1-a/K)^{L_i}$, i.e. $K_i \simeq K/L_i$. Thus $K_1+K_2 \simeq K(1/L_1+1/L_2)$, and $<\vec{S}(0)\vec{S}(r)> \simeq \exp(-L_R/\xi_1)$, with $L_R = L_1L_2/(L_1+L_2)$. More generally, $L_R(r)$ is equivalent to the resistance of the bonds connecting the two spins. Again, this should hold only when $L_R<\xi_1$, and we expect a cutoff at larger distances.

The correlations are given as $\exp[-L_i(r)/\xi_1]$ only if the two spins belong to the same cluster. In a real experiment one should average over all configurations, i.e. multiply $<\vec{S}(0)\cdot\vec{S}(r)>$ by G(r). At $p=p_c$ we thus expect that

$$\overline{\langle \vec{S}(0)\cdot\vec{S}(r)\rangle} \propto \frac{1}{r^{d-2+\eta_p}} \exp\,(-Ar^{\tilde{\zeta}}/\xi_1) \tag{9}$$

for $Ar^{\tilde{\zeta}} < u\xi_1$, and zero otherwise.

We note that a similar result is expected on any fractal structure. The exponent $(d-2+\eta_p)$ is then replaced by $2(d-D)$, where D is the fractal dimensionality of the fractal.

STRUCTURE FACTOR AT PERCOLATION

The structure factor is defined as Fourier transform of the average spin correlation function. Using (9), we write (at $p=p_c$)

$$S(q,T) = \int d^d r\, r^{-(d-2+\eta_p)} e^{-A|\vec{r}|^{\tilde{\zeta}}/\xi_1} e^{i\,\vec{q}\cdot\vec{r}}, \tag{10}$$

with $A|\vec{r}|^{\tilde{\zeta}} < u\xi_1$. It is easy to check that S has the scaling form

$$S(q,T) = q^{-(2-\eta_p)} f(q\xi_1^{\nu_T}), \tag{11}$$

with $\nu_T = 1/\tilde{\zeta}$. For small $y = q\xi_1^{\nu_T}$, $f(y)$ behaves as

$$f(y) = Cy^{(2-\eta_p)}/[1+y^2+\Sigma_4 y^4+\ldots]. \tag{12}$$

The structure factor (12) would be a Lorentzian only if $\eta_p = 0$ and $\Sigma_4 = \Sigma_6 = \ldots = 0$. For an infinite cutoff ($u = \infty$) we find Σ_4=-0.32, -0.22 for d = 2,3 Ising systems and Σ_4=0.09, 0.19 for d=2,3 Heisenberg systems. Since at d=3 one also has $\eta_p \simeq -0.01 \pm 0.09$,[12] it seems that a Lorentzian is a good approximation. However, two dimensional systems show large deviations from a Lorentzian.

To obtain a quantitative feeling for S at d=2, we calculated the integral in (10) with $\tilde{\zeta}=1$ and $u=\infty$. We find[10]

$$f(y) = \tilde{C}y^{(2-\eta_p)} P^o_{1-\eta_p}[(1+y^2)^{-\frac{1}{2}}]/(1+y^2)^{(2-\eta_p)/2}, \tag{13}$$

where P^o_μ is the associated Legendre function.

For large y we thus have

$$f(y) \propto P^o_{1-\eta_p}(y^{-1}) \simeq \frac{\eta_p}{4\pi} + y^{-1} + \ldots . \tag{14}$$

Since $4\pi/\eta_p \simeq 60$, we expect significant deviations from the leading $q^{-(2-\eta_p)}$ already for $q\xi_1^{\nu_T} \lesssim 60$, where $q^{-(3-\eta_p)}$ becomes larger.

CORRELATION LENGTH OFF PERCOLATION

Below p_c, $G(r)$ behaves as $r^{-(d-1)/2}\exp(-r/\xi_p)$. Multiplying by Eq. (9), the average correlation function becomes

$$\overline{\langle \vec{S}(0)\cdot\vec{S}(r)\rangle} \propto r^{-(d-1)/2}\exp(-r/\xi_p - Ar^{\tilde{\zeta}}/\xi_1) \tag{15}$$

with some cutoff on r. If $\tilde{\zeta}=1$, as is almost the case for Ising at d=3 or Heisenberg for d=2, and if the cutoff is infinite, then the Fourier transform of (15) becomes a Lorentzian, $S(q,T,p) \propto (\kappa^2+q^2)^{-1}$, with $\kappa=\xi_p^{-1}+\xi_1^{-1}$. Indeed, this expression for κ is exact in one dimension. More generally, the correlation length is given by

$$\xi^2 = \int d^d r\, r^2\, \overline{\langle \vec{S}(0)\cdot\vec{S}(\vec{r})\rangle} \Big/ \int d^d r\, \overline{\langle \vec{S}(0)\cdot\vec{S}(\vec{r})\rangle}, \tag{16}$$

and we expect the scaling form

$$\xi(p,T) = \xi_1^{1/\tilde{\zeta}}\ X\,(\xi_1^{1/\tilde{\zeta}}/\xi_p). \tag{17}$$

Numerical integration, using (15), with A=1 and u=1, yielded the results shown in Fig. 2. For this cutoff we find that the results for Heisenberg ($\tilde{\zeta}$=1.0, 1.25 at d=2,3) and Ising ($\tilde{\zeta}$=3/4, 1.14) are indistinguishable. Note that for large $\xi_1^{-1/\tilde{\zeta}}$, i.e. high temperature, $1/\xi$ is linear in $\xi_1^{-1/\tilde{\zeta}}$, as indeed fitted in many experiments.[11] However, important deviations occur at low temperatures. In this regime, Eq. (17) may be expanded as

$$\xi^{-1} = a_o\xi_1^{-1/\tilde{\zeta}}\ (1+\sum_{n=1} a_n\ (\xi_1^{1/\tilde{\zeta}}/\xi_p)^n), \tag{18}$$

with the a_n's being combinations of incomplete gamma functions. For A = u = 1, we find $a_1 \simeq 0.11$, 0.08 and $a_2 \simeq 0.015$, 0.011 for d = 2, 3. For u = ∞ the results depend on $\tilde{\zeta}$, and we find a_1 = 5.81, 1.15, a_2 = -0.48, 0.09 for the Ising case, a_1 = 1.8, 0.86, a_2 = 10^{-5}, 0.86 for the Heisenberg case. The strong dependence on the cutoff means that these results should be taken only qualitatively. However, it is clear that the usual approximation

$$\xi^{-1} = a_o\ (\xi_1^{-1/\tilde{\zeta}} + a_1\xi_p^{-1}) \tag{19}$$

is wrong.

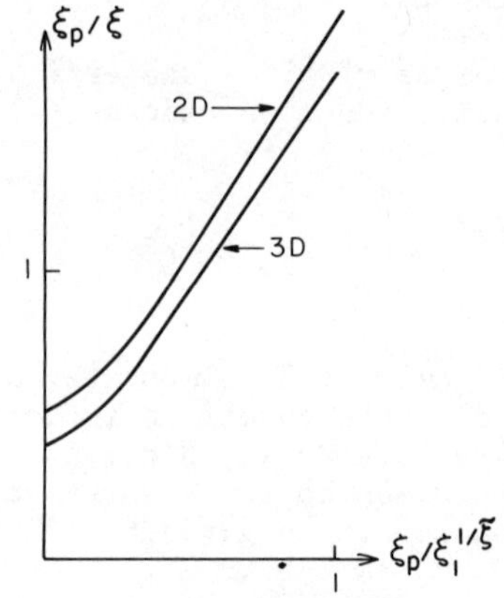

Fig. 2. Inverse Correlation Lengths.

ACKNOWLEDGEMENTS

We enjoyed discussions and collaboration with B.B. Mandelbrot. This work was supported in part by grants from the Israel Academy of Sciences and Humanities and from the U.S. - Israel Binational Science Foundation.

REFERENCES

1. Y. Gefen, A. Aharony, B.B. Mandelbrot and S. Kirkpatrick, Phys. Rev. Lett. 47:1771 (1981).
2. B.B. Mandelbrot and J. Given, Phys. Rev. Lett. 52:1853 (1984).
3. R. Pike and H.E. Stanley, J. Phys. A14:L169 (1981).
4. A. Coniglio, Phys. Rev. Lett. 46:250 (1981); J. Phys. A15:3829 (1982).
5. Y. Gefen, B.B. Mandelbrot and A. Aharony, Phys. Rev. Lett. 45:855 (1980).
6. Y. Gefen, A. Aharony and B.B.Mandelbrot, J. Phys. A16:1267 (1983).
7. Y. Gefen, A. Aharony, Y. Shapir and B.B. Mandelbrot, J. Phys. A17:435 (1984).
8. Y. Gefen, A. Aharony and B.B. Mandelbrot, J. Phys. A17:1277 (1984).
9. Y. Gefen, Y. Meir, B.B. Mandelbrot and A. Aharony, Phys. Rev. Lett. 50:145 (1983).
10. A. Aharony, Y. Gefen and Y. Kantor, J. Stat. Phys. 36:767 (1984).
11. R.J. Birgeneau, R.A. Cowley, G. Shirane and H. Yoshizawa, J. Stat. Phys. 34:817 (1984).
12. J. Adler, Z. Phys. 55:227 (1984).

FRACTAL PROPERTIES OF "DISORDERED SURFACES" AND THE TERMITE PROBLEM

H. Eugene Stanley,* A. Bunde,* A. Coniglio,* D. C. Hong,*
P. Meakin** and T. A. Witten+

*Center for Polymer Studies and Department of Physics
Boston University, Boston, MA 02215

**Central Research and Development Department, E. I. DuPont
de Nemours and Company, Wilmington, DE 19898

+Corporate Research Science Laboratories, Exxon Research
and Engineering Company, Annandale, NJ 08801

In this second lecture, I shall turn to the topic of fractal surfaces, for which knowledge of only "the" fractal dimension d_f is not sufficient to describe the essential physics. In fact, we must distinguish between the entire external perimeter or "hull," with its corresponding fractal dimension d_h, and the <u>unscreened</u> portion of the external perimeter and its fractal dimension d_u. A very simple argument gives rise to an expression for d_u in terms of d_f and d_p, the fractal dimension of the particles used to probe the surface. We shall see that the concept of the unscreened perimeter is at the heart of transport in a two-component random medium made up of superconductors and normal resistors. In particular, the exponent $\tilde{s} = s/\nu$ for characterizing the divergence of the macroscopic conductivity at the percolation threshold p_c is identically equal to d_u.

I am immensely indebted to my collaborators on both the termite project (A. Bunde, A. Coniglio and D. C. Hong) and the disordered surfaces project (A. Coniglio, T. A. Witten, and especially P. Meakin). In addition to these individuals, who kindly consented to join me as co-authors, I wish to thank many--to numerous to mention by name--for helpful remarks and constructive criticism. These include J. Adler, A. Aharony and D. Stauffer (termite project) and R. C. Ball (surfaces project).

We begin with an experimental fact. Consider a beaker full of reactive monomers. If each monomer had only one functional group, the most complex structure that could be formed would be monomers. If each monomer had two functional groups, it becomes possible to form linear chain polymers. If the functionality is three or larger, it one finds branched polymers forming, which are capable of a sort of"amplifed" growth as two large polymers suddenly link up to form a much larger polymer. This amplification manifests itself in mathematical singularities that occur in measured quantities as the system reaches the gelation threshold where there appears a single large molecule spanning the entire system (the "gel" molecule). Among the singular functions is the shear viscosity, which typically can increase by many orders of magnitude as the gelation threshold is approached. The viscosity appears to diverge to infinity with a critical expopnent -s. Adam et al (1981) have found that $s = 0.78 \pm 0.05$.

The puzzle is to explain this exponent. A crucial insight was provided by de Gennes (1980), who noted that the shear viscosity of a system of large ramified polymer molecules was analogous to the electrical conductivity of a large collection of superconducting clusters. Moreover, he predicted that the exponent -s for the shear viscosity should be identical to the exponent, also called -s, for the divergence of the conductivity of a random superconducting network (RSN) in which superconducting bonds are present with probability p and normal bonds with probability 1-p.

Replacing a complicated fluid mechanics problem, with all the messy differential equations, by a simple electrical problem is a majestic feat. However solving the remaining electrical problem is by no means simple. Significant progress is just beginning to occur, and my goal in this talk is to give you a "little taste" of the various approaches being tried. If your appetite is whetted, you will want to consult the original literature which, fortunately, is not too voluminous (Coniglio and Stanley 1984, Adler et al 1985, Bunde et al 1985, Hong et al 1985, Sahimi 1985, Leyvraz et al 1985). An amusing by-product of our work is that we shall learn about how to characterize the surface of a fractal, results with suggestive implications for understanding better catalysis and other surface-related phenomena.

Even an electrical problem, the RSN, is frighteningly complex. How can we adapt the results in Tinkham's classic superconductivity text to a system that is so structurally heterogeneous that it consists of randomly placed superconductng bonds, which in turn form clusters of all shapes and sizes with a size distribution that is not peaked around any particular value of s but rather decays with s according to a power law? This is indeed a challenge. As often is the case, when faced with a challenge it is smart to consider a completely different approach....and that is what we have done. Specifically, we shall treat the general problem of a

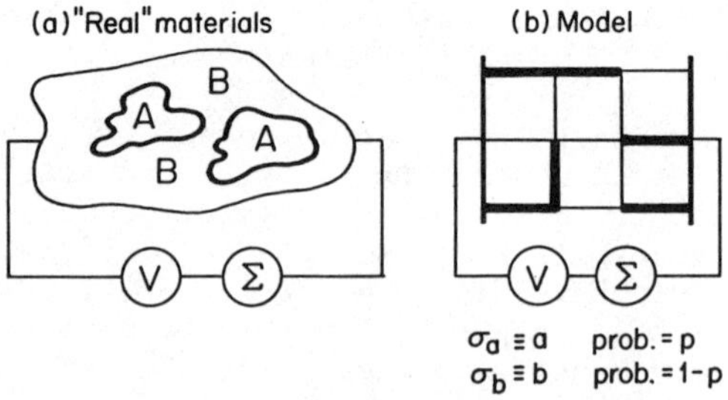

FIG. 1: (a) Schematic illustration of a random two-component composite material before coarse-graining. (b) Replacement by equivalent random network with two conductances $a = \sigma_a$ [probability p] and $b = \sigma_b$ [probability 1 - p].

two-component AB random mixture by replacying the very difficult (and intractable) electrical problem by a very tricky (but tractable!) random walk problem. Then in the limit in which the conductivity of the A component diverges to infinity, we recover the solution to the superconducting limit that was too hard for Tinkham's book.

THE TERMITE PROBLEM

How are the fundamental laws of diffusion and transport modified when the medium in question is a random "AB-mixture" of good and poor conducting regions (Fig. 1a). This question has received a considerable degree of recent attention for two limiting cases:
(i) The random resistor network (RRN)--or pure "ant" limit--for which B, the poor conducting species, has zero conductance, and
(ii) The random superconducting network (RSN) or pure "termite" limit, for which A, the good conducting species, has infinite conductance.

The terms "ant" and "termite" arise from the fact that one can replace the conductivity problem with a diffusion problem using the Nernst-Einstein relation. For the RRN limit, no diffusion can occur on the component with zero conductance, so the constrained diffusion problem is rather like an "ant in a labyrinth." For the RSN limit, the diffusion can occur everywhere since both components conduct, but the fact that the good conductor species has zero resistance means that the diffusion is remarkably different in this region than elsewhere. Some years ago de Gennes invented the term "termite diffusion" to describe this subtle phenomenon. However to this date there has been no clear statement of exactly how to properly define or measure this phenomenon, in contrast to the "ant" limit where the diffusion is simply constrained to one component. There are many reasons for the current upsurge of interest in this problem.

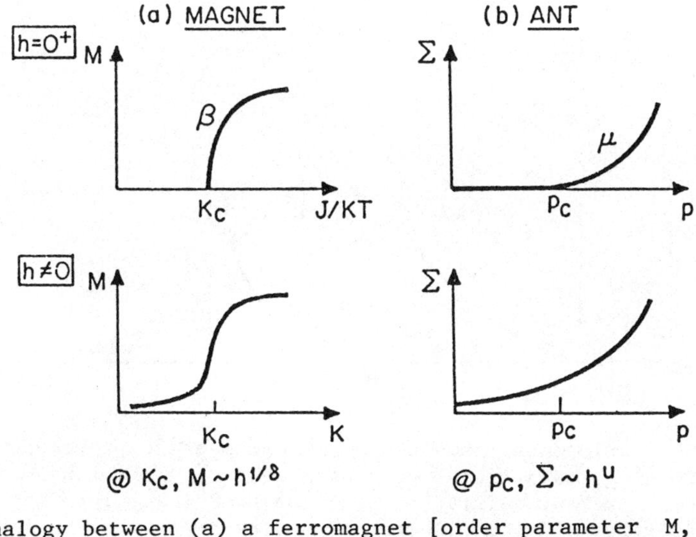

FIG. 2: Analogy between (a) a ferromagnet [order parameter M, conjugate field H] and (b) a random mixture ["order parameter" Σ, "conjugate field" h].

(i) One reason is that there are many experimental systems that are random and inhomogeneous. For example, a rock is composed of tiny grains of different conductivities (to heat, to fluid flow, to electricity). To the extent that such inhomogeneous materials are also random, we may think of using a site-random description of this material: a "lattice-gas" description. One first coarse grains the material and then assigns to each cell one of two conductivities, σ_a and σ_b. Calculations based upon such a straightforward approach have been usefully compared with a wide range of experiments, from conductivities of thin films of lead depositions on an insulating substrate (roughly the RRN limit) to thin films of superconducting material vacuum deposited on a normal substrate (roughly the RSN limit). Moreover, ionic conductors mixed with a dispersed insulating phase represent random heterogeneous materials, where both limits seem to play an important role.

(ii) A second reason is related, perhaps, to the reason why the Ising model has always been of great interest: it is an extremely simple model that captures the essential physics of a realistic system in nature. The analog of the Ising model for random inhomogeneous materials is a mixture of sites (or bonds) randomly distributed on a lattice--see, e.g. , Fig.1b. The sites (or bonds) are assumed for simplicity to have only two possible values of the conductance,

$$\sigma = \begin{cases} \sigma_a & \text{[probability p]} \\ \sigma_b & \text{[probability 1-p].} \end{cases} \tag{1}$$

By convention, we choose $\sigma_a > \sigma_b$, so that the ratio $h = \sigma_b / \sigma_a$ is always less than unity.

Conventionally, one wants to know the macroscopic magnetization of an Ising ferromagnet composed of elements (spins) whose microscopic property is a two-valued variable. Similarly, we now want to know the macroscopic conductivity which depends on all possible configurations of the microscopic elements (conductors) whose property is again a two-valued quantity (σ_a and σ_b). Just as the magnetization couples to a conjugate field H, we shall see that the conductivity couples to a conjugate field h (Fig. 2).

The two limiting cases mentioned above can now be discussed more precisely (Fig. 3):

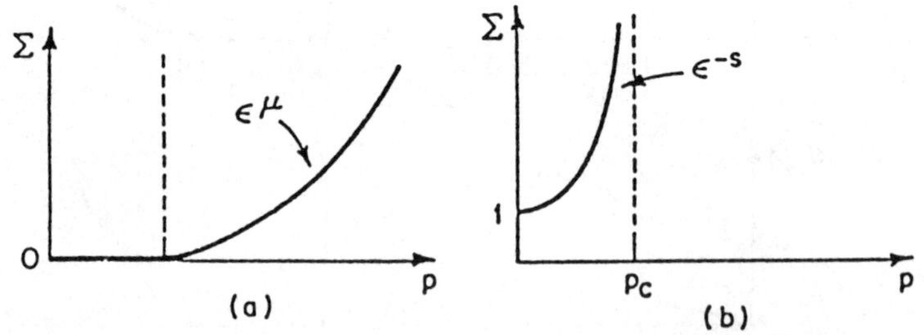

FIG. 3: Schematic dependence on p of the macroscopic conductivity for two limiting cases of the present model: (a) the random resistor network (RRN) or "ant" limit, and (b) the random superconductor network (RSN) or "termite" limit. The corresponding exponents μ and s are equal for d=2.

(a) In the RRN limit, the large conductance is set to unity and the small conductance is set to zero. As the percolation threshold p_c is approached from above, the macroscopic conductivity Σ approaches zero with a critical exponent μ,

$$\Sigma \sim (p - p_c)^{\mu}. \tag{2a}$$

(b) In the RSN limit, the small conductance is set to unity, and the large conductance is infinite. As the percolation threshold is approached from below, the conductivity diverges to infinity with an exponent $-s$

$$\Sigma \sim (p_c - p)^{-s}. \tag{2b}$$

The traditional approach to the RRN limit has been to replace Kirchhoff's laws by an equivalent diffusion problem, where the macroscopic conductivity is related to the diffusion constant D by the Nernst-Einstein relation,

$$\Sigma \sim nD, \tag{3}$$

where n is the density of the charge carriers.

RANDOM WALK MODEL FOR THE CONDUCTIVITY

In this section, we briefly explain the model we have been developing to describe--using random walks--the conductivity of a general two-component random mixture. Perhaps the best starting point is the pure "ant" limit in which the conductance of the poor conducting species tends to zero ($\sigma_b = 0$). In this case, the Nernst-Einstein theorem applied to this case tells us that the macroscopic conductivity Σ (measured, e.g., by a pair of bus bars--see Fig. 1) is proportional to the diffusion constant D.

It is convenient to imagine a randomly diffusing particle which jumps from site to site over potential barriers. Each barrier has the same height if the bond has conductance σ_a, while the barrier is infinite in height if the bond has conductance σ_b (see Fig. 4). Thus when the particle is at a boundary between two bonds with conductivities $\sigma_a = 1$ and $\sigma_b = 0$, it will be "reflected" with probability unity by the zero conductivity bond. We could say that its jump frequency into the high-conductivity region is $f_a = 1$, while its jump frequency into the low-conductivity region is $f_b = 0$. Similar remarks apply to the "termite" or RSN limit, except that now $f_a \to \infty$ and $f_b = 1$. For the case of a <u>general inhomogeneous material</u>, f_a and f_b are both finite constants, different from zero or infinity. The ratio $h = f_b/f_a$ is zero for both the RRN and RSN limits, but for the general inhomogeneous material h is a number between zero and 1. Note that since $f_a \sim \sigma_a$ and $f_b \sim \sigma_b$, h is also given by

$$h = \sigma_b/\sigma_a = f_b/f_a. \tag{4}$$

With the foregoing picture in mind, we can now state clearly the random walk model of a general two-component inhomogeneous material. Place a walker on a d-dimensional lattice made of two kinds of bonds, A and B (for illustration: d=1 here, the general-d case is discussed in Hong et al 1985). The walker carries two coins, weighted and unweighted, and a clock. Without loss of generality, let the origin be well inside a high-conductivity A region. At each tick of the clock, the walker tosses the unweighted coin and moves to the left or right depending on the outcome of the coin toss. When the walker comes to a site on the boundary between the A region and the B region, he tosses the other coin that is weighted with probability

$$P_a = f_a/(f_a + f_b) = 1/(1 + h), \tag{5a}$$

to stay in the A region, and a probability

$$P_b = f_b/(f_a + f_b) = h/(1 + h), \tag{5b}$$

to go outside into the B region. In the event that the walker steps outside the A region, then he must <u>slow</u> down by the ratio f_a/f_b. For example, if the conductivity of the B region is 10 times smaller than that of the A region, then f_b is 10 times smaller than f_a (h = 0.1) and the walker steps <u>only</u> after every 10 ticks of his clock.

Limiting cases of our random walk model are as follows:

(i) h=1. There is no distinction between regions, no reflection on the boundaries ($P_a = P_b$), and no difference in walk speed on and off the A clusters.

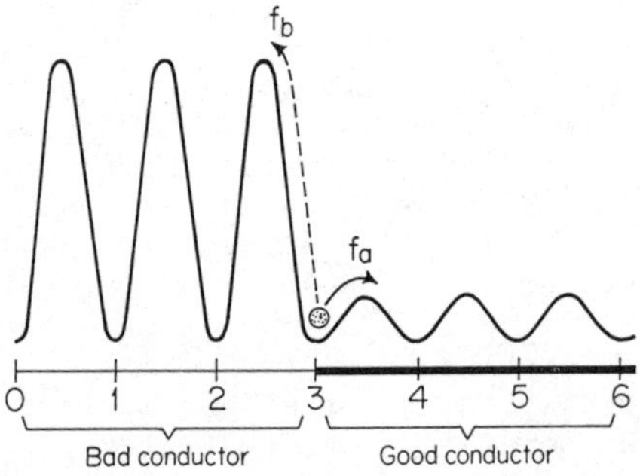

FIG. 4: Schematic illustration of our random walk model for a one-dimensional lattice, showing the presence of a bondary between a good-conductor cluster of conductors σ_a and a poor-conductor cluster of conductors σ_b. The corresponding jump frequencies f_a and f_b determine, through Eq. (2.1), the probability of a particle being reflected at the boundary. After Hong et al (1985).

(ii) $h \ll 1$. The walker now moves at one step per clock tick when he is on an A cluster, and is almost always reflected when he comes to the boundary. Extremely rarely he passes out of an A region and into a B region, whereupon he walks much, much slower--taking a new step only after his clock has made h^{-1} ticks. Statistically speaking, in a very large time $\gg h^{-1}$, the walker peforms $O(f_a)$ moves in the A region and $O(f_b)$ moves in the B region.

Suppose we make a motion picture of the walker's motion. Then we see that the walker is reflected from the walls almost all of the time, and only very rarely--roughly once per h^{-1} trials--will come outside the cluster (see Hong et al 1985). When this does occur, his motion will slow down by a factor of h. If we watch this motion picture, perhaps we become impatient watching the walker in the B region and we speed up the motion picture projector by a factor of 1/h so that the walker is now taking one step per unit of time while in the B region. Then we are no longer impatient while the walker is in the B region. However, when he finally encounters an A cluster, he moves onto it with a high probability, $1/(1 + h)$, and proceeds to move about the A cluster with a motion that is also sped up by the same factor 1/h. Thus the orginal "ant" who stepped normally on an A cluster and extremely slowly on B clusters has suddenly been transformed into a "termite" who moves normally on B clusters and extremely fast on A clusters. Indeed, the only difference between the two domains, "ant" (RRN domain) and "termite" (RSN domain), is the definition of the time scale. This simple observation can be formalized in terms of a rigorous transformation (Hong et al 1985). That transformation in turn forms the basis of the scaling laws for the ant and termite limits of the general two-component random mixture.

To summarize thus far, the two-component random mixture requires for its treatment the understanding of how to handle a diffusion process to which there are two time scales, not one. This problem has not been treated previously and is proving to be quite subtle in many respects. Until quite recently it was widely believed that the physics governing transport near the RSN or "termite" limit was quite different from the physics governing transport near the RRN or "ant" limit. Now we appreciate that these two regions are related by a simple change of time scale (Hong et al 1985).

SURFACES, INTERFACES AND SCREENING OF FRACTAL STRUCTURES

The rest of this talk is devoted to the subtle and fascinating subject of disordered surfaces. At first sight, you might wonder how this new topic is related to the previous topics. In fact, it is fractal surfaces that enable us to unite both previously mentioned subjects: gel viscosity and the conductivity of a random superconducting network.

That the shear viscosity is related to a fractal surface problem is clear if we realize that the screened regions of the polymer carry solvent along with them and hence do not provide "friction" between polymer and solvent. Only the unscreened regions near the external tips of the polymer provide friction and so contribute to the shear viscosity. That the termite problem is related to fractal surfaces is also clear, since the termite can pass from one cluster to another only through the surface.

For the termite problem, the external perimeter sites are not equally effective in producing a macroscopic conductivity across the sample. This is because percolation clusters are deeply invaginated structures, so a termite who escapes the cluster deep inside an invagination may be captured again by the same cluster before having the chance to reach another cluster. This is the reverse of the DLA problem. In DLA, incoming random walkers rarely penetrate a deep invagination and so stick to the unscreened part of the hull at the tips. In the termite problem, exiting random walkers (termites) rarely escape a deep invagination, becoming captured on the walls.

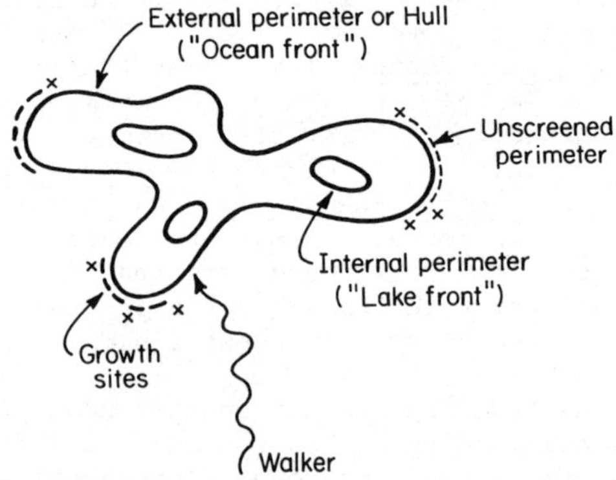

FIG. 5: Schematic illustration of four different fractal surfaces arising in the description of a percolation cluster. (a) The external "oceanfront" parameter or hull has a fractal dimension d_h. (b) The total perimeter has a fractal dimension d_f, equal to that of the total bulk mass of the cluster. Since $d_f > d_h$, it follows that the internal "lakefront" perimeter must have the same fractal dimension d_f of the total perimeter. (c) The unscreened perimeter where an incoming walker is more likely to hit has fractal dimension d_u (heavy solid lines). (d) The growth sites are those perimeter sites that form the living frontier of the cluster. These have fractal dimension d_g, but the nature of the G-site fractal depends on the actual mechanism of how the percolation cluster grows (see, e.g., Stanley et al 1984; Bunde et al 1985a,b; Herrmann and Stanley 1985).

But what do we mean by "the" surface of a fractal object? In fact, we shall see that there are many different surfaces, depending on the physical process in question (Fig. 5). We shall discuss these roughly in order of increasing subtlety.

External Perimeter ("Hull"): d_h

The total number of external surface sites, or "hull," scales with the caliper diameter or radius of gyration L as (Table 1, lecture 1)

$$N_{hull} \sim L^{d_{hull}} \qquad \text{[FRACTAL DIMENSION No. 4].} \tag{6}$$

For d=2 percolation, d_{hull} appears to be about 1.74 ± 0.02 (Voss, this conference; Sapoval et al 1984), thus motivating the conjecture

$$d_{hull} = 1 + d_{red} = 7/4, \tag{7}$$

since $d_{red} = 1/\nu = 3/4$ for d=2.

Total Perimeter: d_f

We know that the total number of perimeter sites $N_{perimeter}$ scales in the same fashion as the total number of cluster sites,

$$N_{perimeter} \sim N_f \sim L^{d_f}, \tag{8}$$

for d-dimensional percolation (Kunz and Souillard 1978) where, for d=2, $d_f = 91/48 = 1.896$. Hence the fact that $d_{hull} < d_f$ means that the ratio of external perimeter sites to total perimeter sites approaches zero at the percolation threshold. As the clusters get larger the internal perimeter sites ("lakefront" sites) completely swamp the external perimeter sites ("oceanfront" sites). For $d > 2$, internal perimeter sites are less commonly seen in finite computer simulations since it takes a lot of clsuter sites to completely surround a "3-dimensional lake." An open question is the value of d_{hull} for $d > 2$: could it be that $d_{hull} = d_f$? Work is underway to test this possibility.

Unscreened Perimeter: d_u

Coniglio and Stanley (1984) introduced the concept of the "unscreened" perimeter to describe that portion of the hull that is effective in termite motion:

$$N_{unscreened} \sim L^{d_u} \qquad \text{[FRACTAL DIMENSION No. 5].} \tag{9}$$

Moreover, they showed that the conductivity exponent for the RSN is simply related to d_u. Recalling the Nernst Einstein relation, we have

$$\Sigma \sim D \sim (R_{cluster})^2 \tau^{-1} \tag{10a}$$

Here the jump frequency τ^{-1} scales as the fraction of cluster sites belonging to the unscreened perimeter,

$$\tau^{-1} \sim [N_{unscreened}/N_f]. \tag{10b}$$

Substituting the Stauffer expression for the mean radius of the finite clusters, $R_{cluster}$, and the definitions for d_u and d_f into (10), we obtain

$$\Sigma \sim L^{2-(d-d_f)} L^{d_u-d_f}. \tag{11}$$

Since $\Sigma \sim \epsilon^{-s}$, we have

$$\tilde{s} = s/\nu = d_u - (d - 2). \tag{12}$$

The conductance between two points scales as $L^{\tilde{\zeta}_{RSN}}$ where $\zeta_{RSN} = \tilde{s} + (d - 2)$. From (12) we obtain the extremely simple result that the conductance exponent is identical equal to the fractal dimension of the unscreened perimeter

$$\tilde{\zeta}_{RSN} = d_u. \tag{13}$$

Let us contrast now the RSN with the RRN, which is well understood. For the RRN, we can also relate the conductance exponent to fractal dimensions characterizing the substrate fractal. The Einstein relation (3) holds, but we must set $n = P_{\infty}$, the probability that a randomly dropped ant will land on the incipient infinite cluster. Now P_{∞} scales as the co-dimension $(d - d_f)$, and D scales as L^2/time. Hence (Gefen et al 1983; Ben-Avraham and Havlin 1982)

$$\Sigma \sim P_{\infty} D \sim L^{d_f - d} L^{2-d_w}. \tag{14}$$

Recalling that $\Sigma \sim \epsilon^{\mu}$ for the RRN limit, we have

$$\tilde{\mu} = \mu/\nu = (d - 2) + (d_w - d_f). \tag{15}$$

The conductance between two point scales as $L^{-\zeta_{RRN}}$, where $\zeta_{RRN} = \mu - (d - 2)$. Hence for the RRN problem, (13) is replaced by

$$\tilde{\zeta}_{RRN} = d_w - d_f. \tag{16}$$

It is convenient to think of the resistance between two points as the "mass" of 1-ohm resistors that we would place in series between the two points in order to have the same resistance (Fig. 6). In this way, we can interpret $\tilde{\zeta}_{RRN}$ as a proper fractal dimension for some specific fractal object (the set of resistors)

$$N_R \sim L^{d_R} \qquad \text{[FRACTAL DIMENSION No. 6]}. \tag{17}$$

From (13) and (16) it then follows that

$$d_R = \begin{cases} -d_u & \text{[RSN]} \\ d_w - d_f & \text{[RRN]}. \end{cases} \tag{18}$$

FIG. 6: One example of how a set of bonds (unit resistors) between sites i and j is replaced by an equivalent number (in this case $N_R = 1$) of unit resistors placed in series. Thus the resistance between i and j is equal to N_R, which in turn scales with the Pythagorean distance L between i and j with an exponent d_R (see Eq. (17)).

From (18) we see that both the RRN and the RSN have resistance exponents d_R that are simply expressed in terms of fractal properties of the substrate. The expressions are completely different, of course, since the mechanism of transport is completely different (Fig. 7). From Fig. 7a, we see that the transport is determined by a "fisherman's net" structure, with a mesh size given by the connectedness length ξ. The strands of the net are made of singly-connected "red" bonds and multiply-connected "blue" bonds, the statistics of which will be described in lecture 3. From Fig. 7b, we see that for the RSN just below p_c transport from one bus bar to the other is determined by the motion of charge carriers from one cluster to another--more precisely out of the unscreened perimeter of one cluster and into the unscreened perimeter of the next.

Thus the clusters in the RSN limit play the role of the nodes in the RRN limit. As one moves close to p_c in the RRN problem, the critical bonds are the singly-connected "red" bonds (the hottest). As one moves close to p_c in the RSN limit, the critical bonds are those bonds on the lattice which--if occupied--would connect two clusters. I have always called these pink (since once occupied they will become red links), but recently Stauffer suggests calling these bonds anti-red because in every sense they are the complement of the red bonds. Thus the red bonds control the physics of the RRN problem just above p_c, while the anti-red bonds control the physics of the RSN problem just below p_c. Coniglio (unpublished) has proved that the anti-red bonds have the same fractal dimension as the red bonds,

$$d_{\text{anti-red}} = 1/\nu . \tag{19}$$

Thus there is a certain symmetry between the RRN and RSN limits, which in some way should follow directly from the homogeneity theorems mentioned above. Work on this important topic is underway, and perhaps at this meeting some of you can help make progress along these lines.

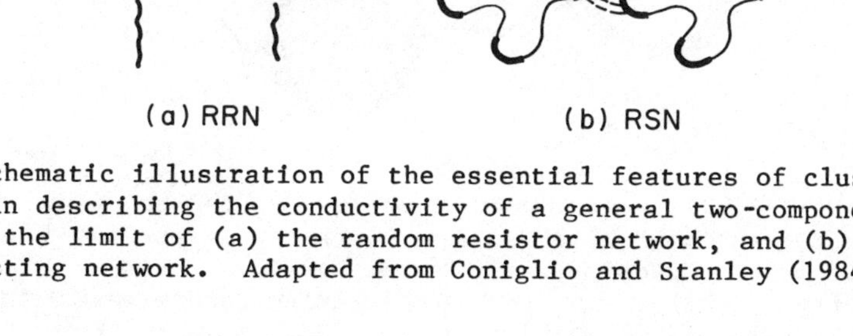

FIG. 7: Schematic illustration of the essential features of cluster structure in describing the conductivity of a general two-component random mixture in the limit of (a) the random resistor network, and (b) the random superconducting network. Adapted from Coniglio and Stanley (1984).

Can we evaluate the fractal dimensions d_u and d_w appearing in (18) in terms of the fractal dimension d_f of the underlying substrate? Some progress along these lines has been made using arguments that require for their validity certain assumptions. In this section we will review a mean-field type argument (Coniglio and Stanley 1984) that

$$d_u = (d_f - 1) + (d - d_f)/d_w \qquad \text{[CONIGLIO-STANLEY]}, \tag{20a}$$

and describe its tests, both by computer and by experiment. In the section below on growth sites we will review a proposal (Aharony and Stauffer 1984)

$$d_w = d_f + 1 \qquad \text{[AHARONY-STAUFFER]}, \tag{20b}$$

and describe the conditions under which this tantalizingly simple relation might hold (Havlin et al 1984, Havlin 1984) or might not hold (Stanley et al 1984).

The argument for d_u is very simple. It begins by sending, say, a million walkers at a fractal one-by-one and recording the perimeter site where they touch before throwing them away. Then if is the mean penetration depth (Fig. 8), we have

$$N_u \sim \lambda \xi^{d_f - 1}, \tag{21}$$

since this is the area of the annulus in Fig. 8. In writing (21) we have used the fact that the area is the thickness of the annulus λ times the circumference, where the circumference scales with dimension $d_f - 1$ (since our d_f-dimensional fractal is cut by a 1-dimensional cookie cutter). To see how λ depends on ξ, we consider the average number of steps

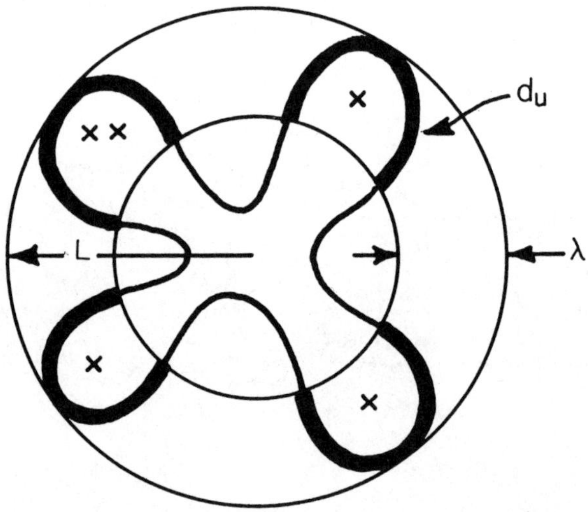

FIG. 8: Schematic illustration of parameters needed to obtain the mean-field expression, Eq. (20a), for the fractal dimension of the unscreened perimeter (Coniglio and Stanley 1984). The symbol x denotes a typical cluster site that is hit by an incoming random walk, while λ is the mean thickness of this band of sites.

$$N_w^* \sim \xi^{d_w}, \tag{22}$$

that a projectile of fractal dimension d_w takes before being absorbed in a shell of width λ. We expect $N_w^* \sim 1/\rho$, where

$$\rho = N_f/\xi^d \sim \xi^{d_f - d} \tag{23}$$

is the number density and N_f is the total cluster mass. Hence

$$\lambda \sim \xi^{(d-d_f)/d_w}. \tag{24}$$

Substituting (24) into (21), we obtain the desired result, (20a). Note that d_u is a function only of the fractal dimension d_f of the substrate and the fractal dimension d_w of the incoming (or exiting) walk. Equation (20a) is easy to remember since the first "cookie-cutter term," $d_f - 1$, would be present for any fractal, while the second term (directly proportional to the co-dimension $d - d_f$ and inversely proportional to d_w) is a measure of how easily the invaginations of the fractal can be penetrated by a walk of fractal dimension d_w.

The important matter is not remembering it but rather learning if it is correct. To this end, we must devise a method of probing the surface of a fractal object. The method we chose (Meakin et al 1985) was to release random walkers, one at a time. When the random walker touched perimeter site i, a counter on site i was incremented by one unit (N_i becomes $N_i + 1$). After typically a million walkers have been released, statistics were done. Our analysis is based on the idea that only a relatively small fraction of the total perimeter will have a large probability of being contacted. Hence to analyze the distribution function N_i ($i = 1,2, \ldots ,P$--where P is the total number of perimeter sites), we formed the moments μ_j defined through

$$[\mu_j]^j = N_T^{j+1}/\sum_i N_i^{j+1} = 1/\sum_i P_i^{j+1} \sim N_f^{j\gamma_j} \tag{25}$$

Here

$$N_T = \sum_i N_i, \tag{26a}$$

is the total number of particles, and

$$P_i = N_i/N_T, \tag{26b}$$

is the probability that an incoming walker will hit site i. The P_i are normalized to unity by virtue of (26a).

First we calculated the γ_j for $j=1-3$, and found that the Coniglio-Stanley mean field relation (20a) was satisfied to within the accuracy of the calculations. We did notice a systematic dependence on j, so to test the possibility that the γ_j depend on j we extended the moment calculation to $j=8$. The hinted dependence from $j=1-3$ became much clearer and so we conclude that there is not a single exponent but rather an entire hierarchy of exponents (Fig. 9).

Why is the Coniglio-Stanley relation wrong? Presumably because it smears out the interface or "active zone" of the fractal into a band, and then assumes that there is an equal probability of capture for all surface sites within this band. In reality, there is a continuous gradation in "temperature," with the outermost tips being immensely hotter and the deepest invaginations being extremely cold. This situation is reminiscent of that found by De Arcangelis et al (1985) for $N(V)$, the distribution of the number of bonds in the backbone across which the voltage drop is V. Here also there is a continuous gradation in temperature from the red bonds (the "hottest" in the sense that the full voltage drop of the entire cluster falls across each red bond) to the very cold bonds arising from the very long loops comprising the blobs.

This discovery of an infinite hierarchy of critical exponents--both in the voltage distribution of the percolation backbone and in DLA--is striking because normally one assumes that two exponents will suffice to describe a critical object. For example, we noted in lecture 1 that y_h ($= d_f$) and y_T ($= d_{red}$) were sufficient to describe percolation. However when we "do something" to the fractal, such as put a battery across it or bombard it with random walkers, we introduce a new measure. Instead of each fractal site having weight 1, each site has a weight that depends on what we are doing to the fractal (e.g., each site has a voltage attached to it, or with each site we associate the number of hits on that site). Several groups (Meakin, Stanley, Coniglio and Witten unpublished; Turkevich and Scher 1985; Halsey et al 1985) are currently seeking to understand the meaning of this new measure and what we can learn from this infinite hierarchy of exponents.

In retrospect, we might have anticipated this infinite hierarchy in advance. This is because for two extreme values of j, $j = -1$ and $j = \infty$, exponents differ by more than a factor of two: $\gamma(-1) = 1$ and $\gamma(\infty) = 1 - 1/d_f$. The first result follows immediately from the definition (25) and the fact that the total surface in DLA scales with exponent d_f just like the total mass. The second follows from the recent theorem (Leyvraz 1985) that P_{max} (the maximum value of all the P_i) scales with cluster mass to the exponent $1 - 1/d_f$. This prediction is confirmed by our calculations for $j = 1$-8. Our hierarchy or spectrum of surface fractal dimensions tends clearly toward a number fairly close to the predicted value $1 - 1/d_f$ (Fig. 9).

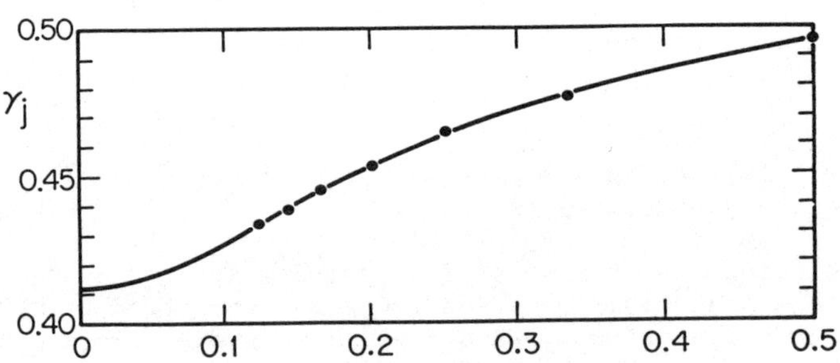

FIG. 9: The exponent γ_j characterizing the behavior of the jth moment of P_i. Here P_i, the basic quantity in surface growth, is the probability that perimeter site i is the next site at which the cluster will grow. In order to see if this apparent hierarchy or "spectrum" of exponents tends in the $j=\infty$ limit toward the expected limit, $1 - 1/d_f$, we have plotted these exponents against $1/j$.

We have just seen that the Coniglio-Stanley mean-field relation (20a) is in rough accord with the calculations of Meakin et al (1985) on the unscreened perimeter of DLA. We conclude this section by comparing with experimental data on the shear viscosity near the gel point. To this end, we must first obtain an expression for the exponent $\tilde{s} = s/\ $ for the RSN, obtained by substituting the expression (20a) into (12). We also need to choose d_w in (20a). For systems with fractal dimension above 2, a random walk will best penetrate the invaginated cluster and we choose $d_w = 2$ (Fig. 10). For $d_f < 2$, a cluster will be more efficient (the co-dimension, which controls penetration, is larger). Hence we choose $d_w = d_f$. Thus we find

$$\tilde{s} = 2 - d + d_u = \begin{cases} 1 - (d - d_f)/2 & [d_f \geq 2] \qquad (27a) \\ d/d_f + d_f - d & [d_f \leq 2]. \qquad (27b) \end{cases}$$

Relation (27a) was recently conjectured to hold <u>for all</u> d_f (not just $d_f \leq 2$) by Kertész (1983), but no argument supporting the conjecture was given. Note that (27) agrees with exact results for d=1,6 and is in rough accord with calculations of $\tilde{s}$ and d_f for d=2-4. This mean field calculation shows the importance of $d_f = 2$ as a sort of "critical dimension" below which the physics of transport on random fractals changes character; $d_f = 2$ also enters as a critical fractal dimension in the work of Aharony and Stauffer (1985). We leave it as an exercise to the student to correct a minor calculational error in the above argument, stemming from neglect of that all important factor P_{oo}.

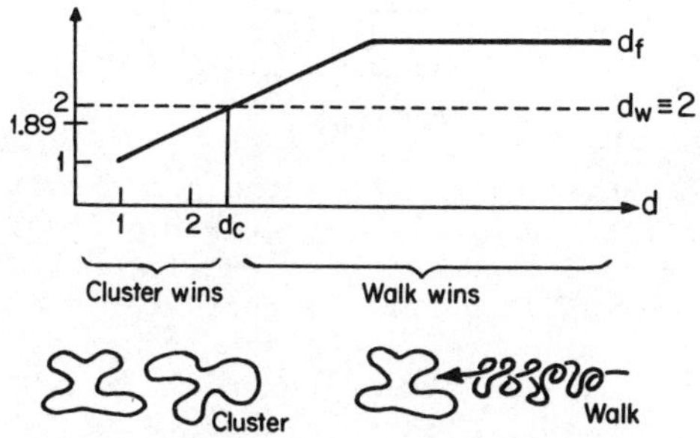

FIG. 10: Dependence on lattice dimension d of two quantities: (a) the cluster fractal dimension d_f, and (b) the fractal dimension of a random walk d_w. For d above some critical dimension d_c, the walk dimension is smaller so a walk penetrates more easily the invaginations of a cluster. For d below d_c, the reverse is true. Above d_c we substitute $d_w = 2$ into (20a), while below d_c we substitute $d_w = d_f$. Numerically, d_c^w is slightly larger than 2.

For d=3 percolation, $d_f = 2.5$ which is larger than two. Hence we use (27a), and predict $s = 3/4$. Since $\nu \cong 0.9$, $s \cong 0.7$ which is within (but just barely within) the error bars of the experimental result, $s = 0.78 \pm 0.05$ (Adam et al 1981). The theory was very rough, as noted above, and the experiment is immensely subtle. Hence we regard the fact that the RSN exponent comes even close to the viscosity exponent to be cause for joy.

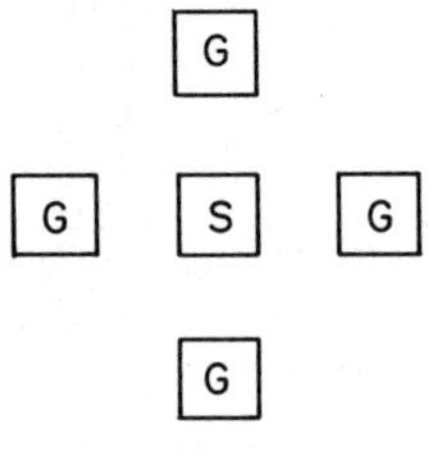

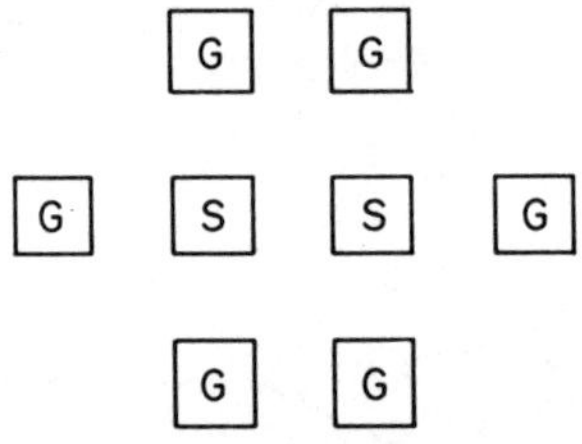

FIG. 11: Precise mechanism of creating the growth site fractal for percolation clusters created by the ant growth mechanism. In (a), we have t=1 and the ant is placed randomly on a Euclidean lattice; all four neighboring sites are available for the next step and hence are called "growth sites." (b) At time t=2, the ant tosses a four-sided coin, which comes up West, so she focusses attention on the growth site on the west. She then tosses a second two-sided coin that is weighted to come up "blocked" with probability $1 - p_c$ and "un-blocked" with probability p_c. Had the coin come up "blocked," then the site on the west would have been blocked forever, and the growth frontier would have been reduced from 4 to 3 sites. In fact, this time the coin came up "un-blocked," so that ant proceeds to move to the new growth site. The incipient cluster fractal now has 2 sites, and the growth site fractal has 6 sites. This process is continued until typically there are 10^4 sites in the cluster fractal and about 200-300 sites in the growth site fractal (see, e.g., the discussion in Stanley et al 1984).

FRACTAL GROWTH

How can we characterize the fashion in which a fractal grows? This is the question that we shall address in this final section. It is important to state at the outset an obvious fact: completely different growth mechanisms can lead, eventually, to the same static fractal object (see, e.g., the discussion in Bunde et al 1985a). For now, let us consider one of the simplest kinetic mechanisms of growth, the ant. Instead of dropping the ant onto a pre-formed fractal structure as in lecture 1, we could instead drop the ant onto a Euclidean lattice but give her a set of rules with which she could form the fractal as she moves. This means that the ant would need the four-sided coin that all walkers have (taking a square lattice for now), but she would also need some dynamical mechanism of generating the ultimate static fractal. For percolation fractals, all she needs is a weighted coin so that each site can be blocked with probability $1 - p_c$. The precise mechanics for the ant motion are described in the caption to Fig. 11.

After some time has evolved, the ant is moving around in a rather interesting region of space that is characterized by 3 sorts of sites (the terminology in brackets suggests an epidemic interpretation of this entire growth problem):

(i) cluster sites already visited ["sick"]
(ii) sites already tested and blocked by the second coin ["immune"]
(iii) neighboring sites that have not been tested ["growth"].

Thus from the reference frame of the ant, only the growth sites are special in the sense that only these enlarge her territory.

Why are the growth sites interesting? Subjectively, they represent the open frontier of the growing fractal. Objectively, this disconnected set of sites has a well-defined fractal dimension d_g in the sense that the number or "mass" of growth sites obeys a scaling relation of the form

$$N_g \sim L^{d_g} \qquad \text{[FRACTAL DIMENSION No. 7].} \tag{28a}$$

Here, as always, L represents the cluster diameter or radius of gyration.

The number of growth sites has been calculated as a function of cluster mass N_f. We predict for the intrinsic exponent

$$N_g \sim N_f^x \qquad \text{with} \qquad x = d_g/d_f. \tag{28b}$$

Stanley et al (1984) obtained the first estimate of x, $x = 0.49$ [d=2 percolation]. Later Stanley (unpublished) invested 200 hours of IBM 3081 time and found $x = 0.493$ [supporting AO--see below]. Demme (unpublished) has recently used an altogether different approach and found a somewhat smaller value of x supporting AS. Efforts to reconcile this small but important difference are underway , with D. Stauffer serving as a mediator.

How do we interpret the value of x? One way is to first derive, for percolation, a relation between d_g and other exponents. Then we can test the calculated value of d_g against what we know for those other exponents. Rammal and Toulouse (1983) have noted that N_f obeys a simple difference equation

$$N_f(N_w + 1) - N_f(N_w) = N_g/N_f. \tag{29}$$

Thus if the ant takes one more step, the number of additional sites added to the fractal is given by the chance of the ant to step on a growth site. Since the ant is assumed to be completely drunk, and is permitted to step only on growth or fractal sites, this probability is simply the right-hand side of (29). If we regard (29) as a differential equation, then the left side becomes dN_f/dN_w, which scales with the exponent $d_f - d_w$. The right side scales with exponent $d_g - d_f$. Hence

$$d_g = 2d_f - d_w. \tag{30}$$

We can now evaluate d_g for the two popular conjectured relations between d_w and d_f. In lecture 1, we discussed the Alexander-Orbach [AO] conjecture $d_w = (3/2)d_f$. In the previous section we mentioned the Aharony-Stauffer [AS] conjecture $d_w = 1 + d_f$. Thus

$$d_g = \begin{cases} d_f/2 = 91/96 = 0.9479 & \text{[AO]} \\ d_f - 1 = 43/48 = 0.8958 & \text{[AS].} \end{cases} \tag{31a}$$

For the intrinsic exponent x, we then predict

$$x = \begin{cases} 1/2 = 0.5000 & \text{[AO]} \\ 1 - 1/d_f = 0.4725 & \text{[AS].} \end{cases} \tag{31b}$$

Thus we see that the calculated values of x fall in between the AO and AS predictions.

Leyvraz and Stanley (1983) considered the conditions under which the AO conjecture might hold, focussing on the need for complete statistical independence of the increments in N_g. They noted that this statistical independence certainly occurs for the Cayley tree, since it is impossible to have correlations on a loopless fractal. In this fashion, they understand the numerical result that AO holds for the Cayley tree ($d_w = 6$, $d_f = 4$). In the asymptotic limit of a truly huge cluster--say the size of Norway--they imagined that correlations in growth sites would begin to vanish even though Norway has loops. To the extent that these correlations eventually drop off, we can expect that in the asymptotic limit the increments in N_g will be statistically independent and hence the distribution will be normal, with $x = d_g/d_f$ exactly 1/2 [AO].

Recently, the subject of growth sites has arisen in various contexts. One concerns a family of epidemic growth models, all of which eventually grow static percolation clusters (Bunde et al 1985a,b; Herrmann and Stanley 1985). Depending on the growth rules, the dynamic critical exponent d_g can differ--even though the static critical exponent d_f is the same for all rules. We considered already one such epidemic model above. A second model is a variation of the above model in which the "ant" visit only growth sites, never re-visiting cluster sites. Clearly this requires the ant to make long-range jumps. In the simplest version of this model, the ant randomly chooses her next move from among all the existing growth sites, weighting them equally. This model can be simulated very rapidly (at least 100 times faster than the "walking" ant). Had we set $p=1$, we would have obtained an Eden cluster, with fractal dimension $d_f = d$. When a fraction $1 - p$ of the growth sites are poisoned, our flying ant acts rather like a butterfly, choosing carefully to land only on a non-poisoned site. A typical cluster at $p = p_c$ is shown in Fig. 12.

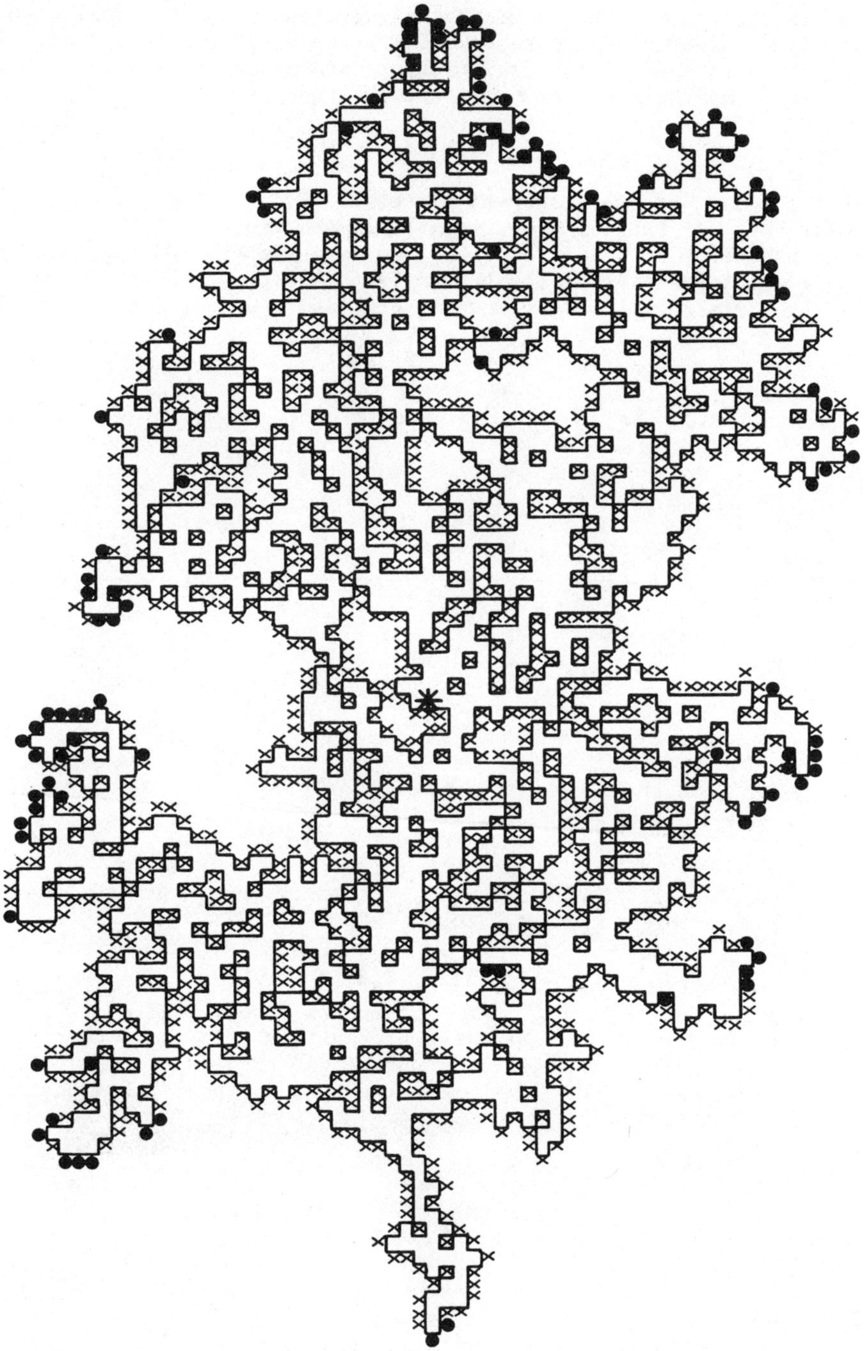

FIG. 12: Typical percolation cluster being grown by the "butterfly" mechanism for the case that each perimeter site has equal weight. This model is equivalent to the Eden model on a "diluted" lattice on which a fraction $1 - p_c$ of the sites have randomly been removed or "poisoned." After Bunde et al (1985a).

An important model in the historical development of critical phenomena was the Fisher-Ma-Nickel long range interaction model (Fisher et al 1972). Here spins are assumed to interact with an interaction parameter that falls off as a power of the separation r between them,

$$J(r) = 1/r^{\alpha}. \tag{32a}$$

What is the analog of the Fisher-Ma-Nickel model for dynamic growth phenomena? Let our butterfly not weight all the growth sites equally (as in the Eden model), but rather weight them as a function of the distance from the present position of the butterfly. Thus the probability for a growth site a distance r to be chosen next is

$$P(r) = 1/r^{\alpha}. \tag{32b}$$

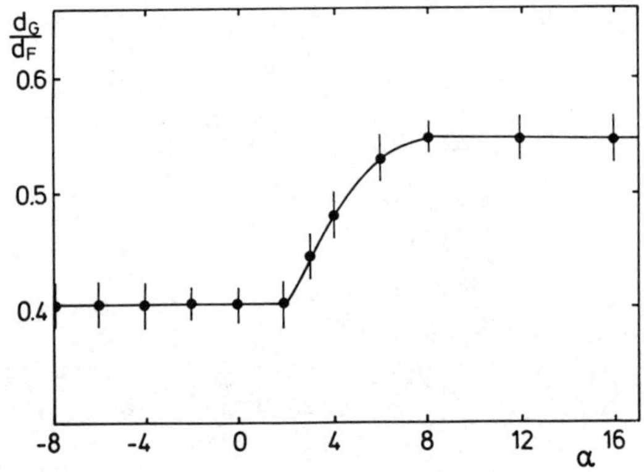

FIG. 13: Dependence on the range parameter α of the intrinsic exponent d_g/d_f governing how N_g, the number of growth sites, scales with N_f, the number of fractal sites. Here the probability that the butterfly at site j chooses growth site i is proportional to $1/r_{ij}$, so small α correponds to a long-range flight and large α to short-range motion. After Bunde et al (1985a).

The Fisher-Ma-Nickel model was of importance because critical exponents dependend smoothly upon the range parameter α, at least over a certain domain of α. This was of initial surprise, since Ising magnets with long-range interactions of long but finite range (say out to 10 neighbor shells) have the same exponents as the nearest-neighbor only case. In retrospect, one could have anticipated this result since we know that in the limit of very long range interactions (small α), one must find mean field exponents while in the limit of very short range interactions (large α) one must recover the normal nearest-neighbor Ising values.

Similarly for the dynamic growth model, in the limit of very long range interactions (small α), we expect to obtain the exponents of a model in which each chemical shell is filled one at a time. Leyvraz (private commmunication) and Bunde et al (1985a) have presented two different arguments that for this limit

$$d_g/d_f = (d_\ell - 1)/d_\ell \qquad \text{[LONG RANGE BUTTERFLY].} \tag{33}$$

For short range interactions (large α), we expect to find exponents corresponding, roughly, to those of ant exploration. Recently Bunde et al (1985a) have carried out extensive simulations for a wide range of α, roughly confirming expectations--except for the fact that the exponent d_g is a little larger than expected in the short range limit (Fig. 13).

We have seen that for some kinetics, the growth sites lie on shells of constant "chemical distance" from the origin of growth. Recently, a model has been developed that can have "forgotten" growth sites lying closer to the origin (Bunde et al 1985b). This model, called the shell model, is similar in some respects to the variable range butterfly except that now the next site to grow is weighted not by its distance from the presently growing site but rather by its distance from the origin. If the weight is chosen so that the nearby growth sites are favored, we have a "banker" who neatly cleans up (occupies) any nearby growth sites before going to further ones. If, on the other hand, the weight is chosen so that the growth sites most distant from the origin are favored, we have a "physicist" who leaves behind many forgotten obligations in his haste to open up new territory. Both extremes are special cases of the weight function used, $P(r) = 1/r^{\alpha}$, which is identical in form to the butterfly of (32b) except that now r is interpreted as the Pythagorean distance of the growth site from the origin, not the distance of the growth site from the previously occupied site. When $\alpha = 0$, the model reduces to the $\alpha = 0$ limit of the butterfly model (also studied recently in the $\alpha = 0$ limit by Family and Vicsek 1984). When $\alpha >$ 0, growth sites at a large distance from the origin are preferred--leaving many "forgotten" growth sites near the origin (Fig. 14).

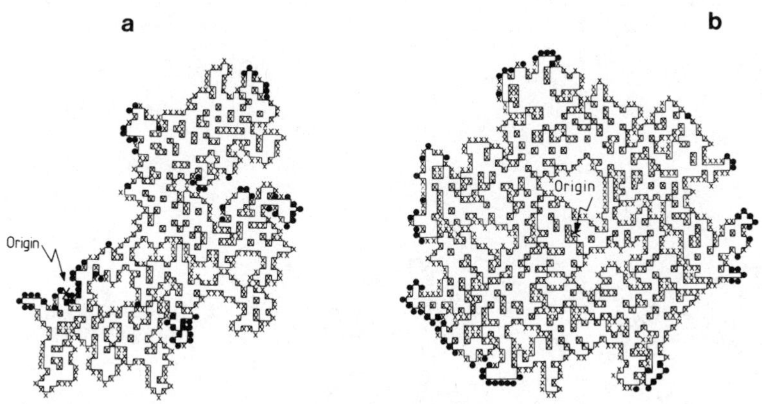

FIG. 14: A percolation cluster in the process of being grown using the shell model, in which the next chosen site has a weight that is solely a function of its Pythagorean distance from the origin, with $P(r) \sim 1/r^{\alpha}$. Shown here are the cases (a) α=-4 and (b) α=12. After Bunde et al (1985b).

The list of growth models could continue , limited perhaps only by the imagination of their inventors. Two models will be mentioned finally: FIFO and FILO. These are not names of pet poodles in Paris, but rather acronyms for "first in first out" and "first in last out". The first is a model for which the condition of occupying sites on successive chemical does apply! The second is the reverse of the first: instead of occupying the earliest-created of the growth sites, one occupies the latest-created. If p were unity, we would get straight line clusters. A FILO cluster at the percolation threshold is shown in Fig. 15. The surprising thing is that this cluster looks like anything but a typical percolation cluster (it more resembles the poodle than the map of Japan). However it must be, since in the long time limit all the sites of this region will have been tested by the growth mechanism--and 59% of them will be randomly occupied.

We conclude this lecture by returning to the important question discussed above: "WHERE ARE THE GROWTH SITES?" In the case of DLA, we had to calculate a "hit probability" P_i by shooting a million projectiles at the fractal. Here the growth sites are well-defined objects, in contrast to DLA where every site is a growth site, "in principle". Hence we not only have but a single exponent d , but we can also calculate the spatial distribution function $P(r,N_f)$ that tells us the probability of finding a growth site a distance r from the origin of a cluster that has a total cluster mass of N_f. This was first done by Stanley et al (1984 and to be published) for the case of the ant growth mechanism. They found that the resulting curves (Fig. 16) resemble those of Plischke and Racz (1984) for DLA except that the characteristic exponent for the width of the growth site distribution function is the <u>same</u> as that for the radius of gyration for ant growth. The Plischke-Racz result for DLA that a different exponent governs the "active zone" has caused considerable excitement since it would imply a second length variable enters in DLA. There has been some concern that possibly the clusters used were not large enough (Meakin and Sander 1985, Jullien and Botet 1985).

We will leave this debate to settle itself, as larger and larger DLA clusters become available. While a large percolation cluster is typically 10 billion (Margolina et al 1984), a large DLA cluster until recently was only 25,000 and even now is typically 100,000--about 5 orders of magnitude smaller. The important point is that one never knows when one has reached the asymptotic limit and, even worse, one knows that many of the problems we are concerned with are "slow crossover problems" in which the basic physics is dominated by extremely rare events. Sometimes we are lucky in

FIG. 15: A percolation cluster being grown using the FILO rule ("first in last out"). After Herrmann and Stanley (1985).

that probing a small system is sufficient to unveil the truth--and sometimes we are not. Recent examples when we have not been so lucky include two with which this speaker has been associated. In the first, we considered a new model of interacting random walks in which each new site visited has a weight factor p (Stanley et al 1983). When p=1 it reduces to a random walk. When p > 1, it interpolates between purely random walks and self-avoiding walks (p=oo). When 0<p<1, the model describes attracting

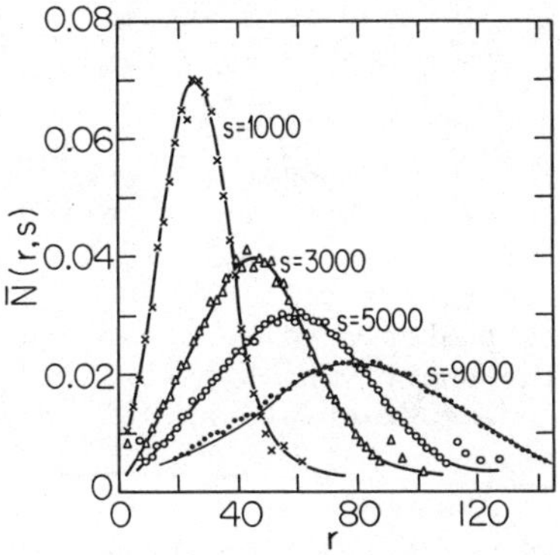

FIG. 16. The normalized probability of finding a growth site a distance r from the origin for clusters whose mass is s. The clusters are grown by an ant mechanism here, but analogous results are found for other growth dynamics such as the variable range butterfly (Bunde et al 1985a), the shell model (Bunde et al 1985b), and the FILO/FIFO mechanism (Herrmann and Stanley 1985). After Stanley et al (to be published).

random walks and also noninteracting random walks on a lattice with static traps (Fig.17). Our initial studies on this model suggested that it shared some of the intriguing fatures of random walks on percolation fractals--e.g., dimension-independent exponents. This initial hope was seen to be false: it turns out that some time ago, the mathematicians Donsker and Varadhan (1975) showed rigorously that for all values of p < 1, the mean number of visited sites varies as $S(N,p) \sim N^{\chi}$ with $\chi = d/(d+2)$. Stanley et al (1983) had estimated $\chi = 2/3$ for all d, the same number found for a random walk on a fractal substrate.

The apparent failure of methods that are fairly standard in physics to come up with results that are in at least moderate agreement with rigorous mathematics is a cause for some concern. This type of problem seems to have plagued a number of recent numerical studies of the trapping problem. To help resolve these discrepancies, Redner and I have extended the series for $Z(N,s)$, the number of N-step random walks that visit s sites, to order $N=18$ $[s=2,3, \ldots, N+1]$ for the square lattice. Figure 18a shows a conventional log-log plot of $S(N,p)$ vs. N for $p = p_c$ and also $p=1$, where $S(N) = N/(\ln N)$. A least squares fit to this data gives $\chi \cong 0.72$ $[\neq 1/2]$ for $p = p_c$, and $\chi \cong 0.82$ $[\neq 1]$ for $p=1$. However, there is significant curvature when we plot the slopes based on successive data points against $1/N$ [Fig. 18b] for $p = p_c$ the intercepts form a decreasing sequence which extrapolate to a value in the vicinity of the Donsker and Varadhan result. We have repeated this analysis for the p values studied in Stanley et al (1983), and find that the series are consistent with the Donsker and Varadhan result; hence we believe that the apparent dependence of χ on p is actually a manifestation of very slow crossover to asymptotic behavior. A random walk must travel for a typical time $t \sim r^2 \sim c^{-2/d}$. For shorter time, the exponent will be close to one (for $d \geq 2$), and our data are consistent with a very slow crossover for much larger times to the asymptotic Donsker and Varadhan value $\chi = d/(d + 2)$.

What can we learn from our discovery? Because of the isomorphism with the trap problem, our model must share its peculiar features. Specifically the survival probability of the random walk is slower than exponential due to relatively rare large trap-free regions. Because of the dominant role of such rare events, the most probable value of an observable and the average value do not coincide. The underlying probability distribution is sufficiently skewed that one must obtain numerical data over a non-zero

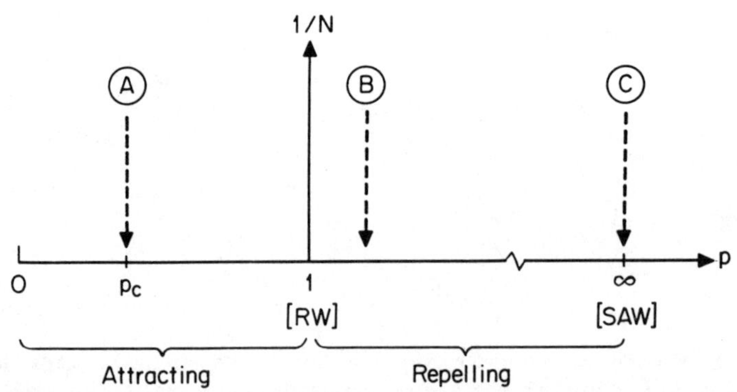

FIG. 17. Phase diagram for the interacting walk model, whosing three typical paths that probe the infinite N behavior. The case p=1 corresponds to noninteracting random walks. For p<1 we have a self-attracting walk, and for p>1 a self-repelling walk (with the special case of infinite p being the self-avoiding random walk).

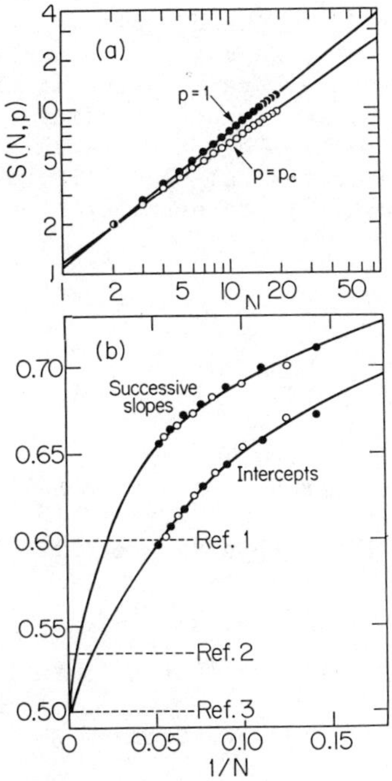

FIG. 18: (a) The mean number visited sites $S(N,p)$ for $p=1$ and $p = p_c$. {From rigorous results for $S(N,p)$ the limiting slopes should be 1 for $p=1$ and 1/2 for $p = p_c$, while the slopes of the straight lines shown are actually 0.82 and 0.72, respectively.} (b) The top curve shows the dependence on $1/N$ of the "successive slopes" line [line segments through successive even (o) or odd (•) pairs of points of part (a)]. The bottom curve shows the intercepts of alternate pairs of points in the top curve. Shown for comparison are the predictions of Refs. 1-3 , which here denote Stanley et al (1983), Meirovitch (unpublished),and Donsker and Varadhan (1975).

fraction of the sample space in order that the average value be correctly measured. Exact enumeration methods, on the other hand, correctly sample even skewed probability distributions and hence should work provided the series are long enough to account for the very slow crossover of the trapping problem.

More recently, an analogous mistake has occurred. Several groups have independently proposed a new random walk model termed the "kinetic growth walk" (KGW) which grows by choosing the next site of the chain from among the empty nearest neighbor lattice sites of the head of the chain (Majid et al 1984, Kremer and Lyklema 1984, Hemmer and Hemmer 1984, Family unpublished). Initially, both exact enumeration and Monte Carlo studies suggested that the fractal dimension of this walk was different from that of the pure self-avoiding walk. We now know, thanks to rigorous work of Peliti (1984), that the extremely rare events in which the walk traps itself are sufficient to change its universality class to be the same as that of the self-avoiding walk. Thus the fractal dimension appears to be constant for a large range of fractal sizes (for d=3, this range is up to about 5000 monomers). Eventually, however, it crosses over to its ultimate asymptotic value.

LITERATURE CITED

Adam M, Delsanti M, Durand D, Hild G and Munch JP 1981 Pure and Appl Chem 53 1489

Adler J, Aharony A and Stauffer D 1985 J Phys A 18 L129

Aharony A and Stauffer D 1984 Phys Rev Lett 52 2368

Alexander S 1983 Ann Israel Phys Soc 5 149

Ben-Avraham D and Havlin S 1982 J Phys A 15 L691

Bunde A, Herrmann HJ, Margolina A and Stanley HE 1985a Phys Rev Lett 55 653

Bunde A, Herrmann HJ and Stanley HE 1985b J Phys A 18 L523

Bunde A, Coniglio A, Hong DC and Stanley HE 1985c J Phys A 18 L137

Coniglio A and Stanley HE 1984 Phys Rev Lett 52 1068

DeArcangelis L, Redner S and Coniglio A 1985 Phys Rev B 31 4725

Donsker MD and Varadhan SRS 1975 Comm Pure Appl Math 28 1, 279, 525

Family F and Vicsek T 1985 J Phys A 18 L75

Fisher ME, Ma SK and Nickel B 1972 Phys Rev Lett 29 917

Gefen Y, Aharony A and Alexander S 1983 Phys Rev Lett 50 77

de Gennes PG 1976 La Recherche 7 919

de Gennes PG 1980 J Phys (Paris) Colloq 41 C3-C17

Halsey TC, Meakin P and Procaccia I 1985 Phys Rev Lett (submitted)

Havlin S 1984 Phys Rev Lett 53 1705

Havlin S, Djordjevic Z, Majid I, Stanley HE and Weiss GH 1984 Phys Rev Lett 53 178

Hemmer S and Hemmer PC 1984 J Chem Phys 81 584

Herrmann HJ and Stanley HE 1985 Z Phys B 36 (1 Sept)

Hong DC, Stanley HE, Coniglio A and Bunde A 1985 Phys Rev B 32 xxx

Jullien R and Botet R 1985 Phys Rev Lett 54 2055

Kertèsz J 1983 J Phys A 16 L471

Kremer K and Lyklema JW 1984 J Phys A 17 L691

Kunz H and Souillard B 1978 J Stat Phys 19 77

Leyvraz F 1985 J Phys A 18 xxx

Leyvraz F, Adler J, Aharony A, Bunde A, Coniglio A, Hong DC, Stanley HE and Stauffer D 1985 J Phys A Lett 18 xxx

Leyvraz F and Stanley HE 1983 Phys Rev Lett 51 2048

Majid I,Jan N,Coniglio A and Stanley HE 1984 Phys Rev Lett 52 1257

Margolina A, Nakanishi H, Stauffer D and Stanley HE 1984 J Phys A 17 1683

Meakin P and Sander LM 1985 Phys Rev Lett 54 2053

Meakin P, Stanley HE, Coniglio A and Witten TA 1985 Phys Rev A 32 (1 Oct)

Peliti L 1984 J de Physique 45 L925

Plischke M and Racz Z 1984 Phys Rev Lett 53 415

Rammal R and Toulouse G 1983 J de Physique 44 L13

Sahimi M 1985 J Phys A 18 1543

Sapoval B, Rosso M and Gouyet JF 1985 J Physique Lett 46 L149

Stanley HE, Kang K,Redner S and Blumberg RL 1983 Phys Rev Lett 51 1223

Stanley HE, Majid I, Margolina A and Bunde A 1984 Phys Rev Lett 53 1706

Turkevich LA and Scher H 1985 Phys Rev Let 55 xxx

EXCITATIONS OF/ON FRACTAL NETWORKS

R. Orbach

Department of Physics
University of California
Los Angeles, California 90024

ABSTRACT

Fractal symmetry (dilation invariance), as opposed to Euclidean symmetry (translation invariance), requires three dimensionalities to contain a physical description of the excitations of a fractal network: d, the Euclidean (or embedding) dimension; D, the Hausdorff (or fractal) dimension; and $\bar{\bar{d}}$, the fracton (or spectral) dimension. The dynamical properties of percolarting networks are examined in this context. The vibrational density of states and the vigrational excitation dispersion law are calculated. The former is shown to be proportional to ω^{d-1} in the phonon or long length scale regime. A crossover is found at frequency ω_c, proportional to $(p-p_c)^{\nu[1+(\theta/2)]}$, where p is the bond occupancy probability and p_c the critical percolation bond occupancy. Here, ν is the correlation length exponent, and θ is the exponent appropriate to the range dependence of the diffusion constant $[D(r) \propto r^{-\theta}]$. At ω_c, a scaling argument shows that a reasonably rapid rise occurs in the vibrational density of states, to which we shall refer as a "fracton edge." For frequencies higher than ω_c, the vibrational density of states continues to rise, but at a slower rate, proportional to $\omega^{\bar{\bar{d}}-1}$. The excitations in this regime are termed "fractons." It is shown that the electrical conductance on a fractal network also depends on D and $\bar{\bar{d}}$. Use is made of Anderson localization scaling theory to show that, for $\bar{\bar{d}} < 2$, the fracton eigenstates are localized. Very recent inelastic neutron diffraction measurements of the vibrational excitations of fused silica are shown to exhibit a density of states which agrees in form remarkably closely with a recent scaling model of phonon to fracton crossover.

The vibrational dispersion law also exhibits the effects of crossover. At long length scales, the dispersion law is linear, with a sound velocity proportional to $(p-p_c)^{\nu\theta/2}$. Near crossover, the dispersion curve reduces its slope, exhibits an inflexion point, and then rises as the $D/\bar{\bar{d}}$ power of the inverse length scale.

These results suggest that conventional electron-phonon interactions may be strongly modified on a fractal network. We have analyzed the two cases of localized electron-one fracton interaction, and a localized electron-two fracton interaction. To do so, we have had to take explicit account of the spatial extent of the vibrational wave function. This has necessitated the introduction of a fourth dimensionality, d_ϕ, which determines the pythagorean range of the localized fracton wavefunction. The electronic relaxation rate differs from electronic site to electronic site. We are able to calculate the probability density for the electronic relaxation rate. The lack of a single relaxation rate results in a significant departure from an exponential time decay of the electronic departure from equilibrium. We find a time dependence which is faster than power law, but slower than exponential or stretched exponential. We suggest that these theoretical results call into question interpretations of phenomena involving electron-vibrational interactions in glassy materials which have been based on extended phonon states. We recalculate the energy and temperature dependence for the inelastic scattering time for extended electron states. Our results exhibit marked departure from those obtained with extended (and localized) phonon states.

We suggest a variety of calculations and experiments which can further elucidate the affect of fracton excitations on physical properties. We believe that fracton excitations may be relevant to some glassy materials above a crossover frequency. These excitations can have a profound influence on the vibrational density of states and the dynamics associated with electron-vibration interactions.

I. INTRODUCTION

It is now generally recognized that self-similar fractal structures[1] require (at least) three dimensionalities for the description of their dynamic properties: the embedding (Euclidean) dimension d; the Hausdorff (fractal) dimension[1] D; and the fracton[2] (spectral[3]) dimension $\bar{\bar{d}}$. In general, $\bar{\bar{d}} \leq D \leq d$, with equality obtaining in the limit that one passes from the self similar regime (exhibiting dilation invariance),[4] to the Euclidean regime (exhibiting translation invariance). We are interested in studying physical systems where a length scale separates the underlying geometry between fractal and Euclidean. An example which has been closely studied is the percolating network. Many physical systems are thought to map onto such a construct [for example, polymers and gels], and it proves to be an effective "laboratory" in which one can study dynamic physical processes. The percolating network is thought to exhibit only a single characteristic length, ξ_p, which we shall abbreviate for conciseness to ξ. For excitations whose length scale is greater than ξ, the network appears uniform and can be described by Euclidean geometry. For excitations whose length scale is less than ξ, the network appears open and "tortured," and can be described by fractal geometry.[5,6] We have conjectured[2] that glasses and other amorphous materials may exhibit qualitatively similar behavior. The precise origin of the characteristic length in such materials is still open to question. However, it does appear that, when an appropriate test for fractal properties is made, such materials follow the theoretical expectations contained in Ref. 2.

The purpose of this paper is to present the underlying theory for the dynamical properties of a fractal network. We derive in Sec. II the vibrational density of states in the short length scale (fracton) regime. The problem of the "connection" of this result with the long length scale phonon density of states has recently been treated using a scaling approach,[7] and we shall reproduce the results here. We shall show that one expects a rapid increase in the vibrational density of states when the phonon and fracton length scales are of order ξ, or, more conveniently, when the vibrational frequency equals the so-called crossover frequency,[3]

$$\omega_c \propto \xi^{-[1+(\theta/2)]} = \xi^{-D/\bar{\bar{d}}} \quad , \tag{1}$$

where θ is the exponent introduced by Gefen et al.[8] to describe the range dependence of the diffusion constant on a fractal network $[D(r) \propto r^{-\theta}]$. This expression is derived in Sec. II.

Section III will treat the vibrational dispersion law directly in the two regimes. We shall use the scaling arguments of Refs. 3 and 7 to generate the framework for our solution. We will then introduce the effective medium approximation to obtain an explicit form for the dispersion relation as a function of the bond concentration p. We shall show that one can predict both the real and imaginary parts of the vibrational response to a plane wave perturbation (e.g., neutron scattering). Finally, we shall discuss the character of the vibrational wave function, showing that one can expect localization in general in the fracton regime.

Section IV will contain some remarkable recent experimental observations which, we believe, demonstrate that the ideas expressed in this Introduction could be relevant to a wide class of "random" materials.

Section V will treat the interaction between electronic states on a fractal network and the lattice vibrations of that network. The "electron-fracton" interaction can have profound effects. We shall compute the consequences of this interaction in Sec. VI for electronic relaxation involving localized electronic states (e.g., paramagnetic impurities, optical centers) and one fracton. We shall find that the relaxation rate differs from site to site. We are able to calculate the probability density for the relaxation rate. We then go on to calculate the time dependence of the electronic states occupation. In Sec. VII we perform the same type of calculation for the localized electron (and nucleus) interacting with two fractons. This has direct relevance to the Raman spin-lattice relaxation rate in both electronic and nuclear magnetic resonance. We again calculate the probability density for the relaxation rate, and the time dependent recovery of the magnetization after saturation. We present the results of a calculation of the inelastic scattering time for an extended electronic state interacting with a single fracton in Sec. VIII. This calculation has direct relevance to electronic localization and to superconductivity. Finally, in Sec. IX we summarize our calculations and suggest future investigations, both experimental and theoretical.

II. VIBRATIONAL DENSITY OF STATES

Gefen et al.[8] have shown that for diffusion on a single percolation cluster at short times (so that the "ant"[9] has not had a chance to reach the edge of the cluster) the mean square distance at time t is proportional to

$$\langle r^2(t)\rangle_s \propto t^{2/(2+\theta)} \quad . \tag{2}$$

Here, the cluster contains s sites within a linear range $r_s \propto s^{1/D}$. The number of distinct sites visited is then

$$N_s(t) \propto t^{D/(2+\theta)} \quad . \tag{3}$$

For conditions where return to the origin is probable (see below), this leads to the probability of remaining at the origin at time t:

$$\langle P_o(t)\rangle \propto [N_s(t)]^{-1} \propto t^{-D/(2+\theta)} \quad , \tag{4}$$

which we have denoted by the autocorrelation function $\langle P_o(t)\rangle$. We now assume that the cluster is part of the infinite network, so as to avoid edge problems. The spectral density of states for the diffusion problem is given by the Laplace transform of Eq. (4), and is of the form

$$N(\varepsilon) \propto \varepsilon^{[D/(2+\theta)]-1} \quad . \tag{5}$$

It is well known that the diffusion problem maps directly onto the vibrational problem (for the most simple vibrational force constant).[10,11] This allows us to obtain the vibrational density of states from the spectral density of states for diffusion by analytically continuing from ε to $-\omega^2$, and multiplying by ω:

$$N(\omega) \propto \omega^{[2D/(2+\theta)]-1} \quad . \tag{6}$$

Noting that the standard Euclidean result is $N(\omega) \propto \omega^{d-1}$, Alexander and Orbach[3] were led to identify the fracton dimensionality,

$$\bar{\bar{d}} = 2D/(2+\theta) \quad . \tag{7}$$

The essential problem which remained [until an effective medium approximation indicated the form of $N(\omega)$ over the entire range] was how to connect Eq. (6) with the Euclidean form, $N(\omega) \propto \omega^{d-1}$, at the frequency where the length scale equaled the percolation correlation length ξ. That frequency is implicitly contained in Eq. (2), by noting that diffusion "time" scales as the inverse square of the vibrational frequency [see the discussion immediately before Eq. (6)]:

$$\omega_c \propto \xi^{-[1+(\theta/2)]} \quad ,$$

where the subscript c means crossover. An early attempt suggested a drop in $N(\omega)$ at ω_c, but this was in the opposite direction suggested by experiment.[12] A subsequent effective medium approximation (EMA) calculation of the vibrational density of states by Derrida et al.[13] exhibited a sharp increase in $N(\omega)$ at ω_c, leading one to ask whether a scaling argument could be developed which would be independent of model. This has recently been accomplished with the following results.[7] We write

$$N(\omega) = A\omega^{p-1} f(\omega/\omega_c) \quad . \tag{8}$$

In the fracton regime, we have just shown [Eq. (6)] that $N_{fr}(\omega) \propto \omega^{p-1}$ with $p = \bar{\bar{d}}$. We choose to normalize the vibrational spectrum per unit mass (i.e., per atom). In Eq. (8), A is a constant and does not depend on ω_c. In the phonon regime, $N_{ph}(\omega) \propto \omega^{d-1}$. The only way this can happen and have Eq. (8) simultaneously satisfied is by requiring

$$N_{ph}(\omega) \propto \omega_c^{p-d}\, \omega^{d-1} \quad , \qquad \omega/\omega_c \ll 1 \quad , \tag{9}$$

so that one is in the phonon regime. This requirement immediately leads to the following conditions on f(x) from Eq. (8):

$$f(x) \to 1 \quad , \qquad x \to \infty \quad ,$$

$$f(x) \to x^{d-p} \quad , \qquad x \to 0 \quad . \tag{10}$$

The ratio of the two extrapolated densities of states at crossover is therefore

$$N_{fr}(\omega)/N_{ph}(\omega) = C \quad , \tag{11}$$

where C does not depend on ω_c (i.e., is non-critical). Because it will turn out in most cases that $1 \leqslant p \leqslant d$, one also notes that $N_{ph}(\omega) < N_{fr}(\omega)$, $\omega \ll \omega_c$.

The normalization condition requires that the integral over $N_{fr}(\omega)$ from 0 to the upper cutoff, $\omega_{fD} = \omega_D(\xi/a)^{\theta/2}$ [see Ref. 2] must be unity. Yet, we are forced to use Eq. (9) at the lowest frequencies by virtue of the crossover to Euclidean geometry at long length scales. And because $N_{ph}(\omega) < N_{fr}(\omega)$ at low frequencies, we are therefore missing vibrational modes. The only way to remedy this is to "fill in" with the missing modes. But $f(x) \to 1$ when x is large ($\omega >> \omega_c$). Hence, f(x) must exceed unity over at least some range of x. It appears that this is most plausible near $x \sim 1$, or in the region of $\omega \sim \omega_c$. Thus, we allow a "bump" in f(x) of size Δf near $x = 1$. This leads directly to an increase in the vibrational density of states:

$$\Delta N(\omega) = A\ \omega_c^{p-1}\ \Delta f \quad . \tag{12}$$

This means that although the *extrapolated* ratio of the fracton to photon density of states is non-critical, the increase in the density of states at crossover scales like Eq. (12) and therefore depends explicitly on the characteristic length ξ through Eq. (1). A suggested form for $N(\omega)$ is plotted below in Fig. 1.

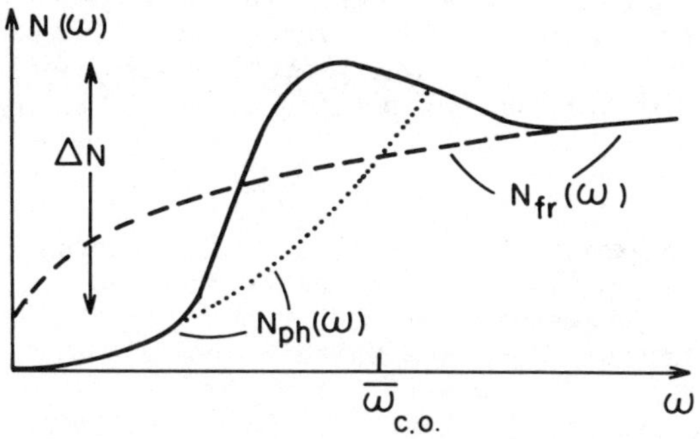

Fig. 1. Crossover of the density of vibrational states (solid line). The dotted (dashed) lines represent the continuation of the phonon (fracton) asymptotic behavior into the crossover regime.

The form from EMA exhibits a sharper structure near ω_c and can be found in Ref. 13.

Recently, purely mechanical central and "bending" forces[14-16] have been introduced. These additional terms, which diminish with increasing length scale more rapidly than the scaler forces of Ref. 3 [S. Alexander, J. Phys. (Paris) Lett. 45, 1939 (1984)], destroy the relationship between the conductivity and elastic modulus exponent.[17-18] These forces result in an "elastic" fracton dimensionality, $\bar{\bar{d}}_E$,[19] which can be less than unity. This results in a vibrational density of states $N(\omega)$ for percolation clusters at $p = p_c$ which is slightly divergent as $\omega \to 0$. A numerical simulation[19] nicely confirms this picture.

III. VIBRATIONAL DISPERSION LAW

The same "connection" problem arises with the vibrational dispersion law as with the vibrational density of states, with one significant additional complication. The propagation characteristics of the two classes of vibrational excitations also differ: the phonons are presumed to be extended (unless localized in the Anderson sense), while the fractons, for $\bar{d} < 2$, are localized (by virtue of the underlying fractal geometry). This was first noted by Rammal and Toulouse,[4] who calculated the range dependence of the conductance on a fractal network. They found,

$$\Sigma(r) \propto r^{(D/\bar{\bar{d}})(\bar{\bar{d}}-2)} \equiv r^{\beta} , \tag{13}$$

which defines β in the Anderson localization sense.[20] We see that β is negative for $\bar{\bar{d}} < 2$, leading to localization of the wave function. In the Euclidean limit for d = 3, $\bar{\bar{d}} \to d = 3$, and β is positive. Hence, the crossover from Euclidean to fractal geometry as the length scale shortens beyond ξ (or the excitation frequency increases beyond ω_c) causes a geometric crossover which forces β to reverse its sign. This shows that it is the fracton dimensionality which plays the role of dimension (and not the fractal) for localization. Indeed, this is the reason why the argument about return-to-the-origin worked in the preceding Section when we calculated the auto-correlation function. For $\bar{\bar{d}} < 2$, the diffusing particle always returns to the origin, and our "trick" can work.

Inherent localization of fractons occurs when $\bar{\bar{d}} < 2$. For the percolating network, the Alexander-Orbach conjecture[3] places $\bar{\bar{d}} = 4/3$, so that the fractons are most assuredly localized. This makes the description of a dispersion law difficult. We shall refer to the wave vector in the fracton regime as the inverse of the characteristic wave function length, which we take to be the fracton localization length. Remarkably, we can obtain a precise expression for this relationship from Eq. (1). We simply write

$$\omega_{fr} \propto \ell^{-D/\bar{\bar{d}}} , \tag{14}$$

where ℓ is the characteristic length of the fracton. In general, the exponent in Eq. (14) is larger than unity so that the linear dispersion law expected for the phonon regime will cross the fracton dispersion law. One expects there may be structure in this crossing regime on the basis of the behavior of $N(\omega)$ at crossover. This is seen explicitly in the EMA solution for the dispersion law, obtained by Entin-Wohlman et al.[21] Because the EMA predicts a strong critical character for the behavior of the density of states near crossover, the effect on the dispersion law is also strong. Thus, for p near p_c, one finds a shape which is nearly roton-like. This is exhibited in Fig. 2.

The reason for Fig. 2 being relevant is associated with the manner in which experiments are done. Neutron scattering involves the momentum change of the neutron. The usual experimental analysis generates an apparent energy width. In reality, the apparent energy width is a momentum width.

The scaling approach[7] can also be useful here for obtaining the velocity of sound for the acoustic branch of a percolating network. We write the dispersion law as,

$$\omega(\ell^{-1}) = \ell^{-a} f(\xi\ell^{-1}) . \tag{15}$$

We use Eq. (1) to note that $a = 1 + (\theta/2)$. It follows immediately that

$$\omega_{ph}(\ell^{-1}) \propto \xi^{-\theta/2} \ell^{-1} , \quad \ell/\xi \to \infty . \tag{16}$$

Hence, the sound velocity, c_s, must scale as,

$$c_s \propto \xi^{-\theta/2} \propto (p-p_c)^{\beta\theta/2} . \tag{17}$$

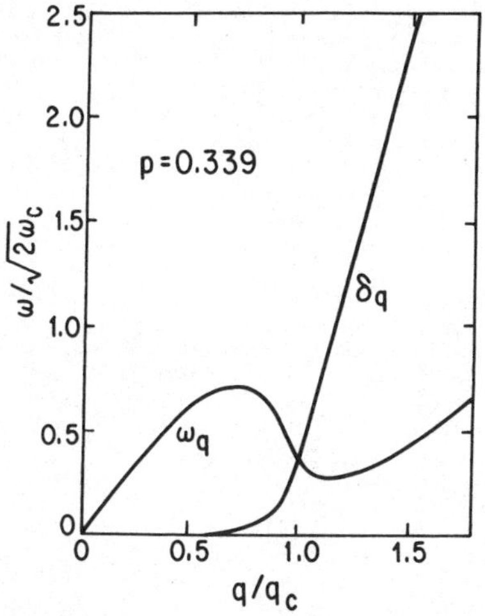

Fig. 2. A plot of the real and imaginary parts of ω, ω_q , and δ_q, respectively, vs. q, for d = 3. Here, p = 0.339, and ω_c = 0.0104. The critical percolation concentration within EMA is p_c = 1/3.

IV. COMPARISON WITH EXPERIMENT

Alexander et al.[2] conjectured that glasses, amorphous solids, epoxy-resin, and neutron irradiated quartz could exhibit a crossover length, such that for short length scale excitations a fractal description would be appropriate. This was based on the near universal character of the specific heat and thermal conductivities of these materials.[22] In addition, the analysis of Zaitlin and Anderson[23] displayed a near universal precipitous drop in the thermal excitation mean free path from thermal conductivity measurements. The two approaches came together in the paper of Kelham and Rosenberg on epoxy resin[12] where they noted that the peak in the specific heat required a jump in the vibrational density of states at precisely the same frequency where the mean free path crossed the "dominant phonon" wavelength. These results were interpreted[2] to mean that a crossover length (frequency) existed in these materials; that the peak in the specific heat was associated with the increase in the density of states at length scales ξ (frequencies ω_c); and that the fracton dimensionality was less than two, so that for shorter length scale excitations (frequencies higher than ω_c), the thermal vibrational states were localized. This was in analogy with our calculations for percolating networks, though there was no reason to restrict the experimental parameters accordingly.

It was difficult to find other experimental evidence for these predictions. The optical data exhibited only small structure[24] which could be interpreted as a "fracton edge," but may not have been convincing.[25]

It is the author's opinion that the situation has markedly changed. A very recent low energy neutron scattering experiment by Buchenau et al.[26] on fused silica, when replotted (courtesy of the authors) on a log-log scale, exhibits what the author believes one would have to regard as a fracton edge (Fig. 3). Comparison with either Fig. 1 of this paper, or the predictions of the EMA result of Ref. 13, is impressive. What is even more convincing is the much older plot of Zeller and Pohl for the thermal

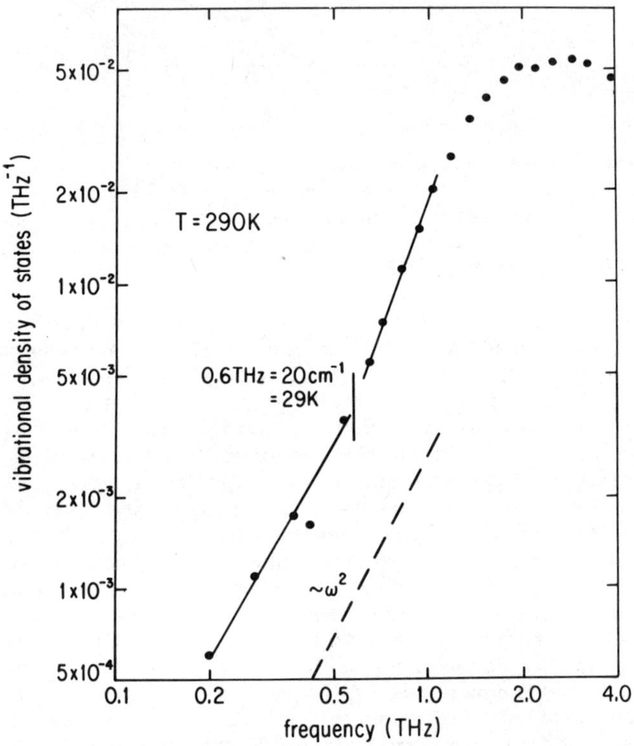

Fig. 3. A plot of the vibrational density of states for fused silica from data replotted by the authors of Ref. 26. A line has been drawn at an energy which could reasonably be identified with the beginning of the fracton edge. The lines through the data are a guide to the eye. Note the similarity in form to the results of a scaling analysis [Ref. 7] exhibited in Fig. 1.

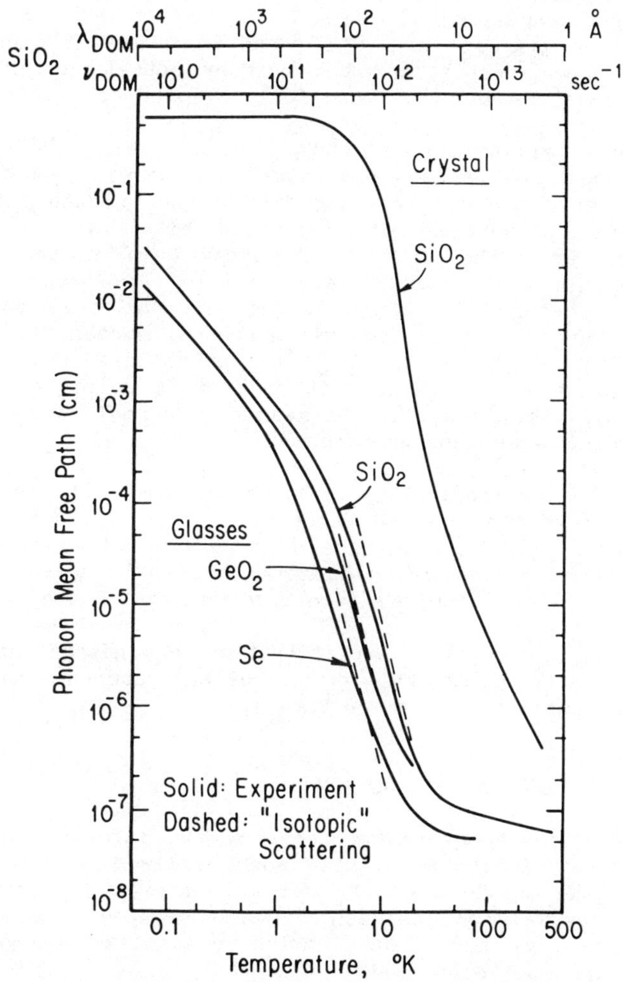

Fig. 4. A plot of the mean free path for the "dominant" phonon from thermal conductivity data for three glassy materials, taken from Ref. 22. Note that the wavelength of the dominant phonon in fused silica is near 20 angstroms when the wavelength equals the mean free path (a condition for localization). Note that this occurs at an energy equal to 20 cm^{-1}, identified in Fig. 3 as the beginning of the fracton (localization) edge.

mean free path for fused silica, which is exhibited here in Fig. 4 as it appeared in Ref. 22.

One sees that the localization condition of Ioffe and Regel,[27] wavelength equalling the mean free path, is satisfied. The fracton edge occurs at an energy appropriate to a lattice vibration whose wavelength is roughly 20 angstroms. This means that only a small number of atoms (say a thousand) are arranged with a fractal geometry. This may account for the rather rounded character of the fracton edge. It would be interesting to see if the softer glasses (e.g., GeO_2 and Se), where the crossover length is expected to be greater, exhibit sharper density of states structure because the fractal geometry is more fully allowed to develop.

An even more telling case is that of epoxy-resin. As indicated above, it was Kelham and Rosenberg's analysis[12] of thermal measurements on this material which gave the original impetus to the fracton interpretation of the vibrational properties of amorphous materials. Very recently, Rosenberg[28] measured the vibrational density of states of epoxy-resin directly using inelastic neutron scattering. The nature of epoxy-resin allowed him to change the length scale (he was able to reduce it by adding three times the stoichiometric amount of hardened). His curves are exhibited in Fig. 5, where the shift in fracton edge from 13.3 K to 15.7 K is apparent. This is consistent with the shift to higher temperatures of the beginning of the plateau in the measured thermal conductivity upon the same change in hardener concentration.[29]

These examples are indicative of the evidence which one can muster from experiment. Though they may be heartening, they are not necessarily compelling arguments for the existence of true fractal (self similar) structures in amorphous solids. For example, one could argue that all of the above evidence "merely" implies localized phonons, rather than fracton excitations. In the next three sections, we explore the electron fracton interaction and argue that explicit time dependence measurements can differentiate between fractal or localized phonon character of the vibrational excitations in a given material.

V. THE ELECTRON-FRACTON INTERACTION

We outline our subsequent treatment of the interaction between localized and extended electronic states, and localized lattice vibrations, in this Section. We shall treat the case of interactions with fractons (as opposed, for example, to localized phonons) because it will prove to be a very simple matter to reduce our results to localized phonons (essentially by passing to the Euclidean limit).

An ulterior motive for this examination of interaction effects involving fractons is the hope to elucidate the consequences for experiment of a fractal lattice. In particular, we shall show that electronic and nuclear vibration-induced dynamics are altered substantially when the underlying atomic arrangement possesses fractal geometry.

We shall show that far more than a change in vibrational density of states is involved when one passes from extended phonons to fractons. The localized character of the fractons, the explicit energy dependence of their localization length, and their peculiar dispersion relation all combine to alter substantially the form of the electronic response. For example, we shall show that the electronic relaxation rate varies from electronic site to site, and we shall be able to calculate the probability density for that rate. The computation of the time dependent reponse of the electronic system will yield very surprising results, but

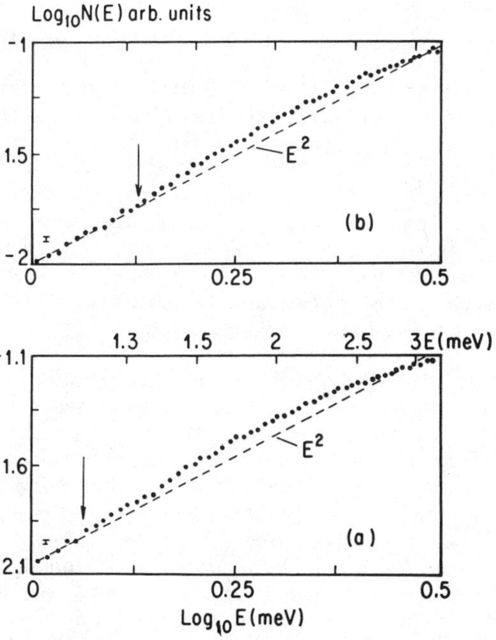

Fig. 5. The dependence of the density of vibrational states, N(E), on the energy for two epoxy-resin samples, derived from inelastic neutron-scattering experiments. The deviation at the arrows from the E^2 dependence (dashed line) should be noted. The bar indicates probable error limits. (a) Specimen of MY 750 hardened with the stoichiometric quantity of ethylene diamine, (b) specimen of MY 750 hardened with three times the stoichiometric quantity of ethylene diamine. This should have shorter structure units than (a) and it should be noted that the breakaway from E^2 occurs at a higher energy than in (a). The data are taken from Ref. 28.

ones which are entirely responsible in terms of the underlying fractal dimensionalities. Thus, observation of the effects predicted in the subsequent Sections of this paper will allow for evaluation of the fractal parameters which describe fracton dynamics. They will also allow one to distinguish between fracton and localized phonon excitations in disordered lattices.

We now explicitly consider the interaction between a localized electronic state and a distortion of the fractal network. We expand the distortion in terms of fracton normal modes. The wave function for the αth mode with energy ω_α is assumed to have the form

$$\phi_\alpha(\omega_\alpha) \propto (\ell_{\omega_\alpha})^{-D/2} \exp\left[-\frac{1}{2}(L/\ell_{\omega_\alpha})^{d_\phi}\right] \tag{18}$$

where ℓ_{ω_α} is the localization length. Here, d_ϕ is an exponent which characterizes the character of localization on a fractal network. It is expected to lie between unity (the Euclidean limit) and d_{min},[30] where d_{min} is defined by $\ell \propto L^{d_{min}}$ where ℓ is the shortest path between two points separated by a Pythagorean distance L.[31] The interaction matrix element is proportional to the spatial derivative of the fracton wave function, and thence to $(\ell_{\omega_\alpha})^{-d_\phi}$. Using the energy dependence of the fracton localization length [Eq. (14)], we find the spatial derivative of ϕ_α proportional to $\omega_\alpha{}^q$ where $q = \tilde{d}(d_\phi/D)$. We shall reduce our expressions to localized phonons by setting $q = d_\phi = 1$, $D = d = 3$ (the dimension of the embedding space), and replacing ℓ_ω in Eq. (18) by a localization length ξ assumed to be frequency independent.

Our treatment of the one-fracton and two-fracton relaxation processes will make use of the Fermi Golden Rule expressions for the relaxation rate experienced by a localized electron at a particular site. Because the fracton states are localized, the usual energy conserving delta function will not be found but rather will be replaced by the combined energy widths of the electronic and fracton states. Eventually, one will sum over the entirety of the (discrete) fracton vibrations interacting with the electronic state. We term this combined width δ_L, where L is the distance between the electronic and fracton states. In general, δ_L will be a function of L. For the one-fracton relaxation case, we shall treat two limits for δ_L. The first is that appropriate to rapidly relaxing fractons (by virtue of anharmonic interactions, for example). If the fracton energy width exceeds the largest electronic relaxation rate, we can safely replace δ_L by a constant, independent of L. A curious opposite extreme, which is of intrinsic interest because of the bizarre result, is the opposite limit where there is no external broadening of the electronic and fracton energy levels. In such a case, δ_L is to be replaced by the relaxation rate itself (obviously L-dependent). This self-consistency will lead to consequences entirely unexpected: some electronic states will not relax at all! This is because of the self-consistency condition leading to a vanishing of the relaxation rate for electronic sites far away from the fracton of required energy. We treat only the former limit when we calculate the two-fracton relaxation process, because of the formidable mathematics encountered even for δ_L a constant.

After formulating the relaxation rate for a particular electronic site interacting with a fracton a spatial distance L away, we next calculate the probability density that that rate is the largest relaxation rate "seen" at that site. It will turn out that the distribution of rates is strongly skewed towards small rates (because of the exponential localization of the fracton wavefunction). This has the happy consequence that the largest rate dominates the sum of all the other rates experienced at a particular site caused by all the other fractons in the fractal network. Were this not the case, we should probably not have been able to obtain a closed expression for the final time dependence of the electronic state occupancy.

After we have calculated the probability density for the electronic relaxation rate, we compute the time profile one would obtain in an experiment which would measure, for example, the initial electronic state occupancy (e.g., the electronic or nuclear moment magnetization relaxation rate). We shall find remarkable time dependences: faster than any power law, but slower than an exponential, for example. This would be appropriate for experiments on electronic (or nuclear) systems in the absence of strong cross relaxation. When rapid communication between electronic (or nuclear) states is present, experiment will measure the average relaxation rate. We compute this as well, both by direct computation of the first moment of the probability density for the relaxation rate, or through a direct Golden Rule calculation of the average relaxation rate.

It should be emphasized that our results will all be in terms of time dependences involving the exponents we have introduced for a fractal network: D, $\bar{\bar{d}}$, and d_ϕ. Thus, measurement will yield these parameters directly, both through the time and/or temperature dependence of our results. When one passes to the localized phonon limit, the exponents will all be Euclidean, and determined entirely by the embedding dimension d. It is in this manner that one can determine whether, in a particular sample, the network is of fractal or Euclidean geometry. This is an important consequence of our calculations: we believe they enable one to make this distinction directly, and thereby to determine the essential underlying geometry of the host lattice.

There are only fragmentary examples of experiments which can test these predictions. It appears that most published data have been obtained under conditions of strong cross relaxation, so that an exponential time dependent return-to-equilibrium has been obtained. Even under those conditions, we shall predict peculiar temperature dependence, and indeed such have been seen.[32,33] However, until the cross relaxation has been overcome (by going to much smaller electronic or nuclear concentrations), the distinctive time profile predictions contained in our expressions will escape detection.

The next Section (VI) will treat the one-fracton interaction; the following (VII) two-fracton interactions, and the final (VIII) the case of extended electronic states interacting with a single fracton. Further details can be found in the full papers listed as Refs. 34 - 38.

VI. LOCALIZED ELECTRON-ONE FRACTON RELAXATION RATE DISTRIBUTION AND DECAY PROFILE

The work reported in this Section has been submitted in preliminary form for publication, Ref. 34, and in a more complete form in Ref. 35. We shall only outline the steps in the calculation here--the philosophy of the calculation having been given in Sec. V.

In the one-fracton relaxation process, the change in electronic energy must equal the energy of the fracton state either emitted or absorbed in the relaxation process. We denote the change in electronic energy by ω_o. We find the transition probability per unit time for fracton emission to be

$$W(\omega_o,L) \sim \omega_o^{2q-1}\, \ell_{\omega_o}^{-D} \coth(\beta\omega_o/2)\, \frac{1}{\delta_L} \exp\left[-(L/\ell_{\omega_o})^{d_\phi}\right] . \tag{19}$$

Here ω_o^{2q} arises from the spatial derivative of the wave function, ω_o^{-1} results from the normal mode expansion, $\ell_{\omega_o}^{-D}$ from the normalization of the wave function, and $\coth(\beta\omega_o/2)$ is the fracton occupation number. The exponential factor in Eq. (19) describes the wave function localization and δ_L the energy width of the fracton state.

We now construct the probability density $\bar{P}(W)$ for a certain relaxation rate at a given electronic site. We then use it to derive the probability density P(W) that W is the largest relaxation rate at that electronic site.

The probability $\bar{P}(W)$ consists of: (i) the probability that there is a fracton level with energy in the range ω_o, $\omega_o + \delta_L$ centered on the fractal at a distance between L and L+dL from the electronic site:

$$\text{(i)} \qquad N_{fr}(\omega_o)\, \delta_L\, D\, L^{D-1}\, dL \quad , \tag{20a}$$

where $N_{fr}(\omega)$ ($\propto \omega^{\bar{\bar{d}}-1}$) is the fracton vibrational density of states, and (ii) a delta function requiring the electronic relaxation rate $W(\omega_o, L)$ caused by the fracton centered a distance L away be equal to W:

$$\text{(ii)} \qquad \delta[W(\omega_o, L) - W] \quad . \tag{20b}$$

Integrating over all L, we obtain

$$\bar{P}(W) = \int dL\, L^{D-1}\, D\, N_{fr}(\omega_o)\, \delta_L\, \delta[W(\omega_o, L) - W] \quad . \tag{21}$$

From Eq. (21) we can construct the probability that there is no relaxation rate larger than W as seen from a given electronic site:

$$\prod_{W \leqslant W' \leqslant W_{max}} [1 - \bar{P}(W')\, dW'] = \exp\left[- \int_W^{W_{max}} dW'\, \bar{P}(W')\right] \quad . \tag{22}$$

Finally, the probability density P(W) that W is the largest relaxation rate is

$$P(W) = \bar{P}(W) \exp\left[- \int_W^{W_{max}} dW'\, \bar{P}(W')\right] \quad . \tag{23}$$

Note that one expects P(W) to be normalized ($\int P(W)\, dW = 1$) while $\bar{P}(W)$ counts all relaxation channels and is not normalized.

As we indicated in Sec. V, there are two limits in which we can evaluate P(W). We treat here that appropriate to rapid relaxation of the fracton state due, for example, to anharmonic interactions. Under such a condition, we can replace δ_L with δ. We insert Eq. (19) with this replacement into Eq. (21) and evaluate it to obtain

$$P_{\delta_L = \delta}(W)\, dW = c_1 (D/d_\phi)\, [\ell n(W_{max}/W)]^{(D/d_\phi)-1}\, dW/W \quad , \tag{24}$$

where

$$c_1(\omega_o, \delta) = N_{fr}(\omega_o)\, \delta (\ell_{\omega_o})^D \quad . \tag{25}$$

We note that the geometrical index d appears only through the combination D/d_ϕ. The value of this ratio will in general depend upon the Euclidean dimensionality, but is always $\geqslant 1$. This will assist us in the derivation of P(W).

Inserting Eq. (24) into Eq. (23), we obtain the probability density for the largest W:

$$P_{\delta_L=\delta}(W) = c_1(\bar{D}/d_\phi)\,[\ell n(W_{max}/W)]^{(\bar{D}/d_\phi)-1}$$

$$\times \exp\left\{-c_1[\ell n(W_{max}/W)]^{D/d_\phi}\right\} / W \quad . \tag{26}$$

This probability density is normalized to unity, exhibiting the fact that all electronic sites relax. This will not be so for the case where the level width δ_L is given by W(L).

Under such conditions (we omit the considerable algebra), we obtain the probability density for the relaxing sits (denoted by a subscript r) for the largest W:

$$P_r(W)\,dW = c_2(D/d_\phi)\,[\ell n(W_{max}/W)]^{(D/d_\phi)-1}$$

$$\times \exp[-c_2(\bar{D}/d_\phi)]\,[\Gamma(D/d_\phi) - \Gamma(D/d_\phi, \ell n\{W_{max}/W\})]\,dW/W_{max} \,, \tag{27}$$

where $\Gamma(\eta,z)$ is the incomplete gamma function, and the subscript r means relaxing. This notation is introduced because the probability density Eq. (27) is not normalized. One has

$$\int_0^{W_{max}} dW\,P_r(W) = 1 - \exp[-c_2(D/d_\phi)\,\Gamma(D/d_\phi)] \quad , \tag{28}$$

which means that a fracton $\exp[-c_2(\bar{D}/d_\phi)\ \Gamma(D/d_\phi)]$ of the electronic states do not relax. Their contribution to the probability density for the largest W is therefore

$$P_{nr}(W)\,dW = \exp[-c_2(D/d_\phi)\,\Gamma(\bar{D}/d_\phi)]\,\delta(W)\,dW \quad , \tag{29}$$

where the subscript nr signifies non-relaxing. These are sites for which the largest channel relaxation rate is zero. The probability density for the largest W is the sum of Eqs. (27) and (29):

$$P_{\delta_L=W(L)}(W) = P_r(W)\,dW + P_{nr}(W)\,dW \quad . \tag{30}$$

We now use these probability densities for the largest relaxation rate, P(W), to calculate the time profile of the electronic state population and the average relaxation rate. The latter is relevant for rapid cross relaxation. We omit the considerable algebra associated with taking the Laplace transform of P(W) to obtain the time profile. We find

$$P_{\delta_L=\delta}(t) \sim [2\pi(D/d_\phi)c_1]^{1/2}$$

$$\times\ (\ell n\ \tau_a)^{(D-d_\phi)/2d_\phi}\ \tau_\alpha^{-c_1[\ell n\ \tau_a]^{(D/d_\phi)-1}} \quad , \tag{31}$$

with $\tau_a = (d_\phi/D)(W_{max}t/c_1)$. With W_{max} the maximum of Eq. (19), and c_1 given by Eq. (25). For the limit of self-consistent broadening, we find

$$P_{\delta_L=W(L)}(t) \sim \exp(-\alpha c_2) + (D/d_\phi)c_2\ \sqrt{2\pi}\ \exp[-(\alpha c_2 + 1)]$$

$$\times\ (1/\tau_b)\ (\ell n\ \tau_b)^{(D/d_\phi)-1} \quad . \tag{32}$$

Here $\tau_b = W_{max}t$, and $c_2(\omega_0)$ is given by,

$$c_2(\omega_o) = 2^{D/d_\phi} N_{fr}(\omega_o) W_{max}(\ell_{\omega_o})^D \tag{33}$$

The forms for P(t) are of interest, first because they are unusual in themselves, and second because they are expressed in terms of the various dimensionalities appropriate to fractal networks. Hence, observation of the predicted time dependences will allow one to determine the parameters which describe the dynamics of the fractal lattice. We see from Eq. (31) that the one-fracton relaxation time profile is faster than any power law in time, but slower than an exponential or stretched exponential decay. The form for P(t) for the self-consistent broadening, Eq. (32), is of special interest because of the fraction of states which do not decay at all (first term).

These dependences can be expected for a wide variety of physical situations: (a) non-radiative decay for an optical center in a glass; (b) excitation transfer between two optical sites in a glass where the energy mismatch is greater than ω_c; and (c) electron-hole recombination in amorphous semiconductors in the band-tail regime where the electronic (and hole) states are localized.[39] It would be interesting to determine if the observed time dependences for recombination follow these predictions.

The calculation of the average relaxation rate is carried out in Refs. 34 and 36. Summarizing these calculations, one finds (to within numerical coefficients of order unity),

$$\langle W\rangle \sim N_{fr}(\omega_o)\, \omega_o^{2q-1} \coth(\beta\omega_o/2) \quad . \tag{34}$$

The dependence of the average relaxation rate upon the magnitude of the change in electron energy ω_o is

$$\langle W\rangle \sim \omega_o^{2q+\bar{\bar{d}}-2} , \qquad \beta\omega_o >> 1 ,$$

$$\langle W\rangle \sim \omega_o^{2q+\bar{\bar{d}}-3}\, k_\beta T , \quad \beta\omega_o << 1 . \tag{35}$$

It is now appropriate to examine the form of the quantities Eqs. (31) and (32), for localized phonons. Making the transition to Euclidean geometry (as outlined in Sec. V) results in

$$P_{\delta_L=\delta}(t) \sim (\ell n\, \tau_a)\, \tau_a^{-c_1[\ell n\, \tau_a]^2} \quad ; \tag{36a}$$

and, for self-consistent broadening, in

$$P_{\delta_\omega=W(L)}(t) \sim \tau_b^{-1} [\ell n\, \tau_b]^2 \quad . \tag{36b}$$

We now see explicitly how one can determine the difference between a fractal network and a Euclidean lattice: the exponents appearing in P(t) are quite different. Observation of the time dependent response of an electronic system prepared out of equilibrium should generate the parameters which will determine which picture is relevant to the particular system under investigation: fractons or localized phonons.

This Section has been concerned with rather large changes in electronic energies. The next Section discusses the opposite limit, a change in electronic energy small compared to ω_c, requiring inelastic scattering of two fractons for effective relaxation.

VII. LOCALIZED ELECTRON-TWO FRACTON RELAXATION RATE DISTRIBUTION AND DECAY PROFILE

We consider in this Section the relaxation of a localized electronic state caused by the absorption and subsequent emission of a vibrational mode, the difference in vibrational energies equaling the change in electronic energy. We obtain, therefore, an explicit expression for the spin-lattice relaxation rate probability density, and for the time profile of the magnetization, for a paramagnetic impurity in a fractal network. As in the preceding Section, we first write down the probability for unit time for a localized electronic state to change its energy by ω_o upon absorbing a fracton of energy ω_α at a distance L away, and subsequently emitting a fracton of energy $\omega_\alpha + \omega_o$ a distance L' away. The overall rate will be proportional to the fourth power of the interaction matrix element (because both fractons enter the expression for the transition probability per unit time). We find

$$W(\omega_\alpha,L,L') \propto \left[(\ell_{\omega_\alpha})^{-2D}\,\omega_\alpha^{4q-2}\exp\{-(L/\ell_{\omega_\alpha})^{d_\phi} - (L'/\ell_{\omega_\alpha})^{d_\phi}\}\right] \times \left[e^{\beta\omega_\alpha}/(e^{\beta\omega_\alpha}-1)^2\right]\frac{1}{\delta(\omega_\alpha,L,L')}\frac{\omega_\alpha^2}{\Delta^2} \quad . \tag{37}$$

The factor $\delta(\omega_\alpha,L,L')$ represents the combined widths of the electronic and fracton vibrational states. The same situation applies for two-fracton relaxation as for one-fracton relaxation if the participating states have combined energy widths which are less than the maximum calculated relaxation rate. However, the computational complexity of the two-fracton case prohibits us from performing the self-consistent calculation, as, for example, we were able to do for one-fracton relaxation. Thus, we shall set $\delta(\omega_\alpha,L,L') = \delta$ for the remainder of the two-fracton calculation. The factor $(\omega_\alpha/\Delta^2)^2$ appears in the relaxation rate of Kramers doublets, and arises from the dynamical breaking of the time reversed symmetry of the ground double.[40] For non-Kramers transitions, it is replaced by $(1/\Delta)^2$. We rewrite Eq. (37) in the form

$$W(\omega_\alpha,L,L') = W_m(\omega_\alpha)\,\frac{\delta(\omega_\alpha,0,0)}{\delta(\omega_\alpha,L,L')}\exp\left[-(L/\ell_{\omega_\alpha})^{d_\phi} - (L'/\ell_{\omega_\alpha})^{d_\phi}\right] \quad , \tag{38}$$

where

$$W_m(\omega_\alpha) = W_o\,\beta^{-\alpha}\,(\beta\omega_\alpha)^a\,e^{\beta\omega_\alpha}/e^{\beta\omega_\alpha} - 1)^2 \quad . \tag{39}$$

Here

$$a = \begin{cases} 4q + 2\bar{\bar{d}} \quad , & \text{Kramers transitions} \\ 4q + 2\bar{\bar{d}} - 2 \ , & \text{non-Kramers transitions} \end{cases} \tag{40}$$

and W_o is a constant divided by $\delta(\omega_\alpha,0,0)$ and $\beta = 1/k_BT$.

We now need to go through the steps followed in Sec. VI leading to the calculation of P(W). We omit the detailed algebra which can be found in Ref. 31. We find the following result,

$$P_1(W)dW = \alpha_1\,\beta\delta\,\frac{1}{a-2}\,(W_1/W)^{1/(a-2)} \times \exp\left[-\alpha_1(\beta\delta)(W_1/W)^{1/(a-2)}\right] dW/W \ , \qquad W > W_c \quad , \tag{41a}$$

$$P_2(W)dW = \alpha_2(\delta/\omega_c)\ 2\eta[\ell n(W_c/W)]^{2\eta-1}$$
$$\times \exp\left[-\alpha_2(\delta/\omega_c)(\ell n(W_c/W))^{2\eta}\right] dW/W\ , \qquad W \ll W_c\ , \tag{41b}$$

where the crossover rate $W_c = W_o\ \beta^{-a}(\beta\omega_c)^{a-2}$, and α_1 and α_2 are the numerical constants $\alpha_1 = \eta B(\eta,\eta+1)(\bar{\bar{d}})^2(a-2)^{2\eta}\ \Gamma(2\eta+1)$ and $\alpha_2 = \eta B(\eta,\eta+1)(\bar{\bar{d}})^2$, and $\eta = D/d_{min}$. The crossover relaxation rate, W_c, specifies two time regimes. The time dependence of the relaxation profile for shorter times can be found by taking the Laplace transform of Eq. (41a). One finds

$$P_1(t) \sim (t/t_1)^{1/2(a-1)} \exp\left[-\alpha_3(t/t_1)^{1/(a-1)}\right], \quad t < t_1(\delta/\omega_c)^{a-1}, \tag{42a}$$

where the time constant t_1 is

$$t_1 = [\alpha_1/(a-2)]\ (1/W_1)\ (1/\beta\delta)^{a-2}\ , \tag{42b}$$

and α_3 is the numerical constant $\alpha_3 = \alpha_1(a-1)/(a-2)$. The time profile for longer times can be found by taking the Laplace transform of Eq. (41b). One finds

$$P_2(t) \sim (\delta/\omega_c)^{1/2}\ [\ell n(t/t_2)]^{n-1/2}$$
$$\times \exp\left[-\alpha_2(\delta/\omega_c)[\ell n(t/t_2)]^{2\eta}\right., \qquad t > t_2\ , \tag{43a}$$

where the time constant t_2 is

$$t_2 = (2\eta\alpha_2/W_c)(\delta/\omega_c)\ . \tag{43b}$$

From these equations, we can see that the long time behavior of P(t) begins as a stretched exponential [Eq. (42a)]. The temperature dependence of the time constant $t_1 \sim \beta^2$. At a time of the order $t_1(\delta/\omega_c)^{a-1}$, the decay changes its character. From Eq. (43a), it is seen that it is slower than stretched exponential, but faster than a power law. In this regime, the temperature dependence of the time constant $t_2 \sim (\beta\delta)^2(\omega_c)^{1-a}$. This time dependence is the same as found for the long time behavior of the one-fracton interaction.

We now quote without derivation the result for the average relaxation rate, appropriate to a situation of strong cross relaxation. Again, the details can be found in Ref. 37. We find

$$1/T_1^{ave} \sim \begin{cases} W_o\ \beta^{1-a}\ \delta\ , & a > 3 \\ W_o\ \beta^{1-a}\ \delta(1/\beta\omega_c)^{3-a}\ , & 3 > a > 2 \end{cases}$$

One should note that W_o is proportional to $1/\delta$, so that the average relaxation rate is independent of δ.

Most reasonable values for the fractal exponents result in a > 3, so that, from above,

$$1/T_1^{ave} \propto T^{2\bar{\bar{d}}[1+2(d_\phi/D)]-1}\ , \qquad \text{for Kramers transitions}\ , \tag{44a}$$

and

$$1/T_1^{ave} \propto T^{2\bar{\bar{d}}[1+2(d_\phi/D)]-3}\ , \qquad \text{for non-Kramers transitions}\ . \tag{44b}$$

Our ansatz for q seems more natural than that of Stapleton et al.[32] in the context of a scaling description. One can, however, also have physical conditions where q = 1 would be correct (as in Ref. 32). Because $\bar{\bar{d}}$, D, and d_ϕ are not known independently, there is at present no way of determining q for the systems investigated by Stapleton et al.[32]

For percolating networks in $d = 3$, we set $d_\phi = d_{min}$ in order to obtain a numerical estimate. Using, $\bar{\bar{d}} = 4/3$,[3] $D = 2.5$,[5,6] and $D/d_{min} = 1.8$,[31] this leads to

$$1/T_1^{ave} \sim \begin{cases} T^{4.63} , & \text{for Kramers transitions} , \\ T^{2.63} , & \text{for non-Kramers transitions} . \end{cases} \tag{45}$$

It is now appropriate to examine the form of the quantities, Eqs. (42a) and (43a), for localized phonons [we omit the details of the probability density for W--they can be found in Ref. 37]. We find

$$P_1(t) \sim (\gamma_2)^{1/2} [\ell n(t/t_1)]^{2d-1} \exp\left[-\gamma_2[\ell n(t/t_1)]^{4d-1}\right] , \tag{46a}$$

$$t_1 < t < t_1 \exp[\beta\Omega_{ph}] ,$$

where the characteristic time t_1 is

$$t_1 = \gamma_2[(4d-1)/W_o]\beta^a , \tag{46b}$$

and

$$P_2(t) \sim (\gamma_3)^{1/2} [\ell n(t/t_2)]^{d-1/2} \exp\left[-\gamma_3[\ell n(t/t_2)]^{2d}\right] , \tag{47a}$$

$$t > t_2 \exp[\beta\Omega_{ph}] ,$$

where the characteristic time t_2 is

$$t_2 = \gamma_3[2d/W_o]\beta^a . \tag{47b}$$

The three quantities γ_i are: $\gamma_1 = d^2(\xi/\beta\Omega_{ph})^{2d}\,\beta$,

$\gamma_2 = [d^3/(2d-1)]\,\frac{(d-1)!d!}{(2d)!}\,(\xi/\beta\Omega_{ph})^{2d}\,\beta\delta$, and $\gamma_3 = \gamma_2(\beta\Omega_{ph})^{2d-1}$. Here, ξ is the phonon localization length (taken to be energy independent), and Ω_{ph} is the phonon Debye frequency. We see from Eq. (46a) that the short time decay is rather rapid, with the characteristic time t_1 increasing with temperature for $d = 3$. At longer times, the decay becomes slower, with the characteristic time t_2 decreasing with increasing temperature.

In both regimes, the time decay obeys a law of the form

$$P(t) \propto e^{-(\ell n\, t)^x} . \tag{48}$$

This is the same form for the time dependence as found in the previous Section for both one-fracton and one-phonon relaxation processes. The reason for the difference between the time dependences for the two-fracton and two-phonon response arises from the difference in the energy dependence of the localization lengths for fractons (proportional to $\omega^{-\bar{\bar{d}}/D}$), and for phonons (set equal to the constant ξ).

The temperature dependences for the average relaxation rates for localized phonons are the same as those found for extended phonon states:

$$1/T_1^{ave} \sim \begin{cases} T^7 , & \text{non-Kramers transitions} , \\ T^9 , & \text{Kramers transitions} . \end{cases} \tag{49}$$

It seems apparent that the temperature dependence of the spin-lattice relaxation rate, even in the regime of strong cross relaxation, may be a clear measure of fractal geometry.

VIII. EXTENDED ELECTRON-ONE FRACTON RELAXATION

In the two previous Sections, we have treated the interaction of a localized electron and localized fracton (or localized phonon) excitations. In this Section, we relax the condition for localization of the electronic wave function, but still keep the fracton states localized. Further, for simplicity, we shall consider only the electron-one fracton interaction, and take $d_\phi = 1$. We relax the latter constraint in future work.[38] The situation we treat may be relevant to inelastic electron scattering in transport measurements. Such scattering rates play a crucial role in, for example, superconductivity and electron spatial localization. When we compare with particular fractal networks in order to obtain numerical estimates, we shall limit ourselves to the infinite cluster of the percolating network.

Consider the scattering of an electron within the Born approximation. The transition probability per unit time to scatter an electron is given by,

$$P_{i\to f} = 2\pi |V_{fi}|^2 \, \delta(E_f - E_i - \varepsilon) \quad , \tag{50}$$

where E_i and E_f are the initial and final system energies of the medium, and ε is the electron energy loss $\varepsilon = E_i - E_f$. The dynamical interaction between the electron and the medium is proportional to the strain components of the medium for short range forces. We therefore set $V_{fi} = \gamma \; \partial u(\vec{r})/\partial \vec{r}$, where γ is the interaction constant. The electron is scattered from a plane wave state of quasi-momentum $\vec{k}_i$ into a state $\vec{k}_f$. Equation (50) becomes,

$$P_{i\to f} \sim \gamma^2 \int_{-\infty}^{\infty} dt \; e^{i\varepsilon t} \int d\vec{r} \; d\vec{r}\,' \times \langle e^{i\vec{q}\cdot\vec{r}} [\partial u(\vec{r},t)/\partial\vec{r}] e^{-i\vec{q}\cdot\vec{r}\,'} [\partial u(\vec{r}\,')/\partial\vec{r}\,'] \rangle$$

$$= \gamma^2 \, q^2 \, P(\vec{q},\varepsilon) \quad . \tag{51}$$

Here, $\vec{q} = \vec{k}_i - \vec{k}_f$, and $P(\vec{q},\varepsilon)$ is the correlation function defined by,

$$P(\vec{q},\varepsilon) = \int_{-\infty}^{\infty} dt \; e^{i\varepsilon t} \int d\vec{r} \; e^{i\vec{q}\cdot\vec{r}} \; P(\vec{r},t) \quad , \tag{52}$$

where,

$$P(\vec{r},t) = \int d\vec{r}\,' \; \langle u(\vec{r}+\vec{r}\,',t) u(\vec{r}\,',0) \rangle \quad . \tag{53}$$

The algebra is tedius, but straightforward, to insert Eq. (18) into Eqs. (53) and (52), and thence into Eq. (51). As stated above, we shall take $d_\phi = 1$ for simplicity. We find [Ref. 38],

$$P(q,\varepsilon) \sim [\omega_c]^{-\theta/(2+\theta)} \; \varepsilon^{-1} \; \delta\left\{q - \varepsilon/[\omega_c]^{\theta/(2+\theta)}\right\} \quad , \quad \varepsilon < \omega_c \quad ; \tag{54a}$$

and,

$$P(q,\varepsilon) \sim \varepsilon^{-2} / \left\{1 + [q/\varepsilon^{2/(2+\theta)}]^2\right\}^{(D+1)/2} \quad , \quad \varepsilon > \omega_c \quad . \tag{54b}$$

We now wish to calculate the electronic inelastic relaxation rate for changes in electronic energy within the fracton energy range ($\omega_c < \varepsilon < \Omega_{fD}$). This time is given by,

$$1/\tau_{\varepsilon_i} = \sum_f \left[P_{f\to i} f(\varepsilon_f) + P_{i\to f}[1 - f(\varepsilon_f)]\right] \quad , \tag{55}$$

where the sum extends over all the final electronic states, and $f(\varepsilon_f)$ is the Fermi function at the final electronic energy. Inserting Eq. (54) into Eq. (55), we find after considerable algebra,

$$1/\tau_{\varepsilon_i} \sim \omega_c^{s-d}\, T^d \quad , \quad \varepsilon_i < T \ll \omega_c \quad ; \tag{56a}$$

and,

$$1/\tau_{\varepsilon_i} \sim T^s \quad , \quad \varepsilon_i < T \quad , \quad \omega_c \ll T \ll \Omega_{fD} \quad , \tag{56b}$$

with,

$$s = (2d - \theta)/(2 + \theta) \quad . \tag{57}$$

This is an interesting result because it coincides with the conventional temperature dependence for electron-phonon inelastic scattering rates at low temperatures [Eq. (56a)], with a specific scaling prediction as a function of crossover frequency. At higher temperatures, one finds quite a different temperature dependence. For example, for percolating networks in d = 2, s ≃ 8/7. This means the temperature dependence of the inelastic scattering rate will be markedly different from that expected phonon states in d = 2. It will be interesting to see if measurements of this rate in thin amorphous films will exhibit a crossover in the power of the temperature from 2 at low temperatures to 8/7 at temperatures above the crossover temperature (energy).

The full details of this calculation, along with the expressions for arbitrary d_ϕ, will be found in Ref. 38.

IX. SUMMARY AND CONCLUSIONS

We have introduced four dimensionalities which appear sufficient to describe dynamical processes of/on fractal networks: d, the Euclidean or embedding dimensionality; D, the fractal or Hausdorff dimensionality which governs the scaling of the mass density; $\bar{d}$, the fracton dimensionality which gives the frequency dependence of the vibrational density of states; and d_ϕ, which determines the Euclidean range of the localized fracton wavefunction.

We introduced the concept of a characteristic length which, in many physical systems, separates long length scale phonon vibrational states from short length scale fracton vibrational states. We showed how this length translates into a crossover frequency which separates the two regimes of the vibrational energy-density of states. We then introduced a scaling approach which showed how one can connected the phonon and fracton density of states in the crossover region.

We discussed the vibrational dispersion law, again using a scaling approach to connect the phonon and fracton dispersion curves. We found that one might naturally expect a "plateau-like" feature in the dispersion law at crossover.

Some remarkable new experimental results were presented which beautifully confirmed this analysis. A "worst case example," that of fused silica, was shown to exhibit a fracton edge in the density of vibrational states determined from neutron scattering.[26] The position of the edge matched precisely with the energy of the "dominant phonon" for which the wavelength equalled the mean free path as derived from thermal conductivity measurements of Zeller and Pohl.[22] This provided strong evidence for the validity of a previous assertion[2] that a fracton edge existed in glasses, and that the inherent geometrical localization of fractons (for $\bar{\bar{d}} < 2$) was the explanation for the plateau in the thermal conductivity found in nearly all glasses. In addition, the value of $\bar{\bar{d}} = 1.4$, extracted from the slope of the fracton density of states in fused silica,[26] is remarkably close to that expected for percolation, 4/3,[3] so that percolation theory may yet prove to be a useful model for glassy dynamics.

An equally strong case was presented for the lattice dynamics of epoxy-resin. Recent neutron diffraction measurements[28] were presented which were interpreted in terms of a fracton edge which moves to higher frequencies as the length scale is reduced by adding hardener beyond stoichiometry.

The remainder of this paper was concerned with the impact of fractal lattice dynamics upon the properties of electronic relaxation. We showed that the localization inherent in fractons leads to an unusual structure for relaxation dynamics. The relaxation time was calculated for localized electrons interacting with one- and two-fractons. We demonstrated that the relaxation rate obeyed a probability density, leading to a strongly non-exponential relaxation time profile. We argued that this profile could be used, in reverse, to establish the dynamical exponents which govern the behavior of fractal networks.

The localized electron-one fracton relaxation rate distribution and time decay profile should be relevant to nonradiative relaxation in glassy materials, and to band-edge electron-hole recombination in amorphous semiconductors. The calculations were performed in two limits: that appropriate to rapid (anharmonic) fracton relaxation, and that appropriate to little lifetime damping of either the electrons or fractons (except, of course, for their mutual interaction). We showed that the former led to a time decay profile which was faster than any power law, but slower than exponential (or stretched exponential). The latter generated the bizarre result that some electronic states did not relax at all, while others relaxed proportionally to the time. This should be observable, especially in amorphouse semiconductors where photoabsorption can monitor directly the initial electronic population.[39]

The localized electron-two fracton relaxation rate distribution and time decay profile should be relevant to spin-lattice relaxation of electronic centers (and nuclei) in glassy materials. The predicted departure from exponential decay should be a signature of fractal structure when cross relaxation is weak. When cross relaxation is strong, the departure from the temperature dependence expected for interaction with extended phonon states should prove a unique signature for a fractal host network.

For extended electronic states interacting with fractons, one recovers the exponential time decay for the electronic state population. This situation should be encountered in many of the materials which have been studied for electronic localization effects. We pointed out that when one also considers the effect of vibrational excitation localization, the dynamics of inelastic conduction electron scattering are profoundly altered. We believe that this effect could alter the interpretation of many properties of amorphous metals.

Future opportunities abound. On the theoretical side, the calculation for the nonradiative decay of a localized electronic state for multi-fracton emission needs to be done. This situation is probably relevant to most band-tail electron-hole recombination processes in amorphous semiconductors. Another calculation which would be related to experiment is the quantity analogous to Mott's variable range hopping, but using fractons instead of extended phonons. Another important topic is anharmonic phonon-fracton interactions. For example, what is the temperature dependence of the acoustic (long length scale) phonon velocity when phonons interact with thermally excited fractons? The great different in the temperature dependence of the sound velocity between crystalline quarts and fused silica calls for such a calculation. Finally, one should be able to calculate the difference in thermal expansion between crystalline and amorphous solids by treating fracton anharmonicity. We are suggesting that it is now important to "re-do" the classical condensed matter calculations for electron-lattice vibrational and anharmonic lattice vibrational interactions, using (localized) fractons in place of the conventional (extended) phonon excitations.

This program will give the experimenter the tools to determine whether measurements of physical processes are relevant to Euclidean or fractal geometries.

On the experimental side, neutron scattering measurements of the vibrational density of states for materials "softer" than fused silica would be very important. As mentioned in Sec. IV, the crossover length in fused silica is only 20 angstroms. Other, softer, glasses should exhibit considerably greater crossover lengths, allowing the fractal geometry of the short length scale regime to more fully develop. It is plausible that this will generate a more profound fracton edge than that exhibited by fused silica. In addition, it would be of great interest to investigate similar properties of polymeric materials, and macromolecules which are thought to be fractal.

Another investigation of interest would be the nuclear spin-lattice decay process for quadrupole relaxation in glassy materials. It would be relevant to examine the time dependence of the recovery of the magnetization for a dilute nucleus (so that cross relaxation is minimized). When cross relaxation dominates, the temperature dependence over a wide temperature range should be measured. The crossover between two-level systems or extended phonons to fracton relaxation should take place at a temperature proportional to the crossover energy. Similar arguments may hold for electronic spin-lattice relaxation in glassy materials.

Finally, it would be interesting to explore the difference between geometrical localization (fracton excitations) and Anderson localization (localized phonons). We have shown how the two differ for electronic relaxation processes and how one can use the time dependence (or temperature dependence) of the decay to tell the difference. It would be interesting to add "dirt" to an otherwise clean system and investigate the dynamical properties of electronic or nuclear centers. As more "dirt" is added, presumably the phonons would localize first, then the overall structure would alter to fractal.

We believe that the fractal properties of amorphous structures has gone beyond being a conjecture to having to be taken seriously. We hope this paper has shown the implications of fractal geometry upon physical processes sufficient to encourage additional investigations, both theoretical and experimental.

This research was supported by the National Science Foundation grants DMR 84-12898 and INT 83-12985.

REFERENCES

1. B. B. Mandelbrot, "The Fractal Geometry of Nature," W. H. Freeman and Co., New York (1983); Ann. Isr. Phys. Soc. 5:59 (1983).
2. S. Alexander, C. Laermans, R. Orbach, and H. M. Rosenberg, Phys. Rev. B28:4615 (1983).
3. S. Alexander and R. Orbach, J. Phys. (Paris) Lett. 43:L-625 (1982).
4. R. Rammal and G. Toulouse, J. Phys. (Paris) Lett. 44:L-13 (1983).
5. A. Kapitulnik and G. Deutscher, Phys. Rev. Lett. 49:1444 (1982).
6. R. F. Voss, R. B. Laibowitz, and E. I. Allessandrini, Phys. Rev. Lett. 49:1441 (1982).
7. A. Aharony, S. Alexander, O. Entin-Wohlman, and R. Orbach, Phys. Rev. B31:2565 (1985).
8. Y. Gefen, A. Aharony, and S. Alexander, Phys. Rev. Lett. 50:77 (1983).
9. P. G. de Gennes, Recherche 7:919 (1976).
10. E. W. Montroll, "Proceedings of the 3rd Berkeley Symposium on Mathematical Statistics and Probability," Univ. of Calif., Berkeley, (1955), p. 209.
11. S. Alexander, J. Bernasconi, W. Schneider, and R. Orbach, Rev. Mod. Phys. 53:175 (1981).
12. S. Kelham and H. M. Rosenberg, J. Phys. C14:1737 (1981); C. I. Nichols and H. M. Rosenberg, J. Phys. C17:1165 (1984); D. E. Farrell, J. E. de Oliveira, and H. M. Rosenberg, "Phonon Scattering in Condensed Matter," ed. by W. Eisenmenger et al., Springer, Berlin (1984), p. 422.
13. B. Derrida, R. Orbach, and K.-Wah Yu, Phys. Rev. B29:6645 (1984).
14. S. Feng and P. N. Sen, Phys. Rev. Lett. 52:216 (1984).
15. Y. Kantor and I. Webman, Phys. Rev. Lett. 52:1891 (1984).
16. D. J. Bergman and Y. Kantor, Phys. Rev. Lett. 53:511 (1984).
17. P. G. de Gennes, J. Phys. (Paris) Lett. 37:L1 (1976).
18. K.-W. Yu, P. Chaikin, and R. Orbach, Phys. Rev. B28:4831 (1983).
19. I. Webman and G. S. Grest, Phys. Rev. B31:1690 (1985).
20. E. Abrahams, P. W. Anderson, D. C. Licciardello, and T. V. Ramakrishnan, Phys. Rev. Lett. 42:673 (1979).
21. O. Entin-Wohlman, S. Alexander, R. Orbach, and K.-Wah Yu, Phys. Rev. B29:4588 (1984).
22. R. C. Zeller and R. O. Pohl, Phys. Rev. B4:2029 (1971).
23. M. P. Zaitlin and A. C. Anderson, Phys. Rev. B12:4475 (1975).
24. U. Strom, J. R. Hendrickson, R. J. Wagner, and P. C. Taylor, Sol. St. Comm. 15:1871 (1974).
25. R. Orbach, J. Stat. Phys. 36:735 (1984).
26. U. Buchenau, N. Nucker, and A. J. Dianoux, Phys. Rev. Lett. 53:2316 (1984).
27. A. F. Ioffe and A. R. Regel, Prog. Semicond. 4, 237 (1960).
28. H. M. Rosenberg, Phys. Rev. Lett. 54:704 (1985).
29. R. Orbach and H. M. Rosenberg, "LT-17," ed. by U. Eckern, A. Schmid, W. Weber, and H. Wühl, Elsebier Science Publishers B. V., Amsterdam (1984), p. 375.
30. Y.-E. Levy and B. Souillard, unpublished.

31. K. M. Middlemiss, S. G. Whittington, and D. S. Gaunt, J. Phys. A: Math. Gen. 13:1835 (1980); R. Pike and H. E. Stanley, J. Phys. A: Math. Gen. 14:L169 (1981); D. C. Hong and H. E. Stanley, J. Phys. A: Math. Gen. 16:L475 (1983); D. C. Hong and H. E. Stanley, J. Phys. A: Math. Gen. 16:L525 (1983); H. J. Hermann, D. C. Hong, and H. E. Stanley, J. Phys. A: Math. Gen. 17:L261 (1984); J. Vannimenus, J. P. Nodal, and C. Martin, J. Phys. A: Math. Gen. 17:L351 (1984); S. Havlin and R. Nossal, J. Phys. A: Math. Gen. 17:L427 (1984); S. Havlin, Z. V. Djordjevic, I. Majid, H. E. Stanley, and G. H. Weiss, Phys. Rev. Lett. 53:178 (1984).
32. H. J. Stapleton, J. P. Allen, C. P. Flynn, D. G. Stinson, and S. Kurtz, Phys. Rev. Lett. 45:1456 (1980); J. P. Allen, J. T. Colvin, D. G. Stinson, C. P. Flynn, and H. J. Stapleton, Biophys. J. 38:299 (1982).
33. J. Szeftal and H. Alloul, Phys. Rev. Lett. 34:657 (1975); J. Non-Cryst. Sol. 29:253 (1978).
34. S. Alexander, O. Entin-Wohlman, and R. Orbach, submitted to J. Phys. (Paris) Lett. for publication (1985).
35. S. Alexander, O. Entin-Wohlman, and R. Orbach, submitted to J. Phys. (Paris) Lett. for publication (1985).
36. S. Alexander, O. Entin-Wohlman, and R. Orbach, submitted to Phys. Rev. B-15 for publication (1985).
37. S. Alexander, O. Entin-Wohlman, and R. Orbach, submitted to Phys. Rev. B-15 for publication (1985).
38. S. Alexander, O. Entin-Wohlman, and R. Orbach, to be submitted for publication, 1985.
39. J. Tauc, private communication.
40. R. Orbach and H. J. Stapleton, "Electronic Paramagnetic Resonance," ed. by S. Geschwind, Plenum Press, New York (1972), p. 121.

LOW FREQUENCY DYNAMICS OF DILUTE ANTIFERROMAGNETS AT PERCOLATION

Timothy Ziman

Institut Laue-Langevin
156X 38042 Grenoble Cedex
France

Dilute magnets with short range interactions provide an interesting laboratory for theories of percolation.[1] Corresponding to the qualitative change of geometry of the percolating cluster as the concentration of magnetic ions decreases to the percolation threshold, the low frequency dynamics are different: coefficients in the hydrodynamic theory of spin-waves are singular in a way that reflects the geometry of the percolating cluster. For dilute isotropic ferromagnets the scaling properties of linearised spin-wave equations are well understood[2]: the spin wave stiffness defined by the low-frequency dispersion relation

$$\omega = D(p)\, q^2$$

is exactly related to the conductance of a dilute random resistor network $\Sigma(p)$ by

$$\Sigma(p) = D(p)\, P(p)$$

where $P(p)$ is the percolation probability. This follows from the equivalence of Kirchoff's equations and the spin-wave equations on the same network. In particular, it implies that the exponent with which D vanishes at percolation is simply derived from that of $\Sigma(p) \sim (p-p_c)^t$:

$$D(p) \sim (p-p_c)^{t-\beta}$$

Furthermore if we accept the conjecture of Alexander and Orbach[3] as being at least approximately correct, then t, and therefore scaling exponents for the long wavelength dynamics, can be found from the geometric exponents ν and β.

The analogous expressions for low frequency modes in the dilute isotropic antiferromagnet are

$$\omega = \pm B(p)\, q$$

where

$$B(p) = \left[\frac{\Sigma(p)}{\chi(p)}\right]^{1/2}$$

The additional coefficient $\chi(p)$ is the susceptibility of the antiferromagnet, at zero temperature, to a uniform magnetic field perpendicular to the antiferromagnetic ordering vector. In mean field theories $\chi(p)$ is finite at percolation: numerically[2] it was found to diverge with critical exponent denoted τ. Physically this enhanced uniform susceptibility, which leads to a relative softening of the low frequency response, is due to the fact that when a two-sublattice antiferromagnet is diluted the sublattices are diluted independently: thus locally there is a net magnetisation that does not exist in the pure crystal. Harris and Kirkpatrick[2] gave a scaling argument to predict a relation between τ and the conductivity exponent t

$$\tau = t - (d-2)\nu - \beta \qquad (1)$$

The relation (1) was re-derived by considering propagation of spin waves on a nodes-and-links picture of the percolating network.[4] As this is now considered to be an over-simplified view of percolation it is important to check the relation (1) numerically.

Recently V. Symons and I have esimated the exponent τ by finite-size scaling: this provides an independent value which given the sizes and numbe of samples simulated, should be more accurate than that given by Harris and Kirkpatrick, who simulated samples of different concentrations above percolation. Our calculation also illustrates, as we shall see, the practical difficulties of extracting accurate exponents from computations on finite lattices.

The assumption of finite-size scaling[5] is that we can determine the exponent τ with which χ diverges for $p \to p_c$ by fixing $p=p_c$ and examining the distributions of $\chi(L)$ for different simulation sizes LxLxL. In practic we calculated $\langle \log\chi°(L)\rangle$ where $\langle\ \rangle$ denotes sample averaging, $\chi°$ is the susceptibility per occupied site of the largest cluster. Then for $L \to \infty$ we expect

$$\chi°(L) = a\, L^y$$

where $y=\left(\frac{\tau+\beta}{\nu}\right)$. Intuitively such a relation can be understood by saying that the dependence on sample size L at criticality is the same as the power-law variation with geometric correlation length for an infinite sample away from criticality. Numerically calculating $\chi°$ involves generating the largest cluster and then solving a matrix equation, whose order is the number of connected sites, which determines the linear response of the individual classical moments to a torque-free magnetic field.[2]

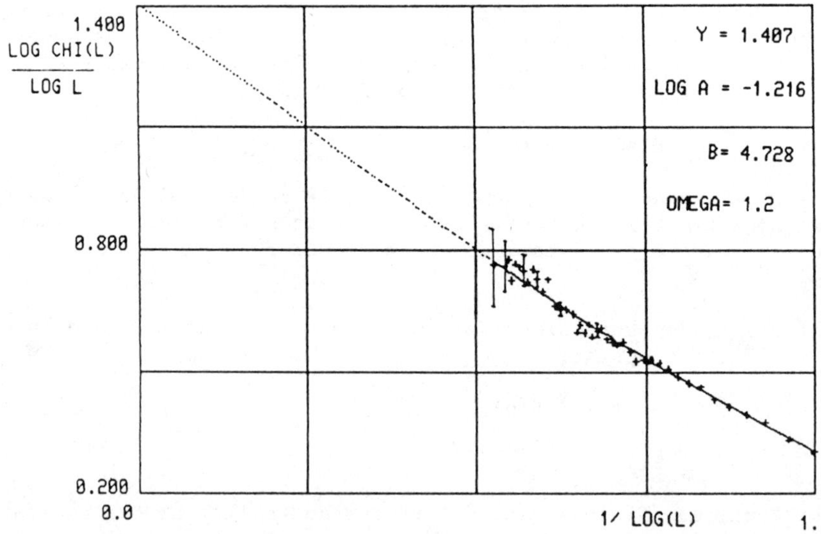

Figure 1. $\log\chi°(L)\,/\log L$ plotted as a function of $(\log L)^{-1}$ for lattices 10x10x10 to 80x80x80 and the fitted curve as defined in the text

In order to extract y we plot in figure 1 $\langle \log\chi^\circ(L)\rangle/\log L$ as a function of $(\log L)^{-1}$. If power-law scaling were exactly obeyed for the lattice sizes simulated, such a plot would be a straight line with intercept y and slope log a. As is apparent from the figure the line is not straight: it is necessary to consider corrections which we assume of the form

$$\chi^\circ(L) = aL^y(1+bL^{-\omega}+\dots)$$

Ideally one would extract b and ω from the data: in practice the data were inadequate for this and the intercept y was estimated by assuming that ω is the same exponent for correction to scaling as appears in other percolating quantities.[6] In addition there may be further corrections analytic in 1/L. It is uncertainties in these correction terms that dominate the uncertainty in our estimate of τ. As a result the analysis is affected more by systematic than statistical error; error bars are to some degree subjective. By varying the fitted range and the exponent ω assumed we find a value

$$y = 1.40 \pm 0.10$$

which corresponds to an exponent

$$\tau = 0.79 \pm 0.10$$

This is rather higher than the Monte Carlo estimate of Harris and Kirkpatrick (0.5 ± 0.2). Equation 1 taken with numerical values either from the Alexander-Orbach conjecture or the two different series expansions of Adler[6] would predict values of τ of 0.66, ($.68 \pm .07$) or ($.72 \pm .07$) respectively. Thus to the accuracy obtained, any discrepancy in equation 1 is explicable in terms of uncertainties in the values of the exponents.

In conclusion it is hoped that interest will revive in dilute antiferromagnets. While this discussion has concentrated on the three-dimensional case in two effects of fluctuation are even stronger: the susceptibility is logarithmically infinite for all dilutions above percolation. The crossover exponent y which we estimate as (1.2 ± 0.1), characterises crossover from divergence as a power to a logarithm of applied field. A theory for the low frequency dynamics taking into account such non-linear effects remains to be elucidated.

REFERENCES

1. R.A. Cowley, R.J. Birgeneau and G. Shirane in "Ordering in Strongly Fluctuating Condensed Matter Systems" T. Riste, ed., Plenum Press, New York (1979)
2. A.B. Harris and S. Kirkpatrick Phys. Rev. B16 542 (1977)
3. S. Alexander and R. Orbach J. Physique (Lettres) 43 L625 (1982)
4. T.A.L. Ziman, J. Phys. C 12 2645 (1979)
5. M.N. Barber in volume 8 of "Phase Transitions and Critical Phenomena" C. Domb and J.L. Lebowitz, eds., Academic Press, New York (1983)
6. J. Adler, M. Moshe and V. Privman in "Percolation structures and Processes (Annals of the Israel Physical Society vol. 5)", G. Deutscher, R. Zallen and J. Adler, eds., Adam Hilger, Bristol (1983)
   J. Adler, J. Phys. A 18 307 (1985)

PHONON-FRACTON CROSSOVER ON FRACTAL LATTICES

B.W. Southern and A.R. Douchant

Department of Physics
University of Manitoba
Winnipeg, Manitoba
Canada R3T 2N2

INTRODUCTION

Recently there has developed a growing interest in the dynamical properties of structures which have a fractal geometry[1]. Alexander and Orbach[2] were the first to point out that three dimensionalities are required to describe excitations on fractals: the embedding (Euclidean) dimension d; the Hausdorf (fractal) dimension D; and the fracton (spectral) dimension $\tilde{d}$. The fractal dimension D describes how the mass of the geometrical object depends on length scale whereas the spectral dimension $\tilde{d}$ characterizes the low frequency behaviour of the density of states. It is well known[3] that percolating networks near the critical threshold p_c have a self-similar(fractal) geometry on small length scales and a homogeneous(Euclidean) structure on larger scales. The length scale which separates these two regions is the percolation correlation length ξ_p which diverges as p_c is approached from above. Orbach et al[4,5,6] have recently used scaling arguments and effective medium approximation (EMA) calculations to predict that there is a sharp increase in the density of vibrational states at a crossover frequency ω_c which corresponds to excitations with wavelengths of the order of ξ_p. They have conjectured[4,5,7] that glasses and perhaps all amorphous materials may exhibit similar behaviour which can be understood in terms of a crossover from phonons to fractons.

In order to elucidate this question concerning the sharp rise in the vibrational density of states (DOS) at a well defined frequency, we have calculated exactly the DOS for a system of masses and springs on a percolating network in both d=2 and d=3. We have used a model for the backbone of the infinite cluster which has a self-similar geometry on length scales from the nearest neighbour lattice spacing up to the correlation length ξ_p and which has a Euclidean geometry at length scales above ξ_p. Our results indicate that a crossover from Euclidean (Debye) to fracton behaviour occurs in the DOS as p_c is approached from above but the change is rather smooth with no apparent sharp increase in the DOS at a crossover frequency.

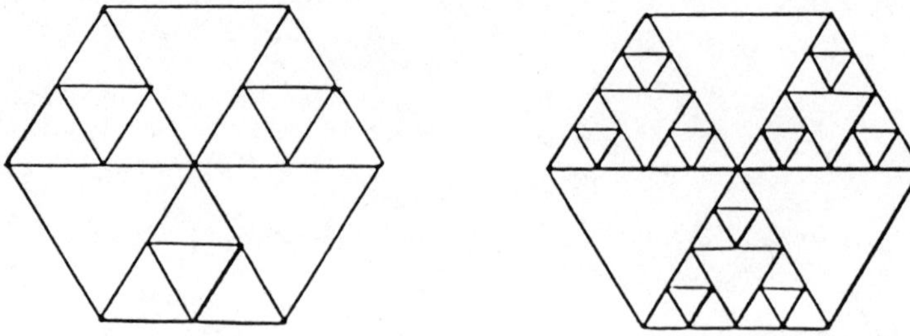

Fig.1. Fractal lattice in d=2 for n=1 and n=2.

FRACTAL LATTICES

Figure 1 illustrates how we construct our fractal lattice in d=2. We begin with a uniform triangular lattice which consists of both upward and downward pointing equilateral triangles. The lattice spacing on this lattice is denoted as L. At each subsequent stage of construction we inscribe downward pointing triangles inside each of the upward pointing triangles and the new lattice spacing is denoted as a . At any level n of construction the Euclidean lattice spacing L is related to a in the simple way $L=2^n a$. The lattice generated in this manner has holes of all sizes from the lattice spacing a up to the Euclidean length L. We identify this length L with the percolation correlation length and since this latter length diverges as p approaches p_c with a characteristic exponent ν, we can relate p to the level n of iteration using $p=p_c+(1-p_c)(L/a)^{-1/\nu}$. Hence n=0 corresponds to a uniform lattice(p=1) whereas $n\rightarrow\infty$ describes the self-similar geometry of the infinite cluster at p_c. A similar method of construction can be used in d=3 if we describe a face-centered cubic lattice in terms of tetrahedra stacked along the cube diagonal with each upward pointing triangle in figure 1 representing the base of a tetrahedron.

VIBRATIONAL DENSITY OF STATES

We now consider the excitation spectrum of these systems when each site has a unit mass with nearest neighbours connected by springs of length a. The spectral properties of this system are calculated exactly using real space rescaling techniques[8,9,10] combined with functional integral methods[11,12]. In the limit L/a=1 (n=0) we have a regular Euclidean lattice and the eigenstates have the Bloch form with the low frequency DOS scaling as

$$N(\omega^2) \sim (\omega^2)^{(d-2)/2} \qquad [1]$$

In the opposite limit $L/a=\infty$ $(n=\infty)$ we have a d-dimensional Sierpinski gasket . The excitation spectrum in this case has been studied previously by several groups[11,13,14] and the low frequency DOS behaves as above with d replaced by the spectral dimension $\tilde{d}$. For both d=2 and d=3 the spectral dimension is given by $\tilde{d}=2\ln(d+1)/\ln(d+3)$ whereas the fractal dimension is $D=\ln(d+1)/\ln(2)$.

Figure 2 shows our results for the DOS in d=2 for n=0,1,2 and 9. As we increase n, or equivalently approach p_c, the continuous spectrum gradually breaks up into a collection of narrower bands separated by gaps with localized modes in the gaps and at the band edges. An alternate way of describing these lattices is to regard the fractal region on length scales from a to L as a unit cell in the Euclidean structure.

For small values of n we can easily calculate the eigenvalues as a function of wavevector and we plot these curves for n=1 and n=2 in figure 3. We have only plotted the dispersion curve along q_x and in this case the zone boundary is located at $4\pi/3L$. Notice that as the fractal region grows many of the modes become independent of q. These modes are localized within the unit cell and are highly degenerate. In fact, for large values of n, the mode at $\omega^2=3$ exhausts 1/3 of the total DOS. The low frequency part of the spectrum consists of a Euclidean band whose weight and width decreases as we approach p_c. The maximum frequency of this lowest band corresponds to a mode with wavelength equal to the correlation length and we identify this as a crossover frequency ω_c.

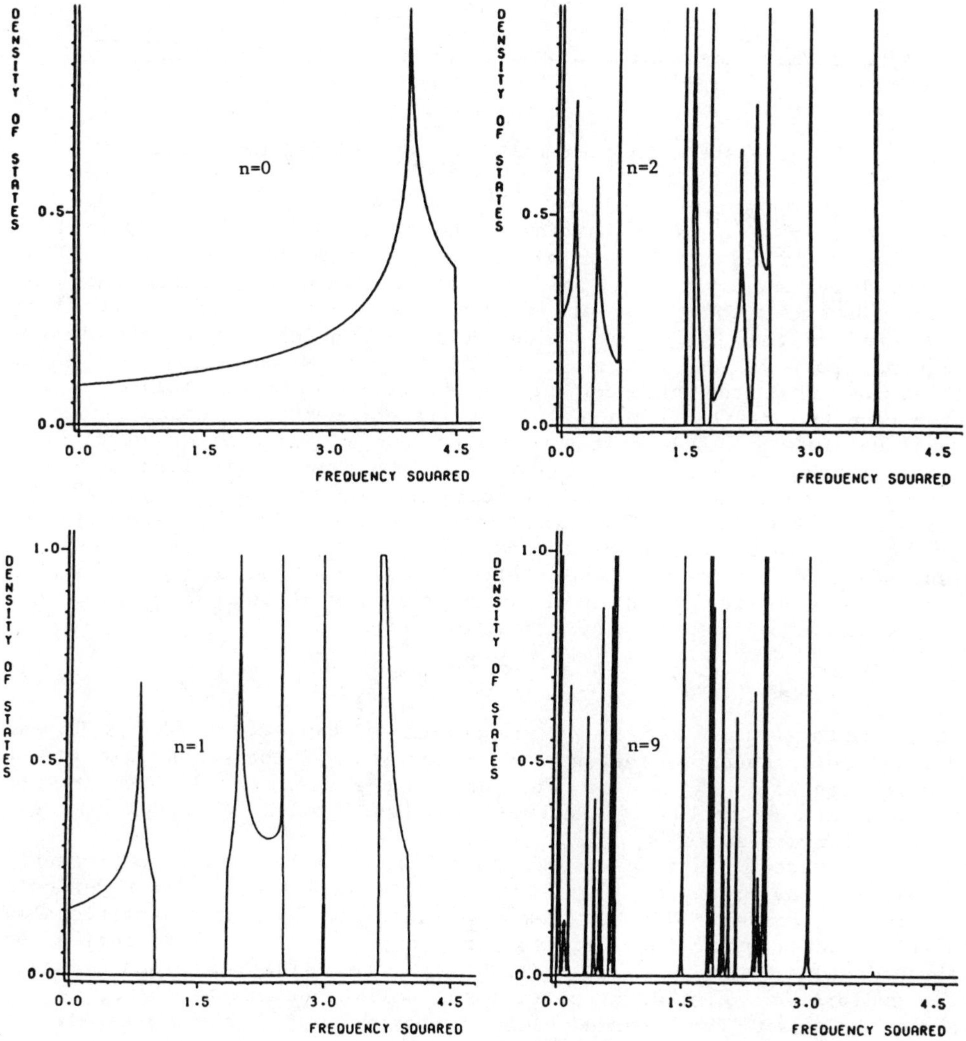

Fig.2. Density of states $N(\omega^2)$ plotted as a function of ω^2 for n=0,1,2 and 9.

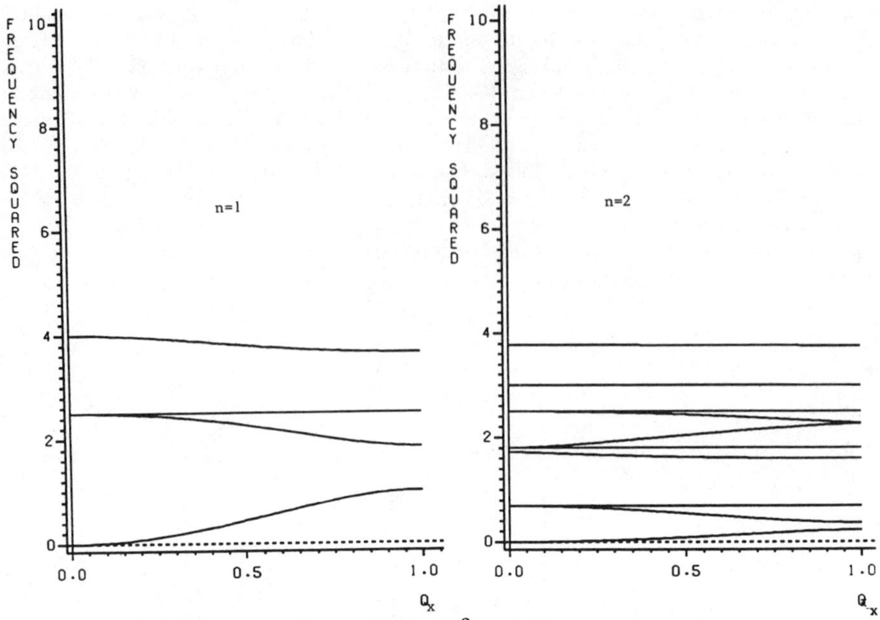

Fig.3. Dispersion curves ω^2 vs. $q_x L$ in d=2 for n=1,2 .

We find that ω_c scales as $(p-p_c)^{\gamma D/\tilde{d}}$ and that the weight of this band scales as $(p-p_c)^{\gamma D}$. As n increases the number of bands and gaps also increases but the weight of these modes decreases and the DOS is dominated by the modes with no dispersion (localized within the unit cell). However, the crossover is rather smooth and there is no sharp structure in the DOS at ω_c. Our results indicate that the higher frequency part of the DOS is composed of both extended and localized states and that the extended states do not simply accumulate at ω_c. These results are also supported by the recent numerical calculations of Grest and Webman[15] who have calculated the vibrational properties of percolation clusters for a simple cubic lattice. They find a smooth crossover between a phonon and fracton regime in the DOS as well as evidence for both extended and localized states at high frequencies.

We have also calculated the specific heat C_V and figure 4 shows our results for d=2 where we have plotted C_V/T^d vs T on a log-log scale. There is a Debye region at low T which gradually shrinks as n increases. The curves can be described by a scaling form

$$C_V/T^d \sim (\xi_p^{-D/\tilde{d}})^{\tilde{d}-d} f(T\,\xi_p^{D/\tilde{d}}) \qquad [2]$$

The prefactor arises due to the reduction in band width of the lowest band and corresponds to the anomalous behaviour of the sound velocity at p_c which scales as $\xi_p^{(1-D/\tilde{d})}$. The peak in the curves at low temperature is due to the deviation of the Euclidean band from a Debye spectrum and its position scales as $\xi_p^{-D/\tilde{d}}$.

The fractal lattices used here are similar to the recently discovered quasicrystals[16,17,18] which are crystals with quasi periodic translational order. Our lattices are neither random nor translationally invariant but they do have orientational order in that all triangles and tetrahedra have the same orientation in space. The real quasicrystals also exhibit self-similar properties not unlike our fractal lattices and we would expect the thermodynamic properties of these materials to exhibit non-Euclidean behaviour at low temperatures.

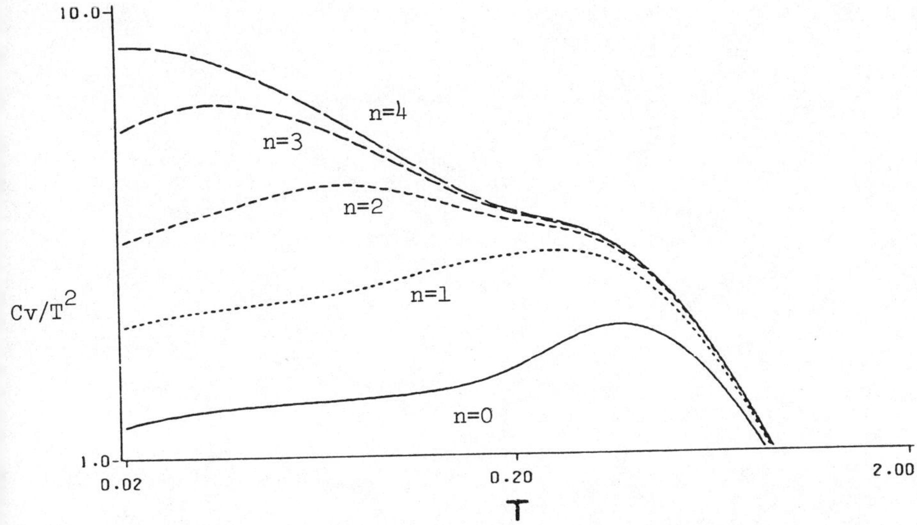

Fig.4. C_V/T^d plotted as a function of T on a log-log scale in d=2 for n=0,1,2,3 and 4.

REFERENCES

1. B. B. Mandelbrot, "The Fractal Geometry of Nature," Freeman, San Francisco (1979).
2. S. Alexander and R. Orbach, J.Phys. (Paris) Lett. 43, 625 (1982).
3. Y. Gefen, A. Aharony, B.B. Mandelbrot and S. Kirkpatrick, Phys. Rev. Lett. 47, 1771 (1981).
4. S. Alexander, C. Laermans, R. Orbach, and H. M. Rosenberg, Phys. Rev. B28, 4615 (1983).
5. B. Derrida, R. Orbach, and K.-Wah Yu, Phys. Rev. B29, 6645 (1984).
6. A. Aharony, S. Alexander, O. Entin-Wohlman, and R. Orbach, Phys. Rev. B31, 2565 (1985).
7. R. Orbach, Geilo Lectures (1985).
8. B.W. Southern, A. Kumar, P.D. Loly, A.-M.S. Tremblay, Phys. Rev. B27, 1405 (1983).
9. B.W. Southern and P.D. Loly, J. Phys. A18, 525 (1985).
10. R. Stinchcombe, Geilo Lectures (1985).
11. A.-M.S. Tremblay and B.W. Southern, J. Phys. (Paris) Lett. 44, 843 (1983).
12. A. R. Douchant, M.Sc. Thesis, University of Manitoba, (1985).
13. E. Domany, S. Alexander, D. Bensimon and L.P. Kadanoff, Phys. Rev. B28, 3110 (1983).
14. R. Rammal, J. Phys. (Paris) 45, 191 (1984).
15. G.S. Grest and I. Webman, J. Phys. (Paris) Lett. 45, 1155 (1984).
16. D. Shechtman, I. Blech, D. Gratias and J.W. Cahn, Phys. Rev. Lett. 53, 1951 (1984).
17. D. Levine and P.J. Steinhardt, Phys. Rev. Lett. 53, 2477 (1984).
18. P. Bak, Geilo Lectures (1985).

GRASSMANN PATH INTEGRAL APPROACH TO TWO-DIMENSIONAL PERCOLATION NEAR THE CRITICAL POINT

Giancarlo Jug

Process Physics Division, AFRC-Food Research Institute

Colney Lane, Norwich NR4 7UA, England

The concept of percolation is a fundamental one in the physics of disordered systems and has been discussed at length at this Institute. Over the last decade, a consistent theory of percolation has emerged, borrowing heavily from the familiar description of critical phenomena. In particular, in two dimensions (2D) a small set of plausible assumptions has led to a conjectured exact theory for the asymptotic behaviour near the percolation threshold, p_c.[1] Central amongst these is the assumption that the critical singularities are in the form of power laws.

Until recently, no systematic field-theoretic approach to 2D percolation was available to check up on the existing theories. Here, I report on a recently developed approach of this kind and on some of the results obtained in this way.[2,3] The new approach relies on a Grassmann Path Integral (GPI) representation for the thermodynamics of the 2D dilute Ising model which has the percolation process of interest as its zero temperature limit. This Hamiltonian formulation of percolation is an alternative to the more widely used q=1 limit of the q-state Potts model.

The new approach to 2D percolation [2] exploits the fact that the singular part of the average free energy of, say, the bond-dilute Ising model on the square lattice can be written as a GPI:

$$-\beta f_{sing} = \lim_{n\to 0} \frac{1}{n} \ln \int D\psi \exp[S_{eff}],$$

where, in momentum space, the effective action reads:

$$S_{eff} = \frac{1}{2} \sum_{k\alpha} \bar{\psi}^{\alpha}(-k) A(k) \psi^{\alpha}(k) + \frac{1}{N} g \sum_{k_1+\ldots+k_4=0} \sum_{\alpha,\beta} \psi_1^{\alpha}(k_1)\psi_2^{\alpha}(k_2)\psi_1^{\beta}(k_3)\psi_2^{\beta}(k_4) + \ldots$$

Here, $\psi(k)$ is a two-component Grassmann lattice field and, with the notation $h(k)=1+pt\exp(ik)$:

$$A(k) = \begin{pmatrix} -h(k_x)^{-1} + h(-k_x)^{-1} & h(k_y) - h(k_x)^{-1} - h(-k_x)^{-1} \\ -h(-k_y) + h(k_x)^{-1} + h(-k_x)^{-1} & h(k_x)^{-1} - h(-k_x)^{-1} \end{pmatrix}.$$

Also,

$$g = \frac{1}{2} p(1-p)t^2 \left(1+4\frac{1-pt}{(1+pt)^3(1-pt^2)}\right)$$

and $t=\tanh(\beta J)$. The effective action thus generated is an expansion in powers of $\bar{\psi}\psi$ and the above expression is exact to order $(\bar{\psi}\psi)^2$. In this approximation, the singular part of the Ising free energy can be calculated perturbatively in g. A first order calculation gives:

$$-\beta f_{sing} = \frac{1}{2N}\sum_k \ln[\det\bar{A}(k)+2gB(k)-\frac{1}{N}\sum_{k'}\frac{B(k')}{\det\bar{A}(k')}],$$

with $\det\bar{A}(k)=h(k_x)h(-k_x)\det A(k)$ and $B(k)=h(k_x)h(-k_x)h(-k_y)-2h(k_x)$. Once $f(p,T)$ is known, percolation's mean number of clusters per site, $K(p)$, follows from the ground state entropy of the Ising model, $s=-\partial f/\partial T$, since $K(p)=s(p,T=0)/\ln 2$. It is then possible, within this approach, to evaluate $T_c(p)$, p_c and $K(p)$ perturbatively. To first order in g, the critical line is given by:

$$p\tanh[J/T_c^{(1)}(p)]=\sqrt{2}-1+\frac{2+\sqrt{2}}{16}(2-1)^2\left[1+\frac{2(\sqrt{2}-1)}{p-(\sqrt{2}-1)^2}p\right]\frac{1-p}{p}.$$

Near p_c, I find the expected behaviour for the critical line, $\exp(-2J/T_c(p))\propto p-p_c$, to all orders in g. Also, $p_c^{(0)}=\sqrt{2}-1$ and $p_c^{(1)}=0.4978\ldots$ show excellent convergence towards the exact value $p_c=1/2$ for bond percolation on the square lattice (BSQ). For the same process, $s(p,0)=K(p)\ln 2$ is reported in Fig.1, both to zero (curve a) and first order (b) in g. The theoretical curves are compared with numerical results for $K(p)$ for a lattice of linear size L=40. Next, the critical behaviour of $K(p)$ near p_c can be investigated by taking the continuum limit of $S_{eff}(T=0)$ and expanding near k=0 and $p=p_c$. The continuum action now reads:

$$S_{eff}=\int d^2x[\tfrac{1}{2}i\sum_\alpha \bar{\psi}^\alpha(m_0+\partial)\psi^\alpha - g_0(\sum_\alpha \bar{\psi}^\alpha\psi^\alpha)^2]+\ldots,$$

where $m_0\propto p-p$ and $g_0\propto g(T=0)$. By inspection, one can verify that, in the renormalization group (RG) sense, higher order operators in S_{eff} are irrelevant whilst the quartic interaction term is marginal in 2D.

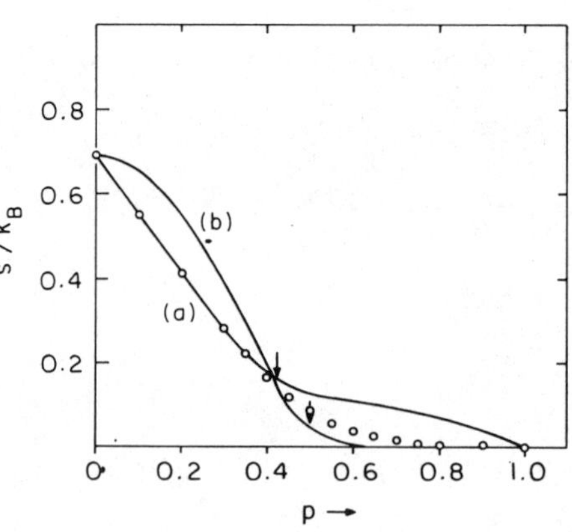

Fig.1 Mean number of clusters $K(p)=s(p,0)/\ln 2$.

Thus, the same marginal RG fixed point, $g^*=0$ in this approach, appears to describe both the dilute Ising and the percolation critical behaviour in 2D.[2] This denies the central assumption of existing theories of 2D percolation, as the following singularity is predicted for K(p):

$$K(p) \sim (p-p_c)^2 \ln|\ln|p-p_c|| ,$$

implying $\alpha=0$, rather than the accepted form $K(p) \sim |p-p_c|^{2-\alpha}$ with $\alpha=-2/3$. At present, only the above prediction for K(p) is available from the GPI approach. Thus, it is not clear whether this merely implies a breakdown of the hyperscaling relation $2-\alpha=d\nu$, where $\nu=4/3$ in the accepted theory, or whether the entire scaling theory of 2D percolation is to be revised.

Evidence in support of the GPI prediction has recently been found [3] through a series expansion and numerical study of the singularity in K(p). From the point of view of series analysis, it is not difficult to show that weak logarithmic corrections generate fictitious exponents if a pure power law singularity is assumed. I have therefore postulated a singularity of the form:

$$K(p) \sim |p-p_c|^h [\ln|\ln|p-p_c||]^{\hat{z}h} ,$$

where $h=2-\alpha$, and analysed the existing series expansions for K(p) [4] to see which value of $\hat{z}$ would fit the input value of h. In Fig.2, I have reported the central group of Padé approximants to the curve $\hat{z}=\hat{z}(h)$ and it can be seen that the GPI prediction, $\hat{z}=1/2$ for h=2, is supported whereas the conventional theory, $\hat{z}=0$ for h=8/3, is ruled out. A similar analysis, for the assumed form

$$K(p) \sim |p-p_c|^h [\ln|p-p_c|]^{zh} ,$$

also disclaims the conventional wisdom (z=0 for h=8/3) and confirms the GPI prediction (z=0 for h=2). Fig.2 refers to BSQ percolation; similar results can be obtained for site percolation on the triangular lattice.

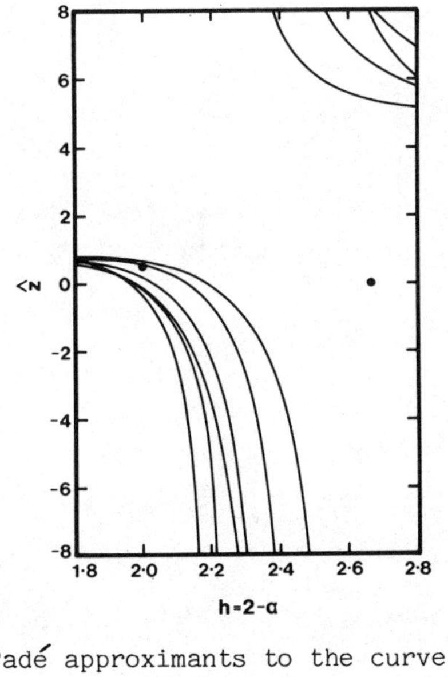

Fig.2 Padé approximants to the curve $\hat{z}=\hat{z}(h)$.

An alternative independent test consists in evaluating numerically the quantity $d^3K(L,p)/dp^3$ for $p=p_c$ as a function of the lattice size L. Finite-size scaling predicts a divergence of the form

$$K'''(L,p_c) \sim L^{\frac{1+\alpha}{\nu}} ,$$

for L>>1. The results for BSQ percolation are reported in Fig.3. K''' is evaluated through a fluctuation formula and in order to achieve good statistics averaging over a very large number of configurations ($>10^7$) is required. This imposes restrictions on the value of L. However, even for relatively small lattices an asymptotic behaviour seems to be observed, with an exponent $(1+\alpha)/\nu$ inconsistent with that of the accepted theory (=1/4). Four sets of data are reported in Fig.3; these refer to different definitions of K(p).[3] On all sets one consistently reads an exponent $(1+\alpha)/\nu$ =1 which agrees with α =0 and, surprisingly, implies ν =1 for the connectedness length exponent. The latter value would fit the naive expectations of the GPI approach, but further theoretical work is required before this value can be accepted confidently.

In conclusion, there is definite evidence that the known predictions of the GPI approach to 2D percolation are correct and, therefore, that this critical phenomenon is described by a marginal RG fixed point.[5] Hopefully, further results will emerge from this promising approach which will help clarify the true nature of this and other subtle phenomena in the physics of disordered systems.

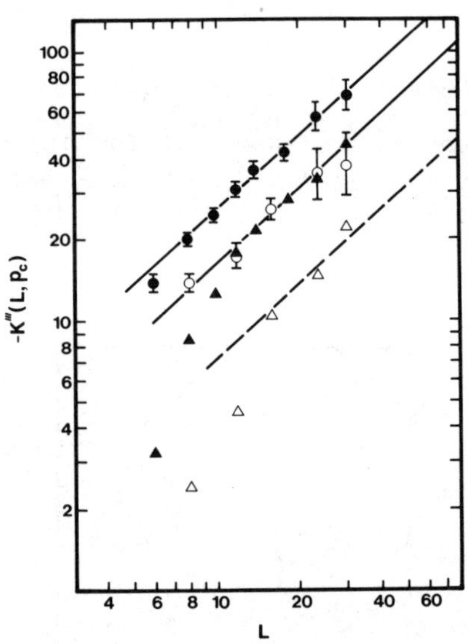

Fig.3 Divergence of $K'''(L,p_c)$ with lattice size L.

REFERENCES

1. B.Nienhuis, J.Stat.Phys. 34, 731 (1984), and references therein
2. G.Jug, Phys.Rev.Lett. 53, 9 (1984)
3. G.Jug, Phys.Rev.Lett. (submitted)
4. M.F.Sykes, private communication
5. See also D.Andelman and A.N.Berker, J.Phys.A 14, L91 (1981)

GEOMETRY AND DYNAMICS OF FRACTAL SYSTEMS

Toshiya Ohtsuki and Thomas Keyes

Department of Chemistry
Boston University
Boston, MA 02215

1. INTRODUCTION

Recently, physics of fractals has been one of the most exciting fields of science.[1,2] A number of studies of percolation clusters, Sierpinski gaskets, lattice animals etc. have been reported and considerable interest has been generated in their anomolous behavior. On the basis of real-space renormalization group methods and/or scaling theories, we have investigated relation between geometrical structure and physical properties of fractal systems. As a result, it becomes evident that various types of critical phenomena are combined into a few universality classes and given unified geometrical interpretations. Conclusions are

$$\text{self-similarity} \rightarrow \text{non-analytic (critical) behavior,} \quad (1.1)$$

$$\text{critical exponents} = \text{fractal dimensionalities.} \quad (1.2)$$

Here we summarize main results.

2. SEGMENT DISTRIBUTION (STATIC CRITICAL PHENOMENA)

Scaling behavior of a distribution function of segments in a self-similar pattern and all static critical exponents are expressed in terms of fractal dimensionalities of the pattern and segments.[3]

Consider a pattern which consists of unit components of length unity and is self-similar in the length scale $1 \ll L \lesssim \xi$. We define a distribution function $\rho(s,\xi)$ of s-unit segments by

$$\rho(s,\xi) = \frac{\text{number of s-unit segments}}{\text{number of units belonging to the whole pattern}}. \quad (2.1)$$

On the basis of a scaling hypothesis, $\rho(s,\xi)$ is assumed to have the scaling form[4]

$$\rho(s,\xi) = s^{-\tau} F(s/s_\xi), \quad (2.2)$$

where s_ξ is a typical segment size dominating all critical behavior and is given by

$$s_\xi \propto \xi^{1/\hat{\sigma}}. \quad (2.3)$$

We introduce other critical exponents as[4]

$$\left[\sum_s \rho(s,\xi) \right]_{\text{sing}} \propto \xi^{-\hat{\alpha}}, \tag{2.4}$$

$$\left[\sum_s s\, \rho(s,\xi) \right]_{\text{sing}} \propto \xi^{-\hat{\beta}}, \tag{2,5}$$

$$\left[\sum_s s^2 \rho(s,\xi) \right]_{\text{sing}} \propto \xi^{\hat{\gamma}}, \tag{2.6}$$

$$\left[\sum_s s\, \rho(s,\xi)\, e^{-hs} \right]_{\text{sing}} \propto h^{1/\delta} \tag{2.7}$$

where the subscript 'sing' means the singular part. The self-similarity of the system leads to the expressions of all critical exponents in terms of a fractal dimensionality d_F of the pattern and that d_F' of segments[3]

$$\hat{\alpha} = d_F, \quad (2.8) \qquad \hat{\beta} = d_F - d_F', \quad (2.9)$$

$$\hat{\gamma} = 2d_F' - d_F, \quad (2.10) \qquad \delta = d_F'/(d_F - d_F'), \quad (2.11)$$

$$\hat{\sigma} = 1/d_F', \quad (2.12) \qquad \tau = 1 + d_F/d_F'. \quad (2.13)$$

The general theory is applied to dead end distribution in percolation clusters and link distribution in percolation backbones.[3,5] It should be emphasized that ordinary thermal and geometrical (percolative) critical phenomena are special cases where the pattern is the Euclidean space. Critical exponents α, β, γ and σ defined through a temperature are related to $\hat{\alpha}$, $\hat{\beta}$, $\hat{\gamma}$ and $\hat{\sigma}$ as $\hat{\alpha} = (2-\alpha)/\nu$, $\hat{\beta} = \beta/\nu$, $\hat{\gamma} = \gamma/\nu$ and $\hat{\sigma} = \sigma\nu$, where ν is a correlation length exponent. From (2.8)-(2.13), therefore, we find that all scaling relations hold. The hyperscaling relation $2-\alpha = d_F\nu$ also holds but in this case, the Euclidean dimensionality d is replaced by d_F. Generally, there exist three independent and foundamental quantities d_F, d_F' and ν in static critical phenomena. In usual cases, $d_F = d$ and non-trivial exponents are only two.

3. ANOMOLOUS DYNAMICS

Unusual dynamics of fractals such as density of states,[6] particle motion[7-9] and wave propagation[10-12] are described by two 'dynamic' fractal dimensionalities: a thickness (fracton or spectral) dimensionality d_T and a length dimensionality d_L.[13]

d_T and d_L represent different geometry and govern different physics. In general, a fractal is connected multiplicatively and have finite thickness. d_T stands for this thickness and is defined by

$$N \propto L^{d_T}, \tag{3.1}$$

where N is a number of units belonging to a fractal and L is a contour length along a fractal. On the other hand, a fractal forms not a smooth but an irregular structure. d_L represents this irregularity and is defined by

$$L \propto R^{d_L}, \tag{3.2}$$

where R is a distance measured in the Euclidean space. Comparing (3.1) and (3.2) with the definition of d_F,[1] $N \propto R^{d_F}$, we have

$$d_F = d_T\, d_L. \tag{3.3}$$

Similarly to thermal critical phenomena, therefore, independent dynamic dimensionality is only one.

Using these concepts, we can derive most of existing results easily. In this picture, a fractal is regarded as an irregularly folded sheet with a fractal dimensionality d_T. Then only d_T is responsible to the connectedness of the structure and the system shows normal d_T-dimensional behavior, for example, for the density of states.[6] Similarly, particle migration and wave propagation along a fractal have no singularity. However, these spatial motions are measured in the Euclidean space. Equation (3.2) informs us that even if a particle or a wave travels L along the structure, it is regarded as moving only $R = L^{1/d_L}$ in the Euclidean space. This causes anomolous slow behavior of a particle or a wave described by d_L[7-12]

$$<X^2>_R \sim <X^2>_L^{1/d_L} \propto t^{1/d_L}, \tag{3.4}$$

$$<X>_R \sim <X>_L^{1/d_L} \propto t^{1/d_L}, \tag{3.5}$$

$$\omega \sim k_L \propto k_R^{d_L}, \tag{3.6}$$

$$\omega \sim k_L^2 \propto k_R^{2d_L}, \tag{3.7}$$

where $<X^2>$ is a mean-square displacement of a diffusing particle, $<X>$ is a mean displacement due to an external force and (3.6) and (3.7) are dispersion relations of acoustic and spin wave propagation, respectively. Here the subscript L denotes the quantity measured along a fractal and R indicates that in the Euclidean space. Note that we use L and associated quantities only as heuristic concepts and definite results are drawn merely to quantities measured in the Euclidean space.

For percolation clusters and backbones, d_T and d_L are expressed in terms of known exponents[6,7,10]

$$d_{TC} = 2(d\nu-\beta)/(t-\beta+2\nu), \tag{3.8}$$

$$d_{TB} = 2(d\nu-\beta_B)/(t-\beta_B+2\nu), \tag{3.9}$$

$$d_{LC} = 1+(t-\beta)/2\nu, \tag{3.10}$$

$$d_{LB} = 1+(t-\beta_B)/2\nu, \tag{3.11}$$

where t is a conductivity exponent, β is a percolation probability exponent, β_B is a backbone probability exponent and subscripts C and B denote dimensionalities of clusters and backbones, respectively. In this case, therefore, we can evaluate d_T and d_L explicitly from available data of t, β, β_B and ν.

Several predictions are also proposed.[14] For instance, we conjecture that if measured along a fractal, a distribution function P(L,t) of a diffusing particle has nothing unusual and

$$P(L,t) \propto t^{-d_T/2} \exp(-C\, L^2/t), \tag{3.12}$$

where C is a constant. In the Euclidean space, we have

$$P(R,t) \propto t^{-d_T/2} \exp(-C' R^{2d_L}/t). \tag{3.13}$$

Here the R-dependence of P is a new prediction expected to be verified by computer simulations etc.

4. SUPER-NORMAL RANDOM MIXTURES

In super-normal random mixtures, a contact area (unscreened perimeter) dimensionality d_{CA} plays a crucial role and describes various physical properties, e.g., conduction,[15] diffusion[16] and thermal critical phenomena.[17]

Just below a percolation threshold p_c, typical size clusters (TSCs) of super components dominate critical behavior. TSCs are linked by channels of normal components. Among these channels, only the shortest paths are considered to be responsible. The number N_c of the shortest paths connecting adjacent TSCs in parallel is given by[15,16]

$$N_c \propto \xi^{d_{CA}}, \tag{4.1}$$

$$d_{CA} = d-2+s/\nu, \tag{4.2}$$

where ξ is a percolation correlation length and s is a super-normal conductivity exponent. Explicit values of d_{CA} are evaluated from known estimates of s and ν.[15]

On the basis of these geometrical pictures, we can clarify the critical divergence of conductivity, diffusivity etc. A self-diffusion coefficient D of 'termites' in super-normal mixtures exhibits the same power law divergence as a conductivity[16]

$$D \sim \xi^{2-(d-d_F)}/\tau_\xi \propto \xi^s, \tag{4.3}$$

because the relaxation time τ_ξ follows

$$\tau_\xi \propto \xi^{d_F}/\xi^{d_{CA}}. \tag{4.4}$$

For conduction in a random network of superconductors and nonlinear conductors whose current (I) - voltage (V) characteristics is given by

$$I = \sum_n g_n V^n, \tag{4.5}$$

we find[15]

$$I(\ell)/\ell^{d-1} = \sum_n \sigma_n (V(\ell)/\ell)^n, \tag{4.6}$$

$$\sigma_n = \xi^{n-1+s/\nu} \; g_n, \tag{4.7}$$

where σ_n is a n-th order conductivity and ℓ is a system size. With increasing n, σ_n diverges more rapidly and enhancement of nonlinearity occurs near p_c. This kind of enhancement of nonlinearity is thought to be observed widely in super-normal random mixtures. In super-normal random ferromagnets, a couping constant J* between adjacent TSCs becomes[17]

$$J^* \sim N_c J_0 \propto \xi^{d_{CA}} \; J_0, \tag{4.8}$$

where J_0 is a couping constant between normal magnets. It follows that a critical temperature T_c^* of the system obeys[17]

$$T_c^* \propto \xi^{d_{CA}} \; T_0, \tag{4.9}$$

where T_0 is a critical temperature of pure normal ferromagnets. Critical behavior of physical quantities are also derived. For example, a magnetization M varies as[17]

$$M(\varepsilon,0;\xi) \propto \xi^{d_F} \; |\varepsilon|^\beta, \tag{4.10}$$

$$M(0,h;\xi) \propto \xi^{d_F+(d_F-d_{CA})/\delta} \; h^{1/\delta}, \tag{4.11}$$

where $\varepsilon = (T-T_c^*)/T_c^*$ is a reduced temperature, h is an external magnetic field and β and δ are critical exponents of pure normal ferromagnets defined as usual.

References

1. B. B. Mandelbrot, "The Fractal Geometry of Nature", Freeman, San Francisco (1982).
2. M. F. Shlesinger, B. B. Mandelbrot and R. J. Rubin eds., Proceedings of a Symposium on Fractals in the Physical Sciences, J. Stat. Phys. 36: Nos.5/6 (1984).
3. T. Ohtsuki and T. Keyes, J. Phys. A17: L495 (1984).
4. D. Stauffer, Phys. Rep. 54: 1 (1979).
5. T. Ohtsuki and T. Keyes, J. Phys. A17: L267 (1984).
6. S. Alexander and R. Orbach, J. Phys. Lett. (Paris) 43: L625 (1982).
7. Y. Gefen, A. Aharony and S. Alexander, Phys. Rev. Lett. 50: 77 (1983).
8. R. Rammal and G. Toulouse, J. Phys. Lett. (Paris) 44: L13 (1983).
9. T. Ohtsuki and T. Keyes, Phys. Rev. Lett. 52: 1177 (1984).
10. C. K. Harris and R. B. Stinchcombe, Phys. Rev. Lett. 50: 1399 (1983).
11. T. Ohtsuki and T. Keyes, J. Phys. A17: L137 (1984).
12. T. Ohtsuki and T. Keyes, J. Phys. C17: L317 (1984).
13. T. Keyes and T. Ohtsuki, submitted.
14. T. Ohtsuki and T. Keyes, Phys. Lett. 105A: 273 (1984).
15. T. Ohtsuki and T. Keyes, J. Phys. A17: L559 (1984).
16. A. Coniglio and H. E. Stanley, Phys. Rev. Lett. 52: 1068 (1984).
17. T. Ohtsuki and T. Keyes, J. Phys. C17: L645 (1984).

QUANTUM PERCOLATION

Yigal Meir and Amnon Aharony

School of Physics and Astronomy
Tel-Aviv University
Tel-Aviv 69978, Israel

A. Brooks Harris

Department of Physics
University of Pennsylvania
Philadelphia, Pennsylvania 19104

INTRODUCTION

Most of the lectures in these proceedings discuss the conductivity of *classical resistor networks*, in which basic resistors are present (with concentration p) or absent (with concentration 1-p)). The conductivity becomes finite above the *geometrical percolation threshold*, p_c, as soon as there exists an infinite cluster, and its properties are uniquely determined by the geometry of the *backbone* of this cluster. Both of these aspects change when one considers a *quantum* particle, whose wave function obeys the Schrodinger equation. Since the wave function may vanish on many sites on the infinite cluster, due to *interference* effects, the *quantum threshold* p_q turns out to be higher than p_c. Quantum conductance also depends drastically on dangling ("dead end") bonds: a wave may be totally reflected by such bonds, yielding no conductance between the two ends of the sample. Quantitatively, these effects imply *localization* of all the electronic wave functions up to p_q.

MODEL

In this paper we describe results of the *tight binding Hamiltonian*

$$\mathcal{H} = \sum_i \varepsilon_i |i\rangle\langle i| + \sum_{\langle ij\rangle} V_{ij} |i\rangle\langle j|, \tag{1}$$

where $|i\rangle$ is the state on the site i of a lattice, and $\langle ij\rangle$ are nearest neighbors. Anderson[1] considered (1) with $V_{ij} \equiv V$ and with the ε_i's randomly distributed in a range W, and found localization of all states for sufficiently large W/V. In our work we consider the model of *quantum percolation*, in which $V_{ij} = V$ (with probability p) or $V_{ij} = 0$ (with probability 1-p),[2] and we let the ε_i's have very small random values (to eliminate unwanted degeneracies of eigenstates). We average our results over 100 samples with these random energies, and the results are not sensitive to this detail.

We consider the limit of small p, when all the connected clusters are finite. For each cluster Γ we solve the Schrodinger equation, and find the eigenstates $\psi_E(i)$. We then define the inverse participation ratio,[3]

$$X(\Gamma) = \sum_E \left(\sum_i |\psi_E(i)|^4\right)^{-1}. \tag{2}$$

Finally we average over clusters,

$$\chi(p) = \sum_\Gamma P(\Gamma)\, X(\Gamma), \tag{3}$$

where $P(\Gamma)$ is the average number per site of clusters Γ. For a totally extended function, $|\psi_E(i)| \sim 1/\sqrt{N_\Gamma}$ (N_Γ is the number of sites on Γ), and $X(\Gamma) \sim N_\Gamma{}^2$, so that $\chi(p)$ diverges at p_c. In practice $X(\Gamma) < N_\Gamma{}^2$, and $\chi(p)$ diverges at $p_q > p_c$, when some of the states become extended. Solving up to clusters of 11 bonds, we derived $\chi(p)$ as an eleven term series for hypercubic lattices in general d-dimensions, and then used the non-homogeneous differential Padé method[4] to extrapolate to the form[5]

$$\chi(p) \sim (p_q - p)^{-\gamma}. \tag{4}$$

Values of γ and p_q are shown in Table 1. Note that we find no sign of extended states at d=2, in agreement with the scaling theory of localization.[6]

An exact solution of our model on a Cayley tree[7] yields $p_q/p_c \simeq 1.4$, and our results are close to this ratio at most dimensions. Quantum

Table 1. Results for the Divergence of $\chi(p)$.

d	p_c	p_q	γ
1	1	1	1
2	½	1 (?)	?
3	0.247±0.003	0.35±0.01	2.2±0.2
4	0.161±0.002	0.220±0.005	1.4±0.1
5	0.118±0.002	0.161±0.005	1.2±0.1
6	0.094±0.001	0.127±0.005	1.1±0.1
7	0.079±0.001	0.105±0.003	1.00±0.05
8	0.068±0.001	0.091±0.003	1.00±0.05

percolation thus behaves as an effective classical percolation, in which there is an effective probability $p_c/p_q \approx 0.7$ to tunnel through each bond. We also note that γ approaches its Cayley tree value of unity at high d, but our results are not accurate enough to identify uniquely an *upper critical dimension*. We plan to address this question more critically in the future.

MAGNETIC FIELDS

The effects of magnetic fields on quantum conduction have recently drawn much attention, especially in the context of the quantum Hall effect.[8] A magnetic flux through a ring changes the interference pattern of the wave function, and may *delocalize* it. We introduce the magnetic field via gauge phase factors,[9]

$$V_{ij} = \exp\left(i \frac{e}{c} \vec{A}\cdot\vec{r}_{ij}\right), \tag{5}$$

where $\vec{A} = \frac{1}{2}\vec{H} \times \vec{r}$ is the vector potential. Such phase terms influence $X(\Gamma)$ only for clusters with loops. For example, Fig. 1 shows $X(\Gamma)$ for the cluster shown in the insert. The field is measured in units of the flux through a unit plaquette, $\phi/\phi_o = Ha^2e/hc$. As might be expected, the results are periodic in ϕ/ϕ_o, with period 1.

Since the magnetic field distinguishes between clusters which have the same topology (but different loop areas), we were able to derive only 8 terms in $\chi(p)$ at finite fields. The results are somewhat uncertain, and Fig. 2 shows p_q versus ϕ from one of our Padé estimates in three dimensions.

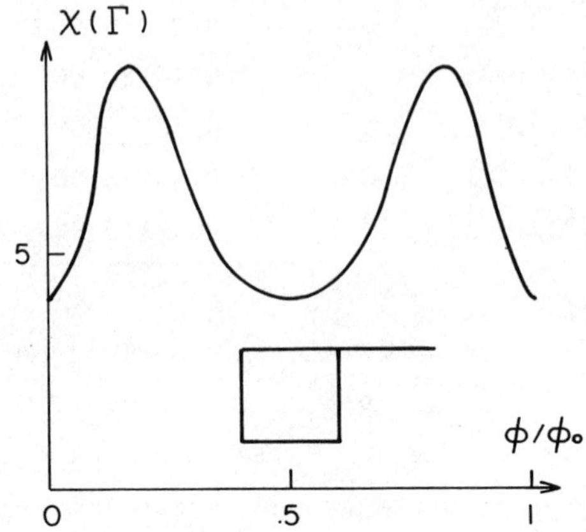

Fig. 1. Inverse participation ratio for graph in insert.

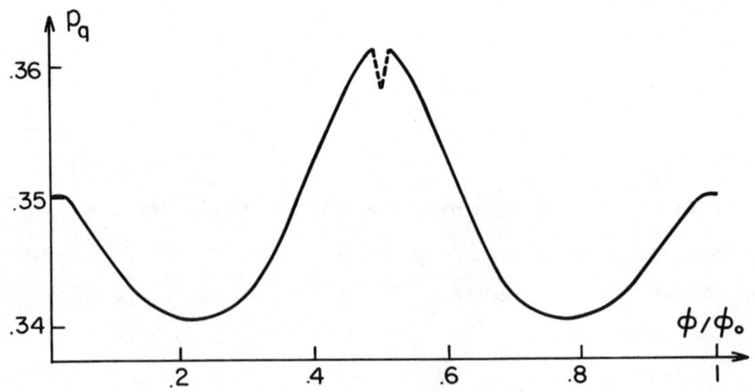

Fig. 2. Quantum threshold in magnetic field, in d = 3.

The results, which are again periodic in ϕ, indicate a lower threshold, i.e. delocalization, for fluxes of order (0.2–0.3) ϕ_o. We find similar indications in two dimensions.

CONCLUSION

The percolation model of localization seems to be an excellent tool to study quantum effects in percolating systems. In particular, we have presented the first quantitative results which show delocalization by a magnetic field.

ACKNOWLEDGEMENTS

This work was supported in parts by grants from the U.S. - Israel Binational Science Foundation and from the Israel Academy of Sciences and Humanities.

REFERENCES

1. P.W. Anderson, Phys. Rev. 109:1492 (1958).
2. Y. Shapir, A. Aharony and A.B. Harris, Phys. Rev. Lett. 49:486 (1982) and references therein.
3. F.J. Wegner, Z. Phys. B36:209 (1980) and references.
4. M.E. Fisher, in *Statistical Mechanics and Statistical Methods in Theory and Applications*, edited by U. Landman (Plenum, N.Y. 1977).
5. Details will be published elsewhere.
6. E. Abrahams, P.W. Anderson, D.C. Liciardello and T.V. Ramakrishnan, Phys. Rev. Lett. 42:673 (1979).
7. A.B. Harris, Phys. Rev. Lett. 49:296 (1982); Phys. Rev. B29:2519 (1984).
8. Y. Gefen, Y. Imry and M. Ya. Azbel', J. Phys. C15:L655 (1982).
9. D. Hofstadter, Phys. Rev. B14:2239 (1976).

NONLINEAR RESISTOR FRACTAL NETWORKS

Rafael Blumenfeld and Amnon Aharony

School of Physics and Astronomy
Tel-Aviv University
Tel-Aviv 69978, Israel

I. INTRODUCTION

Consider a *fractal network* of *nonlinear resistors*, each of which having the characteristic voltage-current relation[1]

$$V = r|I|^{\alpha} \text{ sign } I. \tag{1}$$

For $\alpha=1$ this reduces to the usual resistor network, discussed in detail in other lectures in these proceedings. Other values of α may describe various useful circuit elements, e.g. vacuum diodes.[2] The aim of this seminar is to describe several new results concerning such networks. In particular, we prove that in the limits $\alpha \to 0$ and $\alpha \to \infty$ the resistance of the network reduces respectively to the *chemical-topological-minimal path*[3,4] and to the path of *singly connected "red" bonds*.[4,5] We report explicit calculations on various fractals and discuss the *distribution* of values of the resistivities.

II. LIMITING CASES

Consider for example the graph shown in Fig. 1. It is straightforward to see that if the numbers of unit resistors are L_1, L_2, L_3 and L_4, as shown, then the net resistivity is

$$R = L_1 + (L_2^{-1/\alpha} + L_3^{-1/\alpha})^{-\alpha} + L_4. \tag{2}$$

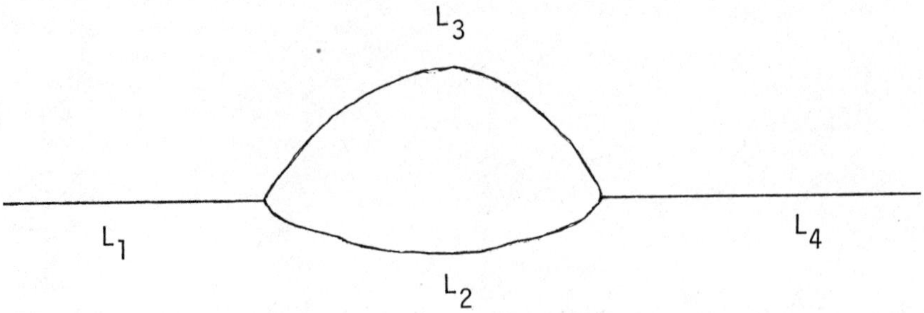

Fig. 1. Example of a graph.

For $\alpha \to \infty$, the middle term disappears, and we have $R = L_1+L_4$, equal to the number of singly connected "red" bonds. In the limit $\alpha \to \infty$, the middle term becomes L_2 if $L_2 \lessdot L_3$, so that $R = L_1+L_2+L_4$, equal to the minimal path. The proof for general networks is slightly more complex, and is given elsewhere.[8]

For a fractal self-similar structure, we expect a power law dependence on the length scale L,

$$R\ (L) \sim L^{\tilde{\zeta}(\alpha)}. \tag{3}$$

We have thus identified $\tilde{\zeta}(0)$, $\tilde{\zeta}(1)$ and $\tilde{\zeta}(\infty)$ with the apparently independent exponents $\tilde{\zeta}_{min}$, $\tilde{\zeta}_R$, and $\tilde{\zeta}_{red}$. For *percolation clusters*, Coniglio[5] proved that $\tilde{\zeta}_{red} = 1/\nu$, where ν is the percolation correlation length exponent. A systematic study of the α-dependence of $\tilde{\zeta}(\alpha)$ may prove useful in finding better values for $\tilde{\zeta}_R$ and $\tilde{\zeta}_{min}$.

III. EXACT RESULTS FOR FRACTALS

Figs. 2 and 3 show the *Mandelbrot-Koch*[6] *curve* (MKC) and the *Sierpinski gasket*[7] (SG), proposed as models for the *infinite cluster* and for its *backbone* for two-dimensional *percolation*. We calculated $\tilde{\zeta}(\alpha)$ for both curves, and the results are shown in Fig. 4, and details are given elsewhere.[8]

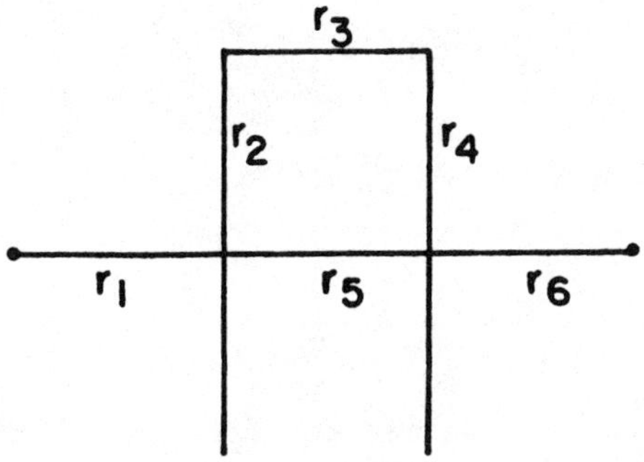

Fig. 2. Mandelbrot–Koch curve.

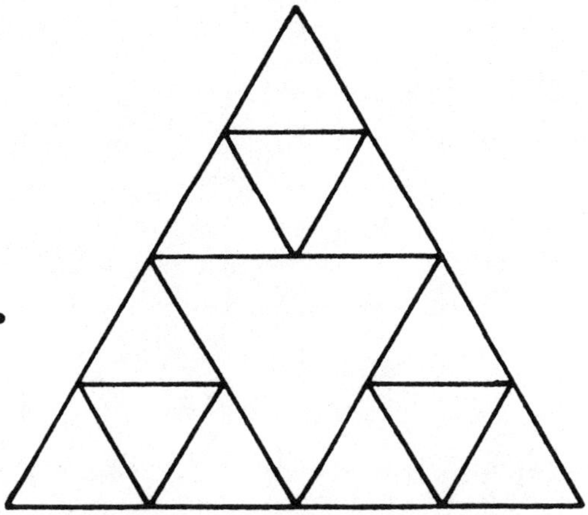

Fig. 3. Sierpinski gasket.

Fig. 4 also shows the exponent $(\tilde{\zeta}_2-\tilde{\zeta})$, describing the *relative width* of a narrow *distribution of resistors*,

$$\Delta R/\langle R\rangle = [\langle R^2\rangle - \langle R\rangle^2]^{\frac{1}{2}}/\langle R\rangle \sim L^{\tilde{\zeta}_2-\tilde{\zeta}} . \quad (4)$$

Note that $\Delta R/R$ always decreases with L for *two-terminal fractals*, like Figs. 1 or 2, but increases with L for the *tri-terminal* SG if $\alpha>1.612$. The exponent $(\tilde{\zeta}_2-\tilde{\zeta})$, measurable e.g. via the 1/f-noise spectrum,[9] seems to be very sensitive to the detailed nature of the fractal (even for the standard case $\alpha=1$).

IV. CONCLUSION

Theoretically, we have found a unifying function $\tilde{\zeta}(\alpha)$ which contains three of the interesting geometrical exponents used to characterize fractals. Practically, our results suggest detailed measurements (or computer experiments) of $\tilde{\zeta}$ and $\tilde{\zeta}_2$, from which one may obtain more insight on the geometry of percolation clusters.

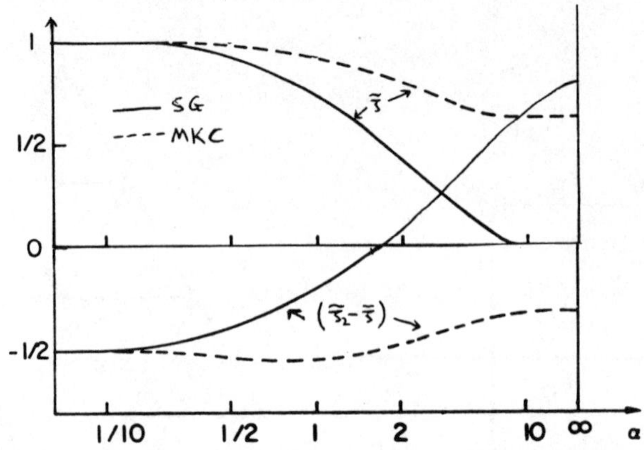

Fig. 4. $\tilde{\zeta}$ and $(\tilde{\zeta}_2-\tilde{\zeta})$ for the MKC and SG. Horizontal axis is $\alpha/(1+\alpha)$.

ACKNOWLEDGEMENTS

We enjoyed discussions with A.B. Harris, Y. Meir and D. Stauffer. This work was supported by grants from the U.S. - Israel Binational Science Foundation and the Israel Academy of Sciences and Humanities. The hospitality of R. Pynn and A. Skjeltorp at Geilo is also very gratefully acknowledged.

REFERENCES

1. S.W. Kenkel and J.P. Straley, Phys. Rev. Lett. 49:767 (1982); J.P. Straley and S.W. Kenkel, Phys. Rev. B29:6299 (1984).
2. I. Langmuir, Phys. Rev. 2:450 (1913).
3. Z. Alexandrowitz, Phys. Lett. 80A:284 (1980); S. Havlin and R. Nossal, J. Phys. A17:L427 (1984).
4. R. Pike and H.E. Stanley, J. Phys. A14:L169 (1981).
5. A. Coniglio, Phys. Rev. Lett. 46:250 (1981); J. Phys. A15:3829 (1982).
6. B.B. Mandelbrot and J. Given, Phys. Rev. Lett. 52:1853 (1984).
7. Y. Gefen, A. Aharony, B.B. Mandelbrot and S. Kirkpatrick, Phys. Rev. Lett. 47:1771 (1981).
8. R. Blumenfeld and A. Aharony, J. Phys. A (in press).
9. R. Rammal, C. Tannous and A.M.S. Tremplay, preprint (1984).

ELASTIC PROPERTIES OF RANDOM SYSTEMS

Yacov Kantor

Corporate Research and Science Laboratories
Exxon Research and Engineering Company
Annandale, N.J.

INTRODUCTION

Recently there has been growing interest in the elastic properties of random systems. Since various elastic model Hamiltonians produce different predictions[1-3] for the effective elastic moduli, it is important to realize that they usually apply to distinct physical systems, or treat different aspects of the same system. Our model Hamiltonian primarily aims to explain the elastic properties of macroscopically disordered systems, such as porous solids or colloidal aggregates in which the size of an elementary grain is at least several atomic radii. However, in certain cases it may also be applicable to disorder on the atomic level.

Let us consider a macroscopically inhomogeneous composite made up of very stiff regions and very compliant regions. (We will actually consider a case when the compliant regions are voids.) Effective elastic moduli of this system decrease with decreasing concentration p of the stiff component and vanish when p approaches the geometrical continuity (percolation) threshold p_c. The dependence of the effective elastic constant κ, e.g. bulk modulus or shear modulus, on p, near p_c is given by $\kappa \sim (p-p_c)^\tau$. This relation reminds the behavior of the effective conductivity of metal-insulator mixture: $\sigma \sim (p-p_c)^t$. The formal similarity of the differential equations governing these problems suggests some analogy in the results. Almost a decade ago de Gennes[4] noted that the rubber elasticity Hamiltonian used to describe the entropic elasticity of gels coincides with the dissipation functional of the conductivity problem and concluded that the critical exponent τ of the entropic elasticity coincides with t of conductivity. For a long time this conclusion has been assumed to be valid for any kind of elasticity (see, e.g. Ref. 5). We will show that in the macroscopically inhomogeneous systems τ is considerably larger than t.

Singular behavior of the elastic moduli of percolating system near p_c is caused by the divergence of the correlation length $\xi \sim (p-p_c)^{-\nu}$, and therefore, on length scales $L \gtrsim \xi$ (homogeneous regime) we can use the relation $\kappa \sim (p-p_c)^{\tau} \sim \xi^{-\tau/\nu} \equiv \xi^{-\tilde{\tau}}$. For $L \ll \xi$ (scale invariant or fractal regime) κ does not depend on ξ, and is proportional to $L^{-\tilde{\tau}}$. Some systems, such as colloidal aggregates, do not have natural correlation length, and therefore the power law $\kappa \sim L^{-\tilde{\tau}}$ is the only way to describe their properties.

CRITICAL EXPONENT OF ELASTICITY OF A PERCOLATING SOLID

In order to investigate the critical behavior of κ we must be able to define a lattice Hamiltonian which adequately describes the elastic properties of the system even in the extremely disordered case. Feng and Sen[1] suggested to describe the elastic behavior of a two-dimensional percolating system by a random spring network on a triangular lattice. The rigidity threshold of such a system in considerably higher than the geometrical percolation threshold, since it has many geometrically connected parts which cannot transmit stress, and the geometrical properties of the "rigid clusters" near the rigidity threshold seem to differ significantly from the properties of usual percolation clusters[6]. Therefore, random spring model does not describe correctly macroscopic disorder. Various finite differences schemes of elasticity also fail in a similar fashion.

Kantor and Webman[2] suggested a lattice Hamiltonian, in which each bond accounts for stretching energy proportional to the square of its elongation, while a pair of ajacent bonds accounts for an energy $^1/_2\, G(\delta\phi)^2$, where $\delta\phi$ is the relative change in the orientation of the bonds: For d=2, $\delta\phi$ is simply the change in the angle between the bonds, while for $d \geqslant 3$ each bond must be treated as a d-dimensional object and $\delta\phi$ represents the d-dimensional solid rotation of one bond in the reference frame of the ajacent bond. Rigidity threshold of a percolating network described by this Hamiltonian coincides with the geometrical p_c.

For a single chain on a lattice the Hamiltonian reduces to a very simple form which is analytically solvable. It can be shown[2] that the force constant k of a long chain (related to the elastic modulus by $\kappa = kL^{2-d}$) does not depend on the bond stretching term in the Hamiltonian (except the case of comparatively straight chains which are stretched along their long dimension), and is given by

$$k = \frac{G}{NS^2} \quad , \tag{1}$$

where N is the number of bonds in the chain and S is its radius of gyration (r.m.s. size) in the direction perpendicular to the applied force.

Elastic modulus of a percolation problem can be bound (in the fractal regime $L \ll \xi$) by the elastic moduli of two single chain problems[2,7]: If we assume that only the singly connected bonds[8] are flexible, while the rest of the infinite cluster is infinitely rigid, we overestimate κ and underestimate τ. If we assume that only the shortest path between two opposite boundaries, which diverges as L^z, supports the stress, while the rest of the cluster is absent, we underestimate κ and overestimate τ. Since each of these problems can be solved using Eq. 1, we can find[2,7] exact bounds on τ:

$$d\nu+1 \leqslant \tau \leqslant (d+z)\nu \quad . \tag{2}$$

The numerical values of the bounds, as well as some Monte Carlo[9] and experimental[10] estimates of τ are depicted in Fig. 1. The uncertainties in the values of the bounds (not shown) are smaller that 0.1 and are caused by uncertainties in the numerical values of ν and z, while the error bars of the Monte Carlo and experimental results (not shown) range from 0.2 to 0.5.

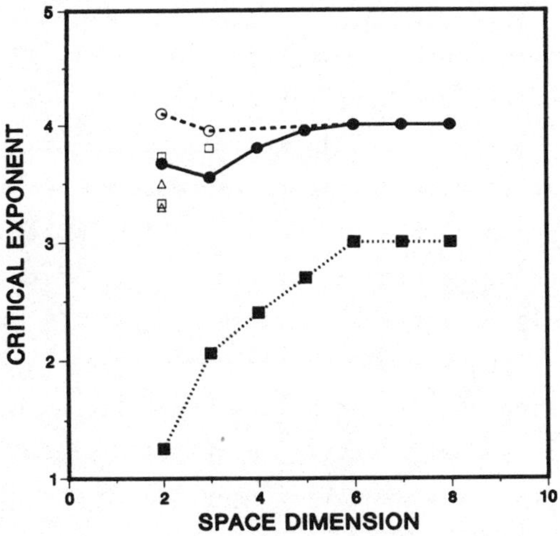

Fig. 1. Lower (full circles) and upper (open circles) bounds on the critical exponent of elasticity τ of percolation in various space dimensions. Full squares depict conductivity exponent t. Open squares show the experimental results[10] for τ, while the open triangles represent the Monte Carlo estimates[9] of τ.

All these results are consistent within the error bars. Note, that τ is close to 4 for any d, and its value is considerably larger than the conductivity exponent t. For d = 6, the two bounds on τ coincide and give $\tau = 4$. This value remains unchanged also for $d > 6$, while the Eq. 2 is not applicable above the critical dimension (d = 6) of the percolation.

MECHANICAL STABILITY OF TENUOUS STRUCTURES

Certain forms of matter, such as colloidal aggregates, have fractal nature: that is the average mass M(L) within a distance L of an arbitrary point varies as L^D, where $D < d$. As the size of the object increases, its effective elastic moduli decrease as $L^{-\tilde{\tau}}$. Thermal fluctuations and gravitational stresses tend to deform the geometry of these structures, and therefore we cannot grow them beyond some maximum size[11].

At temperature T, thermal vibrations induce an average strain ε_T given by $k_B T \approx \kappa L^d \varepsilon_T^2$. Thus in homogeneous solids ε_T fluctuates by amount proportional to $L^{-d/2}$, i.e. overall distortions are smaller in larger objects. In the tenuous structures $\kappa(L) \sim L^{-\tilde{\tau}}$ with $\tilde{\tau} > d$, and therefore the distortions increase with increasing L: $\varepsilon_T \sim L^{(\tilde{\tau}-d)/2}$. Usually aggregates are branched loopless structures, and therefore we can determine their deformation by examining the elastic properties of a single branch[11,12]. If the actual ("chemical") length of a branch N scales as L^{δ}, where L is the end-to-end distance, then from Eq. 1 we find $k \sim L^{-(2+\delta)}$, or $\kappa \sim L^{-(d+\delta)}$, and therefore

$\tilde{\tau} = d + \delta$. Thus, for branched structures $\varepsilon_T \sim L^{\delta/2}$. For sufficiently large L we get $\varepsilon_T \approx 1$ and the structure can no longer be regarded as a rigid body. The exact value L_{max} of this maximal thermally stable L depends on the temperature, the elastic moduli of the material from which the aggregate has been built, the diameter a of the grains from which it is built and the strength of the bonding between the grains. L_{max} strongly depends on a: larger (and thermally more stable) grains increase L_{max}. For most materials at room temperature $L_{max}/a = (Aa/b)^{d/\delta}$, where b is the atomic diameter and $1 < A < 10$. The exponent d/δ for most (three-dimensional) aggregates varies between 2.5 and 3. Typical stability curve is depicted in Fig. 2. Aggregates above the curve are thermally floppy, while the aggregates below the curve are thermally stable.

Another limit on the rigidity of an aggregate comes from gravitational effects. For an aggregate sedimenting under gravity, the weight force is compensated by hydrodynamic friction. For aggregate of weight Mg and size L, the gravitational stress $\sigma_g \approx Mg/L^{d-1} \sim L^{D-d+1}$. This stress induces strain $\varepsilon_g = \sigma_g/\kappa(L) \sim \sigma_g L^{\tilde{\tau}} \sim L^{D+\delta+1}$, and for sufficiently large L, $\varepsilon_g \approx 1$.

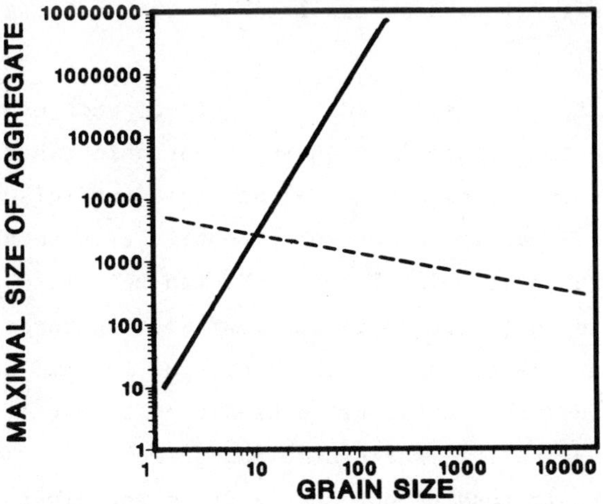

Fig. 2. Double logarithmic plot showing the dependence the maximal stable size of an aggregate L_{max}, measured in grain diameters, on the grain diameter, measured in atomic diameters. Typical lines for thermal stability (solid line) and gravitational stability (dashed line) are shown. For any particular aggregate the location and slope of the curves may slightly vary. An aggregate is stable if it is located below both lines.

As in the thermal case L_{max} mainly depends on a, but now smaller (and lighter) grains are needed to increase L_{max}. For most materials $L_{max}/a = (Bb/a))^{1/(D+\delta+1)}$, where $B \approx 10^{15}$. The exponent $1/(D+\delta+1)$ for most structures is approximately 1/4. Typical gravitational stability line is depicted in Fig. 2. Aggregates above the line are gravitationally saggy, while the aggregates below the line are stable.

Note that the largest stable aggregates can be obtained for $a/b \approx 10 \div 100$. For such grain sizes the size of the aggregate may reach $10^3 \div 10^4$ grain diameters. It appears that aggregates approaching these limits are actually produced in practice, and thus, the prospects for extending the range are distinctly limited.

DISCUSSION

We described a particular model of elasticity of random systems, which primarily intended to explain the properties of macroscopically disordered systems. Critical exponents predicted by this model are considerably larger than the values which could be expected from elasticity-conductivity analogy, i.e. the tenuous structures are considerably softer than the analogy could suggest. This softness leads to such effects as lack of thermal stability.

The applicability of these results to gels depends on the details of the systems. Since the critical exponent of entropic elasticity is smaller than τ predicted by our model, we can expect that sufficiently close to the gel point we will see mainly the entropic contribution to the critical behavior[13]. The crossover point between the two behaviors will be determined by G/k_BT (G is the microscopic bending/twisting force constant), which usually is not a very large number. However, for systems with $G/k_BT \approx 10$ we can expect that mainly the contribution predicted by our elasticity model will be seen experimentally. While some measurements of τ of gels are consistent with the predictions of our model, it is not clear whether this indicates the negligible contribution of the entropy or an inadequate description of the gel geometry by percolation model.

REFERENCES

1. S. Feng and P. N. Sen, Phys. Rev. Lett. 52:216 (1984).
2. Y. Kantor and I. Webman, Phys. Rev. Lett. 52:1891 (1984).
3. D. J. Bergman and Y. Kantor, Phys. Rev. Lett. 53:511 (1984).
4. P. G. de Gennes, J. Phys. (Paris) Lett. 37:L1 (1976).
5. W. Y. Hsu, M. R. Giri and R. D. Ikeda, Macromolecules 15:1210 (1982).
6. M. A. Lemieux, P. Breton and A.-M. S. Tremblay, to be published.
7. Y. Kantor, J. Phys. A17:L843 (1984).
8. H. E. Stanley, J. Phys. A10:L211 (1977).
   A. Coniglio, Phys. Rev. Lett. 46:250 (1981).
   R. Pike and H. E. Stanley, J. Phys. A14:L169 (1981).
9. S. Feng, P. N. Sen, B. I. Halperin and C. J. Lobb, Phys. Rev. B30:5386 (1984).
   D. J. Bergman, Phys. Rev. B31:1696 (1985).
10. L. Benguigui, Phys. Rev. Lett. 53:2028 (1984).
   D. Deptuck, J. P. Harrison and P. Zawadzki, Phys. Rev. Lett. 54:931 (1985).
11. Y. Kantor and T. A. Witten, J. Phys. (Paris) Lett. 45:L675 (1984).
12. T. A. Witten and Y. Kantor, Phys. Rev. B30:4093 (1984).
13. S. Alexander, J. Phys. (Paris) 45:1939 (1984).

ELASTICITY AND PERCOLATION

Shechao Feng

Department of Physics
Harvard University
Cambridge, MA 02138

ABSTRACT

Recent works on the elastic properties of disordered systems near the percolation threshold are reviewed. Results presented show that percolative elasticity critical exponents are very different from those of percolative electrical conductivity.

Until recently it has been thought that the elastic properties of disordered systems near the percolation threshold are essentially the same as the electrical ones in such systems. In particular, it was argued[1] that the elastic constants of a depleted random network near percolation should behave as $K, N \sim (p-p_c)^t$, where p is the fraction of elastic bonds present, and t is the percolative conductivity exponent. Much work since last year, however, has shown that the matter is not so simple. When the randomly assigned elastic bonds on a network are taken to be the ordinary springs, i.e. if the bonds have only stretching forces, it was shown[2,3] that $K, N \sim (p-p_{cen})^{f_{cen}}$, where $p_{cen} > p_c$ and $f_{cen} > t$. The reason behind the first inequality is simply that in this model, many geometrically connected clusters do not resist to mechanical deformations since the singly connected bonds in them can "buckle" without costing any energy. As of now, the two estimates of f_{cen} from ref. 2 and ref. 3 do not agree, and work is still under way in trying to obtain an unambiguous estimate of f_{cen}.[4] When a depleted elastic network has bonds that offer both bond-stretching and bond-bending forces[5], it has been shown by many different theoretical methods, that $K, N \sim (p-p_c)^f$, where p_c is the ordinary geometrical percolation threshold, and $f \approx 3.2-3.7$[6-10] in d=2 and $f \approx 3.6$ in d=3, which is much higher than the corresponding values of t. If one adopts another seemingly quite different 2D depleted elastic network model, in which sites are disks of finite extent which interact via contact forces, it is shown very recently[11] that $K, N \sim (p-p_c)^f$ where f is of the <u>same</u> value as that in the bond-bending force model just mentioned. Recent experiment on the elasticity of 2D (lattice site) hole punched sheets[12] gives $f \approx 3.5 \pm 0.4$, and a 3D experiment on sintered submicron silver powder beams[13] gives $f \approx 3.8 \pm 0.5$, both supportive of the above theoretical estimates. All of these results point out consistantly that percolative elasticity belongs to a different universality class from that of the percolative electrical conductivity.

When a random elastic network consists fraction p of bonds that are completedly rigid (super-elastic) and the rest of unit strength, a super-elasticity exponent c, which is the elastic analog of the percolative super-conductivity exponent s, can be defined through $K, N \sim (p-p_c)^{-c}$ for $p \lesssim p_c$. It has recently been shown[11] that c is close but smaller than s in 2D. This result offers evidence that percolative super-elasticity problem again belongs to a different universality class than that of the superconductivity problem.

If one leaves the discrete percolation networks and consider the elasticity of the "Swiss cheese" type continuum systems near percolation, one finds that since the equivalent bonds in the problem---the narrow "necks" between punched spherical holes---have a distribution in neck width δ which is finite at zero, and since the elastic force constant of a neck depends very strongly on δ, the elasticity exponent $\bar{f}$, defined through $K, N \sim (q_c-q)^{\bar{f}}$ where q is the volume fraction of holes, is even larger than f, by about 3/2 in 2D and 5/2 in 3D.[14]

I am grateful to Dr. B. I. Halperin for his advice and teaching throughout my entire graduate career, during which much of the above mentioned work was completed. I would also like to thank many of my collaborators, among whom are Drs. C. J. Lobb, P. N. Sen, L. M. Schwartz, D. L. Johnson, M. F. Thorpe, and M. Sahimi.

REFERENCES

1. P. G. de Gennes, J. Phys. (Paris) 37, L-1 (1976)
2. S. Feng and P. N. Sen, Phys. Rev. Lett. 52, 216 (1984)
3. M. A. Lemieux, P. Breton, and A.-M. S. Tremblay, J. Phys. (Paris) Lett. 46, L-1 (1985)
4. A.-M. S. Tremblay and S. Feng, private communications; I. Webman and G. S. Grest, private communications
5. J. G. Kirkwood, J. Chem. Phys. 7, 506 (1966)
6. Y. Kantor and I. Webman, Phys. Rev. Lett. 52, 1891 (1984)
7. S. Feng, P. N. Sen, B. I. Halperin and C. J. Lobb, Phys. Rev. B 30, 5386 (1984)
8. D. J. Bergman, Phys. Rev. B 31, 1696 (1985)
9. I. Webman and G. S. Grest, Phys. Rev. B 31 (1985)
10. S. Feng and M. Sahimi, Phys. Rev. B 31, 1671 (1985)
11. S. Feng, submitted to Phys. Rev. B
12. L. Benguigui, Phys. Rev. Lett. 53, 2028 (1984)
13. D. Deptuck, J. P. Harrison and P. Zawadzki, Phys. Rev. Lett. 54, 913 (1985)
14. B. I. Halperin, S. Feng, and P. N. Sen, submitted to Phys. Rev. Lett.

SCALING CONCEPTS IN POROUS MEDIA

Christophe Baudet, Elixabeth Charlaix, Eric Clément, Etienne Gyron, Fean-Pierre Hulin, and Christophe Leroy

Groupe Poreux P.C.

Laboratoire d'Hydrodynamique et de Mécanique Physique
E.S.P.C.I.
10, rue Vauquelin
75231 Paris Cedex 05

INTRODUCTION

Scaling concepts in disordered matter should be of use to the physics of porous media (POM)[1]. The applications of such media are numerous, ranging from hydrogeology, oil industry, chromatography and filtration, and justify the lasting interest for this physics[2]. We will focus our attention towards the effect of disorder in POM and, more precisely, that of multiple scales in the geometry of the fluid(s) penetrating them; but we will also recall some basic properties of POM. Geometrical disorder in POM has several origins :

- There is a strong contrast between the material properties of the solid phase and the fluid one(s) penetrating it.

- The pore space can range from homogeneous to very heterogeneous (e.g. from sintered materials obtained with a relatively uniform particle size distribution to fractured rocks presenting an irregular distribution of cracks of variable size, aperture...). Heterogeneous POM often present a large range of geometrical scales.

- In multiple phase flows, the distribution of the fluid phases contained in the POM introduces additional heterogeneities and multiplicity of scales.

- Local heterogeneities like those due to roughness or of chemical nature (e.g. presence of clays in sandstones) control the wetting properties in polyphasic flows. We will not consider this last class of parameters which, nevertheless, are of dramatic practical importance. Chemical surface effects can also model the properties of walls in random field problems : the advancing front of a wetting fluid on an heterogeneous surface (or, possibly, in a porous material made of wettable and non wettable properties) displays some characteristic features (rough interfaces, hysteresis) of R.F. systems[4] discussed in the article by Villain.

A large amount of information has been obtained on the physics of POM since the original work of Darcy in 1856[5] who expressed the proportionality between the flow rate per unit area, $\vec{Q}$, of a fluid saturating a POM and a pressure gradient $\vec{\nabla}p$ applied to the flow

$$\vec{Q} = -(\kappa/\eta)\vec{\nabla}p \qquad (1)$$

The permeability κ (length^2) is a function of the pore geometry. The properties of the fluid enter in the formula through its viscosity η.

This classical description of monophasic flows through porous media rests on two major assumptions :

- the microscopic velocities vary linearly with the stresses applied to the fluid (Stokes regime at low Re number);

- there exists a length scale ℓ_0 which is large with respect to the pore size or the characteristic size of microscopic variations of the POM but which is small with respect to the macroscopic structures of the medium.

Then, one can define significant macroscopic variables by averaging the microscopic quantities over a representative elementary volume (REV) of linear size ℓ_0. Note that the constitutive law (1) which relates the macroscopically averaged pressure gradient and averaged flow velocity Q is a potential flow equation just because the dissipative effects (the $\vec{\nabla}^2\vec{v}(\vec{r})$ term of the Stokes equation for the local velocity field $\vec{v}(\vec{r})$) are averaged out in the ratio Q/κ. This has important consequences in the description of the average flow lines in a POM and in the dynamics of invasion front in a Hele Shaw cell[6].

In many cases, Darcy's law is not valid even when the flow in the pores is laminar and stationary :

- non linear effects described by the $(\vec{v}\,\vec{\nabla})\,\vec{v}$ term of the Navier Stokes equation occur for moderate values of the Reynolds number $Re \sim v\sqrt{\kappa}\,\rho/\eta \sim 1$ because of the fast local changes of $\vec{v}$. We can expect these effects to be enhanced in multiple scale or poorly connected POM. To our knowledge, the problem has not yet been studied. In the following, we will limit our discussion to linear flow problems;

- the precise definition of a REV is a major problem in heterogeneous media presenting a large range of geometrical scales such as fractured rocks where the structures extend from macrocracks to microfissures or limestones which present a large variety of pore structures (caverns, vugs, inter and intragranular pores);

- it is also a problem in percolation systems such as POM near a permeability threshold. Then, the void space may have a structure qualitatively similar to a percolation infinite cluster.

In both cases, we expect large fluctuations of the permeability factor κ defined by eq. (1) over small scales. A systematic study of these fluctuations for a characterization of heterogeneous POM would certainly be of importance in geological applications.

It should be clear from this introduction that even the simple Darcy problem needs to be revisited using scaling ideas; this will be the object of the following chapter. The chapter entitled "Diphasic flow in porous media" extends these ideas to problems of two phase flows. Original features arising from the presence of interfaces, fronts, will be described and can be connected with geometrical features discussed in the book (cluster statistics, aggregates). The last chapter deals with the dispersion of dyes or small particles present in a fluid flowing at a constant velocity in a POM; the effect of heterogeneities on dispersion is probably even stronger than on permeability.

Before going on with these studies, we mention rapidly a few measurements which can be used to characterize the structures of POM :

- Direct studies of rocks can be made using various microscopic observations. Stereological methods relate the 3 dimensional structure of the solid and void spaces to two dimensional cuts but the long distance connectivity properties cannot be approached as easily as local averages (such as the specific surface area S, or the porosity ϕ) which benefit from mathematical morphology's powerful methods[7].

- A number of nuclear, acoustic, dielectric studies performed "in situ" ("case" studies) as well as on "core" samples can be cross correlated to improve the knowledge of the pore and solid geometries. Many fundamental studies have been carried out recently on these properties. They can be evaluated by extending classical treatments such as E.M.A. and self consistent methods to well connected POM. Direct studies of these properties in the scaling regime are also underway. We refer to the book[3] for a discussion of these properties.

- The electrical conductance of an insulating porous medium filled with a conducting fluid is a most direct way to evaluate the connectivity of the pore space. For example, recent systematic measurements[8] of both conductance Σ and permeability κ of consolidated arrays of monodisperse spheres of variable porosity ($0.023 < \phi < 0.4$) have given the following empirical relations

$$\Sigma \propto \phi^{m}$$

(known as "Archies" law and applied practically to a number of porous materials)

$$\kappa \propto \phi^{m'}$$

(a classical expression for the permeability, the so called "Kozeny-Carman" relation involves in addition to ϕ, the measurement of S and of a tortuosity factor for the flow).

In the study of ref. (8) the relation $m' \sim 2m$ was found. The factor 2 relates to the fact that the conductance of a cylindrical line reduces by a factor of 2 when its area decreases by 2, the permeability of a cylindrical tube by 4 : the dissipative nature of the viscous flow problem within pores leads to a distribution of current different from that of an electrical problem with the same insulating grain geometry (or for the flow of superfluid He_4). The value of m ($\sim$ 2 in the above experiments) is a result of the incoherent potential scattering by the solid grains and can be approached by other potential experiments (like the acoustic 4th sound or slow Biot mode in a POM[9]). The value of m varies widely with the geometry : it is 3/2 for dilute spheres. A value 1.15 was obtained recently for an unconsolidated array of two size spheres[10]. Thus, measurements of conductance are only of empirical and limited value to estimate the connectivity and permeability of POM.

PERMEABILITY IN SINGLE PHASE FLOWS

We consider the effects resulting from the multiplicity of scales and the heterogeneities in POM. We will study them in three sections :

- the effects dealing with the coexistence of structures of different ranges of sizes within the solid phase which can benefit from fractal concepts;

- those found in weakly fractured rocks to which we have applied percolation ideas;

- those related to broad distributions of local permeabilities. General methods introduced in the study of variable range hopping in doped semconductors will be applied to this problem.

A first question concerning these heterogeneous structures is that of the existence and size of REV (Representative Elementary Volume) over which a Darcy permeability can be characterized precisely (beyond local fluctuations). The characteristic scale for such an homogeneous regime can be quite large. Thus it should be of great practical interest to study the size dependence of the mean quantities (pressure losses, flow rates, permeability) as well as of their fluctuations.

Some Examples of POM with a Multiplicity of Geometrical Scales

- The random dense packing of space with objects having a large distribution of sizes is of prime technological interest. In concrete materials, the (undesirable) porosity is empirically known to scale as $(\ell_1/\ell_2)^{1/5}$ (Caquot's law[11]), ℓ_2 and ℓ_1 are the largest and smallest sizes of grains.

This form of Caquot's law is not well understood. We give here a justification of it which assumes that the macrogeometry is self similar with a wide distribution of grain sizes, and an iterative filling of space by grains. A reasonable model for such a structure is the Appolonian iterative packing of spheres such that the nth generation of spheres just fit whithin the voids of the $(n-1)^{th}$ one. The pore space of such a system is a random fractal, and an estimate of its porosity can be obtained from the Minkowski definition of the fractal dimension : the coating of the structure taken over a volume length scale ℓ_2 with smallest ones of radius ℓ_1 leads to a covering volume

$$v \sim \ell_2^{d_f} \ell_1^{d-d_f}$$

Thus the porosity should scale as

$$\phi \sim v/\ell_2^d \sim (\ell_1/\ell_2)^{d-d_f} \quad (2)$$

This result provides an explanation for the Caquot's law but also suggests that the exponent 1/5 should not be universal but should depend on the fractal dimension. This result is also stressed in a recent theoretical analysis by Omnes[11].

- A fractal geometry has been recently presented in real sandstones over a very wide range of scales (from ℓ_2 = 100 μm to $\ell_1 \simeq$ 10 Å) with values of d_f between 2.57 and 2.87, depending on the local geometry, as determined by SEM on cuts[12]. The authors find an excellent agreement with the law (2) They also propose a geometrical determination of the permeability (which they assimilate with the electrical conductance) by expressing d_w [13] in terms of the fractal geometry. The mechanism invoked for the creation of the fractal geometry is "diagenesis" in which rock material is dissolved and next redeposited within pores as multiple scale space filling grains. De Gennes[14] subsequently suggested that the partial filling of this geometry with a liquid film forming films and droplets in the medium could provide useful informations on the geometry of the POM.

- Fracturing of rocks takes place on many scales from megafaults down to microcracks. When such cracks nucleate and grow under applied stresses, a sharp rupture threshold is often observed due to coalescence of fractures. A first R.G. calculation of the problem was given by Madden[15] . In ref. 16 the sharp transition is described in a model in which a cube is divided hierarchically in 8 cubes of half size[15] . At each iteration a fraction p of cubes is stated fragile (a fragile cube is defined as one which has no continuous solid line along any of its axes). The transition from a sound to a fragile rock takes place around a critical value p_c with a transition range which becomes progressively sharper as the number of iterations or of scales of the degradated structure increases.

Permeability Threshold Effects

We describe here some situations where permeability thresholds are met in porous media.

Filtration. The original treatment of percolation by Broadbent and Hammersley[17] was inspired by the study of filters randomly clogged by solid particles, causing a critical transition to an impermeable state. However, filtration involves a collection of particles either in a "cake" on top of the filter surface (directional aggregates), or in "depth" within the medium, mediated by the hydrodynamic field : the closing of pores is neither uniform nor random in space.

Pressure controlled permeability thresholds. An applied pressure field P can close randomly a fraction of fractures in an heterogeneous rock. An abrupt decrease of the permeability for a non zero porosity is often observed around a critical value of applied pressure[18] . On the other hand, the regular closing of porous structures obtained by sintering leads to a smooth continuous decrease of κ with the pressure for low porosities. Recent geometrical models using periodic packings of deformable or overlapping spheres lead to accurate predictions of the variations of $\kappa(P)$ down to a porosity of of a few percent[19]. However, in this limit, the heterogeneity of the narrow passages is probably important and non local effects due to randomness should be present (cf. paragraph "Permeability of fractured rocks"). Indeed very fast decreases of permeabilities are found in sandstones having porosities below a few percent.

Physico-chemical plugging. The swelling of clays, which are often found within sandstones, is a function of salinity and, thus, provides a way to vary continuously the porosity and permeability of a POM. We are developping a model experiment along this line to study 2D and 3D critical permeabilities in random geometries.

Permeability of fractured rocks. Hydrogeological studies indicate large variations of κ (down to zero permeability) in granitic sites where the porosity is mostly due to a network of cracks[20] . Diagenesis can also modify the geometry of sedimentary rocks by dissolving material along certain flow paths and inducing a transition from a regular (intersticial type) to an irregular (fractured type) porosity.

The problem of onset of permeability in fractured rocks is of great importance in hydrogeology and in oil recovery (e.g. in norvegian off-shore oil fields 1). An additional present concern deals with the storage of nuclear wastes.

For weakly fractured rocks, a non zero permeability depends on the existence of a continuous path in the medium. Assuming that the fractures are randomly located in the rock matrix, a percolation model can be used to estimate the permeability threshold of the rock site. If the fractures are modelled by thin disks of radius R (or thin needles of length 2L in a 2D model) with random positions and orientations, a number of predictions can be made :

- The state of connection between fractures being invariant under space dilatation, the percolation threshold is given by a critical value of the dimensionless geometrical number : N_3R^3 (N_2L^2 in a 2D model); $N_{3,2}$ is the number of centers of objects per unit volume or area[21].

- Such a dimensionless scaling can be related to a classical empirical quasi-invariant obtained for percolation threshold on periodic lattices[22] :

. in bond percolation, the critical number of active bonds per site, defined by $B_c = z\, p_{cb}$ (z is the network coordinance) is $B_c \sim 2$ (2D); $B_c \sim 1.5$ (3D)

. in site percolation, the critical volume fraction defined by $\phi_c = f\, p_{cs}$ (f is the volume fraction of spheres centered on the site of the network) is $\phi_c \simeq 0.44$ (2D); $\phi_c \simeq 0.16$ (3D).

Disk percolation can be mapped onto a bond percolation problem on a random lattice : the sites are the centers of the disks; a bond is stated active if the two disks centered on the ends of the bond intersect.

The average number of bonds per site is then the mean number of intersections per disk. In what follows, we estimate B_c from the excluded volume, $<V_e>$[23] which is the mean volume surrounding an object within which the center of another object must lie in order to intersect the first one. $N < V_e >$ is then the mean number of intersections per disk and is equivalent to B. For an isotropic array of disks, $< V_e > = \pi^2 R^3$ (for needles, $< S_e > = 8L^2/\pi$). The correspondance with the values of quasi invariants on periodic lattices gives an estimation for the permeability threshold

$$B_c = N_{2c}\, 8L^2/\pi \sim 2 \qquad \text{(2D)}$$

$$B_c = N_{3c}\, \pi^2 R^3 \sim 1.5 \qquad \text{(3D)}$$

Monte-Carlo simulations made with objects of different shapes[22,24,25] show that this threshold estimate is qualitatively good although the numerical constant depends on the shape of the objects. For the 2D needle percolation, Robinson[24] finds a critical number of intersections $B_c = 3.6$, while for the 3D percolation of squares the value is $B_c = 2.4$.

The method can also give the angular dependence of the permeability threshold in case of an anisotropic fracture field. For example in a 2D model of needles uniformly distributed between 2 limit angles $(-\alpha, +\alpha)$, Robinson finds $B_c(\alpha) \simeq 3.6$ independently of α. The value agrees within 2 % with the numerical value for the isotropic case.

The effect of polydispersity has also been investigated for an uniform distribution of lengths between two extreme values. In the case, the "good" invariant J_c turns out to be the sum of the invariants of each "subnetwork" of fractures of the same length :

$$J_c = \int (8L^2/\pi)\, N(L)\, dL \simeq 3.6 \qquad \text{(2D)}$$

with the same numerical value as for the monodisperse case. This interesting result has to be checked on broader distributions since geostatistical analysis often find large log-normal length distributions.

Besides the percolation threshold, another quantity of interest is the correlation length which gives the characteristic size of the permeability fluctuations. Simulations on random 2D and 3D systems[24,26] confirm the expected scaling $\xi \propto (N-N_c)^{-\nu}$ where ν has the same value as for regular lattices.

The permeability of these systems can be deduced from their electrical conductance, by assuming that the pressure losses occur mostly through the Poiseuille flow within the fractures and that the intersections do not create any appreciable additional dissipation. From the numerical solution in ref.[24] of the conductance of needles of equal size and resistance in 2D we have extracted the (expected) variation in $(N-N_c)^t$ with $t \simeq 1.19$. The domain of validity of this variation ($N < 2\ N_c$) gives a range of fracturation densities N for which critical percolation concepts are useful to describe the hydrodynamic of fractured rocks.

Effect of Local Variations of Permeability

A common statement in the physics of percolating systems is that position disorder is irrelevant compared with composition one. The condition for this to take place ($d\nu - 2 = -\alpha > 0$) expresses that disorder is averaged out over a large enough correlation volume ξ^d where critical fluctuations dominate. The result applies to equilibrium (static) critical properties. Very recently Halperin, Feng and Sen[27] studied the transport properties for a D dimensional swisscheese (solid perforated with overlapping holes of random centers) and found different-larger-dynamical critical exponents than with a periodic lattice. Their argument rests on the following steps :

- narrow paths are characterized by the density distribution $\Pi(\delta)$ of minimum distances between neighbouring spheres, such that $\Pi(\delta) \rightarrow \Pi_o$ as $\delta \rightarrow 0^+$;

- the conductance (electrical, mechanical, hydrodynamical) of a narrow path is equivalent to the one of a channel of width δ and length $\delta^{1/2}$ and varies as δ^x. For example, the equivalent Poiseuille conductance of a neck between three spheres in 3D varies as $\delta^4/\delta^{1/2} \sim \delta^{7/2}$;

- in the node-blob-link description of the percolating cluster, the number L of singly connected (red or sensible) bonds on a mesh size of the superlattice increases as $(p-pc)^{-\zeta}$ near threshold ($\zeta \sim 1$). These bonds form a chain with a distribution fo bond widths. The expectation value δm of the weakest bond of this chain is governed by a scaling relation $\pi_o\ \delta m\ L \sim 1$ (this is a mean field argument within a percolation description);

- finally one computes the transport property of the superlattice, the conductance of each chain of L red bonds being controlled by its weakest bond, which give, as a generalization of Shklovskii-de Gennes argument, a form :

$$\kappa \sim (\xi^{1-d} / \xi^{-1})\ L^{-x} \propto \Delta p^{\bar{e}}$$

with

$\bar{e} = \nu(d-2) + 7/2$ for the permeability in 3D instead of
$e = \nu(d-2) + 1$ for a regular arrangement of spherical grains.

This remarkable prediction of an increase in the value of the dynamic exponents in a random geometry is being tested experimentally on permeability experiments (in Paris and Oslo).

We propose here two extensions of this approach to the study of permeability of fractured rocks with a large distribution of apertures δ. Studies of geological sites[28] give the distribution of fracture apertures by classes (10 to 1 mm, 1 to 1/10 mm...). Log normal distributions are often found in such analyses. In plane Poiseuille flow, the pressure drop across an individual fracture of a given plane geometry (its individual permeability) varies as δ^3, which increases even more the variation of local permeabilities.

i) We can apply directly the result of Halperin et al to this problem if the degree of fracturation characterized by the value of N (see above) is near the percolation threshold N_c. Then even the weakest fractures must carry an appreciable flow. If we assume that the distribution f (δ) of fractures apertures δ is continuous down to $\delta = 0^+$ ($f(\delta) \to f_o$) the premeability of the random network of fractures should vary as $\kappa_o(N-N_c)^e$ with $e = \nu(d-2)+ 3$ (instead of $e = \nu(d-2) + 7/2$ since the length of fractures is assumed to be independent of δ). This critical variation of permeability could be seen between different locations of a geological site where the overall density of fractures N would vary slightly while their aperture distribution would remain the same.

ii) If N is much larger than N_c, we can still construct a percolation solution by analogy with classical treatments of variable range hopping conduction between randomly distributed impurities in extrinsic semiconductors, which are modelled by random lattices with an exponentially large variation of bond resistances[29]. Following the method used to estimate the conductance in this problem, we establish fractures in decreasing order of apertures and we define $\kappa(\delta)$ as the permeability of the medium in which exist only the fractures of aperture larger than δ. For a critical value δ_c such that

$$N \int_{\delta_c}^{\delta_{max}} f(\delta)\, d\delta = N_c$$

the percolation threshold is reached. Up to this point, $\kappa(\delta)$ is zero. If we add a few additional fractures - with apertures just below δ_c, $\kappa(\delta)$ increases continuously and critically from zero. However if the distribution $f(\delta)$ is very wide, we expect a saturation of $\kappa(\delta)$, as sketched on the figure, around a value δ_e not too far from δ_c. Narrower fractures in parallel with this subnetwork will carry practically no current and can be omitted. Thus the actual permeability κ can be known accurately from the knowledge of the bond distribution between δ_c and δ_e. We can estimate δ_e from the following argument :

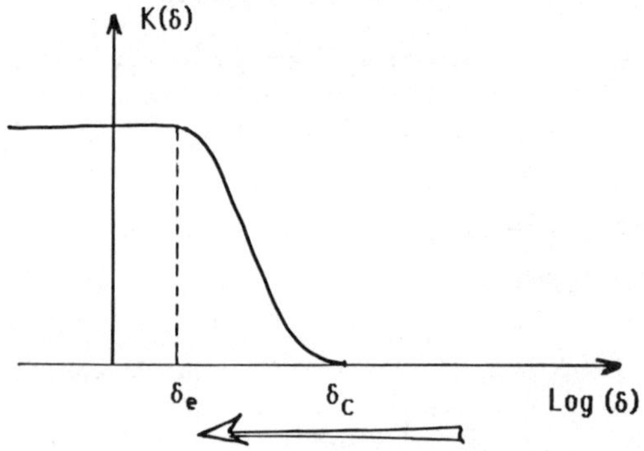

Consider the measurement of permeability κ on boxes of finite size L. The fluctuation of the threshold value is given by the classical finite size scaling result :

$$(\Delta N_c/N_c) \sim (L/\ell_o)^{-1/\nu}$$

where ℓ_o is the size of an individual fracture. In our continuous distribution of bonds the result can be written as

$$\int_{\delta_e}^{\delta_c} f(\delta)d\delta \sim (L/\ell_o)^{-1/\nu}$$

where δ_e will be the aperture of the "weakest" fracture which control the permeability of the box of size L (if $\delta_e = \delta_c/2$ above controlled by δ_e will have a permeability 10 times smaller than one with the average value δ_c). When L increases, the fluctuations on the values of the local permeability decrease and we take as a unit cube size for scaling the global permeability the size L_o of a box such that the mean square fluctuation of the individual box permeability is of the order of unity (which defines a value for $\delta_c / \delta_e \sim \sqrt[3]{2}$). We note that, for a broad distribution of apertures, L is very large : L is, in some sense, a REV for the distribution of fracture.

We can deduce an estimate for the permeability of the media

$$\kappa = \kappa_c/L = \kappa_c \left| \int_{\delta_c}^{\delta_e} f(\delta)\, d\delta \right|^{\nu} /\ell_o$$

κ_c being the permeability of a fracture of aperture δ_c.

On the other hand, we expect that, in a finite size sample, fluctuations of the permeability will increase when the distribution of local permeabilities increases, i.e. when the size of the subnetwork increases. We have analyzed numerical simulations of Robinson[20] on percolation of needles with different distributions of conductance values. For equal resistance needles, we find a critical exponent $t \simeq 1.19$ which agrees with usual percolation. For a uniform distribution of needle widths between two extreme values, the conductance values fluctuate strongly and indicate a faster variation with N_2 although it does not approach threshold close enough to get an estimate of t.

N.B. In practice, correlations often exist between lengths and apertures of fractures. This could modify (enhance) the effects discussed above. We are presently pursuing this particular study using numerical simulations.

DIPHASIC FLOW IN POROUS MEDIA

Percolation concepts have been introduced in this problem a few years ago and a number of recent references can be found on the subject[1,30]; also see the twin paper[31] (chapter 3). We present here a schematic destription of this problem.

We consider two immiscible fluids (one wetting (W) and one non wetting (NW)) filling a porous material. Classical couples of fluids are :

W	NW	function
water	oil	assisted oil recovery (water pushing oil)
water	air	partially saturated soils
Hg vapour	Hg liquid	mercury porosimetry

In the operation of imbibition, the W fluid pushes the NW one out of the POM. This can take place in the absence of an applied pressure gradient under the action of capillary forces. In pratical laboratory "core" evaluation simulating imbibition (water flooding[32]), a sample initially filled with oil is immersed in water. The amount of oil displaced by water is measured as a function of time. The following surprising results are found :

- the recovery is very progressive (over years) and takes place by sudden jumps;

- there is no obvious scaling type dependence of the law of recovery on the oil viscosity (it increases when η_o increases) and on the core sample size.

In fact, imbibition is strongly influenced by local wetting properties of the W phase and we are still a long way from a scaling description taking into account these local effects.

Drainage deals with a non wetting material (say Hg) pushing out of the POM, under an injection pressure Δp, a wetting phase (say Hg vapor). We will consider in more detail this problem by considering first quasi-static injections (infinitesimal pressure step increases). Subsequently we will consider the effect of gravity and of finite flow velocity on the injection profile.

Quasistatic Injection

If we consider a porous material as a set of pores connected by channels of random radii r and a quasistatic injection, the overpressure Δp needed to push an interface meniscus through a given channel is given by the Laplace relation

$$\Delta p = (2\gamma/r)\cos\theta \tag{3}$$

γ is the interfacial tension, θ the wetting angle.

For a given value of Δp applied at an injection face of the POM :

- only pores connected to the sample interface through the invading fluid and such that $r_i > r$ (r being given by eq. (3)) will be invaded;

- the cluster formed by this process is bounded by non invaded pores such that

$$r_j < r$$

The statistical repartition is the same as for a percolation cluster due to the correspondance between the random distribution of pore radii and local capillary pressures. In particular there is a critical injection pressure Δp_c above which there appears a continuous path across the POM (breakthrough) : this pressure value corresponds to a percolation threshold[33].

Above this critical value appears a non-zero permeability for the non wetting phase; it is usually defined by a generalization of the Darcy law :

$$\vec{Q}_{snw} = -(\kappa k_{rnw}/\eta_{nw})\,\vec{\nabla}p_{nw} \tag{4}$$

where $\vec{\nabla}p_{nw}$ is the pressure gradient applied across the non wetting phase, k_{rnw} is called the relative permeability of the NW phase. If one increases the NW fraction (or saturation) by increasing Δp above Δp_c, there exists, in 3D systems, an upper critical value at which the W phase becomes disconnected. One can also define a relative permeability for the W phase, k_{rw}, which should go to zero at this upper pressure value; but local wetting effects and fluid transport by wall films can also play a role at this second threshold, and the application of percolation concepts to it is questionable (let us note that, in two dimensional models, the two thresholds have the same value). We may expect k_{rnw} to vary critically above threshold as $(\Delta p - \Delta p_c)^e$ (e is a critical exponent probably larger than the conductance exponent t) but the present experiments have not considered this critical behavior.

Invasion percolation[34] describes the dynamical invasion problem for a continuous slow increase of pressure or of injected fluid. It describes in particular two differences with classical percolation :

- trapping effects : once it has been surrounded by the invading fluid, a pocket of incompressible residual fluid (e.g. water instead of Hg vapour) cannot be displaced later in the invasion;

- only pores connected to the injection surface by the invading fluid can get filled. Isolated clusters do not appear.

Therefore, the structure of the invading fluid at the percolation threshold is less dense than in classical percolation, leading to a lower fractal dimension.

Other differences with respect to classical percolation appear surface effects. The surface by which the NW phase is injected is "attractive" for percolation in the sense that the order parameter (the fraction of the infinite cluster (relative saturation S_{nw}) is larger that in the bulk[33].

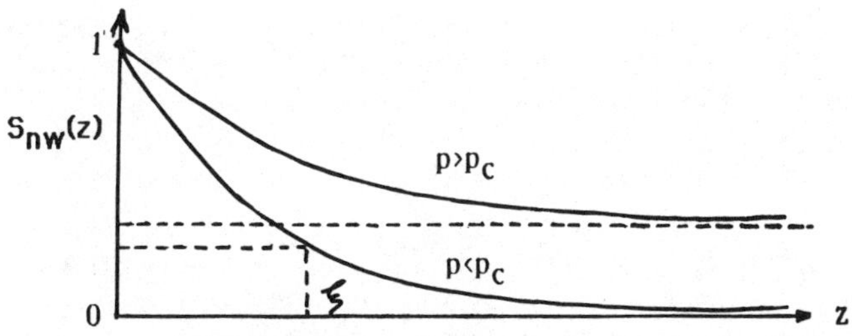

Let us assume that this surface ($\perp$ to z) is very large. Below the threshold for Hg penetration ($\Delta p < \Delta p_c$), the relative saturation to the NW phase decreases to zero for large z.

- At short distances, $z < \xi$ (ξ is the bulk percolation correlation length), S_{nw} decreases as $z^{-\lambda}$. The cluster structure is fractal at this length scale. Both features come from the fact that, at these sort distances, the only reference length is z itself.

- The largest distances there is a residual exponential decay of s_{NW} as $\exp(-z/\xi)$.

An estimate of λ is obtained by expressing that, for $z = \xi$, S_{nw} should scale as $(\Delta p - \Delta p_c)^{\beta}$. Thus $\lambda = \beta/\nu = d - d_F$ as also expected for the fractal dimension of the percolation cluster. Two dimensional experiments on micromodels have tested the fractal description[35].

Above p_c, S_{nw} saturates to the finite bulk value which is the density of the infinite cluster. However porosimetry experiments are done in practice on elongated cylindrical samples. The side walls of the samples are repulsive for percolation since they limit the lateral development of percolation clusters. When ξ becomes larger than the sample width, the residual saturation goes to 0 at large z; the problem becomes a one dimensional one (3D to 1D crossover dimensional crossover) ; there is a unit probability on a sufficiently long sample to find a section across the sample where all the pores will be subcritical). Thus the geometry to study critical injection effects should be that of a long but even wider sample!

Gravity Effects[36]

If we inject from the bottom a non wetting liquid in a vertical POM, gravity tends to reduce the applied overpressure Δp; at any given level h above the injection level the over pressure is :

$$\Delta p - \Delta \rho g h$$

If we consider formula (3) we see that a natural dimensionless parameter to compare gravity and capillarity effects is the Bond number $Bo = \Delta \rho . g\ h^2/\gamma$.

We expect the saturation of the non wetting phase to decrease continuously with h (or Bo). The variation should scale as that of the percolation order parameter P(p). Thus $S_{nw}^{1/\beta}$ should vary linearly with Bo. This result has been found in recent experiments where the NW fluid is Wood metal solidified after a slow injection under gravity[31]. The photograph of plate 1 is a vertical cut of the upper part of the metal block.

Viscosity Effects

Viscosity pressure losses play an important role in fluid invasion characteristics. Their relative importance with respect to that of capillarity can be measured by the capillary number Ca

$$Ca = \frac{\eta v}{\gamma} = \frac{\eta|\vec{\Delta v}|\ell}{\gamma/\ell} \sim \frac{(\eta v/R^2)\ell}{\gamma/\ell}$$

which is a ratio of the viscous pressure drop and of the capillary one for a tube of radius R and length ℓ (in addition to Ca, a dimensionless geometrical ratio ℓ/R appears). In other terms, Ca is a ratio of the local velocity v to a velocity characteristic of the fluid ($\gamma/\eta \sim 10^4$ cm/s for water; the value can be much less for low viscosity interfaces with microemulsions and, consequently, viscous pressure effects will be enhanced). In experiments, the Darcy velocity is used instead of v; this introduces an additional geometrical factor.

It is also known from classical hydrodynamic instability studies in Hele-Shaw cells[6] that the ratio of the viscosity of the invading fluid over that of the displaced one, $M = \eta_1/\eta_2$, controls the stability of the interface.

Consider first the case where η_1/η_2 is larger than unity. The formation of fingers of phase 1 in 2 is strongly inhibited as it would induce strong pressure variations in the injection fluid : those would have to be balanced by transverse flows. Indeed, experiments on 3 dimensional porous media as well as micromodel experiments[37] (see photographs of plate 2) confirm this result and show a front which become progressively flatter as the injection velocity increases. Wilkinson[36] has suggested that Ca could play in this case a role comparable to Bo for gravity as both effects induce a similar reduction of the pressure drop applied to the menisci with respect to the injection pressure. The result in this limit of large M indicates that it will be important to use a large enough velocity to displace another fluid.

Consider now a viscosity ratio M sufficiently smaller than unity. For low flow rates, the experiments[37] indicate a distribution of injected fluid typical of invasion percolation (plate 2). However, as Ca increases one observes strong fingering effects reminiscent of the structures observed in DLA. This can be understood, following Paterson[38]. The flow equation in phase 2 is given by a Darcy law. Thus $\nabla^2 p_2 = 0$. If we neglect the viscous losses in phase 1, the boundary condition on the side 1 of the interface is a constant pressure one, p_1 = cst. If the viscous losses are large enough, the relative pressure drop across the interface is small. The potential problem in phase 2 can be solved by a random walk technique. Thus the progression of the front and the growth rate of a DLA problem can be related if we assume that the field at the interface gives the growth rate in the continuum approach. The description above is crude and does not take into account the fine disordered structure of random pores at the interface. The original Saffman–Taylor instability[36] obtained by displacing a viscous fluid by a less viscous one, the two fluids being contained between two parallel plates, is a limit continuous model of a porous system without local randomness (... provided the local wetting properties on the cell walls are extremely well controlled). It is of fundamental interest to see in a model porous geometry with an uniform pore distribution how much the DLA structure will be different from that with a random pore geometry.

DISPERSION

This chapter deals with the behavior of a tracer in solution in a liquid flowing through a porous medium, adding the effects of molecular diffusion (D_m) and of convection through the random geometry. In general, one looks for a macroscopic dispersion law :

$$\partial c/\partial t + |\vec{U}| \; \partial c/\partial x_{/\!/} = D_{/\!/} \, \partial^2 c/\partial x_{/\!/}^2 + D_\perp \, \partial^2 c/\partial x_\perp^2 \qquad (6)$$

where c is a concentration average over a REV. $D_{/\!/}$ and $D_\perp$ are constant dispersion coefficients giving the spreading along (//) or perpendicular ($\perp$) to the average flow velocity $\vec{U}$. Such a relation describes well the dispersion effects obtained in packed chromatographic columns considered as "good". However this is not always the case and a large variety of (unpublishable) behaviors can be found in the drawers of chemical engineers (shameful curves). Anomalous dispersion also seems to be a rule in hydrogeology ! We will try to relate some of these behaviors to heterogeneous structures in POM. We shall review first schematically the different regimes of dispersion in well connected and homogeneous media and will see how they may be modified in heterogeneous ones. There are only few publications on the last paragraph. One aim of the present experimental effort of our group is to check some conjectures presented here.

Well Connected Materials

Geometric dispersion. Random arrays of monodisperse spheres or of crushed glass give results which agree with eq. (6). For large Peclet numbers, (Pe = Ud/D_m measures a ratio between convective and molecular terms), one observes a linear increase of $D_{/\!/}$ with the applied velocity $|U|$, typically as Ud; the separation between pores, d, is the distance over which the flow rate through pores fluctuates[40].

This dimensional result can be understood from the simple following argument : in a reference frame moving along $x_{\parallel}$ at a velocity U/2, the effect of the variation of velocity from one pore to the next or within a single pore, of the order of U, can be seen as a sequence of brownian steps to the left or right of a mean position : $\Delta x_{\parallel}^2 \sim (U/2\tau_G)^2$, taking place over a time interval $\tau_G \sim d/U$. If the steps are uncorrelated, the central limit theorem leads to a gaussian dispersion with a longitudinal dispersion coefficient

$$D_{\parallel} \sim < \Delta x_{\parallel}^2 > / \tau_G \sim dU \tag{7}$$

The linear dependence in U is compatible with the time reversibility of low Reynolds number flows[41] as demonstrated in visualisation experiments done in Oslo[42] where a line of dye was injected in a fluid flowing back and forth across an array of cylinders. However some dye remains trapped in the stagnant region aft of the cylinders after this process.

Independently we had considered, and we present in the appendix, the effect of the stagnant regions in the flow around a spherical grain (this result can be generalized to random convex objects). A similar problem had been addressed first by Saffman[43] who considered other effects of stagnant flow field :

- along the solid walls of long enough tubes;

- in slow channels nearly transverse to the mean flow field.

These three problems involve a singular distribution of the residence time of the fluid E(t) and a logarithmic divergence of the second moment of the transit time distribution. However, an upper cut off time τ_c is introduced which describes the redistribution of the dye by molecular diffusion out of the stagnant region. After averaging over a random distribution of such structures, one gets a non analytical dispersion

$$D_{\parallel} = Ud \text{ Log } | \tau_c/(\ell/U) | \sim Ud \text{ Log } (U\ell/D_m) \tag{8}$$

(ℓ is a length scale for diffusion out of the stagnant region).

Hydrodynamic dispersion. In a long narrow channel of width a and length ℓ, the joint action of transverse mixing across the channel and of convection along it leads to the so called "Taylor diffusion"[44] with a longitudinal dispersion :

$$D_{\parallel} = D_m + a^2 U^2/D_m \tag{9}$$

This mechanism dominates if the transverse mixing is large enough, i.e. if the transverse time $\tau_H \sim a^2/D_m$ is short enough compared with the convective time d/U.

The surprising form of $D_{\parallel}$ (D_m is in the denominator) can be justified following the same brownian motion argument used for the geometric term. In a referential moving at a velocity $\bar{U}$ (averaged over a parabolic flow profile) the particles move from the central part of the flow field to the side by diffusion at a rate governed by τ_H. Due to the variation of the relative velocity across the channel they execute longitudinal steps $\sim \Delta x_H \sim (U/2)\tau_H$ upstream or downstream in the reference frame moving at a velocity $\bar{U}$. The dispersion coefficient:

$$D_{\parallel} \sim \Delta x_H^2/\tau_H \propto U^2 \tau_H \propto a^2 U^2/D_m$$

adds to the molecular term D_m in eq. (9).

More generally, we expect longitudinal dispersion terms of the form $D_{/\!/} \sim U^2\tau$ to appear due the interplay between convection along the flow and stagnation mechanisms associated with a time constant τ : trapping and diffusive liberation like in the Taylor problem or with adsorption.

The study of the relative importance of geometric and hydrodynamic term as a function of a Pe number is a delicate problem. If the mechanisms of mixing can take place simultaneously like on a fluid line along a tube we may expect that rate τ^{-1} involved in a diffusion equation is the sum of geometric τ_G^{-1} and hydrodynamic τ_H^{-1} rates and geometric dispersion should dominate for large Pe (fastest rate in U/ℓ). On the other hand, in the trapping in dead volumes, it is more sensible to add the geometric and hydrodynamic times and the latter term will dominate large flow (but still laminar) dispersion.

Heterogeneous Media

The form of equation (6) is not obvious and, in fact, is found not to apply well to heterogeneous geometries. In the simplest case of a packed bed of spheres with 2 different radii, Lemaitre[10] has found that the dispersion of a step function variation of tracer injected in the flow departed strongly from the expected sigmoid variation. This is likely to be due to the different classes of pore sizes and of the current distribution. We expect multiple scale and weakly connected media to lead to much stronger effects; dispersion experiments in a partially saturated medium will give us an example of this situation. However, it is likely that, at times long compared with the slow diffusion processes, normal dispersion laws would be recovered.

Multiple scale effects. Iterative Appolonian filling of grains (see section "Some examples of POM with a multiplicity of geometrical scales") is a good example of a statistically self similar geometry, ressembling a Sierpinski gasket. There are loops of all sizes (circuits around grains) but the scale of the elementary flow paths is given by the smallest grain size as that of individual resistors in a gasket. The situation would be quite different for a multiple scale but non iterative filling where voids and pore sizes of all scales can be obtained. Poorly connected, percolation like POM, are another example of self similar geometries at scales smaller than ξ.

- In the absence of flow, we can think of applying the anomalous diffusive ant walk (with $t \sim R^{d_w}$, $d_w > 2$; see Aharony article) to the present problem; the unit time step τ should be taken as a molecular diffusion time step along one short bond.

- If one applies a drift current, the walk is biased along the flow direction and we expect a cross-over from a diffusive behavior to the modes of dispersion discussed in the above section "Well connected materials" as Pe increases[44]. Mitescu and Roussenq[46] have considered the limit of small Pe on a hexagonal lattice such that the probability to exit at triangular node is proportional to the fraction of exit current at this node. Other selection rules can be guessed and should be compared with realistic flow separations in pore geometries (fluidics study). The time of passage in a given bond is taken to be independent of the local flow. On the other hand, for large Pe it is essential to consider the distribution of time steps proportional to the inverse of the local current.

- In the limit $Pe \to \infty$, the dispersion of dye in response to a step injection corresponds precisely to a distribution of first passage times[47] considered in graph problems, if one neglects the effect of channel width. In this limit, we have conjectured[2] that, in a fractal geometry, the dispersion $D_{//}$ should be an increasing function of the distance of spreading L and vary as $D_{//} \propto U(L)L$ because coherent flow patterns around structures L will dominate the convective behavior. The situation can be compared with the problem of turbulent pair diffusion which describes the rate of separation of couples of points separated by L in the inertial turbulent regime where a self similar and fractal eddy structure is obtained[48]. The most efficient eddies to disperse points are those of size $\sim$ L; the largest size eddies merely convect pairs in a coherent fashion. Thus the two particles correlation function gives a richer information for the turbulent structure than a one particle[49]. In a porous medium, the eulerian random velocity field is frozen in time but a lagrangian particle moving with it experiences (in an ergodic way[50]) the randomness of the field. It would certainly be of interest to study the rate of separation of pairs of ants in such a flow $< d < R_{ij}^2(t) > / dt >$ as a function of the distance $L = | \vec{r}_i - \vec{r}_j |$ between pairs of ants.

We also expect that, above the largest fractal scale, the dispersion will reach to a constant value equal twice that for a single particle and taking into account the effect of all the smallest size. This is also obtained above the largest scales of turbulent flow fields.

Dead ends and stagnant region effects. Additional geometrical features are obtained in poorly connected media. De Gennes[51] has considered the effect of dead arms on the geometrical dispersion in percolating systems near their threshold p_c and found that it should give a singular contribution

$$D_{//} = U^2 \tau_{\xi} \tag{10}$$

$\tau_{\xi} = \xi^2/D_a$ is the ant diffusion time over the correlation length (longest arms); U is an average of the flow velocity taken on all the percolating fluid phase (and not only the backbone). Equation (10) should scale as $D_{//} \propto \Delta p^{\mu-\beta-2\nu}$ with $\mu-\beta-2\nu$ negative in 2 and 3 dimensions : the divergence of the stagnation time in dead arms should more than compensate the weak permeability effect near than threshold. No such singular behavior is expected on $D_{\perp}$.

Sahimi et al[52] have made an extensive study of dispersion on the percolating backbone. He also found a large increase of the ratio $D_{//}/D_{\perp}$ near p_c. The increase of $D_{//}$ is probably due to the effect of nearly stagnant zones which are found in the blobs of the percolating cluster.

Let us now present a few experimental facts related to this description.

The experiment of Gaudet[53] shows a beautiful example of the effect of heterogeneity of fluid distribution on dispersion. In a porous medium fully saturated with water, dispersion behaves like described by eq. (6). However, for an unsaturated medium, the distribution becomes dramatically altered. The first exit time (measured from the ratio of the fluid which has flowed through the medium, v_i, from the time of injection normalized to the volume of liquid in the porous medium, v_e) becomes shorter than in a well connected medium because only a fraction of the pore liquid is mobile (on the backbone). The tail in the long time behavior is due to trapping in stagnant regions and dead arms.

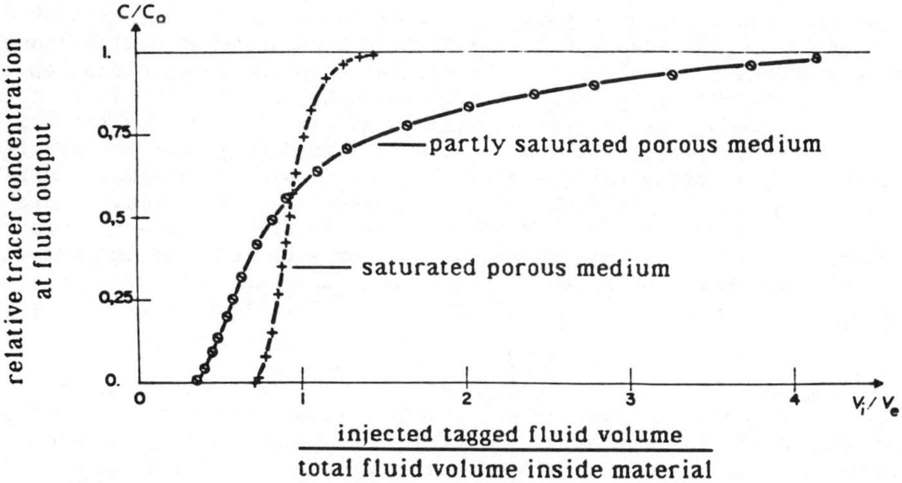

The curves show the variation of concentration c of an ionic marker at the outlet of a porous column. The marker injected at the entrance of a column as a step function (concentration c_o). The results are for a fully saturated porous column and an unsaturated one (relative saturation = 80 %).

We are presently studying the various contributions mentionned in the discussion using 2D micromodels of percolating and fractal channel lattices both by following the paths of colored tracers and from the conductivity response to the injection of saline solutions. We mention our first qualitative findings :

a) there is an extreme heterogeneity in the dye concentration of the channels and only a weak mixing at the nodes (a strong mixing is assumed in the theoretical study of geometrical dispersion). A small number of narrow paths controls the forward fornt of the dispersion curve;

b) long after the injection of a dye step function, the stagnant regions are those where the dye can still be found (the stagnant volumes where the dye was absent at short times are just those which are colored at long times).

Features a) and b) controlled the extreme parts of the unsaturated dispersion curve of Gaudet but the critical behavior of dispersion should be studied experimentally for much smaller residual saturations.

Conclusion

Throughout this chapter we have seen that the dispersion phenomena is extremely sensitive to the heterogeneities of the flow pattern. This is due to the local structure of the hydrodynamic flow but, more importantly, to the extreme heterogeneity of the non local distribution of current through heterogeneous media (the work of Archangelis and Redner[54] gives a Log normal distribution of current in a particular fractal lattice representative of a percolation structure) and the effect of the heterogeneity of the distribution of connecting elements considered in the second section ("Effect of local variations of permeability").

Another important aspect of the dispersion problem deals with extreme distribution of time scales from the shortest ones associated with convection along the shortest paths, τ_m, to the largest ones involving diffusion along the longest dead ends, τ_ξ. The R.E.V. defined as the volume needed to get a normal dispersion law such as eq. (6) is not merely given by the scale ξ as for permeability but should be expressed from the distance, spanned convectively by the fastest particles over a time τ_ξ, which is extremely large and is also a function of the Peclet number. This difficulty has precluded any numerical study up to now. Model experiments are not in a much better state and show anomalous non gaussian behavior even for long columns (length $\sim 10^4$ unit elements). We have to live with it and consider that, in such systems, the asymptotic central limit distribution often is not the physical limit !

On the other hand, hydrogeological systems have a larger range of scales and permit full size investigations provided there are no large scale variations of the mineral structure. We wish to stress that the anomalous features of dispersion in heterogeneous geometries provide a frame of description for two major features in hydrogeology[55] :

- The dispersion coefficient of pollutants is usually found to vary with the distance to source (nearly proportional to the time or distance to injection).

- Anomalously fast as well as slow release times are found. These features have been qualitatively associated with the multiplicity of geological scales and dead arms.

We participate to an extended program of study of permeability and dispersion on granitic sites (in Massif Central) for which the distribution of fractures are documented. It is an open question to see how much the approaches sketched in this review will be of use in a full scale experiment.

APPENDIX : residence time of a particle in a Stokes flow

We compute the residence time distribution E(t) of an idealized fluid particle between two planes $x = \pm \ell/2$ in an axis symmetric flow around a sphere of radius a in the Stokes approximation as a function of the distance y to the flow axis far from the sphere.

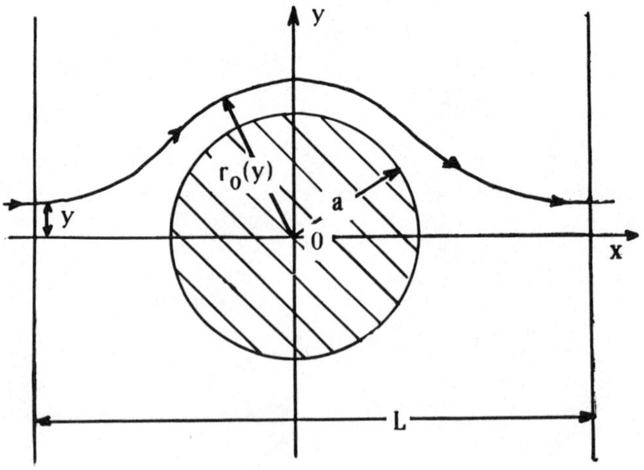

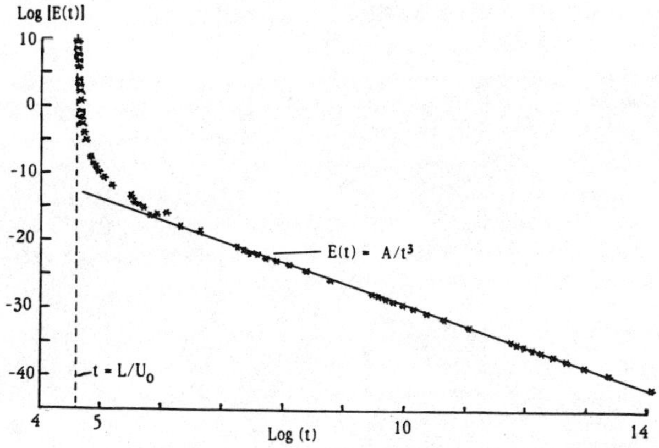

The numerical calculation gives a typical law which can be approximated by

$$t(y) = \frac{\ell}{u_o} \qquad \text{for} \qquad y > \frac{a^2}{\ell}$$

$$t(y) = \frac{a^2}{u_o}\frac{1}{y} \qquad \text{for} \qquad y < \frac{a^2}{\ell}$$

Assuming a uniform concentration of marked particles in the plane $x = -\ell/2$, we calculate the proportion of particles having a passage time $(t < t(y) < t+dt)$ between planes. We get :

$$E(t) = \left(\frac{2}{\alpha^2}\frac{a^2}{u_o^2}\frac{1}{t^3}\right) H\left(t - \frac{\ell}{u_o}\right) + \left[1 - \left(\frac{a}{\alpha\ell}\right)^2\right] \delta\left(t - \frac{\ell}{u_o}\right)$$

if y lies in the range 0 to αa (α should be of the order of ℓ/a). The mean residence time is finite :

$$\bar{t} = \int_0^\infty t\, E(t)dt$$

$$\bar{t} = \frac{2}{\alpha^2}\frac{a^2}{u_o\ell} + \left[1 - \left(\frac{a}{\alpha\ell}\right)^2\right]\frac{\ell}{u_o} \qquad \bar{t} \sim \frac{\ell}{u_o}$$

However the second moment of the distribution diverges due to the dependence of E(t) in t^{-3} for large t. The same behavior of E(t) was obtained by Saffmann in the problem of Poiseuille flow due to the wall slow flow. Here the slow region is that of the vicinity of the stagnation points and should not depend crucially on the detailed geometry of the obstacles[41].

There is a higher time cut off t_c due to molecular diffusion across the flow which can be expressed by equating a convection time and a diffusion time over a distance of order y (the closest approach distance of a flow line to the sphere is proportional to y) y^2/D_m. Thus

$$t_c = \frac{a}{u_o} Pe^{1/3}$$

where the Peclet number $Pe = aU_o/D_m$.

The second moment of the residence time can now be calculated

$$\overline{t^2} - \bar{t}^2 = \frac{1}{\alpha^2}\frac{a^2}{U_o^2}\left[2 \text{ Log}\left(\frac{a}{\ell} Pe^{1/3}\right) + 1\right]$$

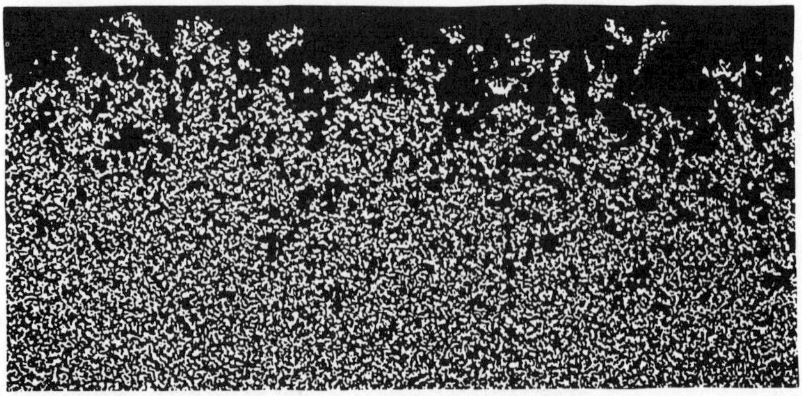

(a) Vertical cut of an injection front (discussed in page 12)

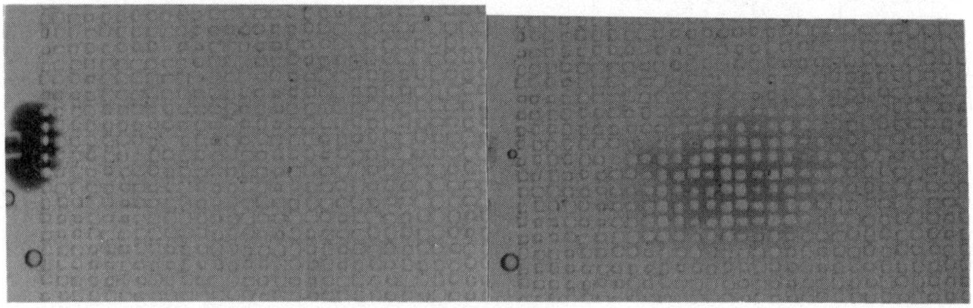

(b) Dispersion in a saturated micromodel : a drop of dye is injected in a stationary flow of water (from the left to the right). The dye spreads with an ellipsoïdal shape.

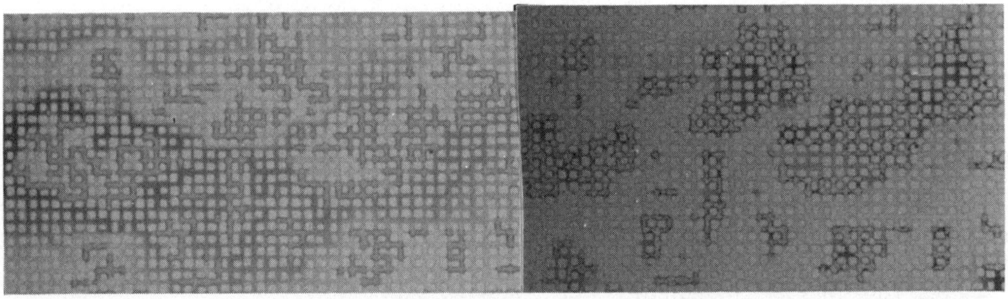

Dispersion in an unsaturated medium : The isolated clusters are air bubbles. In the first stage of its arrival (photograph to the left) the dye (in dark) vizualizes the high velocity channels an avoids dead ends and nearly stagnant regions. Long after the end of its injection (photo graph to the right) some dye remains trapped in the previously avoided regions.

Plate 1

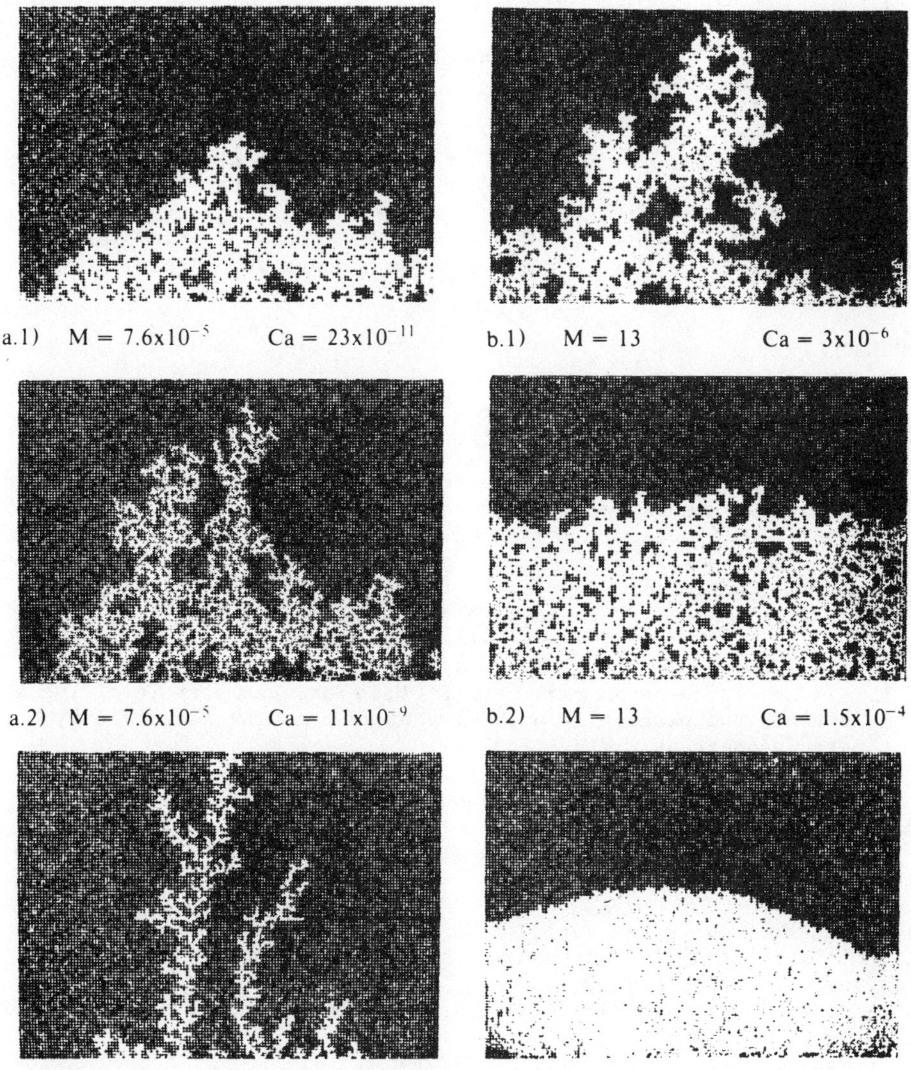

Plate 2 : (courtesy of R. Lenormand). Injection in a 2D micromodel of a liquid 1 (light) pushing a liquid 2 (dark) for various values of the viscosity ratio $M = \eta_1/\eta_2$ and capillary number Ca.

which imposes that the mixing time $aPe^{1/3}/U_o$ be long compared to the convection time ℓ/u_o.

If we consider now a random distribution of such spheres (similar to the fixed bed in Koch and Brady's work[56]) the long time behavior of the dispersion is due to the exploration by the fluid flow of a number of stagnant zones between which the particles are recirculated in the main flowfield. This will lead, following Saffman's calculation, to a relation similar to eq. (8).

ACKNOWLEDGMENTS

We have had many discussions on the content of this paper with J. Koplik, P.M. Sen, D. Wilkinson. This paper has benefited from discussion and communications of unpublished results with many colleagues, in particular B. Halperin, S. Feng and C. Jacquin, R. Lenormand, Y. Pomeau, J. Roussenq, S. Stauffer and J. Vannimenus. A support from Schlumberger-Doll-Research is acknowledged.

REFERENCES

1. F. A. L. Dullien, Porous Media, Fluid Transport and Pore Structure, Acad. Press, New-York (1979).
2. The present review extends and complements the article by E. Guyon, P. P. Hulin, R. Lenormand (in french) in Ann. des Mines, special issue "Ecoulements dans les Milieux Fissurés", 191, (5.6) p. 17 (1984) where a larger number of references can be found.
3. Physics and Chemistry of Porous Media, D.L. Johnson and P.M. Sen edit. A.I.P. Conference Proceed. n° 107, AIP, New-York (1984).
4. For example, see J.F. Joanny, Thèse Paris, p. 135 (1985).
5. H. Darcy, Les Fontaines Publiques de la Ville de Dijon, Dalmont, Paris (1856).
6. L. Sander, in these proceedings. The Hele Shaw geometry involving a flow between two parallel plates distant of b can be described by eq. (1) with a constant $\kappa = b^2/12$.
7. L. A. Santaló, Integral Geometry and Geometric Probability, Encyclopedia of Mathematics, vol. I, Addison Wesley Publishing Cie, Reading Massachusetts.
8. P. Z. Wong, J. Koplik and J. P. Tomanic, to appear in Phys. Rev. B.
9. D. L. Johnson, Appl. Phys. Lett. 37:1065 (1980).
10. R. Lemaitre, thèse Université Rennes (1985).
11. R. Omnes, J. de Phys. 46:139 (1985).
12. A.J.. Katz et A. H. Thompson, Phys. Rev. Lett. 54, 1325 (1985).
13. R. Orbach, these lectures.
14. P. G. de Gennes, Partial Filling of Fractal Structure by Wetting Fluid, to be published.
15. T. Madden, Geophysics, 31:1104 (1976).
16. C.J. Allègre,, J. L. Le Mouel, A. Provost, Nature 297, 5861, 47-49 (1982
17. S. R. Broadbent and J.M. Hammersley, Proc. Camb. Phil. Soc. 53:629 (1957
18. I. Bernabe, W. C. Brace, Mech. of Materials 1:173 (1982).
19. P. Z. Wong, J. Koplik and J. P. Tomanic, to appear in Phys. Rev. B.
20. G. de Marsily, Hydrogeologie Quantitative, Masson, Paris (1981).
21. E. Charlaix, E. Guyon, N. Rivier, Sol. State Com. 50, 11 : 999 (1984). The article also shows that it is possible to derive these parameters form random plane cuts.
22. G. E. Pike and C. H. Seayer, Phys. Rev. B 10:1421 (1974).
23. The same dimensionless quantity describes within a numerical factor the onset of nematic ordering of percolation of arrays of rods (or disks) and that of ordering of calamitic (or discotic) nematics. Indeed both

phases can coexist in molecular systems. See L. Onsager, Ann. N.Y. Acad. Sci. 51:627 (1949).
24. P. C. Robinson, J. Phys. A, 16:605 (1983).
25. I. Balberg, C. H. Anderson, S. Alexander and N. Wagner, Phys. Rev. B 30: 3933 (1984).
26. S. Wilke, E. Guyon and G. de Marsily, Math. Geol. 17:17 (1985).
27. B. Halperin, S. Feng, P.N. Sen, to be published.
28. A. Rouleau, PH.D Thesis, Waterloo, Ontario (1984).
J. C. S. Long, PH.D Thesis, University of California (1983).
29. B. I. Shklovskii, A. L. Efros, Electronic Properties of Doped Semi-Conductor, Springer Verlag, Berlin (1984).
30. J. Bear, Dynamics of Flow in Porous Media, Chapter 9, American Elsevier, New-York (1972).
31. Groupe Poreux P.C., Two components Properties in Heterogeneous Porous Media, Proc. "Physics offinely divided matter", Les Houches, Ed. by Daoud (Springer) (1985).
32. Ch. G. Jacquin, and P. M. Adler, to be published in S. Coll. Int. Sci. (1985).
33. P. G. de Gennes and E. Guyon, J. Meca. 17:403 (1978).
I. Chatzis and F. A. L. Dullien, Journal of Canadian Petroleum Technology, January-March (1977).
R. G. Larson, L. E. Scriven and H. T. Davis, Chem. Eng. Sci. 36:57, Pergamon Press Ldt, Great Britain (1980).
34. R. Lenormand and S. Bories, C.R. Acad. Sc. Paris, 291 B:279 (1980).
35. R. Lenormand and C. Zarcone, to be submitted to Phys. Rev. Lett. (1985).
36. D. Wilkinson, Phys. Rev. A 30:520 (1984).
37. R. Lenormand and C. Zarcone, J. Phys. Chem. Hydr., January (1985).
38. L. Paterson, Phys. Rev. Lett. 52-18:1621 (1984).
39. P. G. Saffman and G. I. Taylor, Proc. R. Soc. Lond. A 245-311 (1985).
S. B. Gorell and G. M. Homsy, S.I.AM, J. Appl. Math. 43-1:79 (1983).
A general condition for marginal stability of the S.T. instability including permeability as well as density difference effect is
$(\mu_2/\kappa_2 - \mu_1/\kappa_1)\ U + (\rho_2 - \rho_1)\ g \cos\theta = 0$
where the subscripts 1,2 refer to the displacing and displaced fluid.
40. J. Bear, Dynamics of Fluids in Porous Media, chap. 10, American Elsevier, N.Y.
We have benefited from several discussions with J. Koplik on chapter 4.1.
41. G. I. Taylor in Low Reynolds Number Flows in Illustrated Experiments in Fluid Mechanics, MIT Press (1982) and film of the N.C.F.M. Films.
42. We thank J. Feder, U. Oxaal and their group for communication of their unpublished data.
43. P. G. Saffman, J. Fl. Mech. 6:321 (1959).
44. R. Aris, Proc. Roy. Soc. A. 235:67 (1956).
The effect of velocity field is quite different from that of an externally applied field considered by Dhar and Barma for a biased ant walk. In this case the local probabilities are determined by the orientation of bonds with respect to the field (like in trickled bed flows in a gravitational field[45]). In the present problem, there is a continuum of particles which respond the local field (the local velocity field) which can be obtained only from a knowledge of the connectivity properties.
45. M. Crine, P. Marchot and G. L'Homme, Chem. Eng. Comm. 7:377 (1980).
46. C. D. Catalin and J. Roussenq, work in progress.
47. J. M. Hammersley and D.J.A. Welsh, Cont. Phys. 21:593 (1980).
48. A recent discussion of the so called "Richardson Pair Diffusion" can be found in S. Grossmann and I. Prococcia, Phys. Rev. A 29:1358 (1984).
49. This correlation function can be measured by forced Rayleigh Scattering M. Cloitre and E. Gyron, to appear in Jour. Fl. Mech.
50. G. Matheron, umpublished.

51. P. G. de Gennes, J. Fl. Mech. 136:189 (1983).
52. M. Sahimi, H. T. Davis and L. E. Scriven, Chem. Eng. Com. 23:329 (1983).
53. J. P. Gaudet, Thèse Grenoble (1978).
54. L. de Arcangelis, S. Redner and A. Coniglio, preprint.
55. A. Dieulin, G. Matheron, G. de Marsily, The Science of Total Environment 21:319 (1981).
56. D. L. Koch, J. P. Brady, to be published.

THE RANDOM FIELD ISING MODEL

Jacques Villain

Institut für Festkörperforschung
Kernforschungsanlage Jülich
5170 Jülich

Abstract

Ising or Ising-like models in random fields are good representations of a large number of impure materials. A domain argument shows that the normal order is not destroyed by weak random fields in more than two dimensions. Domain wall roughening due to random fields produces metastability and hysteresis. Various theories predict a domain size proportional to log t for large times t, and this is confirmed by recent experiments. Metastability is a serious problem both for experiment and for theory! According to a phenomenological theory, and to experiments done on zero-field cooled samples, the critical behaviour is the same in three-dimensions as for t two dimensional systems in zero field.

1. INTRODUCTION

Systems with frozen disorder are interesting for at least two reason: firstly they are frequently encountered in everyday life; secondly they are a source of interesting and fondamental problems in statistical mechanics.

A particular example is the Random Field Ising Model /1,2/. What we have in mind is not necessarily a true Ising model, but any system with two degenerate ground states, which we call + and -. Random fields are generated by random, quenched impurities, when they locally break the symmetry between these two states. Generally, impurities do break symmetry, so that the random field Ising model has a very wide range of applicability in Solid State Physics. A few examples are given below.

i) Commensurate phases. /3/ The simplest cases are probably found in adsorbed layers, e. g. Xenon monolayers on the (110) face of Cu /4/ (Fig. 1). There are two possible sublattices, which may of course be called "+" and "-". An impurity on one sublattice breaks the symmetry between both sublattices, and therefore is equivalent to a random field, the probability of which at any lattice site i is for an impurity concentration c

$$P(H_i) = (1 - c)\ \delta\ (H_i) + (c/2)\ \delta\ (H_i + H_o) + (c/2)\ \delta\ (H_i - H_o) \quad (1.1)$$

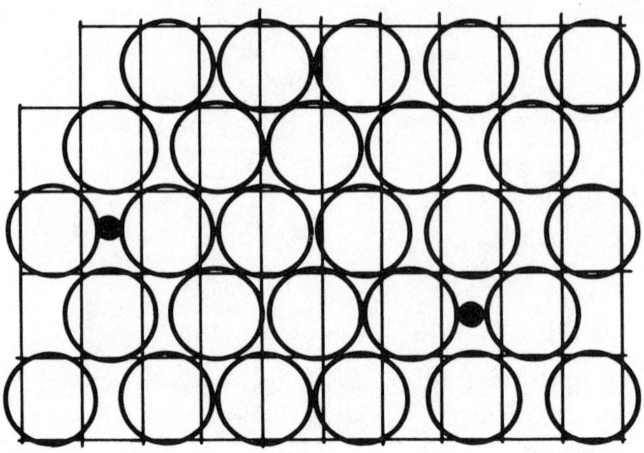

Fig. 1. - A xenon monolayer absorbed by the (110) face of copper has a commensurate phase with two possible sublattices. Big circles represent xenon atoms. Copper surface atoms sit at the vertices of the rectangular lattice. Frozen impurities (small black circles) sitting at easy adsorption sites act as quenched random fields and may result in the formation of domain walls -

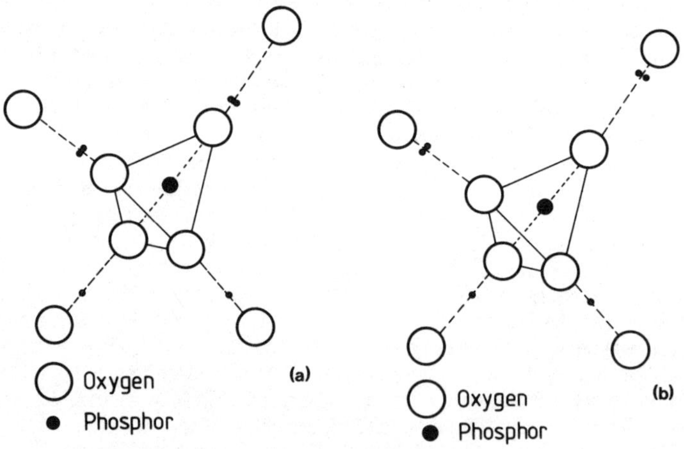

Fig. 2. The ferroelectric material K H_2PO_4 is made of PO_4 tetrahedra bound together by Hydrogen bonds. The replacement of a proton (●) by a deuteron (●●) alters in a significant way the properties of a hydrogen bond. However, a single deuteron does not break the symmetry between the two ground states (a) and (b). Two deuterons are necessary to create a random field.

so that the square average

$$\overline{H^2 = H_i^{\ 2}} \qquad (1.2)$$

is proportional to the concentration

$$H^2 = c\, H_o^{\ 2} \qquad (1.3)$$

H_o is a microscopic constant which depends on the nature of the impurity.

ii) Ferroelectrics. In principle, the natural isotopic disorder is sufficient to produce random fields in a hydrogen bonded ferroelectric like PO_4H_2K (Fig. 2). However, the replacement of a single proton by a deuteron is not enough to destroy the symmetry between the two ground states, and two deuterons are necessary. This is a fairly general feature in materials which undergo a structural phase transition: at least two impurities are necessary to create a local field. Nevertheless, impurity effects are often spectacular near the transition, and random fields are believed /5/ to be responsible for the "central peak" (Fig. 3) observed in neutron scattering experiments /6,7/.

iii) Magnetic systems are exceptional. In zero external field, impurities cannot break the symmetry between + and - because the system is invariant under time reversal. Therefore, random fields can only exist if an external field is applied (Fig. 4). One cannot use a ferromagnet since the external field would destroy the global symmetry and the phase transition. But uniaxial antiferromagnets with non-magnetic impurities in an external magnetic field are very appropriate for a systematic study of the RFIM /8,9/. The advantage is that the random field amplitude, defined by (1.2), can be contiuously varied just by changing the applied field H_o. At low temperature, H is given by (1.3). All recent experimental investigations have been performed on such systems /10, 11, 12, 13/.

Recent reviews of the random field Ising model have been given by Imry /2/ and Toledano /14/ (who also reviews other disordered models) and the 1983 MMM Conference proceedings contain several papers /15, 16, 17/ devoted to this subject.

2. THE LOWER CRITICAL DIMENSION D_c OF THE RANDOM FIELD ISING MODEL

For definiteness, we first consider a "stricto sensu" Ising model with interactions between nearest neighbours, on a D-dimensional cubic lattice. The Hamiltonian is

$$\mathcal{H} = \mathcal{H}_o + \mathcal{H}_1 = -J \sum_{<ij>} S_i\, S_j - \sum_i H_i\, S_i \qquad (2.1)$$

(i,j = lattice sites, $S_i = \pm 1$, <ij> = pair of neighbours). J will be assumed to be positive (ferromagnetic). The case of an antiferromagnet with two sublattices $\mathcal{L}_1$, $\mathcal{L}_2$ can be transformed into a ferromagnetic problem by means of the transformation

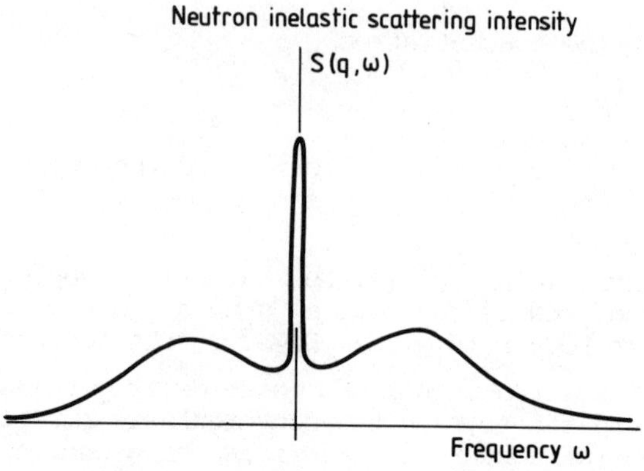

Fig. 3. The "central peak" observed in neutron scattering experiments on $SrTiO_3$ and other materials was the Great Event of the first Geilo Spring Schools /6,7/. Probably a random field effect./5/

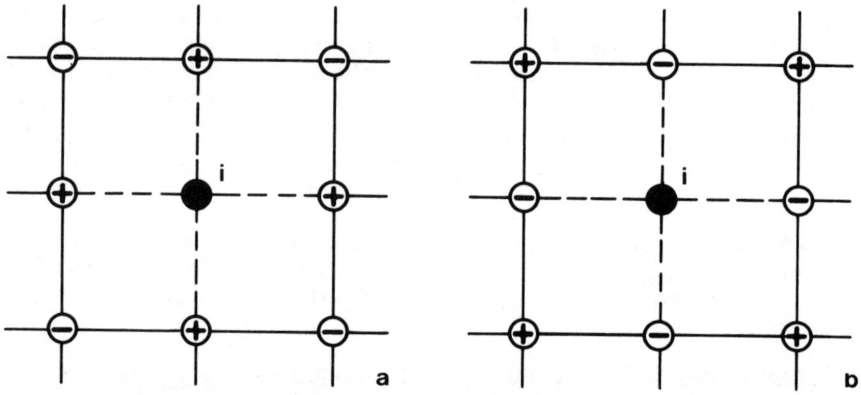

Fig. 4. Effect of non-magnetic impurity (black circle) substituted in an antiferromagnet. Assuming a positive external field H_o, the energy of state (a) is lower then that of state (b) by an amount 2 H_o. Thus, random impurities are equivalent to a random field H defined by (1.3).

$$S_i = \sigma_i \quad (i \in \mathcal{L}_1); \quad S_i = -\sigma_i \quad (i \in \mathcal{L}_2) \tag{2.2}$$

The RFIM is dominated by a conflict between the two terms of (2,1): $\mathcal{H}_0$ favours ferromagnetism at low temperature while $\mathcal{H}_1$ favours disorder. When the random field amplitude H (defined by 1.2) is sufficiently larger than J, the system is expected to be disordered at low temperature, and this guess is confirmed by an exact treatment /18/. Thus, the interesting case corresponds to $H < J$.

If $H \ll J$, the system is expected to order ferromagnetically at low temperature, provided the space dimension D is larger than a lower critical value D_{cl} or D_c.

The standard method to determine a lower critical dimension is to discuss the stability of the ferromagnetic state with respect to the formation of ill-oriented domains. Landau /12/ applied this method to the ordinary Ising model ($H_i = 0$ in 2.1) with short range interactions. If one reverses the spins (Fig. 5) within a domain of size R, the energy cost W_{ex} is proportional to the domain wall area

$$W_{ex} \approx J\, R^{D-1} \tag{2.3}$$

For $D > 1$, the probability of formation of a large domain, proportional to $\exp(-\beta W_{ex})$, is negligible at low temperature $T = 1/K_B\beta$. However, for $D < 1$, domains of very large size are possible and ferromagnetism is unstable for $T \neq 0$. Thus, for $H = 0$,

$$D_c = 1 \tag{2.4}$$

If now random field are introduced, they favour domain formation /1/. Indeed, if one considers the ferromagnetic state, e. g. $S_i = 1$ (Fig. 5), the mean square random field energy W_{RF}^2 inside a region of volume R^D is $HR^{D/2}$ according to the central limit theorem. The energy W_{RF} for a given region may be positive or negative with equal probability 1/2, but it always possible to find a region enclosing any arbitrary point i, such that $W_{RF} > 0$ and therefore it is favourable to reverse the spins inside this region. Adding (2.3), the energy associated to spin reversal in a domain of linear size R is seen to be of order

$$W(R) \approx J\, R^{D-1} - HR^{D/2} \tag{2.5}$$

For $H \ll J$, this energy is positive for $D > 2$, but negative for $D < 2$ if R is large enough. Thus, the lower critical dimension of the RFIM is

$$D_c = 2 \tag{2.6}$$

This result has been confirmed by careful calculations by Chalker /20/ Fisher et al /21/ and Imbrie /22/. Unfortunately, those calculations are performed at zero temperature, but thermal corrections to both therms of (2.5) are probably small at low temperature (in

Fig. 5.
The Imry-Ma argument for the lower critical dimension of the RFIM /1/. Assume a ferromagnetic ground state with positive (for example!) magnetisation. There are regions where negative random fields (± signs on the Fig.) dominate. Is it favourable to reverse the spins in a such a region? This is the question discussed in Section 2.

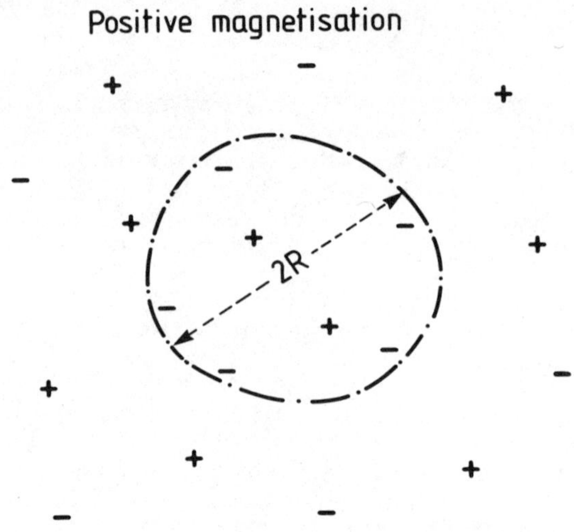

Fig. 6. - If two neighbouring domains of positive (above) and negative (below) magnetisation contain random fields, the corresponding wall has bumps. The dashed curve represents a metastable wall position, and the full curve is the stable position. The Figure applies to the stricto sensu Ising model at low temperature.

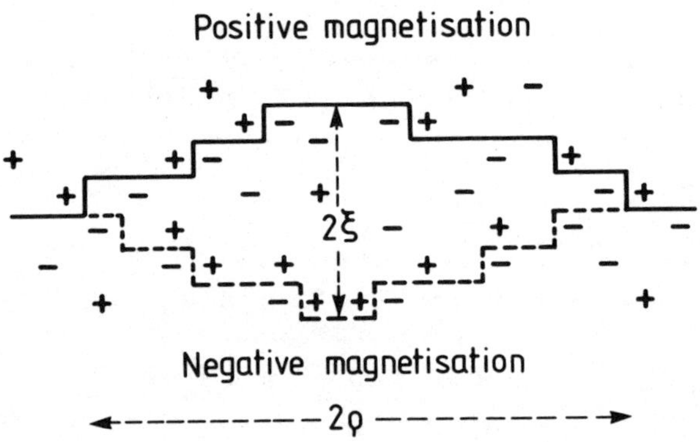

Fig. 7.
Same situation as in Fig. 6, but with continuous domain walls (broader than the interatomic distance).

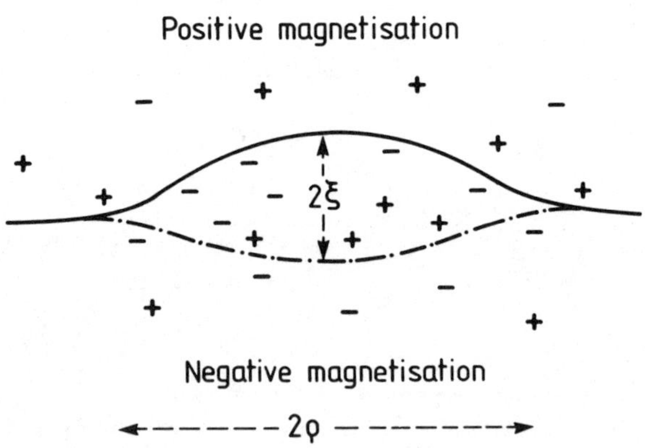

this case, W is a free energy) /3, 23, 24/. Before these calculations were performed, it was suggested /25, 26/ that formula (2.5) might be wrong because domain walls become rough under the effect of random fields. Then, both therms of (2.5) may be underevaluated, and a more careful analysis is necessary. This is the purpose of the next Section.

3. DOMAIN WALL ROUGHENING /3, 23, 24, 27, 28/

We start with the assumption (to be checked at the end of the calculation) that domain walls, if they exist, are smooth, so that they are made of nearly flat parts. This assumption seems to exclude roughenning but, as seen later, this is not completely true. Walls are allowed to make bumps in order to minimize their energy. There are bumps of any radius ρ, but the height ξ of a bump of radius ρ has a well-defined order of magnitude as will be seen now. The energy gain δW_{RF} associated with such a bump, and produced by the random field, is, as in (2.5), proportional to the square root of the bumps.

$$\delta W_{RF} \approx H(\rho^{D-1}\xi)^{1/2} \tag{3.1}$$

This formula is correct for sufficiently large ρ and ξ . The restoring energy due to the increased domain wall area depends on the system. For the "stricto sensu" Ising model (2.1) at low temperature, wall roughening implies the formation of steps (Fig. 6) and the energy increase δW_{ex} is proportional to the total area of these steps /27/.

$$\delta W_{ex} \approx J\rho^{D-2}\xi \tag{3.2a}$$

There are other situations in which domain walls (Fig. 7) are smooth surfaces which can be described by a surface tension g. Then, formula (3.2a) should be replaced by

$$\delta W_{ex} \approx g\rho^{D-1}(\xi/\rho)^2 = g\rho^{D-3}\xi^2 \tag{3.2b}$$

This formula is applicable when the wall thickness is large with respect to the interatomic distance, as it occurs near the transition. It is also applicable to weakly anisotropic Heisenberg models, which often provide a better description of real systems than (2.1). Formula (3.2b) is also appropriate for two dimensional systems. A detailed discussion will be found in Appendix A.

The bump height ξ minimizes the sum of (3.1) and (3.2). The minimum of this sum is the bump energy $\delta W(\rho)$. One finds for narrow walls, using (3.1) and (3.2a):

$$\xi(\rho) \simeq \rho^{3-D}(H/J)^2 \tag{3.3a}$$

$$\delta W(\rho) \simeq -\rho H^2/J \tag{3.4a}$$

In the case of broad walls, formula (3.1) and (3.2b) yield

$$\xi(\rho) \simeq \rho^{(5-D)/3}(H/g)^{2/3} \tag{3.3b}$$

$$\delta W(\rho) \simeq -\rho H\ (H/g)^{1/3} \tag{3.4b}$$

The corresponding energy per unit area $\delta g = \delta W/\rho^{D-1}$ is respectively

$$\delta g(\rho) \simeq -\rho^{\ 2-D}\ H^2/J \tag{3.5a}$$

for narrow walls, and

$$\delta g(\rho) \simeq -H\rho^{\ 2(2-D)/3}(H/g)^{1/3} \tag{3.5b}$$

for broad domain walls.

These expressions are small for small H for D > 2, and constitute therefore a weak correction to (3.2) and also a weak correction to the first term of (2.5). From the present discussion we conclude that the argument of Section 2 is right and that relation (2.6) is correct. Note that relation (3.3) implies $\xi \ll \rho$ for D > 2 and H << J, in agreement with the assumption of nearly flat domain walls, made at the beginning of this Section.

Although the result $D_c = 2$ is supported both by this simple discussion and by recent elaborate treatments /20, 21, 22/, there are other methods /25, 26/ which yield $D_c = 3$, in agreement with the ε-expansion /29, 30/. We believe that these methods are wrong. The ε-expansion is discussed in Appendix B. The replica-trick, which, together with additional approximations /25/, leads to $D_c = 3$, may be a good method in principle, but seems very sensitive to these additional approximations. The reason why Mukamel et al /25/ obtained $D_c = 3$ is probably that "corrections to scaling" are in this case essential /28, 23/. Note that it is extremely difficult to deduce the non-analytic results (3.3b), (3.4b) and (3.5b) from the replica trick, since everything is analytic in H for an integer number of replicas.

The marginal case $D = D_c = 2$ is studied in Section 5. It is argued that order is impossible.

Coming back to (3.3), ξ is found to be infinite for infinite below 3 dimensions for narrow domain walls /27/, and below 5 dimensions for broad walls /25, 3, 23, 24/. In other words, walls are rough, but not enough to destroy order above 2 dimensions. The characteristic dimensions are summarized in Table 1. In this table, the last line refers to a model which is <u>not</u> a random field model, and has been studied by Rice et al /31/, Derrida et al /32/, Feigelmann and Lyuksyutov /33/ and Nattermann /34/. The values of D and D_2 are taken from /34/.

Between D_1 and D_2, interface roughening does not destroy order, but has an important consequence which is metastability. It will be studied in the next Section.

4. METASTABILITY

a) General Idea

As seen in the previous Section, domain walls make bumps in order

Table 1

Characteristic dimensions D_1, D_2, for interfaces (walls) between domains

Dimension	D_1		D_2	
No randomness, $T \neq 0$		Rough interface, but positive surface tension	3	Flat interface
Weak random field, Stepped interface (narrow wall)	2		3	
Weak random field, Continuous interface (broader than the atomic distance)	2		5	
No random field, but continuous interface pinned by impurities	1		5	

to take advantage of the random field. It may occur and it does generally occur that, in a given region of space, two (or more) different bump shapes (Fig. 6 and 7) correspond to relative minima of the energy, so that any continuous deformation of the domain wall from one shape to the other implies passing over a potential energy barrier.

Metastable states and potential barriers are general effects of frozen disorder and already appear for a system of one or two spins (Appendix B). They are most easily analysed in a one-dimensional model and some exact results have been recently obtained in this context by Sinai /35/ and Pomeau and Derrida /36/, but of course in those cases, domain walls are points and cannot deform.

It should also be recalled that hysteresis effects in ferromagnets are attributed to domain wall pinning by impurities, and were studied in detail a long time ago by Néel /33/. The forthcoming analysis is strongly influenced by Néel's work.

b) Continuous Walls

We shall first consider the continuous domain wall picture of Fig. 7, which, as explained in the previous section, holds for broad walls. Again, one can consider a bump with a given basis of radius ρ , and an argument similar to the previous Section shows that the associated maximum barrier height ω is given by (3.4b).

$$\omega = \rho^{(D+1)/3}\, H(H/g)^{1/3} \tag{4.1}$$

The associated bump height ξ is given by (3.3b) and this implies a radius of curvature

$$R \approx \rho^2/\xi \approx \rho^{(D+1)/3}(g/H)^{2/3}$$

or, using (4.1)

$$R \approx \omega\, g/H^2 \qquad (4.2)$$

The domain may have a larger size, but not a smaller one. A bump height smaller than ρ^2/R would not be an acceptable result. This means that, for a given domain size R, only small enough bumps contribute to metastability. We believe it would be a mistake to indentify with the domain size.

Now we follow Néel /37/ or Préjean and Souletie /38/. If the system contains small domains at time t = 0, these domains grow with time, and this implies jumping over potential barriers. After a time t, potential barriers can be jumped if their height ω satisfies the condition

$$(t/\tau)\exp(-\omega/T) \gtrsim 1 \qquad (4.3)$$

This is the Arrhenius law, and is a microscopic time, presumably of order n/J. Inserting the value $\omega \approx H^2R/g$ derived from (4.2), one sees that the minimum domain size after a time t is /39/

$$R_c \approx (gT/H^2)\ln(t/\tau) \qquad (4.4)$$

From the discussion of Section 3, this formula is expected to hold for $2 < D < 5$, and in particular in 3 dimensions.

c) Stepped Walls /40/

Now, what about the narrow, stepped walls of Fig. 6 ? It seems sufficient to use (3.3a) and (3.4a) instead of (3.3b) and (3.4b), so that instead (4.1) one obtains

$$\omega \approx \rho H^2/J \qquad (4.5)$$

Here there is no reason to distinguish ρ from R, so that one recovers relation (4.2), except that g is replaced by J. The same argument as before leads to (4.4) a result obtained in a somewhat more complicated way by Grinstein and Fernandez /40/. The reason of this complication is that the above argument only applies between the critical dimensions. In 2 dimensions, (4.4) does not apply because R_c does not diverge at infinite time, but (4.3) does apply. In 3 dimensions, domains are expected to decay by peeling (Fig. 8). Peeling is a two-dimensional process for which the typical energy barriers are given by (4.3), and then (4.4) may be deduced as before. Since $R^{D-2} = R$ sheets should be peeled, the lifetime of a domain of size R is multiplied by R, but this factor is negligible in comparison with the Arrhenius exponential.

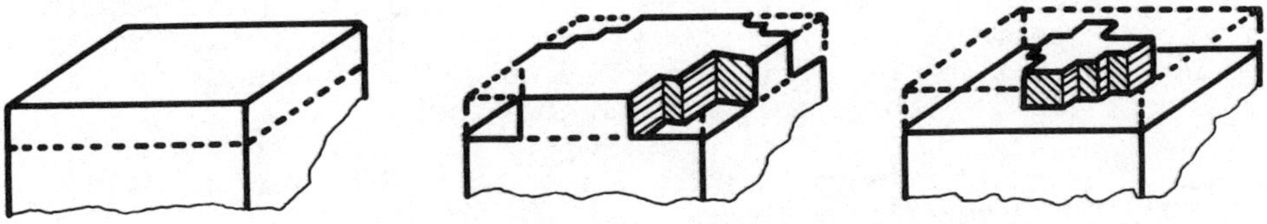

Fig. 8. "Peeling" of a domain, assumed to be cubic for simplicity.

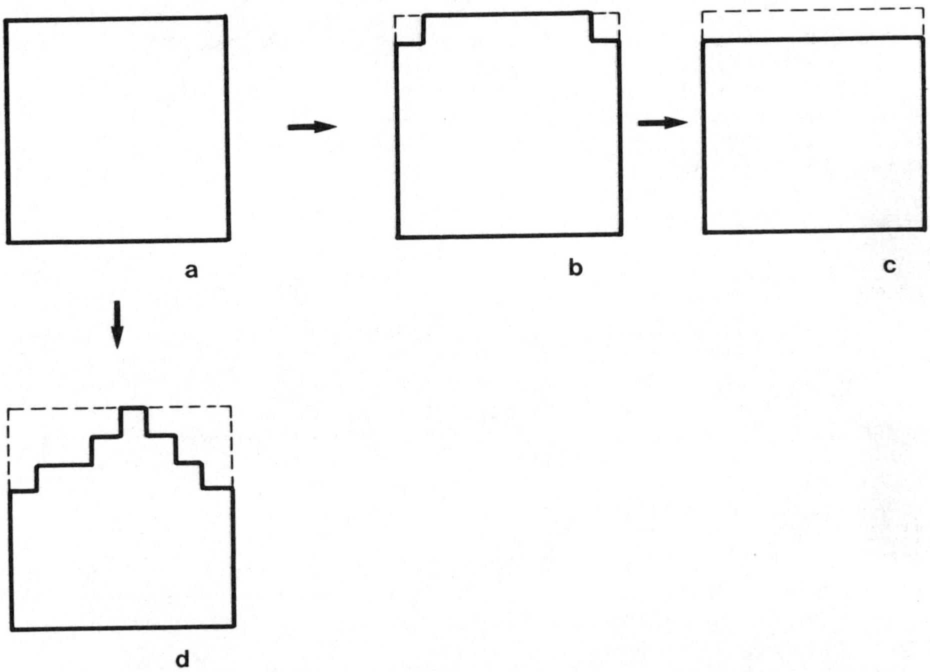

Fig. 9. Decay of a two-dimensional domain, assumed, for definiteness to be a square (a). (b) and (c) show the decay of a small domain which does not satisfy (4.7). Picture (d) applies when (4.7) is satisfied.

d) Short times /40/

Thus, formula (4.4) is expected to hold in 3 dimensions for true Ising models at low temperature as well as for Ising-like models which may be more realistic (see Appendix A). However, in the former case, formula (4.4) applies for very long times only. The reason is that it relies on (3.3a), which is only correct for $\xi >> 1$

(Fig. 9). Inserting this value into (3.1), the typical random field energy for D = 2 turns out to be

$$\delta W_{RF} \approx H\sqrt{\rho} \qquad (4.6)$$

On the other hand the typical exchange energy (3.2a) is $\delta W_{ex} \approx J$, which is larger than $H\sqrt{\rho}$, because if

$$J < H\sqrt{\rho} \qquad (4.7)$$

then (3.3a) is larger than 1 and (3.1) and (3.2a) apply. Then, one might believe that the energy barrier is $\approx J$, the energy necessary to create a kink on a domain wall. This evaluation would be wrong because, in a finite sample, kinks may be formed from the boundary (Fig. 9) without any exchange energy cost. Therefore, the energy barrier height ω is given by (4.6). Insertion into (4.3) yields the typical radius $R_c = \rho$ after a time t /40/

$$R_c \approx (T/H)^2 \ln^2(t/\tau) \qquad (4.8)$$

This value crosses over to (4.4) when (7.5) is satisfied, i.e.

$$\rho = R_c \approx J^2/H^2 \quad \text{or} \quad t \approx \tau \ \exp(4J/T) \qquad (4.9)$$

Instead of 4J, Grinstein and Fernandez found 8J, probably because with their convention, each pair appears twice in (2.1). With our convention, 4J is the energy necessary to create a pair of kinks on a two-dimensional domain wall (Fig. 9).

e) A challenge to simulators !

So far, the long time behaviour (4.4) was not checked by computer simulations. We attribute this failure to the fact that the sample should be large and the time long. probably beyond the possibility of present computers.

"Irreversibility" /41/, "hysteresis" /41, 42/, " domain formation" /42/, "extremely slow" relaxation /43/ have been observed in computer simulations. The formula (4.8) is verified /44/ or at least consistent with the numerical results /45/.

f) Are these considerations relevant ?

In this Section, the decay of a domain structure was investigated. For the experimentalist, the essential question is whether the domain structure is formed when cooling slowly from the paramagnetic region in constant random field. Theoretically, there is no strong reason for this to happen. Indeed, in the region where the phase transition is expected to occur, the domain size R has the same order of magnitude as the domain thickness λ/46, 39/. Therefore, the interaction between the walls of neighbour domains (which has range λ/39/) is appreciable, and one may expect a colective shrinkage of the domains of a given size at a given transition temperature. Experimentally, this seems to occur in sufficiently weak random field /11/, and metastability becomes a problem only in higher field /10/. Our claim of Ref. 39 (a) was too strong.

g) Other point of views

Bruinsma and Aeppli /47/ have proposed a theory analogous to that of ref. /39/ and /40/ but without kinetics. In other words the authors forbid any jump over potential barriers. In addition, the bump size they obtain is much larger than (3.3). The reason for that may be understood from the potential seen by a bump. Assuming this concept to be meaningful, it may reasonably be written /47/ as

$$\mathcal{H}(\xi) = g\,\rho^{D-3}\,\xi^2 + H\,\rho^{(D-1)/2}\sum_{n=0}^{\xi}\eta(n) \qquad (4.10)$$

(Fig. 10) where the first term is (3.2b), and the $\eta(n)$'s are independent random variables which represents the average random field in the n'th atomic layer of the bump. The quadratic average of the second term of (4.10) is (3.1). As argued above (§ 4. b) the highest energy barriers are given by (4.1) and the corresponding ξ is given by (3.3b). We now look for the largest at which one can find energy barriers. If the barrier goes from ξ to ($\xi+\delta\xi$), its height is of order $H\,\rho^{(D-1)/2}\sqrt{\delta\xi}$. It is not erased by the first term of (4.10) if

$$H\,\rho^{(D-1)/2}\sqrt{\delta\xi} > \delta\xi\; d(g\rho^{D-3}\,\xi^2)/d\,\xi$$

or

$$(H/g)\;\rho^{(5-D)/2} > \xi\sqrt{\delta\xi}$$

Since $\delta\xi \gtrsim 1$, the maximum value of ξ at which barriers survive is given by

$$\xi_M \approx (H/g)\;\rho^{(5-D)/2} \qquad (4.11)$$

This is one of the results of Bruinsma and Aeppli /47/. It corresponds, as said above, to a situation where potential barriers are so to speak, "opaque". This might be an appropriate approximation in the case of rapid cooling. On the other hand, this situation is appropriate to computer simulations of molecular dynamics. Nattermann /48/ proposed an interpolation between (4.11) and (3.3b).

5. CRITICAL PROPERTIES

They are, essentially, not understood!

a) Linear response theory for $T > T_{co}$

At temperatures T sufficiently larger than the transition temperature T_{co} of the pure system, the theory is fairly easy /49/. A careful comparison with experimental results was presented by Todedano /14/. In the case of a single impurity at the origin (i = 0) the free energy may be roughly written as

$$F = \frac{1}{2} \sum_{ij} [\chi^{-1}]_{ij} m_i m_j - H_o m_o \qquad (5.1)$$

where $m_i = \langle S_i \rangle$. Introducing the Fourier transform m_q, χ_q one gets

$$F = \frac{1}{2} \sum_q \chi_q^{-1} \left| m_q - H_o \chi_q / \sqrt{N} \right|^2 - (H_o^2/2N) \sum_q \chi_q$$

the minimum of which is obtained for

$$m_q = H_o \chi_q / \sqrt{N} \qquad (5.2)$$

and is equal to

$$F_o = -(H^2/2N) \sum_q \chi_q \approx -H_o^2 \kappa^D \chi/J \approx -(H_o^2/J) \left[(T-T_{co})/T_{co}\right]^{D\nu - \gamma} \qquad (5.3)$$

where $1/\kappa$ is the correlation length for H = 0; ν and $\gamma = \eta(2-\eta)\nu$ are the critical exponents also for H = 0. This expression is small, and therefore the approximation (5.1) is acceptable for 1 impurity (with some minor refinements if H_o is too large). For a finite concentration c, the free energy per unit volume is c F_o if $(T - T_{co})$ is not too small, as will be seen later.

A quantity which can be rather easily measured is the specific heat C, easily deduced from (5.3)

$$C = - cd(dF_o/d\beta)/dT \sim cH_o^2 (T^2/J^2)(d^2/dT^2) \left[(T-T_{co})/T_{co}\right]^{D\nu - \gamma}$$

$$C \sim (H^2/J) \left[(T-T_{co})/T_o\right]^{-\tilde{\alpha}}$$

where $H^2 = c H_o^2$ and

$$\tilde{\alpha} = 2 + \gamma - D\nu = 2-(D-2+\eta)\nu = 2-2\beta \qquad (5.4)$$

The condition of validity of the above relations may be obtained as follows. Random field effects are expected to be correlated within a volume κ^{-D}. Inside this volume the effective field is $\tilde{H} = H^{+D/2}$ (formula 2.5). In the critical region the magnetisation is $\chi\tilde{H}$ for weak field and $(\tilde{H}/J)^{1/\gamma}$ in strong field. The crossover occurs for $\chi\tilde{H} \approx (\tilde{H}/J)^{1/\gamma}$ or $(H/J)^{1-1/\delta} \approx \kappa^{\gamma/\nu} = \kappa^{2-\eta}$ or $\tilde{H}/J \approx \kappa^{(D+2-\eta)/2}$ or

$$H/J \approx \kappa^{1-\eta/2} \tag{5.5}$$

The pair correlation, which is of interest for neutron scattering, consists of two terms

$$\begin{aligned} \overline{\langle |S_q|^2\rangle} &= \overline{|\langle S_q\rangle|^2} + \langle|\delta S_q|^2\rangle \\ &= H^2\chi_q^2 + T\chi_q \end{aligned} \tag{5.6}$$

where S_q is the Fourier transform of S_i and $\delta S_q = S_q - \langle S_q\rangle$. The first term results from (5.2) and (1.3). Note that, when (5.5) is satisfied, both terms are of the same order for $q \lesssim \kappa$. Neglecting the critical exponent η, (5.5) reads

$$(H^2/J^2)(\kappa^2+q^2)^{-2} + (T/J)(\kappa^2+q^2)^{-1} \tag{5.7}$$

The above results are fairly trivial, but they are relevant for the experimentalist. It is remarkable that the specific heat exponent $\tilde{\alpha}$ is larger than the exponent α of the "pure" system (i.e. with H = O). The true exponent α' (which describes the specific heat at equilibrium near the transition temperature T_c of the random field system) has no reason to be equal to $\tilde{\alpha}$. The experimental result /11/ is $\alpha' < \alpha$, and it is not surprising in comparison with other random models /50,51/. Thus, α may have a maximum near T_{co} !

Let an additional remark be made about formula (5.5). It may be shown that, below T_c, the <u>same</u> condition defines the temperature at which the ordered state breaks into domains /39/. Thus, on both sides of T_c, (5.5) defines the limit where random fields becomes important. The temperature region T where the critical exponents of the RFIM can be observed is therefore /8,39/

$$\delta T \approx (H/J)^{2/(2-\eta)\nu} = (H/J)^{2/\gamma} \tag{5.8}$$

<u>b) Upper Critical Dimension, and ε-Expansion</u>

In pure systems, the critical exponents are given by the standard Landau theory /59/ if the space dimension D is larger than an upper critical value D_c. For $\varepsilon \equiv D_{cu}-D > 0$, they can be expanded in powers of ε /59/. $D_{cu} = 4$, and the expansion gives very good results for D = 3 (or $\varepsilon = 1$), although it is divergent, and clever tricks should be used /53/.

Is it so for the RFIM ? As discussed in Appendix B, the Landau theory does not seem to be inconsistent above 6 dimensions, and this implies $D_{cu} = 6$ /25,26/. However, although the case $D = \infty$ is trivial /54/, Bray /55/ found it impossible to carry out a large D expansion. Anyway, since Landau theory is consistent above 6 dimensions, it is reasonable to try to use it to derive an ε-expansion. Then, according to Blankschtein et al /56/, "the ε-expansions of the random field problem in d = 6-ε dimension are the same, to all orders, as those of the non-random problem in d = 4-ε dimension". This simple and powerful result can be derived /26,57/ if one neglects thermal fluctuations, i.e. in the mean field approximation or, in the diagrammatic language, if one keeps only the "tree diagrams". Is that correct ? According to Shapir /58/, "this may be argued to be correct at least for 4 < D < 6." The idea is that disorder induced fluctuations dominate thermal fluctuations (see (5.7)) which can thus be neglected. Then, if the expansion is valid down to the lower critical dimension D_c, the latter should be shifted by the same amount as D_{cu}, i.e. 6-4 = 2. Now, from (2.4), D_c=1 for the pure Ising model. For the RFIM it follows D_c = 1+2 = 3. We believe this result to be wrong because it disagrees with Section 2, and in particular with Imbrie's exact treatment /22/. Not everybody agrees with this statement /59/ because Imbrie's treatment holds at T=0, and at T=0 the lower critical dimension of the pure Ising model is no longer 1, but (at most) 0 : a finite system (D=0) orders at T=0, and we do not want to speculate about what D < 0 would mean.

As argued in Appendix B and already pointed out by Parisi and Sourlas /60/, the reason of this failure of the ε-expansion is probably the existence of metastable states.

Speculatively, we suggest that the ε-expansion might be valid between 5 and 6 dimensions, because potential energy barriers are finite as seen in Sections 3 and 4. The reason is that formula (3.3b) is not applicable if smaller than the interatomic distance. Another possibility is that the ε-expansion is valid between 4 and 6 dimensions, because free energy barriers do not form above T_{co} above 4 dimensions /39/.

c) Theories of the Critical Behaviour /58/

We give here a summary of the attempts which have been made to describe the critical behaviour of the RFIM.

Aharony et al /25/ used the ε-expansion in a modified way and find that the properties of the D-dimensional RFIM are those of the d-dimensional Ising model with

$$d = D - 2 + \eta(d) \qquad (5.9)$$

where $\eta(d) = \eta'(D)$ is the critical exponent of the pure system in dimensions or of the RFIM in D dimensions. Shapir /58,61/ finds

$$d = D - 1/\nu \qquad (\alpha > 0) \qquad (5.10a)$$

$$d = (D + 1/\nu)/2 \qquad (\alpha < 0) \qquad (5.10b)$$

(5.10b) agrees with result (2.6), D_c=2, since then $\alpha=-\infty$, $\nu=\infty$, d=1 fit (5.10b). Relations (5.10) are also satisfied for d=2, D=3, ν=1, and α=0, in agreement with specific heat, birefringence and capacitance experiments /11/. Schwartz rederived (5.9) and also found D_c=2.

Relations (5.10) seem all right for D_c=2, D_{cu}=6 and fit specific heat data for D=3. As commented by Shapir, "the agreement in all three cases is certainly suggestive but should in all three cases be taken with caution before a more formal derivation is achieved." In our opinion, it is not only a question of formalism. Because of metastability, random field systems are essentially different from pure systems and we find it strange that their critical exponents in d dimensions are the same as for the pure systems in d dimensions, as found in refs. /25,58,61,62/, but, on the other hand, in contrast with refs. /71,72/.

6. EXPERIMENT

We will first mention birefringence and capacitance experiments, which are actually an indirect measurement of the specific heat /11/. In two dimensions, no anomaly is found, in agreement with the theoreticallly predicted absence of a transition. In three dimensions, a logarithmic anomaly is found in a weak random field. This is in agreement with the theoretical prediction (2.6), namely $D_c=2$. Moreover, since the zero-field, two-dimensional Ising model has a logarithmic specific heat /52/, it seems that a dimensionality-shift by two accounts not only for D_c but for the specific heat. This is accounted for by Shapir /58/, using somewhat heuristic scaling hypotheses.

Another observation is that the shift $(T_{co}-T_c)/T_{co}$ is proportional to $H^{2/\gamma}$ as expected from (5.8) and refs. /8/ and /63/. These experiments were apparently not perturbed by metastability. However, metastability does occur, as seen from neutron scattering experiments /10,73/. It is most clearly demonstrated by the comparison between samples cooled down in constant random field directly to the temperature T of interest (FC samples) and those cooled in zero field (ZFC) to very low temperatures, and then heated up to T after switching the field on (Fig. 10). At low enough temperature T, three-dimensional FC samples exhibit long range magnetic order, while FC samples have domain structure characterized by a squared lorentzian lineshape (5.7) /10,64/.

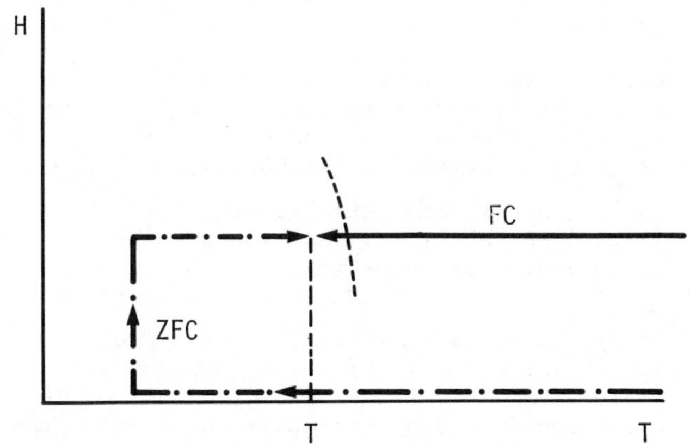

Fig. 10. Field cooled (full line) and zero-field cooled (dash-dotted line) procedure of preparation of samples. The dotted line indicates the onset of metastability.

The squaredlorentzian shape, derived from linear response in Section 5, can also be a result of a domain structure as will now be discussed. This illustrates the fact that the pair correlation function is but an incomplete information! The squared lorentzian can also be derived (exactly) for the spherical model /65/.

Assuming domains of size R_c, the spin pair correlation function G(r) decreases for small r as

$$G(r) \approx G(o) - \mu r/R_c \qquad (6.1)$$

with $\mu \approx 1$. The second term is the probability that two points at distance r are separated by a domain wall. Assuming, for any r,

$$G(r) \sim r^{-\tau} \exp(-\kappa r)$$

the fit with (6.1) yields $\kappa \approx 1/R_c$ and $\tau=0$, and the 3-dimensional Fourier transform is easily seen to be a squared lorentzian.

Since metastability is clearly established, it is of interest to check whether relation

$$R_c \approx (gT/H^2)\ln (t/\tau) \tag{4.4}$$

is satisfied. The H dependence is generally found to be $R_c \sim H^{-\nu_H}$, with $\nu_H \simeq 2.1$ or 2.2, which is not too bad /66,64,73/.

The question is now whether the time dependence observed in experimental systems fits the theoretical law (4.4). There is not a complete agreement among experimentalists, and this is not too surprising in view of the metastability which necessarily generates some irreproducibility. The point of view of two different groups is reproduced hereafter.

According to Birgeneau, Cowley, Shirane and Yoshizawa (preprint) "below $T_N(H)$ all states with length scales between the field-cooled value and infinity are stable for indefinite lengths of times; logarithmic decay of the domain size in time may be ruled out. No existing theory accounts for this transition behaviour."

According to Jaccarino (private communication and ref. /73/) "in three dimensions, the phase transition remains sharp in ZFC experiments for all $H < J$ (birefringence, capacitance and neutron scattering); in the vicinity of $T_c(H)$ new RFIM critical behaviour is observed with $\alpha' \simeq 0$, $\gamma' \simeq 1.75$, $\nu' \simeq 1$, $\eta' \simeq 1/4$ which implies an effective dimensionality $d \simeq 2$; the AF state remains stable for all T and H below $T_c(H)$, whereas the FC state shows irreversible behaviour and, more interestingly, the FC state decays logarithmically with time for $T \leq T_c(H)$. The equilibrium boundary $T_{eq}(H)$ lies slightly above $T_c(H)$ (a necessary condition if $d > d_\ell$) and a metastable configuration is to exist below $T_c(H)$".

In view of so strongly divergent results concerning the logarithmic time dependence, what can the theorist do ? A better derivation of (4.4) would be desirable. Our naive application of the Arrhenius law may be questionable since the number of metastable states is very large. However, the checks in one dimension /35,36/ are favourable, and experimental data on remanent magnetisation in ferromagnets /37/ and spin glasses /38/ show logarithmic laws analogous to (4.4).

Note that the above quotation of Jaccarino, based on ref. /73/, is in agreement with Shapir's phenomenological theory /58/ and disagrees with ref. /72/.

7. THE TWO-DIMENSIONAL CASE

a) One may wonder what the original Imry-Ma argument (Section 2) yields in two dimensions if $H \ll J$. Is it possible to find a region enclosing a given site O, such that energy is lowered by reversing the spins in this region ? For a region of radius R=1, the probability is very low, $p \approx \exp_m(-J^2/H^2)$. Then try with R=2, then R=4, then R=8, ..., $R=2^m$, ... The probability that O is not reversed after m trials is $(1-p)^m \approx \exp(-mp)$, so that the correlation length is $R_c = 2^{m_c}$, with $m_c \approx 1/p \approx \exp(J^2/H^2)$. Thus there is no order in the two-dimensional Ising model in two dimensions.

This result is correct and agrees with experiment /73/, but the above evaluation of R_c is highly overestimated. We believe the correct order of magnitude to be given by a vanishing domain wall energy, as seen below.

b) Continuous Walls /2,23,24/

For continuous walls, formula (3.5b) yields, for D=2,

$$\delta g(\rho) = -B\ H(H/g)^{1/3} \tag{7.1}$$

where B is a constant. (7.1) should be interpreted as the contribution to the domain wall energy of bumps of radius between ρ and λρ, with, say, λ=2. This argument is called coarse-graining. Of course, B depends on λ. The differential form of (7.1) is

$$dg(\rho) = -B_o\ H(H/g)^{1/3}\ d\rho/\rho \tag{7.2}$$

and it is seen that $B(\lambda) = B_o \ln \lambda$. Integration of (7.2) yields

$$g^{4/3}(\rho) = g^{4/3}(1) - \frac{4}{3} B_o\ H^{4/3} \ln \rho \quad . \tag{7.3}$$

This formula is valid until $\xi(\rho)/\rho$ becomes of order 1. According to (3.3b) this implies $g(\rho) \gtrsim H$ and, according to (7.3), $\rho \gtrsim \rho_c$, where

$$\rho_c = \exp\left[K(g/H)^{4/3}\right] \tag{7.4}$$

where $K = 3/4\ B_o$ and $g = g(1)$. The value (7.4) may be identified with the average domain size R_c.

c) Stepped Walls /27,23/

Similarly, formula (3.5a) yields

$$dg = -(H^2/J)\ d\rho/\rho$$

and

$$g = 2J - 2K'^{-1}(H^2/J) \ln \rho$$

Therefore

$$R_c \approx \rho_c \approx \exp(K'J^2/H^2) \tag{7.5}$$

where K' is a constant.

Crossover to (7.4) occurs with increasing temperature. According to Derrida and Shnidman /68/, the ground state of the two-dimensional RFIM shows a Kosterlitz-Thouless correlation function ($\sim R^{-\eta}$). This interesting suggestion requires further check.

Conclusion

It seems well established that the three-dimensional Ising model undergoes a phase transition if the random field is weak enough. A novel critical behaviour is expected below a temperature defined by (5.5), but experimental observation is made difficult by metastability. Possibly, the transition is just first order as recently suggested by A.P. Young. The dimensional reduction suggested by Shapir /61/ and observed by Belanger et al /73/ may hold not too close from the transition, but otherwise it is in contradiction with an exact inequality recently derived by Schwartz.

In addition to the nature of the transition, the time evolution at low temperature is another open question: is relation (4.4) satisfied or not? In this book, the contributions of Andelman, Cowley, Golburg and Satija are also devoted to the random field Ising Model. It may be of interest to make the relation with other topics treated here. Fractals did not appear in our review. For instance, the metastable domain structure at low temperature is not fractal. As noticed in Section 6, the squared lozentzian lineshape of the diffraction peak observed experimentally indicates an approximately uniform (therefore, not fractal) domain size, and reduces for $R_c q \gg 1$ to Porod's law mentioned in their lecture by Kjems and Schofield.

On the other hand, the random field Ising model is related to several systems considered in this book. It may be used for instance, to describe the propagation of a two-liquid interface in a porous medium, adressed in Guyon's lecture. The interface should be smooth in a weak random field and just leaves a few islands or droplets behind it, which correspond to exceptional regions where the random field is strong. Presumably, the interface takes a fractal character in a strong random field or at high velocity. In the latter case, the incompressibility of the liquids and mass conservation are important factors.

APPENDIX A - CONTINUOUS DOMAIN WALLS

The "stricto sensu" Ising model (2.1) represents a system with an infinite anisotropy. Most of real uniaxial magnetic systems have only a finite, and often weak anisotropy. They are then better represented by the following Hamiltonian

$$\mathcal{H} = -J \sum_{\langle ij \rangle} \vec{S}_i \vec{S}_j - A \sum_i {S_i^z}^2 - \sum_i H_i S_i \qquad \text{(A.1)}$$

where $\vec{S}_i$ is a three-dimensional vector (here assumed classical) and A is a positive constant. Omitting an additive constant, one can write

$$\mathcal{H} = (Ja^{2-D}/2) \sum_{<ij>} \int d^D r \sum_{\alpha=xyz} [\vec{\nabla} S_\alpha(\vec{r})]^2 - Aa^{-d} \int d^D r\, S_z^2(\vec{r}) - \sum_i H_i S_i^z \tag{A.2}$$

where a is the side of the cubic cell.

The important point is that the crystal structure does not appear in the first two terms, which are isotropic in the crystal space (not in the spin space, of course). Therefore, the free energy per unit area, g, of a flat domain wall, is independent of its orientation. If the domain wall is not flat, but has a weak curvature, it can be decomposed into approximately flat parts and its free energy is just

$$W' = g\,\mathcal{A} \tag{A.3}$$

where $\mathcal{A}$ is the total wall area. Formula (3.2b) is an approximation of (A.3). From (A.3), one can also deduce (2.3), with J replaced by

$$g \approx \sqrt{J\ A}\ a^{1-D} \tag{A.4}$$

or $g \approx \sqrt{J\ A}$ if $a \approx 1$, as assumed in most of this article.

Formula (A.3) is one of the basis formulae of the physics of interfaces and is by no means special to domain walls.

Weak random fields do not alter formula (A.3). Also, the assumption of a cubic crystal is not essential. (A.2) also holds for the orthorhombic materials provided appropriate length units are chosen on the 3 axes. The wall thickness /46/ is

$$\lambda \approx a\,\sqrt{J/A} \quad . \tag{A.5}$$

APPENDIX B - THE ε-EXPANSION

The ε-expansion relies on a (divergent !) perturbation expansion /52/ in which each term is an integral of type

$$\left[\prod_{q=q_1}^{q_p} \int d^d q\, \langle |S_q|^2 \rangle\right] f_p(q_1 \ldots p_p) \tag{B.1}$$

where p goes from 0 to ∞ and the functions f_p do not play an essential part. $\langle |S_q|^2\rangle$ is given by (5.7). Divergencies arise for κ=0. It is seen from (5.7) that, for $q \approx 0$,

$$d^d q\, \langle |S_q|^2 \rangle_o \sim q^{d-1}\, dq/q^2 \quad , \tag{B.2a}$$

$$d^D q\, \langle |S_q|^2 \rangle_o \sim q^{D-1}\, dq/q^4 \quad . \tag{B.2b}$$

Both factors are the same if $D = d+2$, therefore the condition for convergence of all integrals (which defines D_{cu}) is the same for $H \neq 0$ in D dimensions or for $H=0$ in $(D-2)$ dimensions. Therefore, for $H \neq 0$.

$$D_{cu} = d_{cu} + 2 = 4 + 2 = 6 \quad . \tag{B.3}$$

The complete equivalence of the two ε-expansions (for $H=0$ and $H \neq 0$) is only true if the second term of (5.7) is completely neglected in (B.1), i.e. if only the "most divergent diagrams" are taken into account. Even then, the proof is difficult /30,69/, because the explicit form of f, in (B.1), is important. On the other hand, formulae (5.7) or (B.2) are only correct for $D > D_{cu}$. They can reasonably be used in an ε-expansion, but well below D_{cu}, it seems appropriate to "renormalize the propagators", i.e. to replace $1/q^2$ by $1/q^{2-\eta}$ in (B.2a). After this modification, (B.2a) and (B.2b) are identical if $d = D-2+\eta$. This is relation (5.9).

Let us now see why these ideas are questionable. Neglecting the second term of (5.7) implies neglecting thermal fluctuations around the minimum of the Landau free energy

$$F = \sum_i \varphi(m_i) - \frac{1}{2}\sum_{ij} \gamma_{ij}\, m_i m_j - \sum_i H_i m_i \tag{B.4}$$

where $\gamma_{ij} = \gamma(\vec{R}_i - \vec{R}_j)$'s are coefficients, $m_i = \langle S_i \rangle$ and

$$\varphi(m) = \frac{1}{2}\, Am^2 + \frac{1}{4}\, Bm^4 \quad . \tag{B.5}$$

If F is minimum

$$Am_i - \sum_j \gamma_{ij}\, m_j = H_i - Bm_i^3 \quad . \tag{B.6}$$

An instructive exercise /70/ is to study the case of a single site, i, so that (B.6) reads

$$Am - \gamma m = H - Bm^3 \quad . \tag{B.7}$$

The interesting region corresponds to $A < \gamma$, so that (B.7) has 3 solutions for small H. The natural extension of formulae valid for $A > \gamma$ leads either to the weights 1, -1, and 1, or to 1, 1, and 1. Since a probability cannot be negative and the sum of probability must be 1, both weightings are wrong /70/.

More generally, the correct weighting of all solutions of (B.6) is a difficult, and unsolved, problem.

ACKNOWLEDGEMENTS

It is a pleasure to acknowledge discussions and letters from R. Birgeneau, V. Jaccarino, R. Cowley, and Y. Shapir.

REFERENCES

/ 1/ Y. Imry, S. K. Ma. Phys. Rev. Lett. 35, 1399 (1975)

/ 2/ Y. Imry: J. Stat. Phys. 34, 849 (1984)

/ 3/ J. Villain, J. Physique Lett. 43, L551 (1982)

/ 4/ A. Glachant, M. Jaubert, M. Bienfait, G. Boato, Surf. Sci 115, 219 (1981)

/ 5/ B. I. Halperin, C. M. Varma, Phys. Rev. B14, 4030 (1976)

/6a/ T. Riste, E.J. Samuelson and K. Otnes, in Structural Phase Transition and Soft Modes, edited by E.J. Samuelson, E. Andersen, and J. Feder (Universitetsforlaget, Oslo, 1971)

b) T. Riste, E.J. Samuelson, K. Otnes, and J. Feder, Solid State Commun. 9, 1455 (1975)

c) S. M. Shapiro, J. D. Axe, G. Shirane, and T. Riste, Phys. Rev. B6, 4332 (1972)

/ 7/ R. Blinc, J. Slak, F.C. Sa Barreto, A.S.T. Pires, Phys. Rev. Lett. 42, 1000 (1979)

/ 8/ S. Fishman, A. Aharony, J. Phys. C12, L729 (1979)

/ 9/ T. Cardy, Phys. Rev. B29, 505 (1984)

/10/ R.J. Birgenau, R.A. Cowley G. Shirane, H. Yoshizawa, J. Stat. Phys. 34, 817 (1984)

/11/ Belanger, D.P. King, A.R. Jaccarino, V.: Phys. Rev. Lett. 49, 1050 (1982), J. Appl. Phys. 55, 2383 (1984); Belanger, D.P. King. A.R. Jaccarino, V. Cardy, J.L.: Phys. Rev. B29, 2522 (1983); Ferreira, I.B. King, A.R. Jaccarino, V. Cardy, J.L.: Phys. Rev B28 5192 (1983); S.M. Rezende, A.R. King, V. Jaccarino, J. Appl. Phys. 55 2413 (1983)

/12/ Y. Shapira, N.F. Oliveira, Jr, S. Foner, Phys. Rev. B30, 6639 (1984)

/13/ H. Ikeda, J. Phys, C 16, L21 (1983); H. Ikeda, K. Kikuta, J. Phys. C16, L445 (1983)

/14/ J.C. Toledano, Ann. des Télécommunications 39, 277 (1984)

/15/ G. Grinstein, J. Appl. Phys. 55, 2371 (1984)

/16/ Po-zen Wong, J.W. Cable, P. Dimon, J. Appl. Phys. 55, 2377 (1984)

/17/a) F. Brochard, P.G. de Gennes, J. Physique Lett. 44, L785 (1983)

b) Po-zen Wong, Schlumberger Preprint (1984)

/18/ A. Beretti, J. Stat. Phys. to be published

/19/ L.D. Landau, E.M. Lifshitz, "Statistical Physics" Pergamon Press London (1959)

/20/ J.T. Chalker, J. Phys. C16, 6615 (1983)

/21/ D.D. Fisher, J. Fröhlich, T. Spencer, J. Stat. Phys. 34, 863 (1984)

/22/ J.Z. Imbrie, Harvard preprint (1984) and Phys. Rev. Lett. 53, 1747 (1984)

/23/ Grinstein, G., Ma, S.K.: Phys. Rev. Lett. 49, 685 (1982); Phys. Rev. B28, 2588 (1983)

/24/ J. Villain, B. Séméria, F. Lancon, L. Billard, J. Phys. C16, 6153 (1983)

/25/ E. Pytte, Y. Imry, D. Mukamel, Phys. Rev. Lett. 46, 1173 (9181); K. Binder, Y. Imry, E. Pytte,: Phys. Rev. B24, 6736 (1981); D. Mukamel, E. Pytte,; ibid 25, 4779 (1982)

/26/ H.S. Kogon and D.J. Wallace, J. Phys. A14, L527 (1981)

/27/ K. Binder, Z. Phys. B50, 343 (1983)

/28/ T. Nattermann, J. Phys. C16, 6407 (1983)

/29/ A. Aharony, Y. Imry, S.K. Ma. Phys. Lett. 37, 1367 (1976)

/30/ A.P. Young, J. Phys. C10, L257 (1977)

/31/ T. M. Rice, S. Whitehouse, P. Littlewood, Phys. Rev. B24, 2751 (1981)

/32/ B. Derrida, J. Vanimenus, Phys. Rev. B27, 4401 (1983)

/33/ M.V. Feigelman, J.F. Lyuksutov, Solid State Comm. 48, 397 (1983)

/34/ T. Nattermann, Phys. Stat. Sol. (b) 119, 209 (1983) and preprint submitted to J. Phys. C (1984)

/35/ Ya. G. Sinai, in Proceedings of the Berlin Conference on Mathematical Problems in Theoretical Physics, edited by R. Schrader, R. Seiler, and D.A. Uhlenbrock (Springer Berlin, 1982), p. 12

/36/ B. Derrida, Pomeau, Phys. Rev. Lett. 48, 627 (1982)

/37/ L. Néel, J. Phys. Rad. 11, 49 (1950)

/38/ J.J. Préjean, J. Souletie, J. Physique 41, 1335 (1980)

/39/ J. Villain,(a) Phys. Rev. Lett 52, 1543 (1984) (b) in "Elementary excitations and fluctuations in Magnetic Systems", ed. by S. Lovesey, V. Tognetti and U. Balucani, Springer Verlag (Heidelberg 1985) p. 142

/40/ G. Grinstein, J. Fernandez, Phys. Rev. B29, 6389 (1984)

/41/ C. Ro, G.S. Grest, C.M. Soukoulis, K. Levin, Phys. Rev. B31, 1682 (1985)

/42/ Y. Yoshizawa, D.P. Belanger, Phys. Rev. B B30, 5220 (1984)

/43/ D. Stauffer, C. Hartzstein, K. Binder. A. Aharony, Z. Phys. B55, 325 (1984)

/44/ E.T. Gawlinski, K. Kaski, M. Grant, J.D. Gunton, Phys. Rev. Lett. 53, 2266 (1984), and M. Grant, J.D. Gunton, Phys. Rev. B29, 1521 (1984) and to be published

/45/ E. Pytte, J. Fernandez, Phys. Rev. B31, 616 (1985)
/46/ B. Widom, J. Chem. Phys. 43, 3892 (1985)
/47/ R. Bruinsma, G. Aeppli, Phys. Rev. Lett. 52, 1543 (1984)
/48/ T. Nattermann, preprint (1985)
/49/ A.P. Levanyuk, V.V. Osipov, A.S. Sigov, A.A. Sobyanin, Sov. Phys. J.E.T.P. 49, 176 (1979)
/50/ D.E. Khmelnitskii, Sov. Phys. J.E.T.P. 41, 981 (1975)
/51/ T.C. Lubenski, Phys. Rev. B11, 3573 (1975)
/52/ Simple, but not quite recent textbooks about critical phenomena are G. Toulouse, and P. Pfeuty, "Introduction an groupe de renormalisation et á ses applications" (Presses Universitaires de Grenoble, 1975) and S.K. Ma, "Modern theory of critical phenomena" (Benjamin, London 1976)
/53/ E. Brézin, G. Parisi, J. Stat. Phys. 19, 269 (1978)
/54/ T. Schneider, E. Pytte, Phys. Rev. B15, 1519 (19)
/55/ A.J. Bray, J. Phys. C16, 5879 (1984)
/56/ D. Blankschtein, Y. Shapir. A. Aharony, Phys. Rev. B29, 1263 (1984)
/57/ G. Parisi, N. Sourlas, Phys. Rev. Lett. 43, 744 (1978)
/58/ Y. Shapir, J. App. Phys., to be published (1985)
/59/ V. Krey, J. Phys. C18, 1455 (1985)
/60/ G. Parisi, N. Sourlas, Nucl. Phys. B206, 321 (1982)
/61/ Y. Shapir, J. Phys. C17, L809 (1984), Phys. Rev. Lett. 54, 154 (1985)
/62/ M. Schwartz, J. Phys. C18, 135 (1985)
/63/ A. Aharony, E. Pytte, Phys. Rev. B27, 5872 (1983)
/64/ R.A. Cowley, H. Yoshizawa, G. Shirane, R.J. Birgeneau, Z. Phys. B58, 15 (1984)
/65/ R.M. Hornreich, H.G. Schuster, Phys. Rev. B26, 3929 (1982)
/66/ R.A. Cowley, H. Yoshizawa, G. Shirane, M. Hagen, R.J. Birgeneau, Phys. Rev. B30, 6650 (1984)
/67/ R.A. Cowley, private communication.
/68/ B. Derrida, Y. Shnidmann, J. Physique Lettres 45, L577 (1984)
/69/ G. Parisi, N. Sourlas, Phys. Rev. Lett. 43, 744 (1979)
/70/ J. Villain, in "Unordnungsphänomene in Festkörpern" (KFA Jülich, 1985)
/71/ R. J. Birgeneau, R. A. Cowley, G. Shirane, H. Yoshizawa, D. P. Belanger, A. R. King and V. Jaccarino, Phys. Rev. B 27, 6747 (1983)
/72/ A. Khurana, F.J. Seco, A. Houghton, Phys. Rev. Lett. 54, 357 (1985)
/73/ D.P. Belanger, A.R. King, V. Jaccarino, Phys. Rev. Lett. 54, 557 (1985) and J. Appl. Phys., 15. April 1985

METASTABILITY AND A TEMPORAL PHASE TRANSITION IN THE RANDOM FIELD ISING MODEL

R.A. Cowley

Department of Physics
University of Edinburgh
Mayfield Road
Edinburgh EH9 3JZ, Scotland

R.J. Birgeneau

Department of Physics
Massachusetts Institute of Technology
Cambridge
Massachusetts 02139, USA.

G. Shirane and H. Yoshizawa

Brookhaven National Laboratory
Upton
New York 11973, USA

At the preceding school in this series, we reviewed[1] the then current experimental understanding of the properties of an Ising model in a random field. Since then there have been a large number of both theoretical[2] and experimental studies[3,4,5,6] but as we describe below the problem is still not completely understood. As reviewed in this school by Villain[7], it is now generally accepted that the lower critical dimension in equilibrium, d_ℓ, is 2. Experiments performed by cooling random antiferromagnets in a uniform field to produce a random staggered field, have shown that long range order is not achieved in these experiments and that at low temperatures the properties are history dependent. Recently theories[8] have been developed to reconcile these experimental results with the theoretical prediction that $d_\ell = 2$, by considering the energy barriers to domain wall motion in the presence of random fields. In order to examine this problem in more detail and to test these theories in detail we have performed new measurements on the random d = 3 antiferromagnet $Mn_{0.75}Zn_{0.25}F_2$ and Belanger et al. have performed measurements of the d = 2 antiferromagnet $Rb_2Co_{0.85}Mg_{0.15}F_4$. We choose $Mn_{0.75}Zn_{0.25}F_2$ because the $Mn_xZn_{1-x}F_2$ system is well understood. The spins interact with Heisenberg interactions and weaker dipolar interactions, which are responsible for the uniaxial symmetry. The system therefore belongs to the Ising universality class but there are very many low energy spin waves which might be expected to relax the system towards thermodynamic equilibrium. At low temperatures and applied field H = 0, the spin wave gap is about 8 K and many of our measurements are at about 40 K.

These experiments are an extension of those already reported[4], and were performed on an excellent single crystal supplied by Y. Shapira which was part of the same boule as the crystals used by him for macroscopic measurements[10]. It has a mosaic spread of less than 0.02° and a spread in concentration of less than 0.0015. It was mounted with the c-axis vertical in a variable temperature insert of a superconducting magnet so that the temperature was controlled to ±0.02 K and the applied field to ±0.1 T. The neutron scattering measurements were mostly performed with a triple axis crystal spectrometer at the Brookhaven HFBR. The energy of the incident neutrons was 5 meV and the collimation from reactor to counter was 20', 5', 10', 10' giving an energy resolution of 0.02 meV (HWHM) and a wavevector resolution at the (100) magnetic lattice point of 0.002, 0.0009. 0.015 (HWHM) reciprocal lattice units along the wavevector transfer, $\underline{Q}$, perpendicular to $\underline{Q}$ and vertically. The triple axis configuration for elastic scattering was used to suppress the quasi-elastic fluctuation scattering.

Most of the scans were performed by varying $\underline{Q}$ transversely through the (100) magnetic Bragg reflection as illustrated in fig 1. Experiments were performed by cooling in a field, FC, and by cooling with H = 0 to 8 K, then raising the field and heating, ZFC. Fig. 1 shows that the FC and ZFC procedures give identical results at 44.0 K but quite different results at 43.4 K. This is characteristic of all the measurements that there is a sharp boundary between history independent and history dependent behaviour and that this boundary, $T_M(H)$, is within error at the same temperature for a given field as the magnetic phase boundary found by Shapira et al. from macroscopic measurements[10].

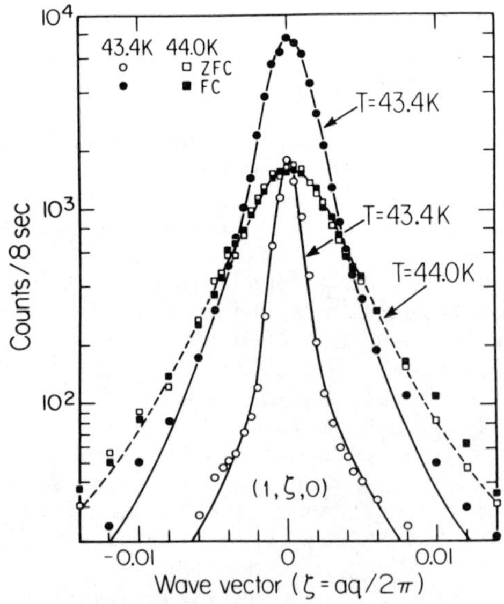

Fig. 1. Wavevector scans (ζ,1,0) at 43.4 K and 44 K using FC and ZFC procedures. The solid lines are fits to Lorentzian squared profiles plus a Bragg Gaussian for the 43.4 K ZFC scan.

The nature of the metastability below $T_M(H)$ has been studied in detail. The width of the FC scans decreases on cooling to about 25 K below which the width increases slightly down to 8.0 K, and the system never attains long range order. If the field is decreased then at any temperature, the domains expand as rapidly as the field is reduced until they reach the same size as they would have achieved by the FC procedure in the final field. When the field is increased the domain sizes are unaltered, as predicted by the theories described by Villain[7]. These theories also predict that the domain sizes should increase with time as $\kappa^{-1} \sim [\ln(t/\tau)]$ with τ a microscopic time constant. We have performed careful scans[4] for H = 7.0 T as a function of time for T = 40.6 K when $T_M(H)$ = 41.0 K. The results of fitting these profiles gave an inverse correlation length κ = 1.46±0.07 (140 secs), 1.39±0.02 (600 secs), 1.37±0.02 (1300 secs). 1.35±0.02 (4300 secs) and 1.38±0.02 (54000 secs) $\times 10^{-3}$ reciprocal lattice units. Since it is unreasonable for τ to be smaller than about 10^{-12} secs the ratio of the κ's at the longest time and at 600 secs is 1.14 whereas the experimental ratio is 1.01±0.03. The relaxation is therefore slower than logarithmic, and yet the system cannot be frozen as it responds to a reduction in the field.

The triple axis profiles, such as those shown in fig. 1, were fitted to a Lorentzian squared form convoluted with the experimental resolution giving $\chi^2 \sim 1$ and the results for the inverse correlation length are shown in fig. 2 for 5.0 T and fig. 3 for 6.5 T. At high temperatures κ decreases almost linearly with $T - T_0$, with T_0 = 43.8 K at 5.0 T and 42.4 K at 6.5 T. Somewhat above T_0, the κ's lie significantly above the line. Below the metastability boundary the FC results yielded a slowly decreasing κ with an almost temperature dependent amplitude of the Lorentzian squared. Figs. 2 and 3 are consistent with there being a break in the slope of κ against temperature at$T_M(H)$, when allowance is made for the small concentration smearing ~0.3 K.

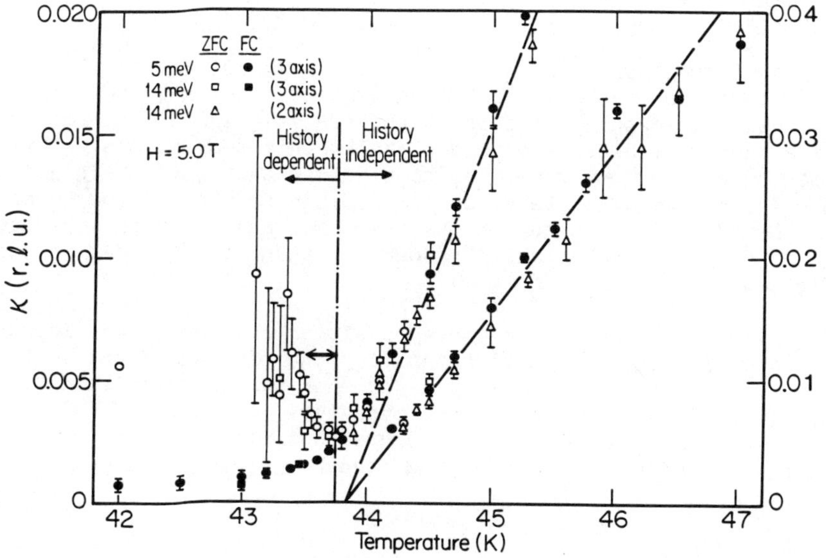

Fig. 2. Inverse correlation length as a function of temperature for H = 5.0 T.

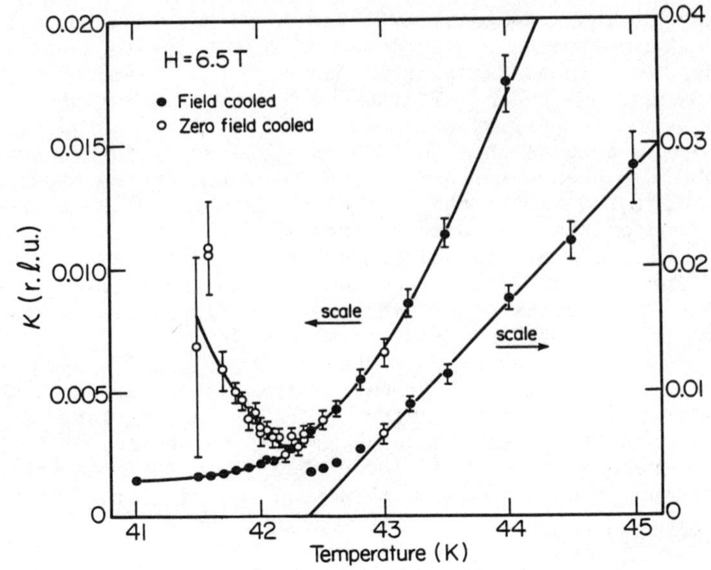

Fig. 3. Inverse correlation length as a function of temperature for H = 6.5 T.

Below the metastability boundary the ZFC results could not be fitted by a Lorentzian squared alone. It was necessary to add a resolution limited Gaussian to represent the long range order scattering. The amplitude of the long range order was then almost independent of temperature below $T_M(H)$ when it dropped abruptly, within the concentration smearing, to zero. The amplitude of the Lorentzian squared was negligible except close to $T_M(H)$, when it rapidly increased, while κ decreased as shown in figs. 2 and 3. These results are consistent with there being a first order transition from the ZFC state when the fluctuations have the same dimensions as those of the FC state, and this transition occurs at the metastability temperature, $T_M(H)$. The former picture is consistent with Villain's discussion[7].

In conclusion it is worthwhile comparing these results with those of Belanger et al.[9] on the d = 2 antiferromagnet. They also found a fairly well defined temperature above which the results were not history dependent while they were history dependent below this temperature. This metastability phase boundary was found to scale so that $T_M(0) - T_M(H) = CH^{2/\phi}$ with $\phi = 1.74 \pm 0.02$. This behaviour is similar to that of the broadened peak in the specific heat but unlike the d = 3 results C is substantially larger for the d = 2 metastability boundary than for the peak in the specific heat.

The time dependence of the decay of long range order was measured close to the vicinity of the metastability boundary and found to be logarithmic in time. No such time dependence was observed in d = 3. Finally on reducing the field at a temperature just below $T_M(H)$, little change in intensity was observed. In contrast, in $Mn_{0.75}Zn_{0.25}F_2$ the same procedure would have led to long range order. Clearly the behaviour of d = 2 and d = 3 are similar in that they both show metastability effects, but the properties are different in detail.

The results we have presented for $Mn_{0.75}Zn_{0.25}F_2$ are consistent with earlier experiments on $Co_xZn_{1-x}F_2$, $Mn_xZn_{1-x}F_2$ and $Fe_xZn_{1-x}F_2$ [1,3-6], but are not consistent with current theories. Figs. 2 and 3 show that on cooling in a field the system begins to develop longer range order, but that a well defined transition occurs to a state with many metastable states. These states are characterised by barriers which give rise to relaxations that are longer than logarithmic in time. These features are not yet understood, and may result because the transition in the random field Ising model is of first order[11] while the low temperature state has very unusual properties, or possibly because the vacancies in the diluted antiferromagnets pin the domain walls more firmly than in the random field ferromagnet[12].

Acknowledgements

We are grateful to Y. Shapira for providing the excellent single crystal used in these experiments. RAC is grateful for the hospitality of Brookhaven where the work is supported by the Division of Material Sciences, US Department of Energy under Contract No. DE-AC02-76CH0016 and at MIT by the NSF Grant No. DMR85-01856.

References

1. R.A. Cowley, R.J. Birgeneau, G. Shirane and H. Yoshizawa, in Multicritical Phenomena, ed. R. Pynn and A. Skeltorp, NATO ASI Series B 106 Plenum (1984).
2. For a review see Y. Imry, J Stat Phys, 34, 849 (1984).
3. For a review see R.J. Birgeneau, R.A. Cowley, G. Shirane and H. Yoshizawa, J Stat Phys, 34, 817 (1984).
4. R.A. Cowley, H. Yoshizawa, G. Shirane, M. Hagen and R.J. Birgeneau, Phys Rev B30, 6650 (1984).
5. R.A. Cowley, H. Yoshizawa, G. Shirane and R.J. Birgeneau, Z Phys B 58, (1984).
6. D.P. Belanger, A.R. King and V. Jaccarino (to be published).
7. J. Villain (this volume).
8. J. Villain, Phys Rev Lett 52, 1543 (1984).
   G. Grinstein and J. Fernandez, Phys Rev B 29, 6389 (1984).
9. D.P. Belanger, A.R. King and V. Jaccarino, Phys Rev Lett 54, 577 (1985).
10. Y. Shapira, N.F. Oliveira Jr. and S. Foner, Phys Rev B 30, 6639 (1984).
11. A.P. Young and M. Nauenberg (to be published).
12. H. Yoshizawa and D.P. Belanger, Phys Rev B 30, 5220 (1984);
    C. Ro, G.S. Grest, C.M. Soukoulis and K. Levin, Phys. Rev. B 31, 1682 (1985).

MAGNETIC EXCITATIONS IN $K_2Co_xFe_{1-x}F_4$: A MIXED MAGNETIC SYSTEM WITH COMPETING SPIN ANISOTROPIES

S.A. Higgins*, W.A.H.M. Vlak†, M. Hagen×□, R.A. Cowley*
A.F.M. Arts‡ and H.W. de Wijn‡

*Department of Physics, University of Edinburgh
Mayfield Road, Edinburgh EH9 3JZ, Scotland

†Netherlands Energy Research Foundation, ECN, P.O. Box 1
1755 ZG Petten, The Netherlands

×Institut Laue-Langevin, BP 156X Centre de Tri
38042 Grenoble Cedex, France

‡Fysisch Laboratorium, Rijksuniversiteit, Utrecht
P.O. Box 80000, 3508 TA Utrecht, The Netherlands

□Present address: Solid State Division, Oak Ridge
National Laboratory, Oak Ridge, Tennessee 37830

INTRODUCTION

$K_2Co_xFe_{1-x}F_4$ is a good physical realisation of a two dimensional mixed magnetic system with competing anisotropies. The end members K_2CoF_4 and K_2FeF_4 have the K_2NiF_4 structure (Birgeneau et al. 1970) in which the magnetic ions are at the corners and body centre positions of a tetragonal unit cell. The a-b planes thus contain quadratic layers of magnetic ions within which the nearest neighbour antiferromagnetic exchange interaction is much greater than that between spins in adjacent layers, leading to quasi two dimensional magnetic properties. In K_2CoF_4 (Breed et al., 1969) the pseudo spin $S = \frac{1}{2}$ aligns along the c-axis due to an anisotropy in the exchange interaction, giving a uniaxial ordered phase. In K_2FeF_4 (Macco et al., 1978) a single ion anisotropy forces the pseudo spin $S = 2$ in the a-b plane perpendicular to the c direction and a small in-plane anisotropy aligns the spins along the a or b axis of the magnetic unit cell. (Rotated 45° about c, relative to the chemical unit cell.) K_2FeF_4 therefore orders in a planar phase.

Thus the system can be modelled by the following Hamiltonian:

$$\hat{H} = \sum_{\langle ij \rangle} [I_{ij} S_i^z S_j^z + J_{ij}(S_i^x S_j^x + S_i^y S_j^y)] + \sum_i D_i (S_i^z)^2 + \sum_i E_i [(S_i^x)^2 - (S_i^y)^2] .$$

where

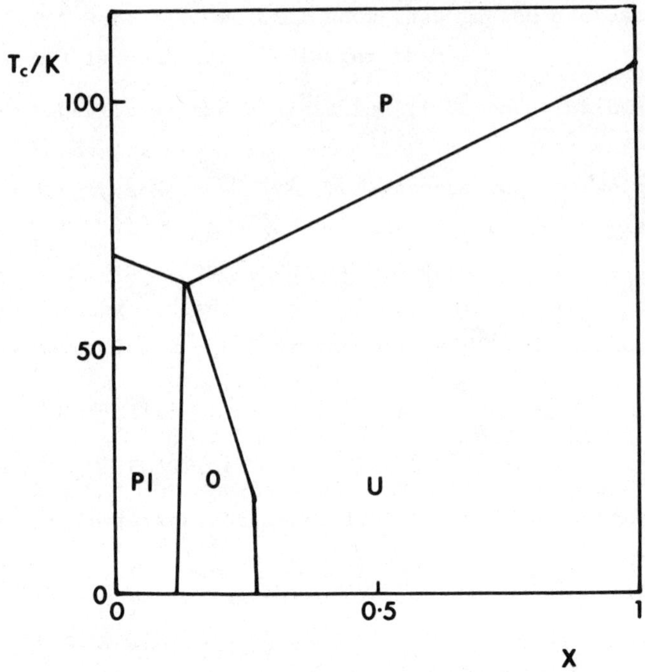

Fig. 1. The phase diagram for $K_2Co_xFe_{1-x}F_4$. (Fendler and Eynatten, 1984). P denotes paramagnetic phase; PL denotes planar antiferromagnetic phase; U denotes uniaxial antiferromagnetic phase and O denotes the oblique antiferromagnetic (OAF) phase.

$I_{ij} > J_{ij}$ for exchange between 2 Co spins.

$I_{ij} = J_{ij}$ for exchange between 2 Fe spins.

$D_i > 0$; $E_i < 0$ gives one domain and $E_i > 0$ the other.

Both mean field theory calculations (Matsubara and Inawashiro, 1977) and renormalisation group theory calculations (Fishman and Aharony, 1978) for systems with competing anisotropy predict a phase diagram with four magnetic phases as a function of concentration, x and temperature, T. The four phases for $K_2Co_xFe_{1-x}F_4$ are shown in figure (1).

EXPERIMENTAL DETAILS

Experiments have been performed with crystals of $K_2Co_xFe_{1-x}F_4$ with x = 0.6 using PLUTO triple axis neutron spectrometer at A.E.R.E. Harwell and with nominally x = 0.2 at Harwell and Risø National Laboratory and the results have already been reported (Higgins et al., 1984). New measurements are reported of the magnetic excitations in a sample with nominally x = 0.27 which was investigated using triple axis neutron spectrometers IN3 and IN8 at I.L.L., Grenoble.

RESULTS

The x=0.6 sample has a uniaxial antiferromagnetic ordering of the spins along the $\underline{c}$-direction below a transition temperature $T_N = 92.2 \pm 0.1$ K. Measurements along the magnetic [1,0,0] direction at T = 4.5 K (figure (2)) indicate two bands of excitations in common with mixed magnetic systems

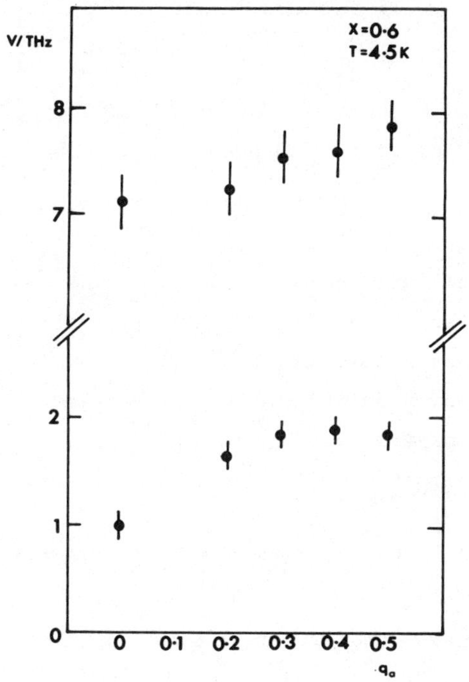

Fig. 2. The two bands of magnetic excitations along [1,0,0] for the $x = 0.6$ sample at $T = 4.5$ K.

without competing anisotropy (Cowley 1983). The upper band which is almost dispersionless can be associated with excitations propagating mainly on the Co ions and the lower band can be associated with excitations propagating mainly on the Fe ions.

In the nominally $x = 0.2$ sample, axial spin components order below $T_N = 66 \pm 1$ K but there is also a second phase transition at $T_N = 32 \pm 2$ K below which the planar spin components order. Thus the sample is of particular interest because it has the Oblique Antiferromagnetic (OAF) structure at low temperatures. Due to a smaller sample size measurements were performed only on the lower band of magnetic excitations. The results of measurements along the [1,0,0] direction are shown in figure (3).

A single branch of excitations was observed attributed to excitations on the Fe spins. (For a fuller discussion of the results for the $x = 0.6$ and $x = 0.2$ samples see Higgins et al., 1984).

To obtain more detailed information on the low lying band of excitations in the OAF high resolution experiments were performed with a sample of nominally $x = 0.27$. The magnetic structure and phase transitions of this sample had been investigated by Vlak et al, 1983 and their results were similar to those for the $x = 0.2$ sample described above but with $T_N = 64.4 \pm 0.6$ K and $T_L = 27 \pm 2$ K.

The first part of the experiment involved measurements of the spin wave dispersion along [1,0,0] at $T = 5$ K (figure (4)). This revealed a splitting of the branch. This is expected in the OAF phase because of the lower spin symmetry compared with the uniaxial phase, the single-ion out-of-plane anisotropy being considerably larger than the in-plane single-ion anisotropy as discussed in the introduction. A split branch of

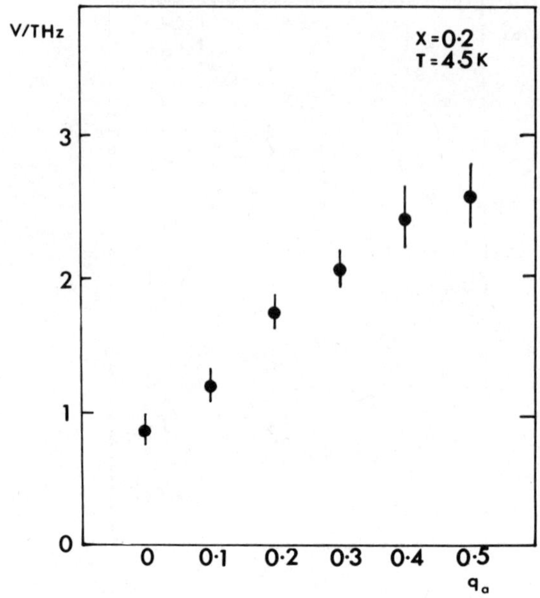

Fig. 3. The lower band of excitations along [1,0,0] for the x = 0.2 sample at T = 4.5 K.

excitations also occurs in K_2FeF_4 (Thurlings et al., 1982) which is in the planar phase below T_N although the energies of the excitations are larger in K_2FeF_4 presumably due to the lack of disorder and the fact that it is in a planar phase rather than in an OAF phase at low temperatures.

The second part of the experiment concerned high resolution measurements of the lower part of the split branch at low wavevector, low frequency transfer as a function of temperature.

Data was collected at T = 5 K, T = 19 K and T = 25 K.

At small wavevector, small energy transfer, the instrumental resolution is important in interpreting the data, therefore at each temperature all scans were fitted to the following $S(q,\nu)$ convoluted with the instrumental resolution function

$$S(q,\nu) = A(n(\nu) + 1)\frac{\nu\Gamma}{(\nu^2 - \nu_0(q)^2)^2 + \nu^2\Gamma^2} \quad .$$

A is a scale factor, ν is the frequency transfer, Γ is an intrinsic frequency width and $\nu_0(q)$ is given by the model dispersion relation.

$$\nu_0(q) = \sqrt{\varepsilon_g{}^2 + (D_s q_a)^2} \quad .$$

Values for ε_g the spin wave gap, D_s the spin wave stiffness, Γ and A are shown in table (1).

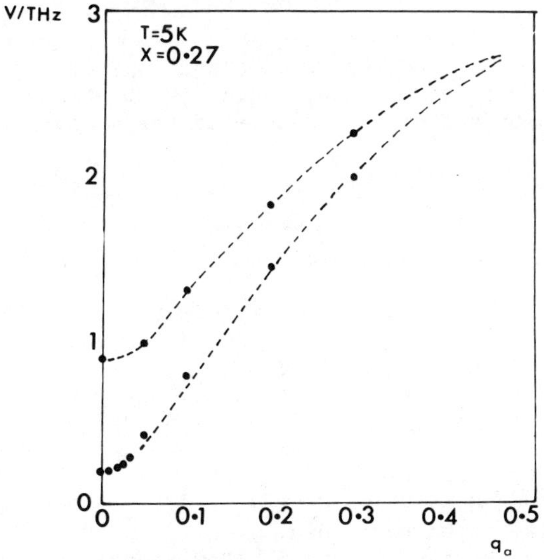

Fig. 4. This figure indicates the splitting of the spin wave dispersion relation along the [1,0,0] direction at T = 5 K. The points indicate peak positions from fitting Gaussian envelopes to the data and the dotted line is a guide to the eye.

ε_g decreases towards a small value (possibly zero) as the temperature increases towards T_L.

D_s decreases slightly and the intrinsic frequency width increases, indicating a decrease in the spin wave lifetimes.

The fact that ε_g becomes small at the boundary between the OAF and uniaxial places is of particular interest since a T = 0 mean-field calculation (Matsubara) indicate that the gap vanishes at the boundary between the OAF and uniaxial phases. A more detailed discussion of the results for the x = 0.27 sample is to be published. (Higgins et al, to be published).

Table (1) Values of ε_g, D_s, A and Γ obtained in fitting the data for the x = 0.27 sample as described in the text. Agreement factor = $\sqrt{\chi^2/N}$ where N = No. of degrees of freedom.

Temperature	ε_g/THz	D_s/THzÅ	A/THz²	Γ/THz	Agreement Factor
T = 5 K	0.196±0.025	5.6±1.4	0.16±0.04	0.068±0.043	1.16
T = 19 K	0.158±0.037	5.0±1.8	0.18±0.04	0.145±0.055	1.16
T = 25 K	0.130±0.049	4.4±1.5	0.18±0.08	0.238±0.12	1.18

ACKNOWLEDGEMENTS

Financial support from the Science and Engineering Research Council is gratefully acknowledged. S.A.H. would like to thank Dr. P.W. Mitchell for useful discussion and help with some of the computing.

REFERENCES

1. R.J. Birgeneau, H.J. Guggenheim and G. Shirane, Phys. Rev. B, 2211 (1970).
2. D.J. Breed, K. Gilijamse and A.R. Miedema, Physica 45, 205 (1969).
3. R.A. Cowley in Excitations in Disordered Systems ed. M.F. Thorpe (New York: Plenum) p373 (1982).
4. K. Fendler and E. Eynatten, Z. Phys. B54, 313 (1984).
5. S. Fishman and A. Aharony, Phys. Rev. B18, 3507 (1978).
6. S.A. Higgins, R.A. Cowley, M. Hagen, J. Kjems, V. Dürr and K. Fendler, J. Phys. C 17, 3235 (1984).
7. S.A. Higgins, W.A.H.M. Vlak, M. Hagen, R.A. Cowley, A.F.M. Arts and H.W. de Wijn, to be published.
8. F. Matsubara, J. Phys. Soc. Jpn 5, 1469 (1981).
9. F. Macco, W. Lehmann, W. Breitling, A.E. Slawska-Wariewska and R. Weber, Solid State Comm. 26, 429 (1978).
10. F. Matsubara and S. Inawashiro, J. Phys. Soc. Japan 42, 1529 (1977).
11. M.P.H. Thurlings, E. Frikkee and H.W. de Wijn, Phys. Rev. B 25, 4750, (1982).
12. W.A.H.M. Vlak, E. Frikkee, A.F.M. Arts and H.W. de Wijn, J. Phys. C 16 L1015 (1983).

SPIN DYNAMICS ON PERCOLATING NETWORKS

G. Aeppli*, H. Guggenheim* and Y. J. Uemura#

*AT&T Bell Laboratories, Murray Hill, NJ 07974

#Brookhaven National Laboratory, Upton, NY 11973

ABSTRACT

We have used inelastic neutron scattering to measure the order parameter relaxation rate Γ in the dilute, two-dimensional Ising antiferromagnet $Rb_2CoMg_{1-c}F_4$ with c very close to the magnetic percolation threshold. Where κ is the inverse magnetic correlation length, $\Gamma \sim \kappa^z$ with $z=2.4^{+0.2}_{-0.1}$. Our results are discussed in terms of current ideas about spin relaxation on fractals.

The static and dynamic behavior of pure magnets near phase transitions is well established, both experimentally and theoretically.[1] There also exists a large body of work dealing with the static critical properties of simple disordered magnets.[2] There are certain random spin systems for which static properties are more difficult to calculate. These systems,[3] where there are couplings between spins or local fields of random sign, are characterized by "frustration", the inability at low temperatures of certain spin clusters to come into unique alignment with respect to other clusters. Much of the qualitatively new behavior of frustrated systems is dynamical, and certain explanations[4] of their apparently static properties use dynamical theories as their starting points, thus reversing the pattern followed in the study of critical phenomena in pure systems. In this context, it is natural to ask whether qualitatively new effects arise in simpler, disordered systems, namely systems without frustration. An indication that interesting dynamical behavior might occur in random systems without frustration comes from studies of diffusion in spaces of non-integral (fractal) dimensionality.[5] For ordinary Euclidean spaces, the mean square distance $<r^2(t)>$ travelled by a particle in time t obeys the law

$$<r^2(t)> \sim t\,. \tag{1}$$

In fractal spaces, signals propagate via more circuitous routes than the straight line paths available in normal Euclidean spaces. Thus, particles will require a longer time to diffuse a certain distance in fractal spaces than in the (Euclidean) imbedding spaces. Correspondingly, the law (1) is generalized to

$$<r^2(t)> \sim t^{2/(2+\theta)} \tag{2}$$

with $\theta > 0$.

The essential idea of our experiments is to probe dynamical effects on fractals by studying the relaxation of spins located on a fractal. Arguably the simplest fractal is the percolating cluster, realized in magnetic compounds containing sites occupied at random either by non-magnetic ions, such as Mg^{2+}, or magnetic ions, such as Co^{2+}, coupled only to their nearest neighbors.[2] The exchange couplings in many of these compounds are such that frustration does not arise; while the randomness has profound geometrical effects, it does not lead to a qualitatively new ground state as in the spin glass problem. For our first study[6] of dynamical effects near percolation, we have chosen $Rb_2Co_cMg_{1-c}F_4$, a system for which the static spin correlations have been studied in detail.[7] Furthermore, there exist experiments[8] on the dynamical critical behavior of the corresponding pure system, Rb_2CoF_4, which is a nearly ideal realization of a two-dimensional Ising antiferromagnet.

We performed our experiments using two triple-axis spectrometers at the Brookhaven high-flux beam reactor. The first of these is located at a thermal beam port and is equipped with a single-pass monochromator. Spectra were collected by varying the final energy E_f with the incident energy E_i fixed at 3.5 meV; the corresponding energy resolution was 64 μeV, full width at half maximum (FWHM). The other spectrometer has a double crystal monochromator and is installed at the cold neutron source of the HFBR. Here, we scanned the incident neutron energy E_i with the final energy fixed at 2.5 meV; the resolution was 37 μeV FWHM.

Our sample was a single crystal of $RbCo_cMg_{1-c}F_4$ with $c=0.58$, which is very close to the nearest neighbor percolation threshold ($c=c_p=0.593$) for a square lattice. We collected inelastic spectra at a reciprocal lattice point lying on the line (in reciprocal space) where the Bragg condition for two-dimensional antiferromagnetic order is satisfied. Thus, our experiment probes the fluctuations in the staggered magnetization, which has zero mean at all temperatures because c is slightly below c_p.

Our spectra are well described by Lorentzian profiles which correspond to simple exponential autocorrelation functions for the staggered magnetization M(t),

$$\langle M(t)\ M(t+\tau)\rangle = \exp - \Gamma|\tau| \,. \tag{3}$$

Because we expect Γ for a classical paramagnet — i.e. a collection of decoupled magnetic ions — to approach 0 according to the law $\Gamma \sim T$ as $T \rightarrow 0$, we show in Fig. 1 the reduced relaxation rate Γ/kT as a function of the inverse correlation length κa established by Cowley and coworkers[7] for the same sample. The data are well-represented by the power law

$$\Gamma/kT = A(\kappa a)^z \tag{4}$$

where $z=2.4^{+0.2}_{-0.1}$.

Our value for z is very high when compared with that ($\sim$1.7) found for pure Rb_2CoF_4.[8] However, the result that z is high is in accord with the qualitative content of Eq. (2), namely that disturbances on ramified, fractal structures require longer times to travel fixed (Euclidean) distances than they do in normal Euclidean spaces. In our first paper on this subject,[6] we suggested that the fractal geometry of percolating networks changes Ising spin dynamics in the same manner as it changes diffusion. Furthermore, we posited that the nonlinearities present in the spin relaxation, but not the particle diffusion problem, are accounted for by the static exponent η, as for the conventional theory of the kinetic Ising model.[9] We consequently guessed that for fractals,

$$z = 2 + \theta - \eta \,, \tag{5}$$

where θ also occurs in Eq. (2). Using θ and η appropriate to the two-dimensional percolation problem, $z=2.47$, in good agreement with the experimental value. Several theories[10,11] of spin relaxation near percolation have subsequently appeared. Most of these[10] use Eq. (4) as a starting point and yield expressions similar to (5) for z. Very recently, however, Henley,[11] Harris and Stinchcombe[12] have made the important observation that Eq. (4) may be incorrect for Glauber models on percolating networks. Instead, they propose a temperature-dependent exponent,

$$z_{eff}(T) = 2Z/T \,, \tag{6}$$

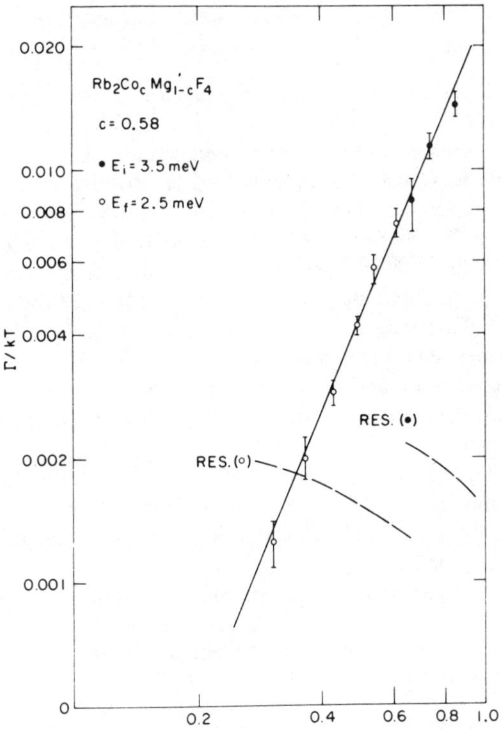

Fig. 1. Normalized relaxation rate Γ/kT plotted against reduced inverse correlation κa. At the dashed curves, the FWHM of the resolution function exceeds Γ by a factor of 4. The open and filled circles correspond to the two different spectrometers described in text.

where Z is a geometrical parameter for the fractal and T is expressed in units of the exchange coupling. The derivation of (6) is based on considerations of the energy barriers dominating the low temperature dynamics. These barriers rise logarithmically with the size L of the system, so that

$$\Gamma \sim \exp(-E_b/kT) \sim L^{Z_{eff}(T)} \exp(-E_o/kT) \qquad (7)$$

where E_o is an L-independent contribution to E_b. According to Eq. (7), there should be curvature in a double logarithmic plot of Γ/kT against κa, since for Ising systems, $\ln(\kappa a) \sim kT$. Curvature of this nature cannot be discerned in Fig. 1. Unfortunately, due to experimental limitations (see the dashed lines in the figure), we could not measure Γ for $\kappa a < 0.3$. However, we anticipate that neutron backscattering experiments[13] currently in progress at Brookhaven will permit a more stringent test of Eq. (7).

ACKNOWLEDGEMENTS

We are grateful to C. Henley for stimulating conversations and a copy of Ref. 11 to publication, and R. B. Stinchcombe for informing us of his work with C. K. Harris. The work at Brookhaven National Laboratory was supported by the Division of Materials Science, U.S. Department of Energy, under contract No. DE-AC02-76CH00016.

REFERENCES

1. See, for example, S.-K. Ma, Modern Theory of Critical Phenomena, (Benjamin; Reading, MA, 1976).
2. A review of experiments on the percolation and random field problems is given by R. J. Birgeneau, R. A. Cowley, G. Shirane and H. Yoshizawa, J. Stat. Phys. 34, 817 (1984). The theory of percolation is reviewed by D. Stauffer, Phys. Rep. 54, 1 (1979).
3. An early overview is Ill—condensed Matter, ed. R. Balian, R. Maynard, and G. Toulouse (North Holland-World Scientific; Amsterdam-Singapore, 1983).
4. H. Sompolinsky and A. Zippelius, Phys. Rev. B25, 6860 (1982).
5. R. Rammal and G. Toulouse, J. Phys. (Paris) Lett. 44, L13 (1983); Y. Geffen, A. Aharony and J. Alexander, Phys, Rev. Lett. 50, 77 (1983).
6. G. Aeppli, H. Guggenheim, and Y. J. Uemura, Phys. Rev. Lett. 52, 942 (1984).
7. R. A. Cowley, R. J. Birgeneau, G. Shirane, H. G. Guggenheim and H. Ikeda, Phys. Rev. B21, 4038 (1980).
8. M. T. Hutchings, H. Ikeda, and E. Janke, Phys. Rev. Lett. 49, 386 (1982).
9. P. C. Hohenberg and B. I. Halperin, Rev. Mod. Phys. 49, 435 (1977).
10. D. Kumar, Phys. Rev. B30, 2961 (1984); J. M. Luscombe and R. C. Desai, preprint.
11. C. Henley, preprint (1985).
12. C. K. Harris and R. B. Stinchcombe, to be published, and R. B. Stinchcombe, this volume.
13. A. Heidemann, J. D. Axe, L. Passell, D. Moncton, H. Guggenheim and G. Aeppli, private communication.

DILUTION-INDUCED CRITICAL DYNAMICS

R. B. Stinchcombe

Department of Theoretical Physics
1 Keble Road
Oxford, OX1 3NP, U. K.

1. INTRODUCTION

This lecture is concerned with the critical behaviour introduced into the dynamics of dilute systems by the divergence of the correlation length at the percolation threshold. In particular, dynamical properties of the infinite percolation cluster, and of non-random fractal models of this geometrical structure will be discussed.

At the percolation threshold[1] (where bond, or site concentration p of the diluted system takes the critical value p_c) the infinite cluster (or its backbone, which is shown in Figure 1) is scale-invariant ("self-similar"). Thus the associated correlation length ξ diverges at p_c, and in the vicinity of the threshold behaves like

$$\xi \propto |p - p_c|^{-\nu} . \qquad (1)$$

The divergence of ξ induces critical behaviour into the dynamics of dilute systems near p_c, causing the softening of modes and the eventual crossover at p_c to anomalous power law dependences characterised by dynamic critical exponents.

This behaviour occurs in the dynamical properties (e.g. of spin waves) in diluted magnets[2,3,4] which is the class of systems emphasised in this lecture. The percolation threshold p_c is already evident in the phase diagram (Figure 2) of diluted magnets as the concentration below which the transition temperature T_c is necessarily zero since for $p < p_c$ there is no infinite cluster to support spontaneous magnetisation. Behav-

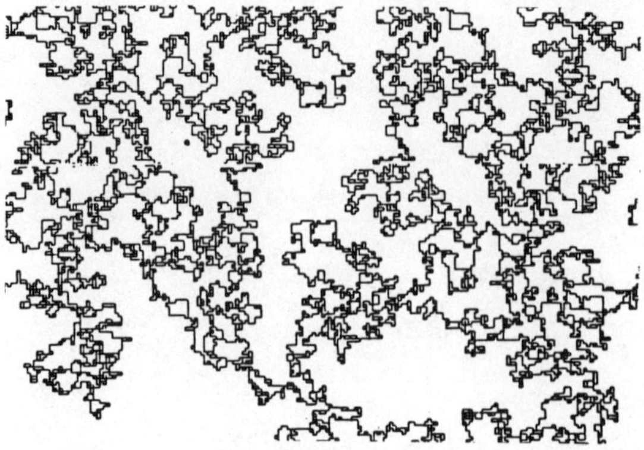

Figure 1. Backbone of infinite cluster of bond-diluted system at the percolation threshold (after Kirkpatrick).

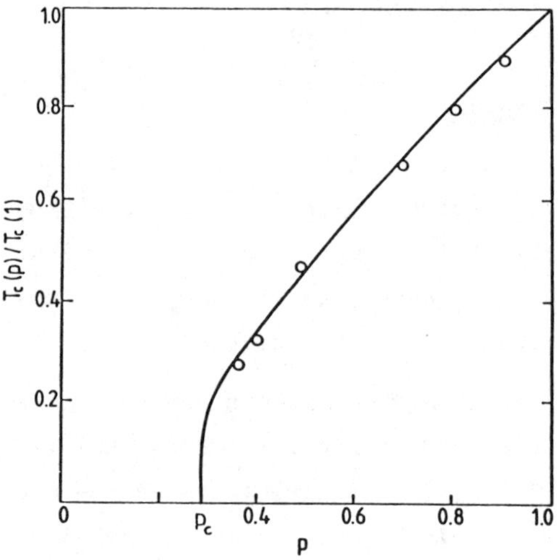

Figure 2. Phase diagram of dilute magnet; p is concentration and p_c is the percolation threshold.

iour on the zero temperature line of the magnetic phase diagram is determined entirely by geometrical (percolative) aspects, and the critical behaviour there is controlled by equation (1), as will be discussed in §2.

Lattice vibrations and diffusion[5,6,7,8] (e.g. "the ant in the labyrinth" problem) on diluted lattices also show the critical softening and crossover mentioned above. As will be pointed out where appropriate, much of our later discussions apply also to such cases; however, there can be additional effects in some cases, e.g. two different thresholds for anomalies in longitudinal and shear modes in vibrating dilute lattices because the shear rigidity disappears at a higher concentration than p_c.

Because the critical behaviour is associated with a diverging length, recursive length scaling treatments are most suitable[9,10] and this lecture will give special emphasis to these methods. These methods are also the ones capable of discussing dynamics on non random fractals[5,6,7,9,10,11,12,13], and this will also be very briefly discussed.

The usual way in which the divergence of ξ causes critical dynamics is presented in Section 2 in the form of a crossover argument starting from the dynamic scaling hypothesis[14]. This hypotheses appears to be almost universally applicable to such situations; however, one case where it appears to fail will be discussed in §9.

2. DYNAMIC SCALING HYPOTHESIS, CROSSOVER

In normal situations, the long wavelength dynamics of acoustic phonons, of ferromagnetic spin waves (or the related problem of diffusion), and so on, is described by a simple power law dependence of frequency on wave vector k (linear for phonons, quadratic for the other cases listed). On dilution, the power law remains unchanged though the coefficient (velocity of sound, spin wave stiffness D, or diffusion constant) decreases and would eventually vanish at the percolation threshold (mode softening[11]). However, as the threshold is approached, the dynamical behaviour crosses over to a new "anomalous" power law dependence characterised by a dynamic exponent z.

This crossover is caused by the divergence of ξ at p. The dynamic scaling hypothesis starts from the observation that ξ could be a relevant length, since it can become arbitrarily large and is therefore capable of causing new features in the dynamics at wavelengths very much greater than the lattice spacing. This could occur through the appearance of a functional dependence on the dimensionless quantity $k\xi$ (i.e. the ratio of correlation length to wavelength). The specific form taken in the usual dynamic scaling hypothesis is[14]

$$\omega = k^z\, f(k\xi) \qquad\qquad k, \frac{1}{\xi} \to 0 \qquad\qquad (2)$$

where z is the dynamic exponent. The different forms taken by the scaling function f(x) in the two limits $x \to 0$, correspond to normal and anomalous (critical) behaviour, which are separated by a region of crossover around $k\xi \sim 1$.

For the rest of this discussion, we take the special case of ferromagnetic spin waves[2,3,4], where the normal behaviour is

$$\omega = Dk^2 \qquad (k\xi \to 0) \tag{3}$$

(where D is, in general, a function of concentration p). This is recovered from (2) in the limit $k\xi \to 0$ provided f(x) behaves for small x like x^{2-z}. This has the implication that, in (3)

$$D(p) \propto \xi^{2-z} \propto (p - p_c)^{(z-2)\nu} \tag{4}$$

where the last step uses (1). Since, in general, $z-2 > 0$, this result implies that the stiffness D vanishes as p approaches p_c (mode softening). However, if k is fixed in (2), $k\xi \to \infty$ as $p \to p_c$ and we recover a finite result (independent of ξ) in the limit if f(x) goes to a constant at large x, making

$$\omega \propto k^z \qquad k\xi \to \infty \quad . \tag{5}$$

This is the anomalous (critical) behaviour (the "fracton" regime) in the spin wave problem at the percolation threshold ($\xi \to \infty$).

Very similar results hold for diffusion, where ω is replaced by a width (inverse lifetime), and a similar argument applies for longitudinal acoustic phonons.

One result of such arguments is a relationship between z and the exponents characterising a static quantity like D(p) (see (4)) or, in this case, through relationships analogous to the Einstein relation between diffusion and conductivity[3], the exponent t characterising the vanishing of the static conductivity[15] of the percolating network as p approaches p_c. Because the percolation probability (whose exponent is β) also enters, the full relationship is

$$z = 2 + (t - \beta)/\nu \ . \tag{6}$$

The dynamic exponent z can therefore be calculated from the static exponents t, β, ν. However, for more details of the dynamics, e.g. the full crossover, or response, a direct attack is required, and the essence of an appropriate (scaling) method is given in the next section.

3. LENGTH SCALING FOR CRITICAL DYNAMICS - SIMPLE ILLUSTRATION

In view of the scale invariance of the dilute system at the percolation threshold, a length scaling method is needed to handle the critical dynamics. The recursive nature of such approaches is required to probe the effects of the diverging correlation length ξ.

The method now presented[9,10,16,17] uses position space, because it is well suited to keeping track of the strong local disorder involved in dilution (removal of bond or site).

For illustration, the method is developed for dynamics of a pure chain. This is trivially soluble by other methods (e.g. k-space), however those methods will not be extendable, as we require, to scale invariant or random situations.

The scaling is carried out on the equation of motion, which we take to be of the form

$$(2 - \Omega)u_n = u_{n+1} + u_{n-1} \quad . \tag{7}$$

Here, u_n is the dynamic variable at site n of the chain. To be specific we think in terms of ferromagnetic spin wave dynamics, where u_n is a transverse component of spin, and then $\Omega = \omega/J$ where J is the exchange constant. However, equation (7), and our subsequent discussion, also apply to lattice vibrations (where $\Omega = m\omega^2/K$, K = spring constant) or to diffusion or Ising Glauber dynamics (where $\Omega = i\omega/J$).

The procedure is to use the equations like (7) but with n increased or decreased by 1 to eliminate u_{n+1}, u_{n-1} from (7). This results in an equation of the same form as (7), but with n ± 1 replaced on the right hand side by n ± 2 and, in place of the factor (2 - Ω), on the left hand side, a 'scaled' form (2 - Ω') where

$$\Omega' = 4\Omega - \Omega^2 \quad . \tag{8}$$

The new equation corresponds to a rescaled lattice in which the nearest neighbours of n are n ± 2, as shown on the right of figure 3. The relationship between u_n and $u_{n\pm2}$ is the same in original and rescaled lattices, so the absolute wave vector q of any periodic excitation of the

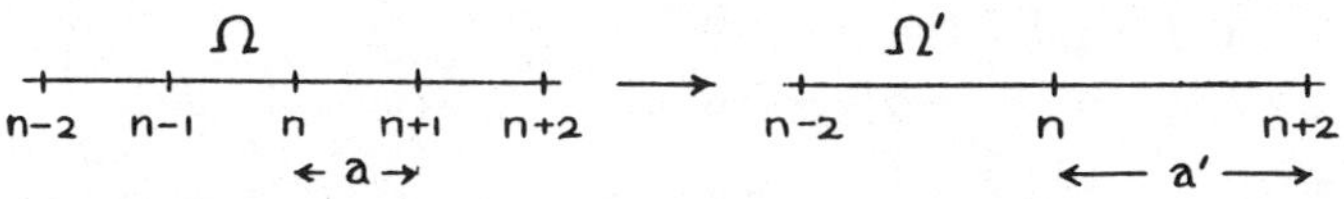

Figure 3. Original and rescaled lattices related by the decimation transformation (8).

lattice is unchanged by the scaling. For our purposes it is more convenient to use a relative (dimensionless) wave vector $k \equiv q\,a$, where a is the lattice constant. Since under the scaling the lattice constant changes from a to $a' = ba$, where in this case the dilation factor b is 2, the transformation of k accompanying (8) is

$$k' = bk\ . \tag{9}$$

The joint transformation (8), (9) can be exploited as follows:

(a) For small Ω (which is in general sufficient to cover the regime in which results like (2) apply) (8) can be replaced by the linearised form

$$\Omega' = \lambda\Omega \tag{10}$$

where $\lambda = 4$. (9), (10) imply

$$\Omega = ck^z \tag{11}$$

where

$$\lambda = b^z \tag{12}$$

and c is a constant. In the present example, $b = 2$ and $\lambda = 4$, so (12) yields $z = 2$. (11) is then the standard quadratic dispersion of long wavelength spin waves. However, the above procedure is generally applicable and shows how a critical exponent (z) is determined by an eigenvalue (λ) of a renormalisation group transformation (equation (8)) linearised (equation 10)) about its fixed point ($\Omega^* = 0$).

(b) The dispersion relation ($\Omega = F(k)$) can be obtained for general Ω from the renormalisation group transformation. Since $\Omega' = F(k')$ where (from (9)) $k' = 2k$, equation (8) gives

$$F(2k) = 4F(k) - F(k)^2. \tag{13}$$

This non-linear functional equation has the solution $F(k) = 2(1 - \cos \alpha k)$, where α is an arbitrary constant. This result is the standard dispersion relation.

We emphasise that the above scaling method is designed to treat situations where elementary methods fail, and especially the case where a diverging correlation length causes real critical behaviour. A closely related case, which we discuss in the next section, is the dynamic behaviour of non-random fractals (which are scale-invariant). This can be treated by the scaling method, while k-space techniques are inadequate.

4. DYNAMICS ON FRACTALS VIA NON-LINEAR MAP

The scaling method discussed in the previous section uses the transformation of dynamical variables resulting from eliminating a fraction of sites from the lattice ("decimation"). Many fractals are built up by the reverse of this elimination process and so, not surprisingly, they can be treated by exploiting the length scaling transformation[5-7, 9-13]. Examples to which this procedure applies include Berker lattices and Sierpinsky gaskets. Dynamical properties of this latter system are now discussed, again taking the specific case of spin wave dynamics.

Figure 4 shows a Sierpinsky gasket constructed by dividing a triangle into four triangles, discarding the central one and repeating (recursively) the procedure on the remaining triangles. This system is self-similar if the process is repeated indefinitely. The sites added at a given stage in the construction can be eliminated from the spin wave equations of motion, in the same way as was shown for the chain in §3, thereby obtaining a relation between the dynamic variables Ω, Ω' of two successive stages in the construction of the fractal. The length scales of these two successive stages differ by a factor $b = 2$, and the (exact) relationship between the dynamic variables turns out to be

$$\Omega' = (d+3)\Omega - \Omega^2 \qquad (14)$$

where d is the (Euclidean) space dimension of the gasket, i.e. $d = 2$ for the triangular Sierpinsky gasket. Equation (14) applies more generally to the whole family of hypertetrahedral Sierpinsky gaskets in arbitrary $d(\geqslant 1)$ space dimensions[11].

Figure 4. Triangular Sierpinsky gasket fractal.

Application of equation (10), (11), (12) then shows that the dynamic exponent for spin waves or diffusion on this family of fractals is

$$z = \ln(d+3)/\ln 2 \qquad (15)$$

(the exponent for phonons is smaller by a factor of 2).

The recursion relation (14) is the quadratic iterative map of the type discussed by Feigenbaum[18] and others. The case d=1 is the only simple one.

For higher d, the density of states in Ω has a fractal structure[19]. No solution exists of the non linear functional equation analogous to (13) (with 4 replaced by d+3) for d > 1.

5. CRITICAL DYNAMICS OF DILUTE HEISENBERG CHAINS: THEORY AND EXPERIMENT

The treatment of random systems requires a generalisation of the scaling technique[9,10,11,16,17,20]. For the specific case of spin wave dynamics of bond-random chains, the basic equation replacing (7) is

$$(J_{n-1\,n} + J_{n\,n+1} - \omega)u_n = J_{n-1\,n}\,u_{n-1} + J_{n\,n+1}\,u_{n+1}$$

where $J_{n\,n+1}$ is the exchange interaction between spins at sites n, n+1. The system thus involves the dimensionless random variables $\{x_n \equiv J_{n\,n+1}/\omega\}$. In the case of bond dilution, the distribution of the random variable x_n can be characterised by the weight (p) and mean ($1/\Omega$) of the non-zero values of x_n. The scaling of the distribution, and hence of these two characteristic parameters, can be obtained (approximately, because of the neglect of correlations which develop) by decimating the equations of motion to achieve a dilation by b=2 in the manner introduced in §3. The result is[20,21]

$$p' = p^2 \tag{17}$$

$$\Omega' = (1-p)^2(3\Omega - \Omega^2) + p(1-p)\Omega(8-9\Omega + 2\Omega^2)/(1-\Omega) + p^2(4\Omega - \Omega^2)$$

$$\equiv \Delta(p, \Omega) \tag{18}$$

$$k' = 2k\,. \tag{19}$$

The procedures introduced in §3 can now be applied to derive the dynamic scaling form (2).

A fixed point of (17), (18), (19) is $(p^*, \Omega^*, k^*) = (1, 0, 0)$. This is the pure situation (at which ξ diverges) in the static, long wavelength limit. (2) is expected to apply in the vicinity of this fixed point, and can be obtained by linearising (17), (18) about the fixed point, which yields

$$\Omega' = \lambda\Omega, \quad \delta p' = \lambda_p \delta p, \quad k' = bk \tag{20}$$

where $\lambda = 4$, $\lambda_p = 2$, $\delta p = p-p^*$, $b = 2$. (20) implies the result

$$\Omega' = k^z f(k\,\delta p^{-\nu}) \tag{21}$$

where $z = \ln\lambda/\ln b = 2$, $\nu = \ln b/\ln\lambda_p = 1$. This is a form of (2) in which the explicit dependence of the correlation length on δp occurs.

(17) has a second fixed point, $p^* = 0$ corresponding to the empty chain

($\xi = 0$). For the two fixed point values of p ($p^* = 0, 1$) the frequency scaling (18) reduces to

$$\Omega' = \lambda\Omega - \Omega^2 \tag{22}$$

where $\lambda = 3, 4$ respectively. These two cases are again special cases of the quadratic iterative map, associated with cyclic and with chaotic behaviour of the iterated variable. These two behaviours are related respectively to (localised) dynamics on the finite chain segments into which the very dilute chain decomposes and (extended state) dynamics on the pure chain[20].

A more detailed (exact) discussion of the dynamic response of the diluted chain is possible by superimposing the contributions from the separated finite segments which make up the diluted chain[21,22]. This results in a dynamic scaling form for the response function, which is shown in Figure 5 as a function of the two scaling variables $k\xi$, ω/k^z. The main features of the theoretical response function are exhibited in results from neutron scattering experiments on a diluted chain Heisenberg anti-ferromagnet[23].

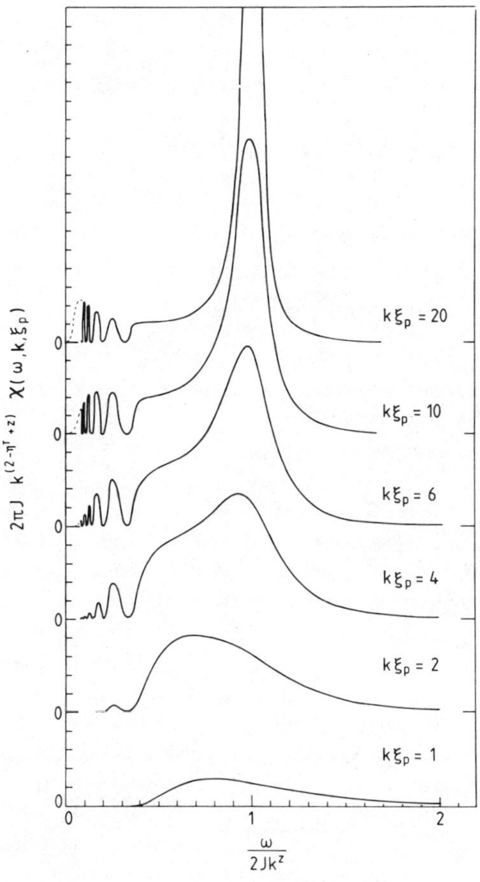

Figure 5. Universal scaling plot of dynamic response function for diluted ferromagnet.

6. CRITICAL SPIN WAVE DYNAMICS AT PERCOLATION THRESHOLD: DYNAMIC EXPONENT

The application of the scaling technique to dilute systems in more than one dimension involves the smae principles as discussed in the preceding section (scaling of variables characterising a distribution). Typically the equations of motion are transformed by eliminating the sites on a sublattice. In more than one dimension, this usually results in a proliferation of couplings on the rescaled lattice. This effect can be suppressed by the use of bond moving techniques. In this way, the transformations of the wave vector k, bond concentration p, and characteristic reduced frequency Ω(see §5) can be obtained for a specific dilatation, and are typically of the form

$$p' = R(p) \tag{23}$$

$$\Omega' = S(p, \Omega) \tag{24}$$

$$k' = bk. \tag{25}$$

A particular calculation of this type[11], using a $b = \sqrt{3}$ dilation of the honeycomb lattice, has been used to obtain the spin wave dynamic exponent z at the percolation threshold from the eigenvalue

$$\lambda = \frac{\partial S}{\partial \Omega} = 4.56 \ (=b^z) \tag{26}$$

where the derivative is evaluated at the fixed point $(p, \Omega) = (p_c, 0)$ of the transformation. This results in $z = 2.76$ for the two dimensional case, which is in good agreement with the result obtained by inserting values for the static exponents t, β into (6) and with Monte Carlo results for the ant in the labyrinth problem[8]. Work is presently in progress on the dynamic exponent for the dilute layer antiferromagnet at the percolation threshold, since that system is likely to be more amenable to experimental investigation.

7. ANOMALOUS DENSITY OF STATES AT PERCOLATION THRESHOLD

The spin wave density of states $\rho(\omega)$ for a pure two dimensional (d=2) ferromagnet is constant at small ω because the volume in k space (dk^2) is proportional to $d\omega$ because of the quadratic dispersion. The modified dispersion law (5) which applies at the percolation threshold will obviously change the density of states there. There is a further reason for a change in $\rho(\omega)$namely the reduced ("fractal") dimensionality of the system at p_c. For the infinite cluster at p_c, this fractal dimensionality is

$$d_f = d - \beta/\nu \tag{27}$$

where β, ν are percolation exponents (β/ν is the "anomalous" dimension of percolation). (27) can be derived by noting that the number of infinite cluster sites per unit volume is proportional to the percolation probability which, by definition of β, is proportional to $(p - p_c)^\beta$ and hence (using (1)) to $\xi^{-\beta/\nu}$. A dilation of the infinite cluster by b thus scales the number of sites by $b^{d-\beta/\nu}$, leading to the effective dimensionality (27). It follows that the density of (spin wave) states on the infinite cluster at p_c diverges at small frequency like[5,6,12,24]

$$\rho \propto \omega^{\bar{d}-1} \tag{28}$$

where the "spectral" dimension defined by (28) is

$$\bar{d} = \frac{1}{z}\left(d - \frac{\beta}{\nu}\right). \tag{29}$$

A direct investigation of the spin wave density of states in a two dimensional system at the threshold has been carried out by using the Lanczos algorithm to diagonalise the dynamic matrix for specific realisations of the random infinite cluster[24]. The result for the associated exponent ($1-\bar{d} = 0.32 \pm 0.01$) agrees within the error with those obtainable with the use of (29) with z calculated directly (as in §6) of via the further scaling relation (6).

The crossover from the behaviour (28) (which applies where $k\xi >> 1$) to the normal behaviour $\rho \propto \omega^o$ (for $k\xi << 1$) has also been considered, and it has been proposed[25,26,27] that the crossover region $k\xi \sim 1$ is marked by a well-defined edge (the "fracton edge") in the plot of ρ versus ω. The edge is certainly necessary unless a special relationship holds between the various exponents introduced above[28]. Whether or not the relationship does hold, making the fracton edge unnecessary, is not at present clear. In any case, neutron inelastic scattering exponents on resins seem to confirm the existence of an edge[29].

8. ISING CRITICAL DYNAMICS AT THE PERCOLATION THRESHOLD: EXPERIMENT

Up to now, the critical dynamics of soft spin and similar problems has been discussed. This and the next section are concerned with Ising dynamics at the percolation threshold.

Dilution-induced Ising critical dynamics has been the subject of a recent neutron scattering experiment[30]. The system investigated, $Rb_2Co_pMg_{1-p}F_4$, is an Ising layer antiferromagnet (d=2) and the concentration used (nominally p = 0.58) is so close to the critical concentration (p_c = 0.593 ...) for the square-lattice layers that ξ/a attains values of about 250 at very low temperatures. The low temperatures at which the neutron scattering is carried out are such that the correlation length (though still large) is more than ten times the very low temperature value, so it is essentially the thermal correlation length $\xi_T \equiv \xi(p_c, T)$.

The results for neutron width Γ versus a/ξ are shown as a log-log plot in Figure 6. The data points fall close to a straight line, suggesting the following behaviour for the characteristic time

$$\tau \equiv {}^1/\Gamma \propto \xi_T{}^z \tag{30}$$

where $z = 2.39 \begin{smallmatrix} + 0.15 \\ - 0.04 \end{smallmatrix}$. The dynamic exponent for the pure system Rb_2CoF_4 at the critical temperature is $z_{pure} \sim 1.7$.

The experiment thus demonstrates a new dilution induced critical behaviour. The theoretical discussion given in the next section shows however that the dynamic scaling form (30) does not apply in this situation and has to be replaced by an entirely new relationship between τ and ξ_T. A consequence is that if (30) is used, the effective exponent z will be seen to be temperature dependent (diverging as $T \to 0$) if the range of temperatures investigated is extended further down.

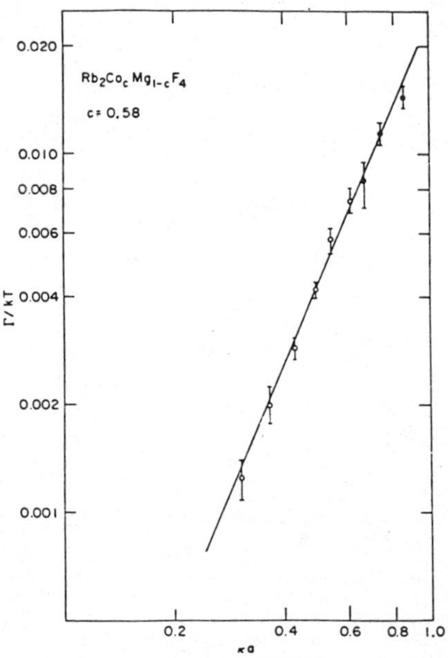

Figure 6. log-log plot of neutron width versus inverse correlation length for diluted d=2 Ising antiferromagnet at the percolation threshold.

9. ISING CRITICAL DYNAMICS AT PERCOLATION THRESHOLD: THEORY

A theoretical discussion[31,32] is now presented of low temperature Ising Glauber dynamics at p_c. The dynamic mechanisms and the influence of the self similar geometry of the infinite cluster are first discussed (in (i) below). These indicate the ingredients of a scaling description (given in (ii) below) leading to a breakdown of dynamic scaling and a new relaxational critical dynamics.

(i) Diffusion, activation and "branchings on branchings"

Dynamics of Ising models at low temperatures involves the motion of domain walls separating up-spin regions from down-spin ones (the discussion of this section is given for the ferromagnet, but applies with obvious modifications to the antiferromagnet). This wall motion is the essential relaxation mechanism. On a pure chain, it costs no energy to move the wall one "step" along (or back) by flipping a spin. The motion is therefore diffusive (c.f. the solution of Glauber dynamics on the Ising chain).

In a dilute Ising system at p_c, the low temperature dynamics is modified by the geometry of the infinite cluster. The backbone of the infinite cluster (Figure 1) includes chain sections along which the domain wall diffuses until it meets a site (site A of Figure 7) at which the chain branches into two or more chains (which occur within a more connected part of the cluster). To flip the spin at A and move the wall into the branched region costs energy 2J, and the rate for this activation process is therefore proportional to

$$\varepsilon \equiv e^{-2J/kT} \quad . \tag{31}$$

In a more general situation in which there is initially an excess of m neighbours parallel to the flipping spin, the rate suppression factor is ε^m, which is very small at low temperatures.

Inspection of typical sections of the infinite cluster backbone shows that a wall encounters a succession of "branchings on branchings", due to the self-similarity of the system at p_c. This self-similarity can be treated by scaling methods, one of which is given immediately below, which obviously must also incorporate the other essential ingredients of the low temperature Ising dynamics at p_c: diffusion and activation (especially the rate suppression factors ε^m).

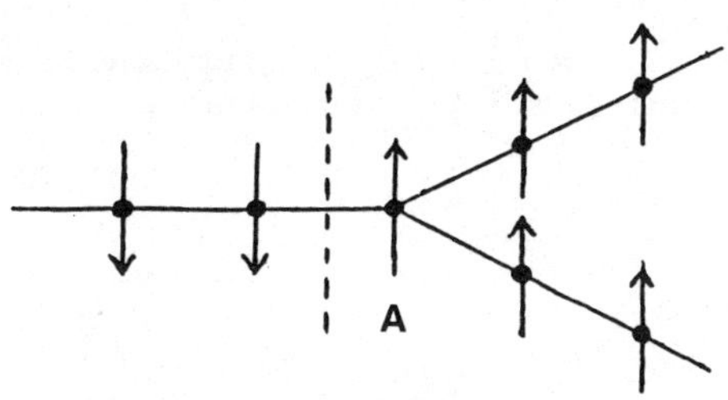

Figure 7. Ising domain wall (dashed line) approaching branching site A on infinite cluster backbone.

(ii) Recursive treatment: breakdown of usual dynamic scaling

Three recursive treatments have been applied to the problem[31]. The one given here involves scaling on a random cluster, and provides a simple framework in which intuitive ideas can be retained without losing the essential ingredients.

If universality ideas apply (see later) contact with the layer system investigated in the experiment can be made by treating any two dimensional dilute lattice and using bond dilution in place of site dilution. The honeycomb lattice is simplest to use because of its low coordination number. A length scaling by b=2 can be achieved by decimating sites to replace the original lattice (full lines in Figure 8) by a rescaled lattice (dashed lines).

One bond of the rescaled lattice represents a hexagonal cell of the original lattice. Each bond of such a cell is present with probability p or absent with probability (1-p) where p is the bond concentration. For our purposes it is necessary to distinguish between configurations of the hexagonal cluster with no paths across, with a single path across (configurations 1a, 1b in Figure 9), and with two paths across (configuration 2 in the figure).

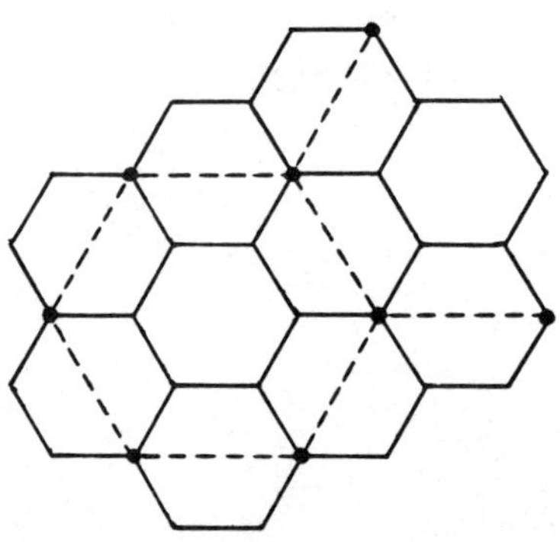

Figure 8. Original and scaled honeycomb lattices (b=2).

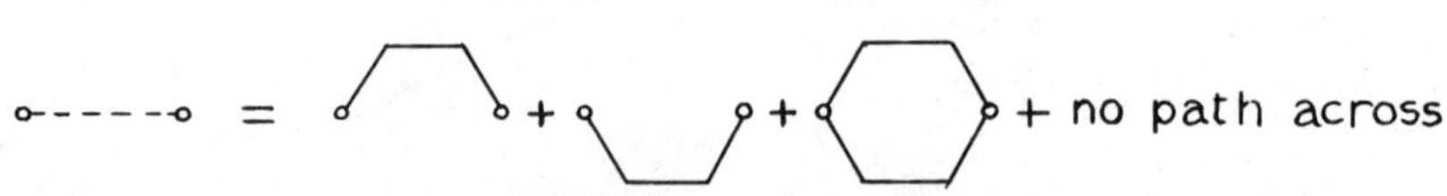

Figure 9. Configurations of a random cluster representing a single renormalised bond of a rescaled honeycomb lattice.

The respective probabilities of configuration 1 (= 1a+1b), 2 are $2p^3(1-p)^3$, p^6, and the sum of these is the rescaled bond probability

$$p' = 2p^3(1-p^3) + p^6 . \tag{32}$$

Relative probabilities of the connecting configurations 1, 2 are

$$Q_1 = 2p^3(1-p^3)/p' \quad ; \quad Q_2 = p^6/p' \tag{33}$$

and these are related to the probabilities of proceeding through the random cluster without or with branching.

The critical dynamics can be obtained from the transformation of the variable τ_o, defined as the time taken for the domain wall to traverse one bond of the lattice (i.e. for one diffusion step). Its scaled value, τ_o', is the time to traverse a renormalised bond, i.e. to traverse configurations 1 or 2. The respective times for these two configurations are

$$\tau_1 = 9\,\tau_o \tag{34}$$

$$\tau_2 = 9\,\tau_o\, c/\varepsilon \ . \tag{35}$$

τ_1 involves only diffusion, and the factor 9 occurs because the diffusion time for N steps (here N=3) is proportional to N^2. τ_2 again involves diffusion across a length of three bonds, but also a slowing factor c/ε (where c is a constant) from the rate suppression associated with activation onto the (branched) cluster.

Under scaling, τ_o therefore develops into a random variable. A good approximation to its scaling can be obtained by averaging the outcome (34), (35), which yields

$$\tau_o' = 9Q_1(p)\tau_o + 9Q_2(p)\, c\varepsilon^{-1}\, \tau_o . \tag{36}$$

At small temperature, τ_2 completely dominates τ_1, because of the activation factor, and it is therefore sufficient to discard the first term on the right hand side of (36). This result appears more convincing if one considers the roles of τ_1, τ_2 (with $\tau_2 >> \tau_1$) in the combined random processes involved in traversing the section of the original random lattice represented after several scalings by a single bond of the renormalised lattice.

We now specialise to the situation at the percolation threshold, $p = p_c$, where p_c is the fixed point of (32). There, the simplified low temperature scaling equation becomes

$$\tau_o' = 9Q_2(p_c)\, c\, \varepsilon^{-1}\, \tau_o \qquad (b=2). \tag{37}$$

This has to be taken together with the equation for the scaling of the correlation length at p_c. Here the correlation length is the thermal length and we use ξ_T to represent its absolute value divided by the lattice spacing. The required scaling is therefore

$$\xi_T' = \xi_T/b. \tag{38}$$

The fixed point of interest is more fully $p = p_c$, $T = 0$. A result of the form (30) (usual dynamic scaling) would be obtained there, by arguments like those leading from (9), (10) to (11), if at the fixed point (37) reduces to $\tau_o' = \lambda\tau_o$. However, this reduction does not occur because the prefactor of τ_o on the right hand side of (37) diverges as $T \to 0$ because of the activation factor ε^{-1}. Dynamic scaling therefore fails.

The new critical dynamics replacing (30) can be obtained as follows. One of the static properties of dilute Ising magnets at the percolation threshold is

$$\xi_T \propto \varepsilon^{-\nu} \tag{39}$$

where ν is the percolation exponent used before; that is, the percolation-thermal crossover has unit crossover exponent. Thus (37) can be re-written in the form

$$\tau_o' = \alpha\, \xi_T^{1/\nu}\, \tau_o \tag{40}$$

where α is a constant. The combined scaling (38), (40) can easily be seen to leave the following quantity invariant under scaling

$$I(\tau_o , \xi_T) \equiv \ln \tau_o + A(\ln \xi_T)^2 + B \ln \xi_T \tag{41}$$

provided $A = [2\nu \ln b]^{-1}$, $B = (1/2\nu) + \ln \alpha/(\ln b)$. That is, if $\tau_o(\ell)$ is the time to traverse a bond, or renormalised bond, of length ℓ,

$$I(\tau_o(a), \xi_T) = I(\tau_o(ba), \xi_T/b). \tag{42}$$

By putting $b = \xi_T$, we can therefore recover from (42), (41) the time for the domain wall to move a distance of the order of the absolute correlation length:

$$\tau \equiv \tau_o(\xi_T a) = \tau_o\, \xi_T^{A\ln\xi_T + B}. \tag{43}$$

This is the characteristic time associated with the critical dynamics, and is proportional to the inverse width measured in the low temperature neutron scattering experiment at p_c.

The dependence (43) of the characteristic time on the thermal correlation length ξ_T at the threshold is quite different to (30) as expected from the breakdown of the step which would have led to usual dynamic scaling. However, in a restricted temperature region where $\ln\xi_T$ varies rather little, (43) can approximate to (30) with an effective exponent z dependent on the region of temperature and this can account for the fit of (30) to the experiment. According to (43), a log-log plot of τ versus ξ_T(c.f.Figure 6) should be quadratic, but for the values of ξ_T occurring in the experiment, the curvature should be quite small because A is (according to the present calculation) only about 0.54. It would clearly be desirable to see if the "effective exponent" z associated with a fit of experimental (or simulation) data to the form (30) diverges as the temperature is lowered towards zero, as predicted by (43).

Universality was invoked at the outset of this discussion. It is by no means clear that the exponents A and, expecially, B are universal in the usual sense, and a real space calculation of this type is unable to decide the point. The value of A quoted above must therefore be accepted with caution. The universality question does not however affect the analytic form (43) arrived at, which should apply to low temperature Ising dynamics on any lattice of dimensionality greater than one at the threshold.

10. CONCLUSION

An attempt has been made to provide a survey of recent work on dilution-induced critical effects in dynamics of mainly soft spin systems, and to provide the length scaling methods required to treat such effects. The anomalous behaviour occurring near p_c is analogous to that discussed elsewhere for quasi scale-invariant phonon and other systems. Extensions being presently considered include non-linear and quantum effects. In addition, section 9 gives an account of low temperature critical dynamics of Ising systems at the threshold, leading to the replacement of the usual dynamic scaling by a new form, which awaits experimental confirmation.

REFERENCES

1. J. W. Essam, Rept. Prog. Phys. 43: 833 (1980)
2. R. B. Stinchcombe, Dilute Magnetism, in: "Phase Transitions and Critical Phenomena", C. Domb and J. L. Lebowitz, ed., Academic Press (1983)
3. S. Kirkpatrick, Models of Disordered Materials, in: "Ill-Condensed Matter", R. Balian, R. Maynard and G. Toulouse, ed., North Holland (1979)
4. I. Ya. Korenblit and E. F. Shender, Usp. Fiz. Nauk. 126: 233 (1978) [Sov. Phys. Usp. 21: 832 (1978)]
5. S. Alexander and R. Orbach, J. Phys. (Paris) Lett. 43: 625 (1982)
6. R. Rammal and G. Toulouse, J. Phys. (Paris) Lett. 44: L13 (1983)
7. A. Aharony, STATPHYS. Proceedings, J. Stat. Phys. (1984)to be published: and Geilo Lectures (1985)
8. R. B. Pandey and D. Stauffer, Phys. Rev. Lett. 51: 527 (1983)

9. R. B. Stinchcombe, in: "Highlights of Condensed Matter Physics", F. Bassani, F. Fumi and M. P. Tosi, eds., Proceedings of the Varenna Summer School, North Holland (1983)
10. R. B. Stinchcombe, Dynamic Scaling in Pure and Dilute Systems, in: "Static Critical Phenomena in Inhomogeneous Systems", A. Pekalski and J. Sznajd, ed., Lecture Notes in Physics, vol. 206, Springer (1984)
11. C. K. Harris and R. B. Stinchcombe, Phys. Rev. Lett. 50:1399 (1983)
12. R. Rammal and G. Toulouse, Phys. Rev. Lett. 49: 1194 (1982)
13. H. E. Stanley, Geilo Lectures (1985)
14. B. I. Halperin and P. C. Hohenberg, Phys. Rev. 117: 952 (1969)
15. R. B. Stinchcombe and B. P. Watson, J. Phys. C9:3321 (1976)
16. C. E. T. Goncalves da Silva and B. Koiller, Solid St. Commun. 40: 215 (1981)
17. B. W. Southern, A. A. Kumar, P. D. Loly and A-M. S. Tremblay, Phys. Rev. B27: 1405 (1983)
18. M. Feigenbaum. J. Stat. Phys. 19: 25 (1978); and 21: 669 (1979)
19 E. Domany, S. Alexander, D. Bensimon and L. P. Kadanoff, Phys. Rev. B28: 3110 (1983)
20. R. B. Stinchcombe, Phys. Rev. Lett. 50: 200 (1983)
21. R. B. Stinchcombe and C. K. Harris, J. Phys. A16: 4083
22. A. C. Maggs and R. B. Stinchcombe, J. Phys. A17: 1555
23. J. P. Boucher, C. Dupas, W. J. Fitzgerald, K. Knorr and J. P. Renard, J. Phys. (Paris) 39: L86 (1978); Y. Endoh, G. Shirane, R. J. Birgeneau and Y. Ajiro, Phys. Rev. B19: 1476 (1979)
24. S. J. Lewis and R. B. Stinchcombe, Phys. Rev. Lett. 52:1021 (1984)
25. R. Orbach, Geilo Lecture (1985)
26. B. Derrida, R. Orbach, Kin-Wah Yu, Phys. Rev. B29: 6645 (1984)
27. Ora Entin-Wohlman, S. Alexander, R. Orbach and Kin-Wah Yu, Phys. Rev. B29: 4588 (1984)
28. A. Aharony and S. Alexander, to be published; A. C. Maggs and R. B. Stinchcombe, to be published
29. H. M. Rosenberg, Phys. Rev. Lett., to be published
30. G. Aeppli, H. J. Guggenheim and Y. J. Uemura, Phys. Rev. Lett. 52: 942 (1984)
31. C. K. Harris and R. B. Stinchcombe, to be published
32. A slightly different version of this approach was first given in C. K. Harris, D. Phil. Thesis, University of Oxford (1983); and it is understood that a fractal model with similar ingredients has been discussed by C. Henley.

SCALING IN POLYMER PHYSICS

R. C. Ball

Cavendish Laboratory
Cambridge
England

INTRODUCTION

Scaling ideas form a central part of our understanding of the behaviour of long, chain-like, flexible macromolecules. As the chain length or molecular weight M becomes sufficiently large, chemically different polymers exhibit universal enhancement of various molecular properties, e.g. coil dimensions and relaxation times. It is the enhanced properties that make polymers such important practical materials - the viscosity of an oil, the elasticity of a rubber, the setting of gels, the stabilization of suspensions, the toughness of 'plastics' and many more. There is, for certain properties at least, an underlying thermodynamic critical point, via an analogy where the conventional role of $T-T_c$ is played by $1/M$. It is explicitly the incidence of high molecular weight that corresponds to critical and scaling behaviour and not a narrow range of parameters.

The outline of these notes is as follows. We will begin with some conformational properties of an isolated polymer chain, where scaling laws can be underpinned by mathematical results and scaling in magnetic critical phenomena. In treating polymer solutions which are not strictly dilute we will depend more on physical arguments and scaling as an ansatz. Finally in addressing the dynamics of polymers and in particular entangled behaviour, the interplay of scaling with phenomenology, albeit of a sophisticated variety, becomes crucial.

THE STATICS OF A SINGLE COIL

The first problem in polymer physics is the single polymer coil of N monomer units connected flexibly in a linear sequence. As well as being connected, the units will also have interactions which are not specific to adjacent members of the sequence: the excluded volume constraint that two monomers cannot be at the same place, Van der Waals forces (longer ranged and canonically attractive) between monomers, and finally in solution there are energies of solvation, monomer-monomer contacts differing in energy from monomer-solvent ones.

If we retain only the primary element of connectivity, the ensemble of conformations of the polymer is just the ensemble of random walks of N-1 steps. The freedom of choice at each step is governed by the flexibility of the molecule, but at large enough scales ($>\ell$) simple gaussian behaviour will ensue. Thus the mean square separation of two monomers n units apart down the chain is given by

$$\langle R^2(n)\rangle = na\ell$$

where a is the (literal) length of a monomer unit and ℓ is the persitence length introduced by Kuhn. Typically ℓ might correspond to the lenght of 8 or so monomers. Furthermore the probability distribution of $R(n)$ is a simple gaussian

$$P_n(R) \sim \exp -d/2\ R^2/na\ell$$

and beyond scales of length ℓ the chain is a statistically self similar fractal of dimension $d_f=2$; then all (unswollen) polymers look alike.

Restictions on the local bonding affect only the value of ℓ, but the remaining interactions can and generally do change the overall fractal structure. The mean density of monomers c_{int} in a large gaussian coil is low (in >2 dimensions),

$$c_{int} \cong N/R^d \sim R^{2-d}$$

so that it is appropriate to treat the additional interactions between monomers U in terms of a monomer-monomer virial coefficient,

$$v = \int d^dr(1 - e^{-U/kT})$$

Both this quantity and the persitence length ℓ must be renormalized to take into account the correlations due to local connectivity. Everything then depends on the sign of the resulting effective excluded volume v : if it is positive (repulsive) a large coil will be swollen (see below) to a fractal dimension $d_f<2$; if it is negative (attractive) the coil will collapse to a ball ($d_f=d$) and only for $v=0$ is the coil essentially gaussian. A temperature $T=\Theta$ at which $v=0$ is known as a Flory or Θ Temperature, and corresponds to an unstable fixed point in the sense that for long enough chains any small net v will drive the coil to swell or collapse. In $d>3$, $v=0$ is sufficient to give a gaussian coil, but in lower dimensions (e.g. $d=2$) it is necessary to consider three monomer effects.

Coil Swelling

In polymer physics it is conventional to write $R \sim N^{\nu}$, where it is understood that R is a r.m.s. radius of gyration and $\nu=1/d_f$. For the ideal gaussian chain we have $\nu=\frac{1}{2}$ and we will now give Flory's argument for ν in the repulsive interaction case. The gaussian distribution applied to the end to end radius of the coil corresponds to a free energy of extension $d/2\ kT\ R^2/Na\ell$ to which we add a term for the monomer repulsions with the assumption that each monomer sees the mean density inside the coil. This gives a total free energy

$$F(R)/kT \cong d/2\ R^2/Na\ell + N\ vN/R^d$$

where we assume that the end to end radius and coil density will both scale according to the radius of gyration. For $v>0$ this has a minimum with

$$R = R_F \sim N^{3/(d+2)}$$

whereas for $v<0$ we would have to look to higher order terms in the density to find the collapsed solution. The generalized Flory formula for ν,

$$\nu = {}^{3}/_{d+2}$$

is extraordinarily successful: exact in d=4, 2 (probably), 1 and within 3% in d=3. It correctly identifies the upper critical dimension where $\nu=\frac{1}{2}$ as d=4, above which the repulsions are irrelevant to the exponents, but no-one has succeeded in making the Flory estimate the starting point of a systematic exact approach.

The 'n=0 trick' introduced by de Gennes provided the first formal approach to calculating swollen polymer exponents, by mapping the partition function of the polymer into the spin-spin correlation function of a classical ferromagnet with n->0 spin degrees of freedom. The end to end vector of the polymer is fixed and the chain length allowed to range over a chemical ensemble giving

$$Z(R,\mu) = \sum_N e^{-\mu N} Z_N(R)$$

where $Z_N(R)$ is the partition function of a chain of N segments spanning vector R . Then

$$Z(R,\mu) = < S_1(0)\ S_1(R) >_{n=0}$$

for a ferromagnet with Hamiltonian density

$$H = {}^{\ell}/_{6}(\nabla S)^2 + \mu S^2 + vS^4$$

where $S^2 = \sum_{\alpha=1}^{n} S_\alpha S_\alpha$ and the correlation is evaluated between the 1 components of the spin vectors before the limit n=0 is taken. Hence it can be seen that N is conjugate to μ equivalent to $T-T_c$. The correlation length divergence $\xi \sim |T-T_c|^{-\nu}$ then gives $R_F \sim N^\nu$ motivating our notation for the polymer exponent and identifying it with the n->0 limit of the correlation length exponent in the n vector model. Thus for example in expansion in $\varepsilon=4-d>0$ we have

$$\nu = \tfrac{1}{2}(1 + \varepsilon/8 + \text{order } \varepsilon^2)$$

indicating that the corresponding expansion of the Flory formula is wrong. All renormalization calculations, direct or indirect, support this conclusion.

The magnetic susceptibilty exponent γ governs the (power law part of the) polymer partition function,

$$Z_N(R) = \int d^dR\ Z_N(R) \quad \sim \quad e^{\mu_c N}\ N^{\gamma-1}$$

and the entire distribution of the end to end vector has a scaling form inferred from that of the spin-spin correlation function,

$$P_N(R) = Z_N(R)/Z_N = R_F^{-d}\ f_P(R/R_F) \quad .$$

From the magnetic specific heat divergence we get the statistical weight for the chain to be closed into a ring,

$$Z_N(a) \sim e^{\mu_c N}\ N^{-2+\alpha} \sim e^{\mu_c N}\ R_F^{-d}$$

where a is a monomer size, and from this it follows that $f_P(x) \sim x^{(\gamma-1)/\nu}$ as x->0 . The magnetic scaling laws $\alpha+2\beta+\gamma=2$, $\alpha=2-\nu d$ and $\gamma=\nu(2-\eta)$ have been used to restrict to two independent exponents γ and ν .

The large x behaviour of $f_P(x)$ can be found by a scaling argument for the polymer: we anticipate the form $e^{-F/kT}$ where the reduced free energy cost F/kT is proportional to N at fixed imposed span per monomer R/N . Hence

$$f_P(x) \sim \exp -x^\delta \quad \text{where} \quad (R/R_F)^\delta \sim (R/N)^\delta N \quad \text{giving} \quad \delta\nu=\delta-1$$

so that

$$\delta = {}^{1}/_{1-\nu} \ .$$

COIL OVERLAP AND THE POLYMER MELT

Large polymer coils are such open structures that they rarely occur in the dilute state, which here demands that the coils are not significantly forced to overlap. In d=3 the threshold is at

$$c = c^* \doteq M/(4/3\,\pi\, R_F^3) = c_{int}$$

where the overall concentration c matches that within an isolated coil. For swollen coils in general

$$c^* \cong M\, R_F^{-d} \sim M^{1-d\nu} \sim M^{-4/5} \text{ in } d=3.$$

To understand what happens beyond this overlap concentration we consider first the extreme case, the polymer melt; subsequently we will address intermediate cases by scaling.

In his classic textbook Flory argued that the coils in a polymer melt must be gaussian because each chain is no more repelled by itself than by all the others around it, and so has no preference to swell. S.F. Edwards gave a diagrammatic justification in terms of screening of the repulsions, and de Gennes pointed out the way the process could be visualised in terms of a correlation hole. There is a deficit in the density of all other chains exactly corresponding to the density profile of the chain of interest. Any doubts were finally dispelled by small angle neutron scattering from a small proportion of labelled (deuterated) chains in an ordinary polymer melt. See the paper by Cotton et al. (1974).

Semi-Dilute Solutions

A polymer solution is said to be semi-dilute if the coils are overlapping but the monomer concentration is still low. We can now understand this regime in terms of a blob scaling picture and a crossover to the screened gaussian behaviour of the melt at large scales. Consider each chain as N/g blobs of mass g . The polymer can be fully swolen inside these blobs provided that we do not exceed

$$c/g\, \xi^d \cong 1$$

where $\xi \sim g^\nu$ is the radius of a blob. These blobs then fill space without interpenetrating, and

$$\xi \sim c^{-1/(d-1/\nu)} \sim c^{-3/4} \text{ in } d=3.$$

Beyond the scale of a blob the coils must interpenetrate (overlap) and we expect a 'melt' of blobs with gaussian behaviour. Thus for example the radius of gyration will obey

$$R^2 \cong N/g\, \xi^2 \cong N/c\, \xi^{2-d}$$
$$\sim N\, c^{-(2-1/\nu)/(d-1/\nu)} \sim N\, c^{-1/4} \text{ in } d=3,$$

as shown experimentally in the paper of Daoud et al.

In general one expects a scaling function

$$R^2 = R_F^2\, f_R(c/c^*)$$

where R_F is the radius of a coil in dilute conditions and we have $f_R(0)=1$ and

$$f_R(x) \sim x^{-(2\nu-1)/(d\nu-1)} \quad \text{as } x\to\infty.$$

Physically, the blob size ξ may be identified as a correlation length of the polymer density fluctuations, characterizing a crossover from the fractal correlations of the swollen coil to the mean density of the solution. It was also shown by des Cloizeaux that the magnetic analogy of de Gennes can be extended to the semi-dilute solution case by the introduction of an external magnetic field, whereupon the blob size becomes directly the correlation length of magnetic order.

This semi-dilute behaviour does not extend all the way to the polymer melt because eventually ξ comes down to the scale of the persistence length ℓ. Screening cannot then further reduce coil dimensions and we have a '(semi-)concentrated regime' where the chain conformations are fairly indpendent of concentration. Strictly it is to this regime that the diagrammatic screening calculations apply exactly.

A quantity of direct importance in polymer solutions is the osmotic pressure Π, given in terms of the Helmholtz free energy F by

$$\Pi = -(\partial F/\partial V)_T$$

for a fixed mass of (polymer) solute in volume V of solution. We again expect a scaling form in dilute and semi-dilute solutions,

$$\Pi = kT\ C\ f_\Pi(c/c^*)$$

where C is the number of chains per unit volume and $f_\Pi(0)=1$ yields the standard colligative law for dilute solutions. For strong coil overlap $c>c^*$, we expect chain ends to be unimportant and Π to be a function of c=CM only. This requires that

$$f_\Pi(x) \sim x^{1/(d\nu-1)} \quad \text{as} \quad x\to\infty$$

and so for $c>c^*$

$$\Pi \cong kT\ \xi^{-d} \sim c^{d/(d-1/\nu)} \sim c^{-9/4} \text{ in } d=3,$$

a law for which there is considerable experimental evidence. See for example the data in the paper by Ferry.

In the concentrated regime binary contacts alone would give $\Pi = \frac{1}{2}vc^2$ but as the monomer concentration becomes large higher order virial effects take over.

POLYMER DYNAMICS

The most important feature of polymer dynamics is the occurrence of long relaxation times. Not only are polymer coils large, but since their equilibrium shapes are primarily governed by entropy their principle mechanism of recovery from deformation is diffusion. Polymer coils also have the intriguing property that since they contain a great many internal degrees of freedom they can sustain very large deformations in their overall shape whilst not being strongly perturbed at fine scales: this gives them characteristic and universal non-linear behaviour.

Dilute Coils

Locally monomers must comove with the fluid flow so that a coil in a steady shear flow must impede the flow in its own vicinity to avoid being indefinitely stretched out. At low enough shear rates the distortion of the coil shape will be negligible and so flow is impeded over a region spanned by the unperturbed coil, a volume of order R_F^3, where here as below we specialise to three space dimensions. This then realizes the Einstein behaviour for hard spheres in suspension and we have a viscosity increment from η_o for pure solvent to

$$\eta \cong \eta_o(1 + 5/2\ 4/3\ \pi\ R_\eta^3\ C)$$

which defines a viscosimetric radius of the coils R_η. In practice R_η/R_F is a universal constant for large enough swollen coils. Thus measurements of the 'intrinsic viscosity'

$$|\eta| = \lim_{c\to 0} (\eta - \eta_o)/c$$

give directly a quantity proprotional to $1/c^*$ and can be calibrated as a molecular weight determination.

The coil diffusion coefficient likewise realizes a Stokes limit

$$D = kT/\ 6\pi\eta_o R_H$$

defining a hydrodynamic radius R_H which through D can readily be determined by quasi-inelastic light scattering. In principle R_H/R_F should also approach a universal constant for swollen coils, but in practice corrections due to finite molecular weight are quite significant here. Then at least qualitatively we have the scaling behaviour $R_H \cong R_\eta \cong R_F$.

Still on hydrodynamic scaling grounds we expect the coil to have a longest relaxation time of order

$$\tau_R \cong R^2/D \cong \eta_o R^3/\ kT$$

with an associated modulus of order $G \cong kT\ C$ which yields the Einstien viscosity increment. For shear rates $\dot\gamma$ which are not small we then expect a scaling function

$$\eta(\dot\gamma) = \eta_o(1 + C\ R_F^3\ f_\eta(\eta_o R_F^3 \dot\gamma/\ kT)\ )$$

Although dilute coils cannot dominate the viscosity of a solution, they do give it a power law time dependence. Here we consider a step strain applied to the system at time 0 and the stress $\sigma(t) = G(t)\ \gamma$ measured after maintaining γ for time t. Assuming the law $\tau_r \cong \eta_o r^3/kT$ to apply to the interior structure of the coils we have after time t, structure equilibrated up to a scale

$$\xi(t) \cong (kT\ t/\eta_o)^{1/3}$$

for which each chain has $(R_F/\xi(t))^{1/\nu}$ blobs. We thus expect a time dependent stress relaxation modulus

$$G(t) \cong C\ R_F{}^{1/\nu}(\eta_o/\ kTt)^{1/\ 3\nu}$$

for times $t<\tau_R$. Since $3\nu>1$ the viscosity increment is still dominated by the longest relaxation time.

Overlapping Coils

It is when the coils overlap that the polymers dominate the dynamics of the whole solution and just as with the statics we must first establish what happens in the polymer melt case.

There are three extra effects to consider in a polymer melt compared to a dilute solution, of which the first is just that the coil dimensions are screened and gaussian. Secondly there is hydrodynamic screening, that the propagation of hydrodynamic disturbances is impeded by the polymer. On the large scale this can be simply quantified in terms of the viscosity increase: the Oseen tensor which gives the flow response to a point force is inversely proportional to viscosity. On the short scale it results in the friction on the coil being a sum of individual monomeric friction forces, rather than a hydrodynamically cooperative Stokes friction, because the local friction is not increased by high molecular weight whereas the Stokes term increases in proportion to the overall viscosity.

Hydrodynamic screening changes the scaling of relaxation times to $\tau_m \sim m^2$ where m is a molecular weight.

The final and most drastic effect of coil interpenetration is the constraint of entanglement, namely that two sections of chain cannot cross through each other. Doi and Edwards suggested that even when the chains confining a given chain are themselves equally free to move, the result is that every chain is constrained to move within a tube.

Beyond the scale of the tube diameter b , the only freedom the chain has is then to snake or reptate along its tube, which it will do with a diffusive motion. As the 'head' of the polymer pushes out from the end of the tube it has a free choice of path through the melt, but all along the rest of the chain the monomers must faithfully follow the tube.

The curvilinear drag coefficient for the chain to move down its tube is a sum of monomeric frictions and hence proportional to M , giving a curvilinear diffusion coefficient $D_c \sim 1/M$ by an Einstein argument. To renew its conformation the chain must diffuse along the tube by a distance of order the tube length L , again proportional to molecular weight, so that one arrives at an estimate of a conformational relaxation time of order

$$\tau_d \cong L^2/D_c \sim M^3 .$$

The motion of the coil centre of mass over this time is of order the coil size so that the true coil diffusion coefficient is expected to vary as

$$D \cong R^2/\tau_d \sim M^{-2}$$

which has been verified by Klein and Briscoe.

The corresponding prediction for the viscosity is an essentially single relaxation time behaviour with each section of tube giving a (molecular weight independent) contribution to the modulus, so that the viscosity depends on molecular weight only through the relaxation time $\tau_d \sim M^3$ above. Qualitatively the experimental data are in excellent agreement, both for the single relaxation time feature and most particularly for the predicted nonlinear strain dependence, but as the data of Berry and Fox show there are two outstanding problems.

The first problem with entanglement is the scale at which it sets in. Intuitively one might expect it to be effective at the scale of monomeric contact, whereas experimentally it is characterised by a molecular weight between entanglements M_e of the order of 100-200 monomer units, with a correspondingly wide tube diameter. On shorter scales in polymer melts the simple scaling due to hydrodynamic screening is observed.

The second problem is the molecular weight exponent of the viscosity. Notwithstanding the difficulties of measuring high viscosities, there is no doubt that at least for the first decade of truly entangled molecular weight $\eta \sim M^{3.4}$ rather than $\eta \sim M^3$ as per the theory and as the observed diffusion law would suggest. Doi has suggested that this is merely a very slow crossover phenomenon; if not it is a fundamental puzzle.

Entanglement in Semi-Dilute Solution

Despite the difficulies surrounding the interpretation of the dynamics of a polymer melt there has been success with scaling the entangled semi-dilute behaviour to it, as with the diffusion results of Leger et al. The variation with molecular weight shows the same law $D \sim M^{-2}$ as observed by Klein in the melt, and the concentation dependence follows from a simple blob scaling where we take the tube diameter to scale with the semi-dilute correlation length ξ and the drag coefficient on each blob to have its local Stokes value $6\pi\eta_o\xi$. Then we expect

$$D \cong (R_F/\xi)^{-2/\nu} kT/\eta_o\xi \sim c^{-(2-\nu)/(\nu d-1)} \sim c^{-7/4} \text{ in } d=3,$$

as observed.

CONCLUDING REMARKS

These notes are only a tiny sketch of scaling in polymer physics, which itself is only a small part of our understanding of polymers. Indeed there is a sense in which those phenomena which we do understand by scaling are the trivial ones, bound by other known properties of the system. Nevertheless, polymer science is one of the subjects where scaling concepts have had most impact on technology.

Scaling in practice always begs two questions: 'what are the values of dimensioned parameters by which we must reduce our data to scaling form?' and 'how do the scaling functions behave?'. The example of polymer melt dynamics is one where apparently there are answers to these questions, but we have yet to firmly establish what they are.

REFERENCES

de Gennes PG 'Scaling Concepts in Polymer Physics', Cornell 1979.
Ferry JD 'Viscoelastic Properties of Polymers', 2nd Edn, Wiley 1970.
Flory PJ 'Principles of Polymer Chemistry', Cornell 1971.
de Gennes PG, Phys. Lett. 38A, 339 (1972).
Edwards SF, Proc Phys Soc (London) 88, 265 (1966).
des Cloizeaux J, J Phys (Paris) 36, 281 (1975).
Cotton JP et al, Macromolecules 7, 863 (1974).
Daoud et al, Macromolecules 8, 804 (1975).
Ferry JD, Macromolecules 13, 1719 (1980).
Doi M & Edwards SF, J Chem Soc Far Trans II 74, 1789, 1802, 1818 (1978)
Klein J & Briscoe B, Proc Roy Soc London A365, 53 (1979).
Berry GC & Fox TG, Adv Pol Sci 5, 261 (1968).
Leger et al, Macromolecules, December 1981.

ADENDUM

The author regrets that the figures originally prepared were not clear enough for the publisher.

DYNAMICAL SCALING IN POLYMER SOLUTIONS

D. Richter

Institut für Festkörperforschung, KFA Jülich
D-5170 Jülich, Fed. Rep. Germany

B. Ewen

Max Planck Institut für Polymerforschung
D-6500 Mainz, Fed. Rep. Germany

I. INTRODUCTION

Long flexible polymers constitute physical systems with many internal degrees of freedom. At large enough scales, various molecular properties exhibit universal behavior such as the radius of gyration Rg which scales as M^{ν}, where M is the molecular mass and ν is an universal exponent (see also lecture notes by R.C. Ball). Similarely the diffusion coefficient D of linear chains scales with $D \sim M^{-\nu}$ in dilute solutions while in melts this relation is altered to $D \sim M^{-2}$ /1-3/.

This lecture discusses our experimental efforts to look into dynamic scaling properties of linear chains on a molecular level. After a brief outline of neutron spin echo we address basic molecular models for chain relaxation. Thereafter we present experimental results on dilute solutions which proof the validity of the so called Zimm model - it states the dominance of hydrodynamic interactions between chain systems - down to microscopic scales. Then the influence of chain concentration on their dynamics is studied: The collective behavior of such a many chain system is characterized by a cross over from single chain Zimm dynamics at short scales to collective diffusion at larger scales, while the single chain relaxation is dominated by the screening of the hydrodynamic interaction. Finally, in concentrated solutions and melts the macroscopic data appear to favour a simple Rouse like relaxation while the macroscopic behavior clearly cannot be understood in these terms.

II. NEUTRON SPIN ECHO

The relationship between neutron energy and its wavelength allows to access the internal relaxation modes of polymer chains on appropriate length and time scales simultanously.

Furthermore the possibility to lable polymer chains by deuterati· permits an observation of single chain dynamics even in concentrated systems and melts. Still, since the coil relaxation modes are very slow, it was only the invention of ultra high resolution neutron spin echo spectroscopy which opened up this field to neutron scattering /4/.

The basic feature of neutron spin echo (NSE) is the decoupling of energy resolution and band width of the incident neutron spectrum $I(\lambda)$. Using the Larmor precession of the neutron spin in an external guide field as an internal clock, it measures the velocity change during the scattering process for each neutron individually. This allows ultrahigh resolution $\Delta E/E \sim 10^{-5}$ at acceptable intensities. The measured quantity is the final neutron polarization P_z as function of the applied guide field H. P_z is given by

$$P_z = \int_0^\infty I(\lambda)\,d\lambda \int_{-\infty}^{+\infty} S(q,\omega)\cos(\omega t(\lambda))\,d\omega \tag{1}$$

with

$$t(\lambda) = \frac{N_0}{h\lambda_0} m \lambda^3$$

where $N_0 \sim H$ is the number of Larmor precessions in the applied field H, λ_0 the center of $I(\lambda)$, and m the neutron mass. According to Equ. (1), P_z is the cosine Fourier transformed of $S(q,\omega)$ or the real part of the intermediate scattering function S(q,t), averaged with respect to $I(\lambda)$ and normalized to one. Though $\Delta\lambda/\lambda$ is in the order of 10 %, the averaging does not cause difficulties, since in the case of diffusive motion $S(q,t)/S(q,0)$ is always a function of q^2t^x and the remaining λ dependence is weak. Other than in normal neutron spectroscopy where the measured quantity is a convolution of instrumental resolution and scattering function, in Fourier space the resolution comes in as a product and can be divided out with no complication. The time range covered by the NSE-spectrometer IN11 at the ILL is determined by the accessible guide fields and for the selected wavelengths of 8.3 Å amounts to $0.5 < t\,[10^{-9}\,s] < 12.7$. The energy resolution corresponds to 40 neV FWHM .

III. BASIC MODELS

As the most simple model for chain relaxation we consider a Gaussian chain in heat bath (Rouse-model /5/). The building blocks of such a chain are so called Kuhn segments which consist of several monomers such that its end to end distance follows a Gaussian distribution. Their conformations are denoted by vectors $\underline{a}_n$ along the chain. The whole chain then is a sucession of freely jointed segments of length σ . Thereby, we neglect any excluded volume interactions. We are interested in internal relaxation modes on a length scale $\sigma < r < Rg$. They are solutions of a Langevin equation

$$\ddot{\underline{a}}_n + \zeta_0 \dot{\underline{a}}_n + \nabla_n F(\{\underline{a}_i\}) = \underline{A}_n(t) \tag{2}$$

ζ_0 is a friction coefficient, $\underline{A}_n(t)$ the stochastic force on segment n and $F(\underline{a}_n)$ is the free energy. For segmental motion the force $\nabla_n F(\underline{a}_i)$ is dominated by the entropy part in F given by $S = k \ln W(\{\underline{a}_i\})$ where

$$W = \prod_{i=1}^{N} \exp\left(-\frac{3}{2}\frac{a_i^2}{\sigma^2}\right) \tag{3}$$

is the probability of a chain conformation $\{\underline{a}_i\}$. We note that the resulting entropic force is a speciality of macromolecular systems, where the conformational entropy generates a force stabilizing the most favourable conformation. This force is also the basis for rubber elasticity.

Using periodic boundary conditions and neglecting the inertial term, Equ. (2) is solved by $a_p = a_p^o e^{-\Gamma_p t}$. For small $p = 2\pi n/N$ (n = 1...N) we have

$$\Gamma_n = \frac{6\pi^2 kT}{\langle R_g^2\rangle^2}\left(\frac{\sigma^2}{\zeta_o}\right) n^2 \tag{4}$$

Thus, the physical parameter of the Rouse model is the friction coefficient ζ_o per segment length squared.

In the next step we consider a more realistic model for chain relaxation in solutions and take into account the hydrodynamic interaction between different segments of a coil. A moving segment creates a backflow $\underline{v}$ in the solvent commonly described by the Oseen tensor $T \sim 1/\eta_o r$ (η_o: solvent viscosity). It gives rise to an additional force term in Equ. (2):

$$K_n = \zeta_o \underline{v}(\underline{r}_n) \tag{5}$$

Thereby $\underline{v}(\underline{r}_n)$ is the result of the forces exhibited by <u>all</u> other segments on the solvent. The inclusion of the hydrodynamic interaction results in a long range coupling between all segments alterning the characteristic relaxation times significantly:

$$\Gamma_n = \frac{12\sqrt{\pi}\, kT}{\langle R_g^2\rangle^{3/2}\eta_o} n^{3/2} \tag{6}$$

Other than in the Rouse model here in the so called Zimm model /6/ no parameter describing specific properties of a segment appears and Γ_n depends in an universal way on the radius of gyration Rg and the viscosity of the solvent allone.

In dense polymer solutions and melts the presence of other chains imposes topological constraints on the motion of a given chain. The most popular model in this field, the so called reptation model /7-9/, assumes that these constraints or chain entanglements effectively restrict the chain motions to a tube along the chain profile. - The chain reptates like a snake. Now we assume Rouse like diffusion inside the tube $D_t = kT/\zeta_o N$. In order to loose memory of its original tube the chain has to move over a tube length

$$2D_t \tau_t = N^2 \sigma^2 \tag{7}$$

The time τ_t is the longest possible relaxation time which governs the viscosity. It scales

$$\tau_t \sim \eta \sim N^2 \sigma^2 / 2D_t \sim N^3 \tag{8}$$

During τ_t in real space the chain has moved a distance $R = \sigma N^{1/2}$. Thus, for the diffusion coefficient we get

$$D_{rep} \simeq \frac{R^2}{\tau_t} \sim \frac{1}{N^2} \tag{9}$$

While $D_{rep} \sim 1/N^2$ is also found experimentally, the experimental exponent for η is more like 3.4 /30/.

IV. SCATTERING FUNCTIONS

As outlined in Section II NSE measures the intermediate scattering function S(q,t) which is determined by the correlation function

$$S(q,t) = \frac{1}{N} \sum_{ij} b_i b_j \langle \exp(-i\underline{q}\underline{r}_i(0)) \exp(i\underline{q}\underline{r}_j(t)) \rangle \tag{10}$$

b_i is the scattering length density of segment i. In Gaussian approximation one gets

$$\langle \exp[i\underline{q}(\underline{r}_j(t) - \underline{r}_i(0))] \rangle = \exp\left[-\frac{q^2}{6}\langle(\underline{r}_j(t) - \underline{r}_i(0))^2\rangle\right] \tag{11}$$

In the case of incoherent scattering (i = j) Equ. (11) shows that S(q,t) measures the mean square displacement of a segment. It can be shown that in general this also holds for coherent scattering. In order to calculate $\langle \Delta r(t)^2 \rangle$ the solutions $\underline{r}_j(t)$ of the respective Langevin equations are needed. The following results are obtained /7,10/

$$\begin{aligned} &\text{Rouse:} && \langle \Delta r_n(t) \rangle^2 \simeq \sigma \sqrt{kT/\zeta_0 t} \sim t^{1/2} \\ &\text{Zimm:} && \langle \Delta r_n(t) \rangle^2 \simeq \sigma^2 (kT/\eta_0 t)^{2/3} \sim t^{2/3} \\ &\text{Reptation:} && \langle \Delta r_n(t) \rangle^2 \simeq \sigma (kT/\zeta_0 t)^{1/4} \sim t^{1/4} \end{aligned} \tag{12}$$

For the three models the different time exponents are characteristic. For normal diffusion we had $\langle \Delta r_n^2 \rangle$ t. The normalized intermediate scattering functions for the Rouse and Zimm models are exponentially decaying functions depending on <u>one</u> scaling variable x

$$\begin{aligned} &\text{Rouse:} && x = \frac{q^2\sigma^2}{3}\left(\frac{3kT}{\zeta_0} t\right)^{1/2} \\ &\text{Zimm:} && x = \frac{q^2\sigma^2}{3}\left(\frac{kT}{6\sqrt{2}\pi\eta_0} t\right)^{2/3} \end{aligned} \tag{13}$$

The characteristic frequencies corresponding to the half width of $S(q,\omega)$ for the coherent scattering function is evaluated to:

$$\omega_c^R = \frac{1}{12}\,\frac{kT}{\zeta_0}\,\sigma^2 q^4$$
$$\omega_c^Z = \frac{1}{6\pi}\,\frac{kT}{\eta_0}\,q^3 \qquad (14)$$

For the reptation model recently De Gennes has calculated an expression for $S(q,t)$ which should be valid in the so called regime of local reptation ($d_t \ll r \ll R_g$, where d_t is the tube diameter) /11/

$$S(q,t)/S(q,0) = \left(1 - \frac{q^2 d_t^2}{36}\right) + \frac{q^2 d_t^2}{36}\exp\left(\frac{x^2}{36}\right)\left(1 - \mathrm{erf}\left(\frac{x}{6}\right)\right) \qquad (15)$$

where x is the scaling variable of the Rouse model.

V. EXPERIMENTAL RESULTS IN DILUTE SOLUTIONS

Measurements on dilute solutions offer a direct test of the Zimm predictions. Since besides some estimates for the initial slope /12/ theoretical calculations cannot treat the effect of the excluded volume interaction, we present data obtained from a polymer solution under Θ-conditions where the excluded volume interaction also vanishes in the real world /13/. We investigated PDMS $[-SiO(CH_3)_2-]_n$ in C_6D_5Br which at 84 C is a Θ system. Fig. 1 displays the results of this experiment.

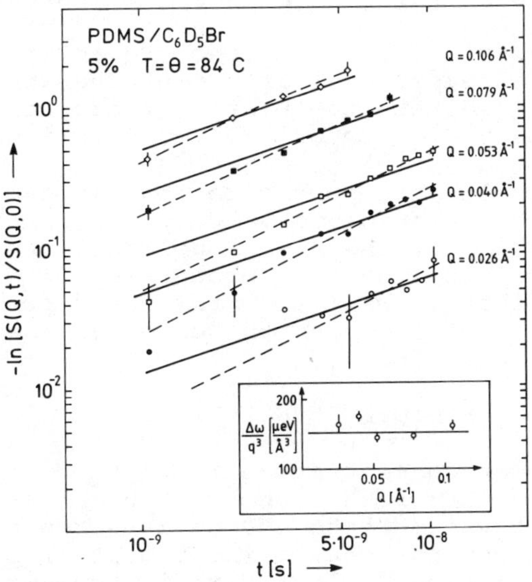

Fig. 1. Results of neutron spin-echo experiments on the dilute Θ system PDMS/C_6D_5Br at Θ = 84 °C. Solid lines result from a fit with the asymptotic Zimm scattering function, and the dashed lines display the fit with the full Zimm function. The insert presents the Q dependence of the characteristic frequency /13/.

The logarithms of the corrected final neutron polarization which resemble the normalized scattering function are plotted double logarithmically vs the guide field H or the Fourier time t. In this manner the time exponent in S(q,t) which asymptotically should assume n = 2/3 for the Zimm case is visualized most easily. In the following we term these exponents lineshape parameter. The dashed lines in Fig. 1 are the result of a fit with the scattering function of the Zimm model while the solid lines indicate the asymptotic n = 2/3 behavior. It is obvious that the Zimm scattering function describes the experiment very well. The asymtotic regime is not yet reached in the covered q-t regime. With increasing q, however, a gradual tendency toward the asymptotic form is observable. We note that the lineshape parameter approaches its asymptotic value from above. The insert in Fig. 1 presents the q-dependence of the characteristic frequency. ω_c/q^3 follows a horizontal line showing the predicted scaling behavior. A common fit of the spectra yields an exponent $\gamma = 3.0 \pm 0.1$.

According to Equ. 14 the absolute value of ω_c should depend only on the viscosity of the solvent. On a first glance the thus determined viscosity $\eta_o = 0.65 \pm 0.03$ cP agrees well with the macroscopic viscosity $\eta^o = 0.61$ cP. However, Equ. 14 is based on a preaveraged Oseen tensor leading to a 17 % too small value of ω_c resulting in a 20 % discrepancy between theory an experiment. Similar deviations have also been observed with light scattering on PS-solutions using two orders of magnitude smaller q-values /14/.

Regarding experimental results in good solvents it was found that within experimental errors the lineshape agrees with that in Θ-solutions /13,15/. The solvent character only affects the prefactor in Equ. 14 which is enhanced in good solvents. Concerning the absolute value of the prefactor again deviations between theory and experiment are seen. Changing gradually from a good - to a Θ - solvent the prefactor crosses over in accordance with the blob picture of polymer solutions /16/.

Finally we'd like to emphasize that in the spite of some difficulties with the absolute value of ω_c, applying the concepts of hydrodynamics, the Zimm model does describe microscopic and macroscopic dynamical properties of dilute polymer solutions astonishingly well and provides a prevailing description the case of PDMS from length scales in the order of 10 Å and time scales above 10^{-9} s.

VI. EXPERIMENTS IN SEMIDILUTE SOLUTIONS

As the monomer concentration is raised different chains start to overlap. As a result above a characteristic length ξ_c the excluded volume interaction is screened out. At distances $r > \xi_c$ the chains behave like a melt of blobs and exhibit Gaussian statistics while for $r < \xi_c$ the swollen conformation prevails. Furthermore the hydrodynamic backflow force acting on a submit $\underline{r}_n$ due to the motion of the submit at $\underline{r}_m$ is disturbed by the movements of all other chains. For distances $|\underline{r}_n - \underline{r}_m|$ larger than a characteristic hydrodynamic length ξ_H we expect the hydrodynamic interactions mediated by η_0^{-1} to be screened out. Finally the collective response of the transient network has to be considered. At short times the overlapping chains react as a rubber like network of mesh size ξ_c and exhibit additional elastic rigidity leading to faster relaxation of long range fluctuations ($r > \xi_c$).

Experimentally the effect of hydrodynamic screening can be observed on single labeled chains in semidilute solutions (tracer diffusion), while the collective behavior is seen on unlabeled systems (collective diffusion). We fist turn to the collective behavior and thereafter treat the single chain relaxation.

VI. 1. Collective Relaxation in Semidilute Solutions

From the analogy with permanent networks swollen by a solvent, the relaxation spectrum for $r > \xi_c$ in the many chain regime is predicted as

$$\omega_c = D_c q^2 \tag{16}$$

where D_c is the collective diffusion coefficient of the pseudo gel. For spatial dimensions $r < \xi_c$ or $q\xi_c > 1$ the network is not yet effective and Zimm relaxation (Equs's 13, 14) should prevail Spatial cross over between both regimes is expected for $\xi_c q \doteq 1$. The condition of a smooth cross over implies $D_c \sim \xi_c^{-1} \sim c^{3}/4$ in good solvents.

Fig. 2 presents experimental results on PDMS/C_6D_6 at 70 C, where Benzene is good solvent for PDMS /17/.

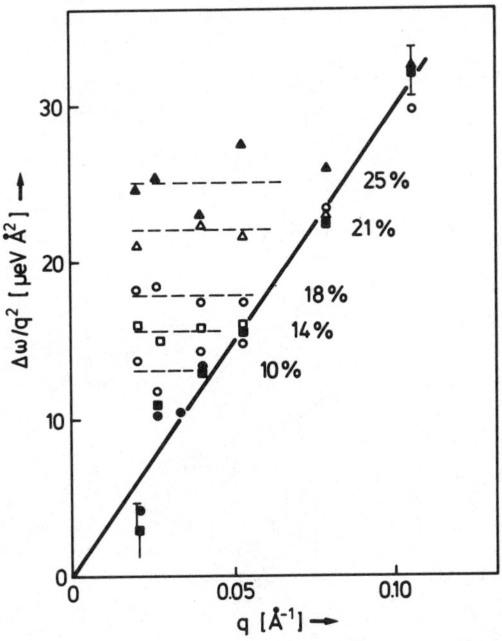

Fig. 2. Concentration crossover in PDMS/d-benzene. Quasielastic line widths ω_c/q^2 for different concentrations at 70°C as a function of momentum transfer q; (----) result of fitting the low-concentration spectra (c = 2 and 5%) simultaneously with $S_{DG}(q,t)$. (●) 2%, (■) 5%, (o) 10 %, (□) 14 %, (o) 18 %, (△) 21 %, (▲) 25 % /17/

The quasielastic linewidths ω_c/q^2 are plotted vs q for several monomer concentrations. At small concentrations (c = 2, 5%) $\omega_c \sim q^3$ is observed in the entire q-range of the experiment. With increasing concentration a cross over from $\omega_c \sim q^3$ to $\omega_c \sim q^2$ occurs at increasing q-values. The data provide a direct experimental observation of the dynamic spatial cross over between Zimm like relaxation at small dimensions and collective diffusion at larger dimensions in space. This is also seen in the spectral lineshape displayed in Fig. 3.

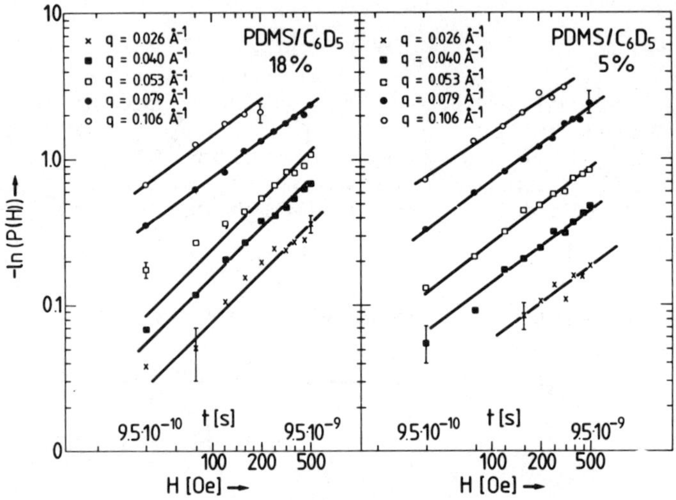

Fig. 3. Line-shape analysis for PDMS/d-benzene at c = 5 and 18 %. Double-logarithmic plot of -ln S(q,t) vs. time t or magnetic guide field H. (x) q = 0.026 $Å^{-1}$, (■) q = 0.040 $Å^{-1}$, (□) q = 0.053 $Å^{-1}$, (●) q = 0.079 $Å^{-1}$, (o) q = 0.106 $Å^{-1}$. /17/

While for c = 5 % the lineshape exponent is close to n = 0.7 (Zimm relaxation) for all q-values, for the 18 % sample a change in slope from about 0.7 at large q to about unity at small q's is evident. The position of the cross over point q*(c) is a direct measure of the correlation length $\xi_c = 1/q^*(c)$; the plateau height at small q reveals the cooperative diffusion coefficient D_c. A detailed data analysis taking into account the lineshape changes yields $D_c = (1.3 \pm 0.1)\ 10^{-5}\ c^{0.67 \pm 0.05}$ cm^2/s and $\xi_c = (3.2 \pm 0.7) * c^{-0.71 \pm 0.07}$ Å. The observed exponents are close to the scaling prediction of 3/4. The slightly smaller values are in agreement with light scattering measurements of Dc on Polystyrene solutions /18/ and discussed in detail by Weill and Des Cloiseaux /19/, who attribute the deviations to slowly converging dynamic exponents.

VI. 2. Hydrodynamic Screening

The calculation of eigenmodes of single polymers in concentrated solutions from first principles is a demanding task requiring the solution of a complicated many body problem /20-29/. It was found out that hydrodynamic screening and interchain interactions responsible also for the screening of the excluded volume interaction are intimately related /27/. On this basis De Gennes suggested the identification $\zeta_H(c) = \zeta_c(c)$ which is not generally accepted however /21,28,29/.

In order to incorporate hydrodynamic screening into the Zimm model we take an empirical approach and replace the solvent viscosity in the Oseen tensor by an exponentially screened quantity

$$\eta_{eff}^{-1} = (\eta_0^{-1} - \eta(c)^{-1})\exp(-r/\xi_H) + \eta(c)^{-1} \tag{16}$$

Equ. (16) defines the hydrodynamic screening length discussed above. We emphasize, that even for $r \gg \zeta_H$ η_{eff}^{-1} does not decrease to zero: rather η_{eff} may become the time dependent viscosity of the polymer solution $\eta(c)$ /13/. Incorporation of Equ. 16 into the Langevin equation yields three dynamical regimes: (i) at large q:$q\zeta_H > 1$ we recover the Zimm limit and expect unscreened Zimm relaxation (ii) for $q\zeta_H < 1$ the characteristic frequency becomes

$$\omega_c = \frac{\eta_0}{\eta(c)}(q\xi_H)^3 + (q\xi_H)^4\left(1 - \frac{\eta_0}{\eta(c)}\right) \tag{17}$$

Depending on the magnitude of $q\zeta_H$ Equ. (17) yields Rouse like relaxation ($\omega_c \sim q^4$) for $q\zeta_H > \eta_0/\eta(c)$ or (iii) screened Zimm relaxation for $q\zeta_H < \eta_0/\eta(c)$ ($\omega_c \sim q^3$). Scaling analysis shows that the third regime is not universal. It only exists if $q\zeta_H < \eta_0/\eta(c)$ is reached before qRg becomes smaller than one.

The observation of single chain dynamics in semidilute solutions requires contrast matching and labelling. This can be achieved using deuterated PDMS containing a 5 % protonated fraction dissolved in C_6D_5Cl. Since the scattering length densities of d-PDMS and C_6D_5Cl are practically equal, for all monomer concentrations the dynamics of the single protonated chains can be observed. Fig. 4 presents the q-dependence of the lineshape parameter n(q) for the intermediate monomer concentrations c = 18 % and c = 45 % /13/. In the 18 % case n(q) starts around 1/2 for small q and rises to $n \sim 2/3$ at larger q's. Obviously at small q the polymer relaxation is dominated by Rouse modes (Equ. (13)) while at larger q Zimm modes take over. Qualitatively, this is the behavior we expect for the cross over from unscreened Zimm to enhanced Rouse-relaxation as discussed above. A first estimate for ζ_H is obtained from the cross over condition $q\zeta_H = 1$ yielding $\zeta_H \approx 20$ Å for this concentration. At 45 % the behavior is inverted. Now n(q) decreases from 0.8 at small q to about 1/2 at large q. Apparently Zimm modes are active over long distances while at shorter distances Rouse like modes dominate. This result is in qualitative agreement with the idea of incomplete screening, where at large ces the effective viscosity of the polymer solution mediates hydrodynamic coupling though the backflow of the pure solvent is already screened.

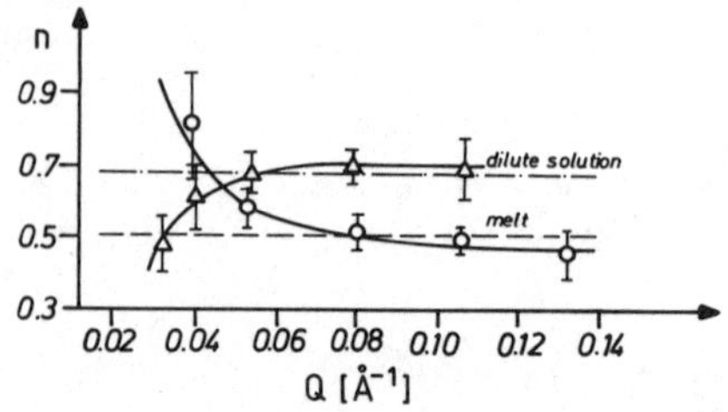

Fig. 4. Q dependence of the line-shape parameter n(Q) for intermediate polymer concentrations of PDMS in C_6D_5Cl ((△), c = 18 %; (o), c = 45 %). The point-dashed line and the dashed line indicate the asymptotic values of n for dilute solutions (Zimm) and melts (Rouse), respectively /13/.

A quantitative data analysis employing the explcitly calculated scattering function for incomplete hydrodynamic screening demonstrated the consistency of the approach and revealed results for ζ_H (c) and η (c). ζ_H and η(c) were obtained by a common fit of 25 experimental spectra varying only two parameters. Fig. 5 displays the concentration dependence of ζ_H and compares with ζ_c determined from the collective response of semidilute PDMS solutions. The magnitude of both lengths nearly coincides. The concentration dependence of ζ_H appears to be close to the mean field prediction of $\zeta_H \sim c^{-1}$ while for ζ_c an exponent near the scaling result of 3/4 was obtained. However, the data on ζ_H are yet not sufficient to answer the question after the concentraion exponent definitively. Fig. 6 presents the concentration dependence of η (c) which mediates the screened Zimm relaxation. Measurements at two different molecular weights (M_w = 60.000, 178.000) revealed no influence of Mw on η(c). Thus η (c) cannot be identified with the macroscopic viscosity of the polymer solution. The result, however, can be understood, if we compare the time scale of the spin echo experiment ($5 \ 10^{-10} < t < 1.3 \ 10^{-8}$ s) with the time needed to relax the viscosity towards equlibrium. For M_w = 178 000 we estimate $\tau_{Zimm} \sim 10^{-6}$s (see Equ. (6)) 2 to 3 orders of magnitude longer than the experimental time window. Thus for large M_w, on our experimental scale the macroscopic viscosity cannot establish itself. The molecular weight, where the relaxation time of the macroscopic viscosity and the time scale of our experiment coincide, is about $M_w \simeq 800$. We now postulate that this viscosity should be indentified with our microscopic result. In order to support this idea, we measured

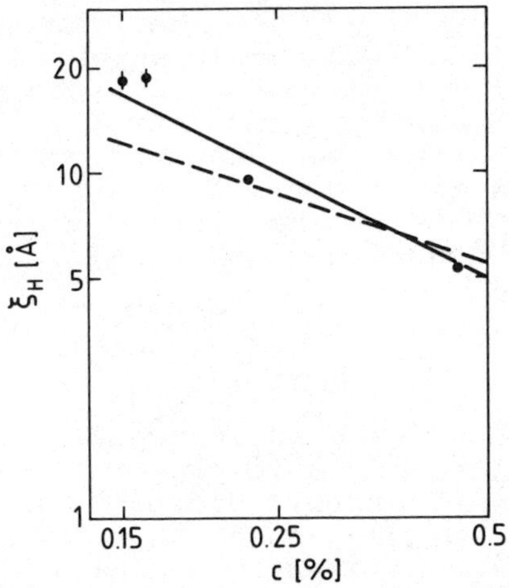

Fig. 5. Concentration dependence of the hydrodynamic screening length ξ_H (●, results of fits at each q separatlely). Solid line represents the result of the simultaneous fit, and the dashed line the excluded volume screening length ξ_c /13/

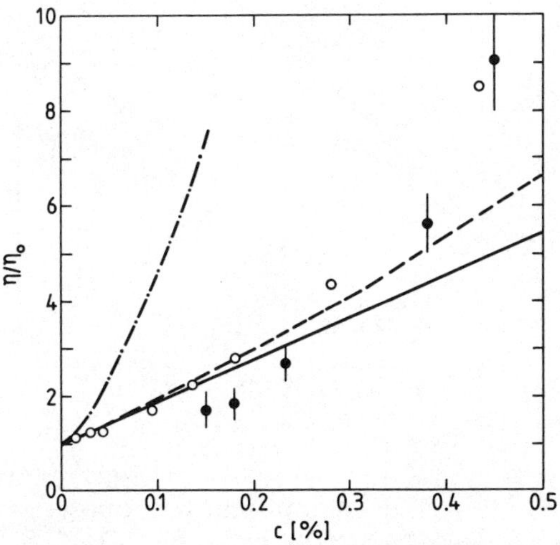

Fig. 6. Concentration dependence of $\eta(c)/\eta_o$ (●), result of separate fits; (o) result of viscosity measurements on PDMS solutions (M_w = 7400). The result of the simultaneous fit considering the linear term in

$$\eta(c) = \eta_o [1 + [\eta]c + K_H[\eta]^2c^2] \quad /33/$$

is given by the solid line; the inclusion of a quadratic term leads to the dashed line. The point-dashed line indicated the macroscopic viscosity for M_w = 60.000 /13/.

the zero shear viscosity of PDMS solutions under the same conditions as in the scattering experiment, taking a molecular weight of M_w = 7400. Figure 6 compares the macroscopic with the microscopic results, and the agreement is startling. Evidently, both sets of data have not only the same concentration dependence but agree nearly quantitatively, supporting the idea of an effective time dependent viscosity describing the screened Zimm interaction. For illustration we also show the concentration dependence of the macroscopic viscosity of M_w = 60.000 (dashed-pointed line). It disagrees completely.

VII. POLYMER RELAXATION IN MELTS

In dense polymer systems the influence of entanglements is supposed to govern their dynamic properties /7-9/. The experimental results on macroscopic diffusion and zero shear viscosity support the reptation idea, though the experimental viscosity expoent is consistently larger (~ 3.4) than the reptation prediction of 3 /30/. Here we examine microscopic relaxation data from labeled PDMS melts. Fig. 7 presents the respective NSE-spectra from a M_w = 60.000 melt /31/.

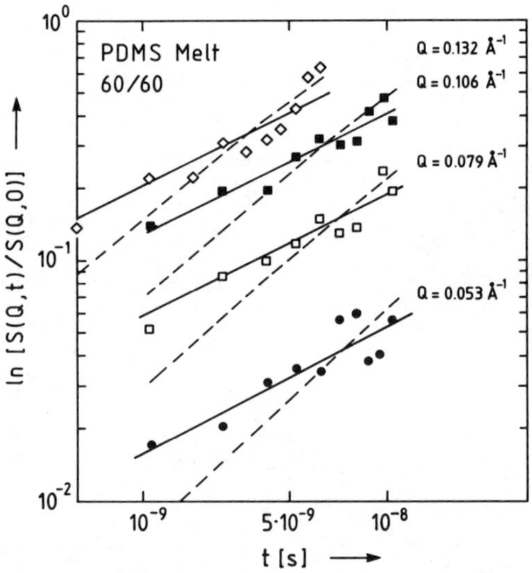

Fig. 7. NSE spectra from the PDMS melt at 100 C (M_w^H = 60.000, M_w^D = 60.000). Solid lines result from a fit with the asymptotic Rouse function, dashed lines display a fit with the full Rouse function.

In a first step the data are analyzed in terms of the Rouse model. The solid lines are a result of a fit with the asymptotic scattering function (equ. 12) while the dashed lines give the result of a fit with the full Rouse scattering function. Other than in the dilute solution where the full Zimm scattering function was an excellent description of the line shape, here serious deviations between $S_{coh}^{Rouse}(q,t)$ and the experimental spectra are obvious. These discrepancies are not restricted to the large molecular weight but occur as well for the small M_w. While $S_{coh}^{Rouse}(q,t)$ predicts line-shape parameters of $n \sim 0.7$-0.8 in our q-t regime, experimentally $n = 0.5$ is observed. Thus, the Rouse model does not reach the same quality of data description for polymer melts as the Zimm model does for dilute solutions. Nevertheless, the asymptotic form of the Rouse model is a reasonable description of the scattering data over a very wide range of time and q. From the characteristics frequences (Equ. 14) the adjustable parameter $\sigma^2/\zeta_0 = (1.9 \pm 0.2)\ 10^{-6}\ cm^2 sg^{-1}$ of Rouse model can be extracted. It may be compared with $1.4 \times 10^{-6}\ cm^2 sg^{-1}$ found in mechanical measurements /32/. The agreement is of the same order as that between the microscopic and macroscopic viscosity in dilute solutions. It demonstrates that there is some basic truth in the molecular foundations of the Rouse model. But of course, the Rouse model can explain neither the law $\eta \sim M_w^{3.4}$ nor the diffusion constant $D \sim M_w^{-2}$.

In a second step, we examine the PDMS data in terms of the scattering function (Equ. 15) which recently De Gennes has proposed for local reptation /11/ Fig. 8 presents a joint fit of Equ. 15 to the PDMS-data.

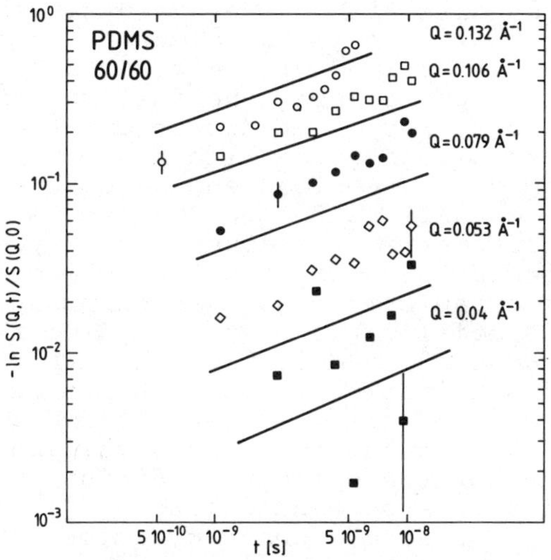

Fig. 8. Result of a fit with the local reptation model to the NSE spectra from the (60K/60K) PDMS melt (solid lines).

Thereby the Rouse rate and the tube diameter were adjustable parameters. They were evaluated to $d_T = 40 \pm 4$ Å and $W\sigma^4 = 1.6 \pm 0.8\ 10^{13}$ Å^4 and have to be compared with $d_t = 51$ Å from viscoelastic data /13/ and $W\sigma^4 = (3 \pm 0.3)\ 10^{+13}$ from the Rouse fit

We draw attention to the following features exhibited by the fit: (i) apparently the line shape of the theoretical scattering function (Equ. 15) is in better agreement with the data than that of the Rouse model (cf. Figure 7). The line-shape parameters of the local reptation model are in the range $0.4 < n < 0.5$ and close to the experimental values of the order of 0.5.

This observation coincides with result of ref. /34/ on long-chain PTHF which also shows strong deviations from the Rouse model and is closer to n = 1/2. (ii) The q dependence of the experimental spectra cannot be reproduced by the local reptation model while the Rouse model could account for it well. If one allows for a variation of the q exponent in Eq. (15) 0.75 instead of 2 is obtained. (iii) Also for a sample with M_w = 3800 well below the molecular weight in between entanglement points ($M_e \sim 8100$ for PDMS) the lineshape is virtually identical with that of the M_w = 6000 sample displayed in Fig's 7 and 8. The failure of the local reptation model to reproduce the experimental q dependence together with a line-shape behavior unaffected by molecular weight implies that the spin-echo data cannot be taken as evidence for de Gennes' local reptation idea. At this stage the data probably do not disapprove this model either, since the accessed q- and t-ranges likely were too small. For further discussion theoretical calculations in the crossover range between Rouse relaxation and local reptation would be helpful and show whether the reptation mechanism leads to a better understanding of the experiment.

VIII. CONCLUSION

This lecture dealt with universal dynamical properties of linear polymer solutions on a molecular level. It presented experimental results obtained with ultra high resolution neutron spin echo spectroscopy covering the full concentration range of polymer solutions. For dilute solutions the experiments showed that segmental diffusion within polymer coils is governed by hydrodynamic interactions and described well by the Zimm model for Θ as well as for good solvents. Only the magnitude of the characteristic frequency deviates from the Zimm prediction. In semidilute solution the collective as well as the single chain dynamic response was discussed. In the collective response a spatial cross over from Zimm relaxation at short scales to collective diffusion at larger scales was observed. It manifests itself (i) in a change of the scaling behavior of the characteristic frequency from $\omega_c \sim q^3$ to $\omega_c \sim q^2$ and (ii) in a change of spectral lineshape. The concentration exponents for the collective diffusion coefficient D_c and the excluded volume screening length ξ_c are close to scaling predictions. The single chain dynamics exhibit the features of incomplete hydrodynamic screening. The experimental spectra allowed an evaluation of the hydrodynamic screening length ξ_H. Its magnitude is close to that of the excluded volume screening length ξ_c as required by the scaling ansatz of only one relevant length. More experiments are necessary in order to determine the concentration exponent of ξ_H reliably. The intermetiate viscosity $\eta(c)$ which appears in incomplete screening is not yet interpreted on a microscopic level. It exhibits startling similarities to the macroscopic viscosity of an equivalent nonentangled PDMS solution of smaller molecular weight. In polymer melts the Rouse model provides a general description of the dynamical behavior on length scales up to 40 Å and time scales to the 10^{-8} s region. Deviations of the experimental line shape from that predicted by the Rouse model cannot be attributed to local reptation. An influence of entanglement constraints appears not to be visible on the experimental length and time scales.

ACKNOWLEDGEMENT

We thank Dr. D. Lehnen for providing the PDMS and Dr. J.B. Hayter for assistance during experiments. The authors greatly benefitted from numerous discussion with Prof. Binder and Dr. Baumgärtner.

REFERENCES

(1) Flory, P.J. "Principles of Polymer Chemistry"; Cornel University Press: Ithaca, NY 1971

(2) Yamakawa, H. "Modern Theory of Polymer Solutions"; Harper and Row: New York, 1971

(3) de Gennes; P.G. "Scaling Concepts in Polymer Physics"; Cornell University Press: Ithaca, NY, 1979

(4) Mezei, F; Ed. "Neutron Spin Echo"; Springer-Verlag: Heidelberg, Berlin, New York, Tokio, 1980

(5) Rouse, P.E. J. Chem. Phys. 1953, 21, 1272

(6) Zimm B.H., J. Chem. Phys. 1956, 24, 269

(7) de Gennes, P.G. J. Chem. Phys. 1971, 55, 572

(8) Edwards, S.F.; Grant, J.M. V.J. Phys. A. Math, Nucl. Gen. 1973, 6, 1169, 1186

(9) Doi; M.; Edwards, S.F. J. Chem. soc., Faraday Trans. 2 1978, 74, 1789, 1802, 1818; 1979, 75, 38

(10) Dubois-Violette, F., De Gennes, P.G., Physics 1967, 3, 1981

(11) de Gennes, P.G., J. Chem. Phys. 1980, 72, 4756; J. Phys. (Paris) 1981; 42, 735

(12) Akcasu, A.Z.; Benmouna, M.; Han. C.C. Polymer 1980, 21 866.

(13) Richter, D., Binder, K., Ewen, B., Stühn B., J. Phys. Chem. 1984, 88, 6618

(14) Han, C.C.; Akcasu, A.Z., Macromolecules 1981, 14, 1980

(15) Richter, D., Hayter, J.B., Mezei, F., Ewen, B., Phys. Rev. Lett., 1978, 41, 1984

(16) Richter, D; Ewen, B.; Hayter, J.B.; Phys. Rev. Lett. 1980, 45, 2121.

(17) Ewen, B., Richter, D., Hayter, J.B., Lehnen B., J. Polymer Sc. Lett. 1982, 20 233

(18) Adam, M., Delsanti M., Macromolecules, 1977, 10, 1229

(19) Weill, G., Des Cloizeaux J., J. Phys. 1979, 40, 99

(20) Edwards, S.F.; Freed, K.F. J. Chem. Phys. 1974; 61, 1189

(21) Freed, K.W.; Edwards, S.F. J. Chem. Phys. 1974; 61, 3626

(22) Freed, K.F., Edwards, S.F., J. Chem. Soc. Faraday Trans 1, 1975, 71, 2025; J. Chem. Phys. 1975, 62, 4032. Freed, K.F. In "Progress in Liquid Physics"; Croxton, C.A., Eds.; Wiley: London, 1978; p. 343.

(23) Freed, K.F; Metiu, H., J. Chem. Phys. 1978, 68,4604

(24) Muthukumar, M, Freed, K.F. Macromolecules 1977, 10, 899; 1978, 11, 843

(25) Freed, K.F. Ferroelectrics 1980, 30, 277

(26) Perico, A.; Cuniberti, C; J. Polym. Sci., Polym. Phys. Ed. 1983, 17, 1983

(27) Freed, K.F., Perico, A. Macromolecules 1981, 14, 1290

(28) Muthukumar, M., Edwards, S.F. Polymer 1982, 23, 345

(29) de Gennes, P.G. Macromolecules 1976, 9, 587, 594

(30) Graessley, W.W. Adv. Polym. Sci. 1982, 47, 68

(31) Richter, D., Baumgärtner, A.; Binder, K. Ewen, B.; Hayter, J.B., Phys. Rev. Lett. 1981, 47, 109, 1982, 48 1695

/32/ Ferry, J.D. "Viscoelastic Properties of Polymers"; Wiley, New York, 1980

/33/ Price, F.P. Martin, S.G. Bianci, J.P., J. Polym. Sci. 19865, 22, 41

/34/ Higgins J.S., Roots J.F. Faraday Discuss. Chem. Soc. in press.

SCALING DESCRIPTION OF POLYMERIZATION KINETICS

Zoran B. Djordjević
Boris Kidrič-Lab 030
Institut u Vinči
Belgrade, Yugoslavia

Zorica V. Djordjević
Institute of Physics
Theoretical Physics Lab
Zemun, Yugoslavia

INTRODUCTION

Applications of scaling ideas to the analysis of geometric features of polymeric molecules has been numerous and very succesful.[1] Recently, it has been realized that these ideas can be extended further and applied to yet another domain of polymer chemistry,[2] the kinetics of branched polymer formation. This opened a new and exciting field in which geometric, fractal, nature of polymers imposes itself on the dynamics and in which reaction time (age of the system) appears as a scaling variable.[3]

We are all aware that the distribution of monomers in a branched polymer can be described by a correlation function which decays as a power of the distance from the polymer center.[4] This is true for all kinds of fractals. However, there is something peculiar about real branched polymers in a chemical solution. They are made of many monomers, each of which carries several reactive chemical groups and only a fraction of these is reacted. Many more are still ready to create new bonds if appropriate collisions bring them in contact. Therefore, polymers are "reactive" fractal objects and characteristic mechanism of their formation and growth is their coalescence through formation of new chemical bonds between smaller molecules. Randomness of this process assures that a distribution of polymer sizes invariably develops.

Our aim is to discover how the fractal character of polymers influences the above distribution and the way in which it evolves with reaction time.

KINETIC EQUATIONS FOR REACTIVE SYSTEMS

We consider a dilute solution of initially unreacted polyfunctional monomers. They diffuse, collide and bind through formation of chemical bonds. Gradually, aggregates with one or more chemical bonds become the more typical feature of the solution than the single monomers. Reaction proceeds until the distribution is shifted towards large clusters and it ends with formation of a micro-gel. A standard laboratory method of monitoring a chemical (polymerization) reaction is to follow the number of chemical bonds created in the system by one of the spectrocopies, like Infra Red, for example. To be able to describe such measurements it is natural to characterize polymers by the number of chemical bonds they contain. A full description of the reaction medium would then be provided by a set of functions $\{ N_b(x,t),\ b=0,1,2\ldots\}$ specifying the local concentrations of polymers with all possible values of bond number b.

Unfortunately, it appears rather difficult to formulate a consistent Lagrangian theory which would simultaneously account for the random shape of polymers, their diffusive motion and their reactivity. Since we can not do better, we resort to the rate equations which, in a typical mean field manner, describe the time development of the spatially averaged concentrations of polymer molecules $\{ N_b(t)\}$.

Collision of two polymer molecules can result in the formation of one or more new bonds. However, only one of them is essential "cluster building bond". Others can be treated as internal bonds created after the collision, by the rearrangement of the molecular structure. Cluster building bonds form the basic tree of the polymer molecule, while every internal bond adds a closed loop to it. By classifying bonds in this way, we have separated bond formation processes into: biomolecular and unimolecular and effectively restricted every reaction to the formation of a single new bond. This makes it possible for us to express the change per unit time of the polymer number $N_b(t)$ over the average rates of all possible bimolecular and unimolecular reactions in which b-sized molecule is formed or lost:

$$dN_b(t)/dt = \frac{1}{2}\sum_{b'=o}^{b-1} K(b',b-b'-1)N_{b'}N_{b-b'-1} - \sum_{b'=o}^{\infty} K(b',b)N_b N_{b'} + L(b-1)N_{b-1} - L(b)N_b. \tag{1}$$

Reaction kernel $K(b',b")$ gives the rate of bimolecular processes in which two polymers with b' and b" bonds each coalesce into a polymer with $b=b'+b"+1$ bonds. Kernel $L(b)$ provides the probability of formation per unit time of a single new bond in a cluster with b bonds. Positive terms of the above equation correspond to the processes in which b-sized clusters are formed and negative to the processes into which these clusters enter and disappear.

It is easy to see that equation (1) does not conserve the first moment of the distribution function

$$I_1(t) = \sum_{b=o}^{\infty} bN_b(t). \tag{2}$$

This is of course expected, since $I_1(t)$ is changing with time and we measure that change readily, by I.R. for example.

There are two essential points through which the fractal nature of polymers enters into the reaction rates. First, for large polymers, the radius of gyration scales as power of polymer index b

$$R_g(b) \sim b^{1/D_f}, \tag{3}$$

where D_f is the fractal dimensionality. Radius of a polymer determines the cross section for collision with other polymers and is obviously a rate determining parameters. Secondly, diffusion coefficient $D(b)$ of a polymer scales as a power of polymer radius or index b. The diffusivity exponent α is necessarely a function of fractal dimensionality D_f and space dimension d but we will not enter into these details here. We will rather treat α as an independent exponent and write

$$D(b) \sim b^{\alpha} . \tag{4}$$

Relying on equations (3) and (4) and using standard arguments for the derivation of the Brownian kernel one can show that for large values of b´ and b", K(b´,b") scales as

$$K(b',b") \quad \sim \quad (b'^{1/D_f} + b"^{1/D_f})^{d-2}(b'^{\alpha} + b"^{\alpha}). \tag{5}$$

K(b´,b") is a homogeneous function of its arguments which satisfies the relation

$$K(\lambda b',\lambda b") = \lambda^{2\omega'} K(b',b"). \tag{6}$$

Exponent $2\omega'$ is called the degree of homogeneity of the bimolecular kernel and is equal to

$$2\omega' = (d-2)/D_f + \alpha . \tag{7}$$

The reader should understand that form of kernel K(b´,b") depends on the density of system, types of chemical interactions and other details. The kernel we have presented is in no way universal and serves just as an illustration of the general ideas.

Unimolecular reaction kernel L(b) has never been examined in detail and we will at this point present only its crude estimate. We assume that every cluster lives sufficiently long so that most of its internal bonds are created before it enters in the reaction with another cluster. Upon formation of the new cluster, in a bimolecular reaction, new internal bonds will primarely be formed through the reaction of groups which belonged to the different parent clusters. Geometric distance Λ between two such groups is a finite fraction of the radius of gyration of the new cluster and we can safely write $\Lambda \sim R_g(b)$. The shortes chemical path between the two groups then has $b_\Lambda \sim R_g{}^2(b)$ elements. The probability per unit time that the two groups will meet and react is proportional to the number of states in which a polymer chain of length b_Λ closes on itself. Taking finally into account the number of possible reaction pairs we come to the estimate for L(b):

$$L(b) \sim b^{-d/D_f+D_f-1} \tag{8}$$

Unimolecular kernal is also a homogeneous function with the degree of homogeneity equal to

$$\omega'' = -d/D_f+D_f-1 \tag{9}$$

SCALING ANALYSIS OF KINETIC EQUATIONS

The above kinetic equations were formulated with respect to specific units of scale. Unit of the polymer index b was one chemical bond and unit of time was one tick of a laboratory clock. If resolution of our I.R. spectrometer degrades for some reason, we will not be able to distinguish between two clusters which are only one bond apart, any more. Under such circumstances the unit of the cluster index scale effectively becomes larger by a factor, let us say λ. With worse instrument one usually has to measure longer, in order to "accumulate statistics". This means that experimental time now scales in units which are, for example μ times longer than previously. It can be shown rigorously that time factor μ must be a power of the index factor λ. Let us not wait for the proof of this but simply assume that $\mu = \lambda^{1/z}$, where z is a scaling exponent.

Let us now perform a formal simulation of this unfortunate degradation of our experimental abilities and examine its consequences on the results of our measurements.

Since we can not distinguish clusters which are but λ bonds apart we bin all clusters with index b in the range $\{k\lambda,(k+1)\lambda-1\}$ into one group and label their number by $M(k\lambda,t)$. Obviously

$$M(k\lambda,t) = \sum_{b=k\lambda}^{(k+1)\lambda-1} N_b(t) \tag{10}$$

We will also treat the time axis as discrete and write

$$dN_b(t)/dt \simeq N_b(t+1) - N_b(t) \tag{11}$$

Since we sample the distribution every μ time ticks, we need to calculate the coarse-grained time derivative:

$$M(k\lambda,(j+1)\mu) - M(k\lambda,j\mu) = \sum_{t=j\mu}^{(j+1)\mu-1} \sum_{b=k\lambda}^{(k+1)\lambda-1} (N_b(t+1)-N_b(t)) \tag{12}$$

We understand that $\mu<<j\mu$ and $\lambda<<k\lambda$, i.e. the coarsening of the scales is not such as to obscure the essential variations in properties as functions of time and cluster index.

Application of the double sum of equaition (12) to kinetic equation (1) results in a difference equation for quantities $M(k\lambda,j\mu)$

$$\begin{aligned} M(k\lambda,(j+1)\mu)-M(k\lambda,j\mu) &= \\ &1/2\mu \sum_{k'=o}^{k-1} K(k'\lambda,(k-k'-1)\lambda)\, M(k'\lambda,j\mu)\, M((k-k'-1)\lambda,j\mu) \\ &-\mu M(k\lambda,j\mu) \sum_{k'=o}^{\infty} K(k'\lambda,k\mu)\, M(k'\lambda,j\mu) \\ &+\mu\lambda^{-1} M((k-1)\lambda,j\mu)\cdot L((k-1)\lambda) - u\lambda^{-1} M(k\lambda,j\mu)\, L(k\lambda) \end{aligned} \tag{13}$$

We would like to avoid keeping track of the actual size of new steps and treat them as unities. This change might make it necessary to rescale all cluster numbers by a factor and so we introduce the rescaled cluster numbers $\tilde{N}_k(j)$, by the equality

$$M(k\lambda,ju) = \xi\tilde{N}_k(j) \tag{14}$$

Using the homogeneity of $K(b',b")$ and $L(b)$ and substituting equation (14) in equation (13) we obtain

$$\begin{aligned} \tilde{N}_k(j+1) - \tilde{N}_k(j) &= \\ &1/2\xi\lambda^{2\omega'+1+1/z} \sum_{k'=o}^{k-1} K(k',k-k'-1)\tilde{N}_{k'}(j)\tilde{N}_{k-k'-1}(j) \\ &-\xi\lambda^{2\omega'+1+1/z}\tilde{N}_k(j) \sum_{k'=o}^{\infty} K(k,k')\tilde{N}_{k'}(j) \\ &+\lambda^{\omega"+1/z-1} L(k-1)\tilde{N}_{k-1}(j) - \lambda^{\omega"+1/z-1} L(k)\tilde{N}_k(j) \end{aligned} \tag{15}$$

This equation would be very similar to our original equation (1) if ξ were a power of λ , i.e. $\xi = \lambda^{-\theta}$, and if the following scaling law

$$1+2\omega' = \theta - 1/z, \tag{16}$$

between the above exponents, would hold. Then, using continuous time t, instead of the discrete j, we can write:

$$\begin{aligned} d\tilde{N}_k(t)/dt = 1/2 \sum_{k'=o}^{k-1} & K(k',k-k'-1)\tilde{N}_{k'}(t)\tilde{N}_{k-k'-1}(t) - \\ - \tilde{N}_k(t) \sum_{k'=o}^{\infty} & K(k,k')\tilde{N}_{k'}(t) \\ +\lambda^{\omega''+1/z-1}L(k-1)\tilde{N}_{k-1}(t) & -\lambda^{\omega''+1/z-1}L(k)\tilde{N}_k(t) \end{aligned} \tag{17}$$

This is an equation for the rescaled (coarsened) cluster numbers $\tilde{N}_k(t)$ and apart from a prefactor infront of the unimolecular rates, it looks exactly the same as equation (1) for original cluster numbers. Exponent $\omega''+1/z-1$ is typically negative and therefore the degradation of the instrumental resolution results in the reduction of importance of unimolecular reactions. Succesive lowering of that resolution will, through the cumulative action of factors $\lambda^{\omega''+1/z-1}$ complitely eliminate the effect of these interactions. What will be left is an equation invariant under the change instrumental resolution, or what is equivalent, the change of cluster index and time scales. Solution of this equation would satisfy the equality

$$N_k(t) = \lambda^{\theta}M(k\lambda,t\lambda^{1/z}) \tag{18}$$

which, with an appropriate choice of λ, can be written as

$$N_k(t) = k^{-\theta}M(1,t\ k^{-1/z}) = k^{-\theta}\ f(k/t^z) \tag{19}$$

This is basically the final award of our analysis and it demonstrates that for large clusters the polymer size distribution follows a power law of polymer size. This equation also shows that the distribution grows with a characteristic size which is a power of reaction time. The characteristic exponent z must, however, be calculated by a separate analysis. The only essential ingredient on which this result rests is the "assumption" on the fractal nature of branched polymers.

The reader has undoubtedly noticed that our construction is a rudimentary renormalization group procedure[5], applied to the kinetic equations. Its simplicity, in this particular case comes from the absence of any statistical weight. The disappearance of the unimolecular reactions is, in R.G. terminology, equivalent to their scaling irrelevance. In the case of application of R.G. to thermodynamic systems, the fixed point of the transformation was a particular Hamiltonian. In our case, fixed point is a kinetic equation which contains only the bimolecular reactions.

Finally we should say that degradation of the instrumental resolution has no effect on the results of measurements of the scale invariant phenomena.

EXPERIMENTAL VERIFICATION OF SCALE INVARIANCE

It is a simple excersize to show that scale invariant form $N_b(t) = b^{-\theta} f(b/t^z)$ of the polymer size distribution implies that moments of the distribution grow as powers of reaction time. The first moment $I_1(t)$, the total number of bonds, should grow as $I_1(t) \sim t^{(2-\theta)z}$. The second moment, defined as

$$I_2(t) = \sum_{b=o}^{\infty} b^2 N_b(t) \tag{20}$$

should grow as $I_2(t) \sim t^{(3-\theta)z}$. $I_1(t)$, as we know, is readily measurable. However, in assessing the second moment we must rely on the inequality which relates s, number of monomers, and b, number of bonds in a polymer. Namely, if monomers are q-functional then

$$s-1 < b < (q/2)\ s. \tag{21}$$

Since, in our experimental system, q equals 3, the bounds on ratio b/s are rather tight and in the scaling sence quantities s and b must be equivalent.

Another common laboratory technique, small angle light scattering, measures the second moment of the distribution $N_s(t)$ in which polymers are indexed by the number of monomers they contain. The above scaling equality of bond and monomer index implies that this technique effectively measures and $I_2(t)$ of equation (20).

To examine the predictions of our scaling theory we have performed experimental measurements of the first and second moment of the polymer size distribution in dilute solutions (2% by volume) of three-functional aziridine molecules and two functional sebacic acid. A set of results of measurements of the number of chemical boinds, $I_1(t)$, as a function of time is presented in figure (1). Results for the second moment, obtained by light scattering, are presented in figure (2). Both moments grow as powers of reaction time, as confirmed by the straight line appearance of the results in the log-log diagram. The slope of the curve in figure (1) is equal to $(2-\theta)z = 0.8 \pm 0.1$ and slope of the second, $(3-\theta)z = 2.7 \pm 0.2$. From these two values the basic scaling exponents

$$z = 1.9 \pm 0.2 \qquad \text{and} \qquad \theta = 1.6 \pm 0.1 \tag{22}$$

are obtained.

Let us only briefly make a final connection of these values with the single polymer exponents D_f and α. A separate analysis[3] demonstrates that time exponent z can be expressed as:

$$1/z = 2/d - d/D_f - \alpha + 1 \tag{23}$$

From equations (23), (16) and (7) we can deduce a similar expression for exponent θ:

$$\theta = 2 - 2(1/D - 1/d) \tag{24}$$

The last two equations can be solved for exponents D_f and α and their numerical values obtained:

$$D_f = 1.84 \pm 0.1 \qquad \text{and} \qquad \alpha = -0.47 \pm 0.1 \tag{25}$$

These numbers tell us that three dimensional kinetically grown branched polymers are highly ramified objects which diffuse almost like Stokesian particles.

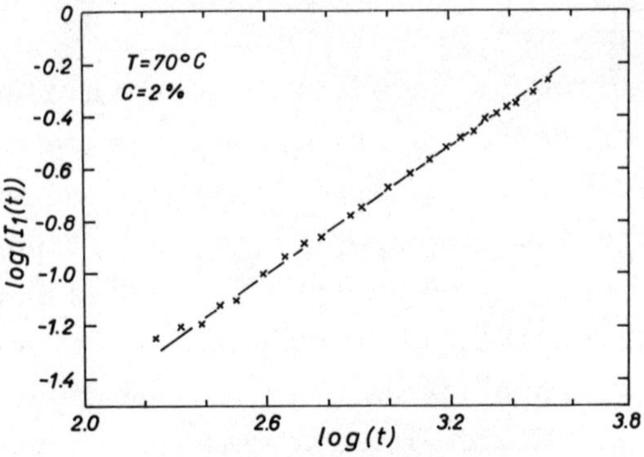

Fig. 1. Time evolution of the number of bonds $I_1(t)$ formed in the system, followed by infrared spectroscopy.

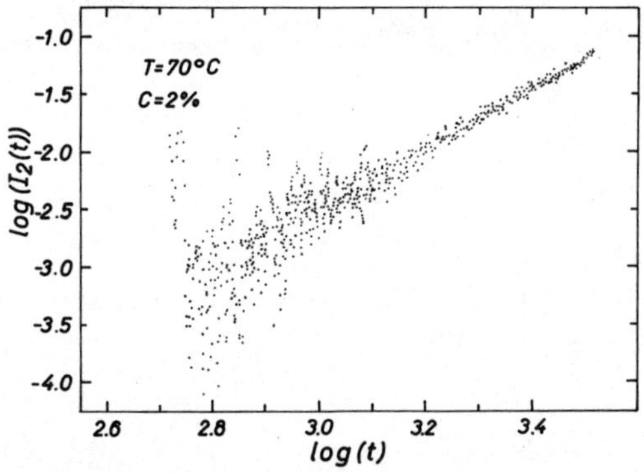

Fig. 2. Double logarithmic plot of the intensity of scattered light as a function of time.

CONCLUSION

Fractal nature of branched polymers manifests itself in the reaction rates which are homogenous functions of participating polymer sizes. This homogeneity is further reflected in the existence of a scale invariant polymer size distribution, characterized by a polymer size which grows as a power of reaction time. Validity of this scaling picture is confirmed by our experiments. Value of the fractal dimensionality of branched polymers, deduced from this work, falls in a very natural range, which is a proof of the consistency of our theoretical reasoning.

REFERENCES

1. P.G. de Gennes, "Scaling Concepts in Polymer Physics", Cornell University Press, Ithaca (1979)
2. P.J. Flory, "Principles of Polymer Chemistry", Cornell University Press, Ithaca (1953)
3. R. Botet and R. Jullien, J. Phys. A17, 2517 (1984)
   Z.B. Djordjević, "Branched Polymers and Sol-Gel Transition", Ph.D. Thesis, M.I.T., Cambridge (1984)
4. T.A. Witten and L.M. Sander, Phys. Rev. Lett. 47, 1400 (1984)
   P. Meakin, Phys. Rev. Lett. 51, 1119 (1983)
5. K.G. Wilson, Rev. Mod. Phys. 47, 773 (1975)

SWELLING OF BRANCHED POLYMERS

Mohamed Daoud

Laboratoire Léon Brillouin*
C.E.N. Saclay
91191 Gif-sur-Yvette Cedex, France

I. INTRODUCTION

In what follows, we would like to discuss the configuration of randomly branched polymers in different environments. This configuration is accessible experimentally through light or neutron scattering experiments for instance, and some experimental results will be given. Before we come to this discussion, some words have to be said about the synthesis of these polymers, and about the influence of the experimental method itself on the conformation of the polymers. Although the detailed chemistry of the synthesis may be very involved, it is most probable that only a limited number of different classes only may be actually met. These may be characterized by the distribution of molecular weights. Hence, we will be concerned only with one such classes, which corresponds to the random percolation model[1]. This will be done in section II, and will provide the distribution function. Our approach to the conformation of the polymers then will make use of a Flory - de Gennes theory[2,3]. Although it is not exact, this usually provides a fairly good approximation of the fractal dimensions. Thus we will consider the percolation problem from an energy point of view rather than from the usual geometric approach : between any two monomers there is an interaction potential V(r), that may or may not be screened depending on the conditions. The case of a single polymer - or animal - will be considered in section III, whereas section IV deals with a polymer in the presence of many others in the reaction bath. With these results in hand, we may discuss the experimental results. In section V, we consider a sol, made of large but finite polymers below the threshold. Here the solution is heavily diluted before light scattering experiments are performed. We will argue that the screening of the interactions, which is important in the reaction bath, disappears upon dilution leading to a swelling of every individual polymer of the sol.

*Laboratoire commun CEA - CNRS

II. THE SOL-GEL TRANSITION

Consider a vessel which is filled by three-functional monomers. Each of these may react by three different functional units. We will assume that each of these units has an equal reactivity, even when the monomer belongs to a molecule, so that one of its units is already reacted. Then the remaining units have still the same reactivity. It is easy to realize that when time proceeds, larger and larger molecules -or polymers- are present in the vessel. As in the percolation problem, it is also important to realize that after some time, there is a large distribution of molecular weights. A model for this polycondensation is just the percolation model : every monomer is a site of a three dimensional lattice. The bonds are the chemical links, which are present with probability p. Note the limitations of the model.

(i) no motion is allowed for any of the species present in the reaction bath. This implies that only concentrated solutions of reacting monomers might be considered.

(ii) the equal reactivity of the functional units may also be criticized. In particular when diffusion is important, we know that unreacted units deep inside the polymers are screened and cannot react.

There are however many systems that obey these restrictive conditions[4,5]. One such systems, to be discussed in section V below is the following : suppose we start with a semi dilute solution of linear polymers. These are highly interpenetrating and have many contact points. This solution is irradiated by rays, which break double bonds at random. Upon recombination, these double bonds may cross-link differents molecules, thus creating a randomly branched structure. This vulcanization procedure has been used by Leibler and Schosseler[5] with polystyrene linear chains dissolved in Cyclopentane. Here the probability that a link is present is related to the time of irradiation.

As is well known in percolation, there is a threshold p_c. Below this threshold, only finite -possibly very large-polymers are present. This is the sol phase. Above the threshold, in addition to the sol, there is an infinite molecule spanning the sample : the gel.

The most important property to be stated at this level is the distribution of molecular weights. Let $P(N,\varepsilon)$ be the number of polymers made of N monomers per site at a distance $\varepsilon \equiv p-p_c$ from the threshold. From percolation we know that

$$P(N,\varepsilon) \sim N^{-\tau} f(\varepsilon N^{\sigma}) \qquad (1)$$

where τ and σ are the usual exponents and depend on the space dimension d. In the same way, the gel fraction, which is the probability for a monomer to belong to the gel starts at the threshold as

$$G \sim \varepsilon^{\beta} \qquad (2)$$

In what follows, we will mainly focus on the conformation

of the polymers and of the gel. We will argue that whereas screening of the interactions between monomers is important in the reaction bath, it is irrelevant when solvent is added in excess both below and above the threshold. This is done within an approximate Flory-de Gennes formulation.

III. THE SINGLE POLYMER

Let us first consider a single polymer made of N three functional units which are randomly branched. All configurations are equally weighted, and we wish to know the radius of a polymer. In order to do this, we may write a free energy for the polymer in the Flory[1,2] approximation.

$$\frac{F}{kT} = \frac{R^2}{R_o^2} + v\,\frac{N^2}{R^d} \qquad (3)$$

where the first term is an elastic contribution and the second one the interaction energy. R_o is the ideal radius, when no interaction is present and was calculated some time ago by Zimm and Stockmayer :

$$R_o \sim N^{1/4}\ell \qquad (4)$$

where ℓ is the length of a monomer.

R is the actual swollen radius when interaction is present. Finally, we approximate the actual interaction potential V(r) by a point like interaction $v\delta(r)$ where v is the binary cluster integral

$$v = \int [1 - e^{-V(r)/kT}]\; d\underset{\sim}{r} \qquad (5)$$

Minimizing the free energy with respect to R gives the fractal dimension

$$N \sim R^{D_A} \qquad (6a)$$

$$D_A = \frac{2^{(d+2)}}{5} \qquad (6b)$$

Equation (6) is the Flory approximation to the random animal fractal dimension, and is exact in three dimensions[7]. It gives the behavior of a randomly branched polymer in a very dilute solution in a good solvent, when only the repulsive part of the potential is relevant.

IV. NEAR THE GELATION THRESHOLD

In a sol there are two main differences with the previous case.

(i) As discussed above, there is a very broad distribution of molecular weights, given by equation (1).

(ii) The sol is a concentrated solution, where there is a screening of the excluded volume interaction. This was discussed by Edwards[8] in the context of linear polymer chains, and is also important here.

We may write down a free energy for a typical polymer in the same approximation as above

$$\frac{F}{kT} = \frac{R^2}{R_o^2} + \frac{v}{N_w}\frac{N^2}{R^d} \qquad (7)$$

where the first term is the same as above, and the second term is the screened interaction term. This screening show up in the presence of the weight average molecular weight N_w

$$N_w = \frac{\int N^2 P(N,\varepsilon)dN}{\int N P(N)dN} \qquad (8)$$

Finally, we have to define the "typical"polymer size N, which may be extracted from the distribution function, equation (1)

$$N \sim \varepsilon^{-(\beta+\gamma)} \qquad (9)$$

Within the Flory-de Gennes approximation, both N and N_w are assumed to have a mean field behavior, so that there is a simple reaction between them : From (8) and (9) we get

$$N_w \sim N^{1/2} \qquad (10)$$

Minimizing the free energy (7) we get

$$N \sim R^{D_p} \qquad (11a)$$

$$D_p = \frac{d+2}{2} \qquad (11b)$$

Relation (11b) is in good agreement with the numerical results for $D'_p = d-(\beta/\nu)$ for all space dimensions.

V. DILUTION OF A SOL

Light scattering experiments are made in three successive stages :

(i) The sol is synthesized as discussed in section I. Then the reaction is quenched.

(ii) Excess solvent is added, so that the experiments are made on a very dilute solution rather than on the one considered above.

(iii) light scattering experiments provide an average radius of giration of the polymers over the distribution :

$$\langle R^2 \rangle_z = \frac{\int R^2(N)\, N^2\, P(N,\varepsilon)dN}{\int N^2\, P(N,\varepsilon)dN} \qquad (12)$$

The distribution function $P(N,\varepsilon)$ in (12) corresponds to the distance $\varepsilon \equiv p - p_c$ where the reaction has been quenched, after step (i)

We argue that because of the dilution in step (ii), the screening effects present in the sol vanish with dilution, so that in the diluted sol, the radius of each individual polymer is the same as in a dilute solution, equation (6). As a result, the averaging is mixing the behavior of a single polymer with the percolation distribution function.

Using equations (6, (1) and (12), and eliminating with relation (8) we get[8]

$$\langle R^2 \rangle_z \sim N_w^{(2/D_A)(1+\frac{\beta}{\gamma})} \qquad (13)$$

where both $\langle R^2 \rangle_z$ and N_w may be measured by elastic light scattering experiments. These were performed by Leibler and Schosseler[5]. Their results are shown on figure 1, and are in good agreement with equation (13) above. The system is the one described in section II above.

Quasi-elastic light scattering experiments on a different system by Candau et al[4] lead to a satisfactory agreement for the average diffusion coefficient.

We conclude this section by emphasizing that these light scattering experiments show that the molecular weight distribution is the same as in percolation and that the dilution has a dramatic effect on the conformation of the polymers.

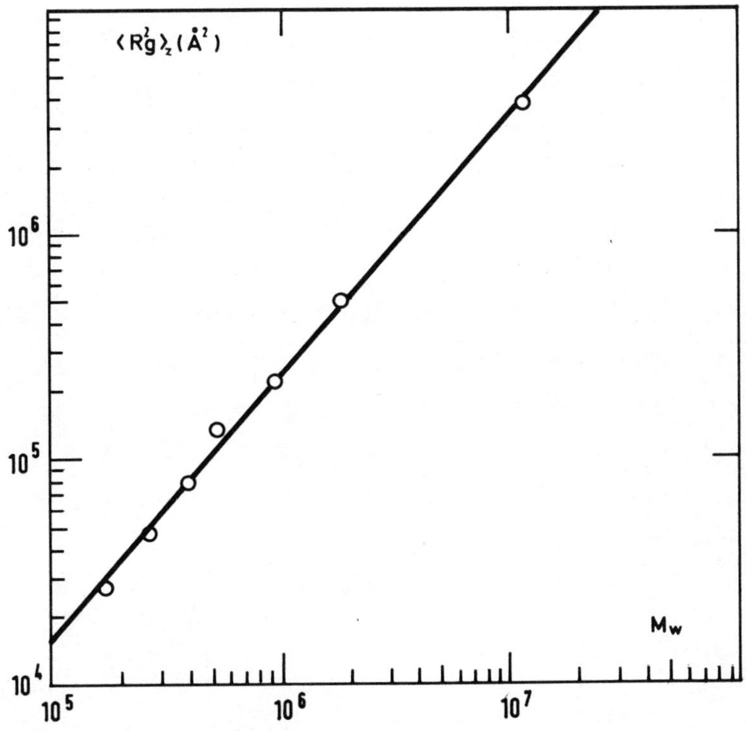

Figure 1.

REFERENCES

1. D. Stauffer, Phys. Rep. 54, 1, (1979). See also this volume.
2. P.J. Flory, Principles of Polymer Chemistry, Cornell U. Press (1953).
3. P.G. de Gennes, Scaling Concepts in Polymer Physics, Cornell U. Press (1979).
4. S.J. Candau et al, Proc. 27th Microsymposium on Macromolecules, Prague (1984).
5. F. Schosseler, L. Leibler, J. Physique Lett. 45, 501 (1984)
6. B.H. Zimm, W.H. Stockmayer, J. Chem. Phys. 17, 1301 (1949).
7. G. Parisi, N. Sourlas, Phys. Rev. Lett. 46, 871 (1981).
8. M. Daoud, F. Family, G. Jannink, J. Physique Lett. 45, 199 (1984).

SIMULATIONS OF POLYMERS IN CONFINED GEOMETRIES

K. Binder

Institut für Physik, Johannes Gutenberg Universität Mainz
D-6500 Mainz, Postfach 3980, Fed. Rep. Germany

K. Kremer *

Exxon Research and Engineering, Corporate Science Lab.
Clinton Township, Annandale, N.J. 08801

I. INTRODUCTION

The properties of flexible polymers moving inside porous structures are believed to be relevant to practical problems such as filtration, gel permeation chromatography, heterogeneous catalysis, oil recuperation, etc.[1]. Similarly the adsorption of macromolecules on interfaces plays an important rôle for problems such as adhesion, flocculation and stabilisation of colloid particles, biological membrane function, artificial organs in medicine, etc.[2]. Aside from this eventual practical application, the configurational statistics of polymers in such confined geometries is a challenging problem of theoretical physics. The present brief review will be concerned with the study of a single long flexible linear macromolecule, consisting of N effective units of size ℓ, which may move inside a tube of diameter d_T, or between two parallel planes a distance d_T apart, or which is fixed at a flat impenetrable surface with one (or both) chain ends.

If instead we were to consider chains in dilute solution, where their configurations are not restricted, for good solvents these chains would form "swollen" random coils. The coil size is measured by the gyration radius (brackets denote an average over the various configurations)[3]

$$< R^2 > \propto N^{2\nu}, \ \nu \approx 0.59 \ (d=3) \text{ or } \nu = \frac{3}{4} \ (d=2) \tag{1}$$

The chain partition function Z_N involves a second exponent[3,4]

$$Z_N \propto N^{\gamma-1} q_{eff}^N, \ \gamma \approx 1.16 \ (d=3) \text{ or } \gamma = 43/32 \ (d=2) \tag{2}$$

q_{eff} being a non-universal constant.

---

* Permanent address : Inst. für Physik, Univ. Mainz, D-6500 Mainz, Postf. 3980.

We now wish to analyse how the laws Eqs. (1,2) have to be modified when we consider polymer chains in a confined geometry. In Sec. II we shall discuss the chain dimensions for chains in tubes or between parallel plates. Sec. III is devoted to the configurational statistics near a free surface, and Sec. IV treats polymer adsorption at surfaces when attractive interactions between polymer and surface are present. As an example of how these considerations can be extended to dynamical properties, we shall discuss chain movements in a tube in Sec. V. This problem is closely related to the reptation concept of polymer dynamics in dense solutions and melts[5]. Sec. VI finally contains some conclusions.

II. LINEAR DIMENSION OF CHAINS TRAPPED IN TUBES OR SLITS

If we consider tubes of diameter d_T such that $d_T > R \equiv \sqrt{\langle R_G^2 \rangle}$ {Eq.(1)} the tube constraint would have little effect. But if $d_T < R$, the chains are deformed into long "cigars" of length $R_{/\!/}$ and diameter d_T. The structure then can be understood[1,6,7] as a one-dimensional succession of blobs of size d_T containing about g units of size ℓ each, with $d_T \propto \ell g^{\nu}$ [Eq.(1)]. Since there must be N/g such blobs, the "cigar" length is

$$R_{/\!/} \propto d_T(N/g) \propto d_T^{1-1/\nu}\, \ell^{1/\nu}\, N. \tag{3}$$

The crossover from Eq.(1) to Eq.(3) as d_T decreases hence is described by a crossover scaling function $\tilde{R}(x)$ where $x = \ell N^{\nu}/d_T$.

$$R_{/\!/} = \ell N^{\nu}\, \tilde{R}\,(\ell N^{\nu}/d_T), \quad \tilde{R}(x \to 0) = \text{const}, \quad \tilde{R}(x >> 1) \propto x^{1/\nu-1} \tag{4}$$

Similarly, for the chain trapped between two plates a distance d_T apart we again have N/g blobs but they arrange to a two-dimensional random coil configuration.

$$R_2 \propto d_T(N/g)^{\nu(d=2)} \propto d_T^{1-\nu(d=2)/\nu(d=3)}\, \ell^{\nu(d=2)/\nu(d=3)}\, N^{\nu(d=2)} \tag{5}$$

More detailed information on the chain configuration is obtained from the coherent structure factor

$$S_{coh}(\vec{k}) = \frac{1}{N} \sum_{ij} \langle \exp\,[i\vec{k}.(\vec{r}_i - \vec{r}_j)\,] \rangle\ , \tag{6}$$

the summations running over the subunits of the chain. While for unconfined chains $S_{coh}(\vec{k})$ is isotropic and scales as

$$S_{coh}(k) = N\,\tilde{S}(kR_G), \quad \tilde{S}(x << 1) \approx 1 - \frac{x^2}{3}\ , \quad \tilde{S}(x >> 1) \propto x^{-1/\nu} \tag{7}$$

For confined chains $S_{coh}(\vec{k})$ depends on the orientation of $\vec{k}$ (parallel or perpendicular to the tube, or planes, respectively). A scaling analogous to Eq.(4) reads

$$S_{coh}(k_{/\!/}) = N\tilde{S}_{/\!/}\,(k_{/\!/} d_T,\ \ell N^{\nu}/d_T), \quad S_{coh}(k_{\perp}) = N\tilde{S}_{\perp}(k_{\perp} d_T,\ \ell N^{\nu}/d_T) \tag{8}$$

Expanding Eq. (6) for small k shows $S_{coh}(k_{/\!/}) = N(1 - k_{/\!/}^2 R_{/\!/}^2/3)$. Thus $S_{coh}(k_{/\!/})$ starts to decrease for $k_{/\!/} \sim R_{/\!/}^{-1}$ already. Conversely, for $k_{/\!/} d_T >> 1$ the structure factor measures correlations over distances much smaller than d_T and hence should reduce to Eq.(7) there. Thus Monte Carlo simulations have been performed both for the tube problem and for the slit[8], and the scaling ideas have been confirmed in both cases.

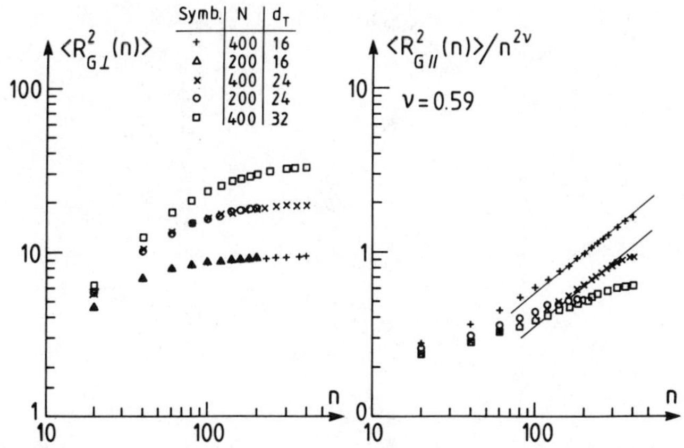

Fig. 1 . Mean square gyration radius parallel ($<R^2_{G//}(n)>$, right part) perpendicular ($<R^2_{G\perp}(n)>$, left part) to the tube axis plotted vs. the number of subunits, n. Various values of total chain length N and tube diameter d_T are shown. Straight lines in the right part of this figure indicate a behaviour $<R^2_{G//}(n)> \propto n^2$. Data were obtained by dynamic Monte Carlo method (combination of 3-bond and 4-bond motions) on the tetrahedral lattice. From Ref. 7.

In the following we briefly describe the results for chains inside tubes.

Fig. 1 shows internal average square distances parallel and perpendicular to the tube, between points n links apart. Indeed, for small n there is little difference between $<R^2_{G//}(n)>$ and $<R^2_{G\perp}(n)>$, both are proportional to $n^{2\nu}$. For larger n, one observes crossover of $<R^2_{G\perp}(n)>$ to constant plateau values, which depend on the tube diameter, of course, but not on the total length N of the chain. In the same regime, the parallel component shows the expected one dimensional behaviour, $<R^2_{G//}(n)> \propto n^2$. Of course the crossover between these two behaviours at the blob size g is quite smooth, and hence the precise definition of g is rather arbitrary.

Fig. 2 checks then the scaling law, Eq.(4). The predicted behaviour is indeed clearly verified. The behaviour of the static structure factor, however, is more complex (Fig. 3). It is seen that for small enough d_T it is non-monotonic, has a minimum at about $k_{min} = 2\pi/R_{//}$, and a subsequent maximum at $k_{max} = 3\pi/R_{//}$. These interference effects mean that the one-dimensional chain of blobs is rather rigid and the fluctuations in its length $R_{//}$ must be small.

III. POLYMER CHAINS ATTACHED TO SURFACES

If a polymer chain is attached to a hard wall with one or both ends, but otherwise there is no interaction between the chain and the wall, we expect that again the components of the gyration radius parallel and perpendicular to the surface will no longer be equivalent. Nevertheless, Eq.(1) still holds for both components with $\nu = \nu$ (3d). However, if there is

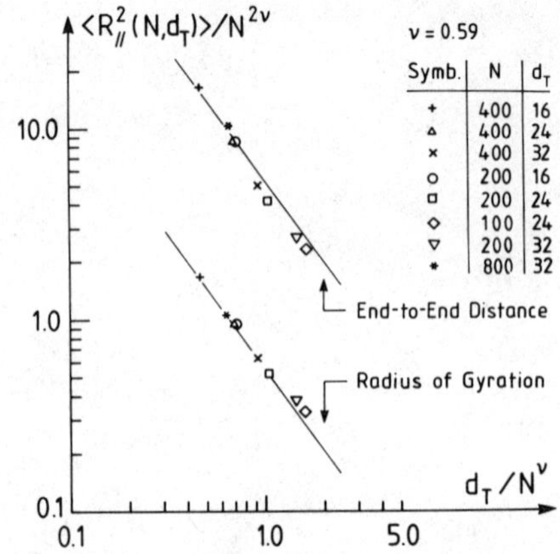

Fig. 2 . Log-log plot of gyration radius and end-to-end distance parallel to the tube axis vs. d_T/N^{ν}. The radii are normalized by $N^{2\nu}$ to show the scaling behaviour. Straight lines indicate the slopes resulting from $\langle R_{//}^2\rangle \propto N^2$. Different symbols denote various values of N and d_T as indicated in the figure. From Ref. 7.

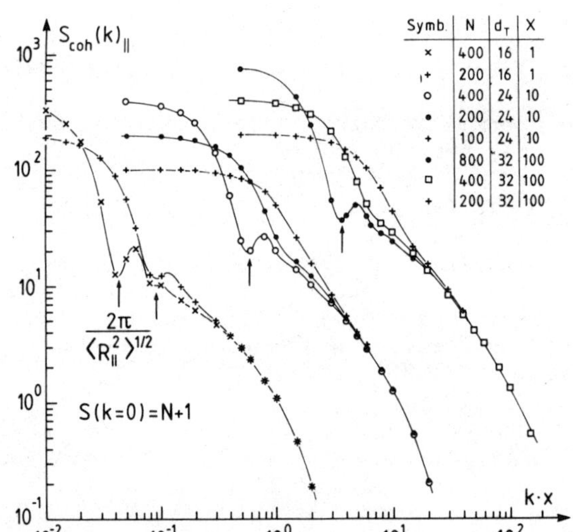

Fig. 3 . Log-log plot of $S_{coh}(k_{//})$ vs. $k_{//}$. The abscissa is displaced by an arbitrary factor x for different values of the tube diameter to avoid confusion. From ref. 7.

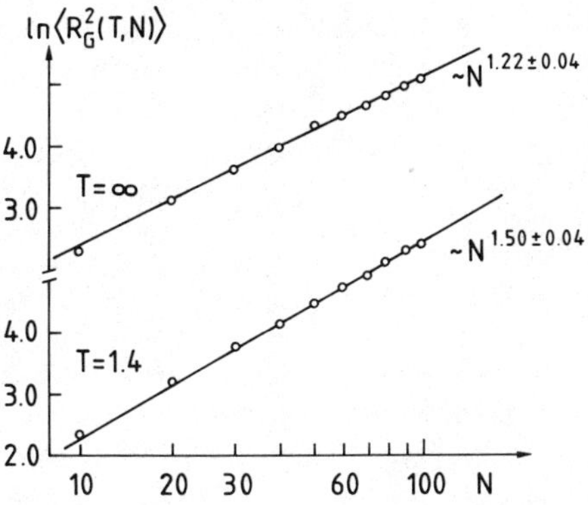

Fig. 4 . Log-log plot of the total gyration radius for $T = \infty$ {non-adsorbed region, $\nu = \nu$ (3d)} and for $T = 1.4$ (temperature measured in units of ε/k_B) {adsorbed region, $\nu = \nu(2d)$}, for chains at the tetrahedral lattice with one end attached to the surface. In this model the adsorption temperature is $T_a = 2.28$. From Ref. 9.

an attractive energy between monomers of the chain and the wall, there occurs a transition at a temperature T_a to an adsorbed state, such that for $T < T_a$ the chain is basically localized at the surface and then Eq.(1) holds with $\nu = \nu$ (2d), Fig. 4. This adsorption transition will be studied in more detail in Sec. IV.

But already in the non-adsorbed region there is need to modify Eq. (2) for the partition function $Z_N^1 (Z_N^{11})$ for a chain with one (both) ends attached to the surface,

$$Z_N^1 \propto N^{\gamma_1 -1} q_{eff}^N \; , \; Z_N^{11} \propto N^{\gamma_{11}-1} q_{eff}^N \tag{9}$$

These exponents have been estimated from Monte Carlo techniques[9] (Fig.5), exact enumerations[10] and expansions to second order in $\varepsilon = 4-d$[11] (the latter yielding $\gamma_{11} \approx -0.377$, $\gamma_1 \approx 0.695$, in fair agreement with the Monte Carlo results). These two exponents γ_1, γ_{11} are not independent of each other but related to the bulk exponents via a "surface scaling law"[12] $2\gamma_1 - \gamma_{11} = \gamma + \nu$. We also emphasize that the distinction between γ, γ_1 and γ_{11} is already necessary for ordinary random walks where excluded volume interactions are neglected (and variants thereof, such as the non-reversal random walk [NRRW]), for which $\nu = 1/2$ independent of dimensionality, and hence questions such as the crossover from $d = 2$ to $d = 3$ (Eq.(5)) cannot sensibly be asked. In the adsorbed region ($T < T_a$), on the other hand, we have $\gamma_1 = \gamma_{11} = \gamma(2d)$, while right at the adsorption transition itself γ_1 and γ_{11} adopt special values reflecting multicritical behaviour.

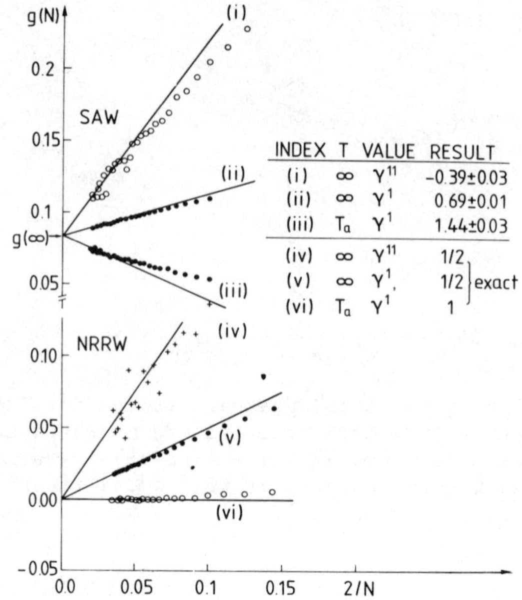

Fig. 5. Plot of $g(N) = \ln Z'(N)/Z'(N+2)$ vs. $2/N$, $Z'(N)$ being the partition function $Z_N^1(Z_N^{11})$ normalized by $Z_{NRRW} = q_0^N = 3^N$, for the case $T = \infty$, Z_N^1(ii), and $T = T_a$, Z_N^1 (iii), for self-avoiding walks on tetrahedral lattice ; analogous data for NRRW's are shown for $T = \infty$, Z_N^{11}(iv), Z_N^1(v), $T = T_a = 2.84$, Z_N^1 (vi). Data were obtained by a biased simple sampling method from Ref. 9.

IV. SCALING BEHAVIOUR AT THE POLYMER ADSORPTION TRANSITION

While the "thickness" ξ of the adsorbed polymer (measured, for instance, by $\xi = \sqrt{\langle R_G^2 \rangle}$) in the non-adsorbed phase for $T > T_a$ behaves as $\xi \propto N^{\nu}$, i.e. ξ is infinite for $N \to \infty$, we have a finite thickness (independent of N) in the adsorbed phase for $T < T_a$. In this case, the fraction N_1/N of monomers right at the surface tends to a non-zero finite limit, while $N_1/N \to 0$ for $N \to \infty$ and $T > T_a$. In order for these behaviours to be compatible at T_a, we expect

$$\xi \propto (1 - T/T_a)^{-x} \; , \quad \frac{N_1}{N} \propto (1 - \frac{T}{T_a})^{y} \tag{10}$$

Before we discuss the exponents x, y defined in Eq. (10), we mention that it makes sense to generalize Eq.(9), defining a partition function for chains with one end at the surface and the other end at a distance r apart, $Z_N(\vec{r})$, and to consider the number of monomers $M_N(z)/Z_N^1$ in a layer a distance z from the surface. For $T > T_a$ one has the simple scaling forms (the index // and ⊥ refer to direction of $\vec{r}$ parallel or perpendicular to the surface)

$$Z_N^{\perp,//}(r) \propto N^{-1-\nu(d-2+\eta^{//,\perp})}\, \tilde{Z}_{\perp,//}(r/N^{\nu}) \xrightarrow[r \ll N^{\nu}]{} r^{2-\eta^{\perp,//}} N^{-\nu d-1} \tag{11}$$

$$M_N(z) \propto N^{\gamma_1-\nu}\, \tilde{M}(z/N^{\nu}) \xrightarrow[z \ll N^{\nu}]{} z^{-1+1/\nu} N^{-1+\gamma_1} \tag{12}$$

the exponents $\eta^{/\!/}, \eta^{\perp}$ being related to γ_{11}, γ_1 via [12], $\gamma_{11}/\nu = 1 - \eta^{/\!/}$, $\gamma_1/\nu = 2 - \eta^{\perp}$. Now the structure of $Z_N^{\perp,/\!/}(r)$, $M_N(z)$ as well as of the other quantities of interest has been clarified [9] by involving the analogy between polymer statistics and the $n \to 0$ limit of the n-vector model of magnetism[3]. By this analogy, the adsorption transition is related to the multicritical behaviour of a magnet with enhanced surface exchange, such that the surface starts to order before the bulk[12]. We are not going through the details of the argument here, but merely quote some of the results. It turns out that the relative distance $1 - T/T_a$ from the adsorption temperature comes in with a scaling combination $(1 - T/T_a)N^{\varphi}$, φ being a crossover exponent, and the exponents γ_{11}, γ_1 (or $\eta^{/\!/}$, $\eta^{\perp}$, respectively) are replaced by their multicritical values γ_{11}^{SB}, γ_1^{SB} (or $\eta_{/\!/}^{SB}$, $\eta_{\perp}^{SB}$). Thus Eqs.(11),(12) are replaced by

$$Z_N^{\perp,/\!/}(r) \propto N^{-1-\nu(d-2+\eta_{SB}^{\perp,/\!/})} \tilde{Z}_{\perp,/\!/}^{SB}\left\{r/N^{\nu}, \quad (1-T/T_a)N^{\varphi}\right\} \tag{13}$$

$$M_N(z) \propto N^{\gamma_1^{SB}-\nu} \tilde{M}_{SB}\left\{z/N^{\nu}, \quad (1-T/T_a)N^{\varphi}\right\} \tag{14}$$

while right at $T = T_a$ we have for $r \ll N^{\nu}$ or $z \ll N^{\nu}$ (note that the scaling functions are singular at this limit),

$$Z_N^{\perp,/\!/}(r) \propto r^{1-\eta_{SB}^{\perp,/\!/} - \varphi/\nu} N^{-\nu d-1+\nu+\varphi} \tag{15}$$

$$M_N(z) \propto z^{-1+(1-\varphi)/\nu} N^{-1+\varphi+\gamma_1^{SB}} \tag{16}$$

Eq.(16) shows that the monomer density at T_a decays with a power law z^{-a} as a function of distance z from the surface (- a is called the "proximal exponent" in Ref. 13). Eq.(9) generalizes as follows (here and already in eqs.(11),(13),(15) we choose a normalization where $q_{eff} \equiv 1$)

$$Z_N^1 \propto N^{\gamma_1^{SB}-1} \tilde{Z}^1\left\{(1-T/T_a)N^{\varphi}\right\}, \quad Z_N^{11} \propto N^{\gamma_{11}^{SB}} \tilde{Z}^{11}\left\{(1-T/T_a)N^{\varphi}\right\} \tag{17}$$

From Eqs.(16),(17), we see that the number of monomers $N_1 = M_N(1)/Z_N^1$ in the first layer is given by the crossover exponent φ, $N_1 \propto N^{\varphi}$. In the general case it is again ruled by a dependence on the variable $(1-T/T_a)N^{\varphi}$, namely

$$N_1 = N^{\varphi} \tilde{N}_1\left\{(1-\frac{T}{T_a})N^{\varphi}\right\}, \quad \tilde{N}_1(t \to +\infty) \to t^{\frac{1}{\varphi}-1}, \quad \tilde{N}_1(t \to -\infty) \to t^{-1} \tag{18}$$

Hence we conclude that the exponent y defined in Eq.(10) simply is $(1/\varphi)-1$. A similar scaling behaviour as in Eq.(18) also is found for the linear dimensions

$$\langle R_G^2 \rangle_{\perp}^{1/2} = N^{\nu}\tilde{R}_{\perp}\left\{(1-T/T_a)N^{\varphi}\right\}, \quad \tilde{R}_{\perp}(t \to \infty) \to t^{-\nu/\varphi}, \quad \tilde{R}_{\perp}(t \to -\infty) = \text{const.} \tag{19}$$

$$\langle R_G^2 \rangle_{/\!/}^{1/2} = N^{\nu}\tilde{R}_{/\!/}\left\{(1-T/T_a)N^{\varphi}\right\}, \quad \tilde{R}_{/\!/}(t \to \infty) \to t^{[\nu(d=2)-\nu]/\varphi}, \tilde{R}_{/\!/}(t \to \infty) = \text{const.} \tag{20}$$

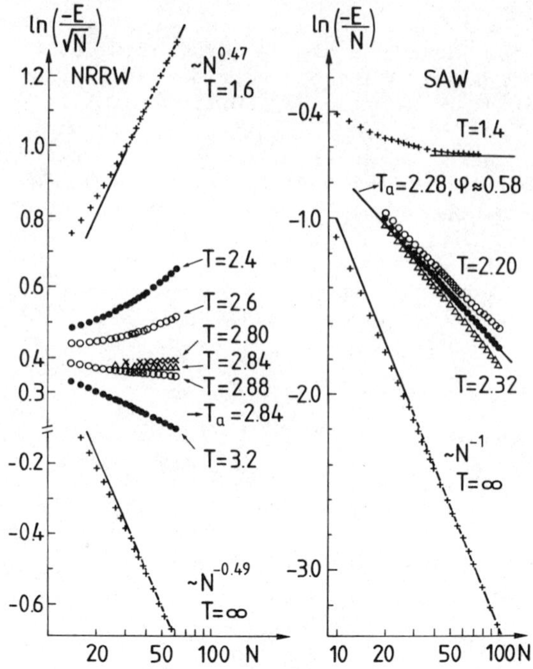

Fig. 6. Log-log plot of $-E/\sqrt{N}$ vs. N for NRRW's (left part) and (-E/N) vs. N for SAW's (right part). For estimating T_a and φ more data were used than the figure shows, which have been omitted for the sake of clarity. This leads to the accurate estimates of T_a and φ shown. From Ref. 9.

From Eq.(19) we conclude that the exponent x defined in Eq.(10) is $x = \nu/\varphi$. Hence it is of primary interest to estimate the exponent φ.

Fig. 6 shows numerical data for the energy E (which is proportional to the monomer number N_1 at the surface). Again the NRRW is treated as a check and the exact result $\varphi = 1/2$ for ordinary random walks[14] is confirmed nicely. For the SAW the estimate $\varphi \approx 0.58$ is smaller than the renormalization group estimate[11] ($\varphi \approx 0.67$) but distinctly higher than the value of $\varphi = 1 - \nu \approx 0.41$ of the de Gennes[15] conjecture. The present estimate is also confirmed by recent exact enumerations[16]. In two dimensions a real space renormalization group study[17] yielded $\varphi \approx 0.55 \pm 0.15$, again in agreement with enumerations[18], and in disagreement with the de Gennes conjecture[15] $\varphi = 1 - \nu = 0.25$. One can also check the scaling of the data in Fig. 6 directly (Fig. 7), which confirms our previous conclusions. The failure of the conjecture $\varphi = 1 - \nu$ now seems generally accepted[13]. One can also use the scaling function of Fig. 7 to extrapolate E(T,N)/N to the limit $N \to \infty$, which yields ($T < T_a$) [9] $E(T,N)/N = 0.966\ (1 - T/T_a)^{0.695}$ (Fig. 8).

Similar scaling is observed also for the chain linear dimensions $<R_G^2>_{\parallel}$, $<R_G^2>_{\perp}$, Fig. 9, where the scaling function can be used to estimate the inverse square thickness $\xi^{-2} = \lim_{N\to\infty} <R_G^2>_{\perp}^{-1}$ of the adsorbed layer, for instance $\{\xi^{-2} \approx 1.68\ (1-T/T_a)^2\}$. Hence we see that the questions asked in Eq.(10) have been fully answered.

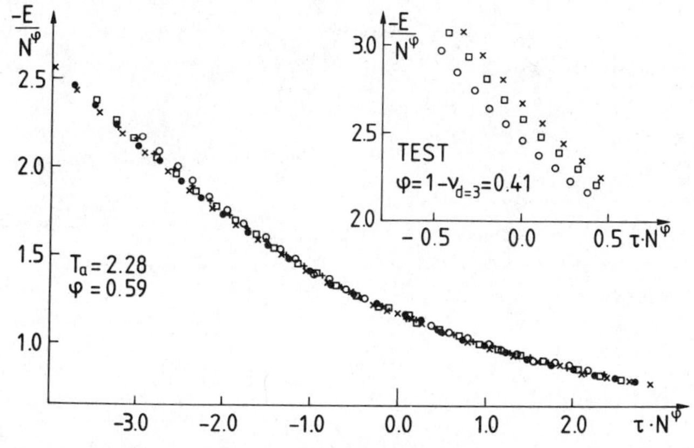

Fig. 7. Plot of the scaling function $-E/N^{\varphi}$ vs. the scaling argument $\tau N^{\varphi} \equiv (1-T/T_a)N^{\varphi}$ with $T = 2.28$, $\varphi = 0.59$. The insert gives a test of the de Gennes conjecture[15] $\varphi = 1 - \nu$.

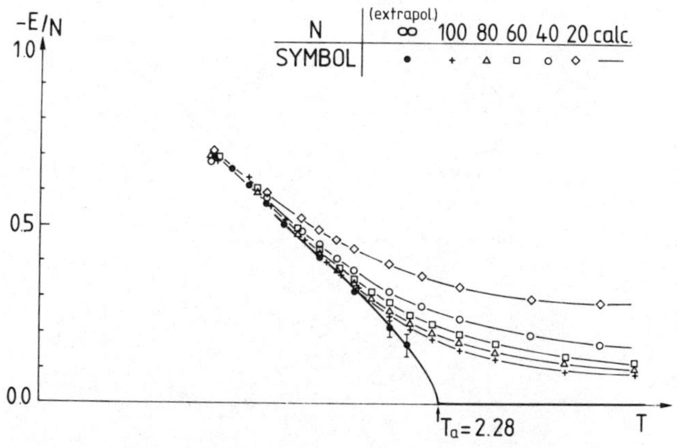

Fig. 8. Plot of $-E/N$ vs. T for different chain lengths including a simple extrapolation (dots) to $N \to \infty$. Thin lines are only drawn to guide the eye while the thick line is the asymptotic law quoted in the text. From Ref. 9.

V. DYNAMICS OF CHAINS CONFINED TO TUBES

Dynamical scaling has been developed for the case[19] where the dynamics of the free chain is governed by hydrodynamic interaction (Zimm model[3]) and for the free draining limit[7] (Rouse model[3]) ; only the latter case is studied here. The mean square displacement of an inner monomer is found as follows (W is a rate for monomer reorientation)[7]

$$\langle \vec{r}^2(t)\rangle \propto \ell^2 (Wt)^{1/(1+1/2\nu)} \quad , \quad 1 \lesssim Wt \lesssim (d_T/\ell)^{2(1+1/2\nu)} \tag{21}$$

This is the standard Rouse regime. The monomers do not yet feel the tube constraint. For longer times, $\langle r^2(t)\rangle \propto d_T^2$, independent of t, while

$$\langle r^2_{/\!/}(t)\rangle \propto d_T^2\ (Wt)^{1/2} (\ell/d_T)^{(1+1/2\nu)} \quad , \quad (d_T/\ell)^{2(1+1/2\nu)} \lesssim Wt \ll Wt_N, \tag{22}$$

where the largest relaxation time t_N of the chain is $t_N^{-1} = W(\ell/d_T)^{2-1/\nu} N^{-2}$. The final regime is a diffusion of the chains as a whole,

$$\langle r^2_{/\!/}(t)\rangle \approx 2D_N t, \qquad t > t_N \ , \qquad D_N = W\ell^2/N. \tag{23}$$

Similar predictions can also be made for incoherent and coherent dynamic structure factors. Fig. 10 shows that the simulations confirm the scaling predictions : the first crossover (seen in $\langle r^2\rangle_\perp$ only since the distinction between $t^{0.54}$ and $t^{1/2}$ is not resolved) occurs at a time depending on d_T, but independent of N. Conversely, the diffusion constant is independent of d_T, and proportional to N^{-1}.

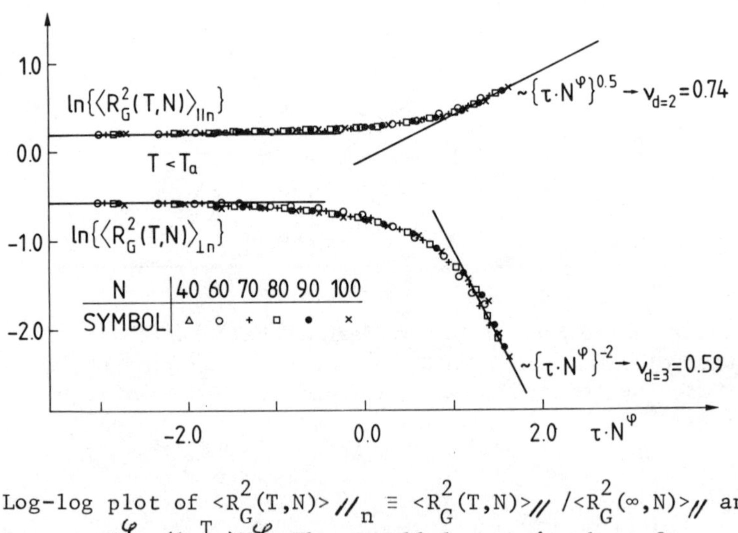

Fig. 9. Log-log plot of $\langle R_G^2(T,N)\rangle_{/\!/ n} \equiv \langle R_G^2(T,N)\rangle_{/\!/} / \langle R_G^2(\infty,N)\rangle_{/\!/}$ and $\langle R_G^2(T,N)\rangle_{\perp n}$ vs. $\tau N^{\varphi} \equiv (1 - \frac{T}{T_a}) N^{\varphi}$. The parallel part is shown for $T < T_a$, the transverse for $T > T_a$, and the asymptotic behaviour is indicated. From Ref. 9.

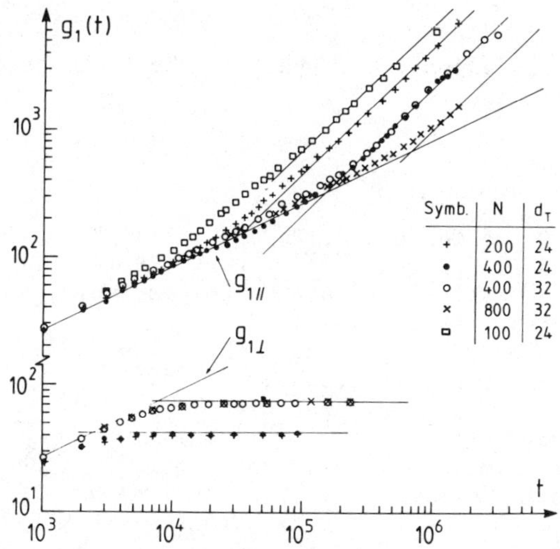

Fig. 10. Log-log plot of parallel (upper part) and perpendicular components (lower part) of mean square displacements $g_1(t) \equiv \langle r^2(t) \rangle$ of inner monomers versus time, for various tube diameters and chain lengths. From Ref. 7.

Other parts of the results, however, are not yet fully understood: one finds pronounced end effects, the monomers near the free ends of the chains can make much larger excursions relative to the center of gravity of the chains than the inner ones; also the structure factor $S_{coh}(\vec{k},t)$ seems to reflect some of the interference effects already seen in the statics (Fig. 3) and does not seem to agree with predictions of de Gennes[20]. While the latter difficulty hampers the interpretation of inelastic neutron scattering from melts in terms of the reptation model, the former problem (enhanced end mobility) should have an important effect on the so-called tube renewal mechanism[20].

VI. CONCLUSIONS

In this lecture, it has been shown that scaling analysis is a powerful tool for the understanding of the properties of polymers in confined geometry, although some quantitative problems still remain (e.g., the exponent $\gamma_1^{SB} \approx 1.44$ according to Monte Carlo (Fig. 5), but $\gamma_1^{SB} \approx 1.31$ according to the ε-expansion[11]. Unfortunately, none of these results referring to the single-chain limit has been open to experimental test so far; studies of the thickness δ of adsorbed layers as a function of molecular weight in the Θ-regime (where $\delta \propto N^{1/2}$)[21] and in good solvents (where $\delta \propto M^{0.4}$)[22] do not reflect Eqs. (15), (16), which are valid for $T < T_a$ for $z \ll \xi$, but rather reflect the concentration profile in the regime $\xi \ll z$. In this regime scaling arguments [13,14,22] simply predict $\delta \propto N^{1-\nu}$, i.e. none of the surface exponents enter. There is now a variety of theoretical results for polymers at surfaces, due to simulations and various renormalization group calculations[23], awaiting experimental tests. With respect to dynamics of polymers in tubes or slits or near surfaces, however, much less is known. A fascinating area of study will also be the extension to types of polymers other than linear: branched chains, rods, polyelectrolytes, block copolymers, etc. While some data on adsorption at surfaces are available for all these systems[2], here the theoretical analysis clearly lags behind.

ACKNOWLEDGEMENTS

This research was supported in part by the Deutsche Forschungsgemeinschaft, Sonderforschungsbereich 41.

REFERENCES

1. M. Daoud and P.G. de Gennes, J. Phys. (Paris) 38:85 (1977).
2. A. Takahashi and M. Kawaguchi, Adv. Polym. Sci. 46:1 (1982).
3. P.G. de Gennes, "Scaling concepts in polymer physics", Cornell University Press, Ithaca (1979).
4. J.C. Le Guillou and J. Zinn-Justin, Phys. Rev. B21:3976 (1980); B. Nienhuis, Phys. Rev. Lett. 49:1062 (1982).
5. L. Leger and P.G. de Gennes, Ann. Rev. Phys. Phys. Chem. 33:49 (1982).
6. L. Turban, J. Phys. (Paris) 45:347 (1984).
7. K. Kremer and K. Binder, J. Chem. Phys. 81:6381 (1984).
8. I. Webman and J.L. Lebowitz, J. Phys. (Paris) 41 (1980); A.T. Clark and M. Lal, J. Chem. Soc. Faraday Trans. Part II, 77:981 (1981).
9. E. Eisenriegler, K. Kremer and K. Binder, J. Chem. Phys. 77:6296 (1982).
10. M.N. Barber, A.S. Guttmann, K.M. Middlemiss, G.M. Torrie and S.G. Whittington, J. Phys. A11:1833 (1978); T. Ishinabe and S.G. Whittington, J. Phys. A14:439 (1981).
11. H.W. Diehl and S. Dietrich, Phys. Lett. A80:408 (1980); Z. Phys. B42:65 (1981); Phys. Rev. B24:2878 (1981); S. Dietrich and H.W. Diehl, Z. Phys. B49: 315 (1981); J.S. Reeve and A. Guttmann, Phys. Rev. Lett. 45:1581 (1980); J. Phys. A14:3357 (1981); J.S. Reeve, Phys. Lett. A 81:237 (1981).
12. K. Binder, "Phase Transitions and Critical Phenomena", ed. by C. Dowb and J.L. Lebowitz, Academic Press, New York (1938), Vol. 8, p. 1.
13. P.G. de Gennes and P. Pincus, J. Phys. (Paris) Lett. 44:L-241 (1983).
14. R.J. Rubin, J. Chem. Phys. 43:2392 (1965).
15. P.G. de Gennes, J. Phys. (Paris) 37:1445 (1967); Macromolecules 14:1627 (1981).
16. T. Ishinabe, J. Chem. Phys. 76:5589 (1982).
17. K. Kremer, J. Phys. A16:4333 (1983).
18. T. Ishinabe, J. Chem. Phys. 77, 3171 (1983).
19. F. Brochard and P.G. de Gennes, J. Chem. Phys. 67, 52 (1977).
20. P.G. de Gennes, J. Phys. (Paris) 42:735 (1981).
21. A. Takahashi, M. Kawaguchi, H. Hirota and T. Kato, Macromolecules 13:884 (1983).
22. M. Kawaguchi and A. Takahashi, Macromolecules 1 :1464 (1983).
23. E. Eisenriegler, J. Chem. Phys. 79:1052 (1983); 81:4666 (1984); and preprint; K.F. Freed, J. Chem. Phys. 79:3121 (1981); A.M. Nemirovsky and K.F. Freed, preprints.

NOTE ADDED : For two-dimensional self-avoiding walks attached to a one-dimensional surface, the exponent γ_1 has been found exactly from conformal invariance[24] as $\dot{\gamma}_1 = 61/64$.

24. J.W. Cardy, preprint.

THE DIMENSIONS OF STRANGE ATTRACTORS: A SURVEY OF EXPERIMENTAL STUDIES

Tormod Riste and Kaare Otnes

Institute for Energy Technology

P.O. Box 40, N-2007 Kjeller, Norway

ABSTRACT

Attractors are objects in multidimensional phase space. Their topology may even so be reconstructed from time-series observations of a single variable. With available techniques, one may also from such observations deduce a dimension for characterization of the attractor. In this paper we review experimental methods and results of such studies for systems that are nonlinearly driven into the chaotic regime. The strange attractors that develop are mostly identified as fractal objects of a single-valued dimension in the range 2-5. We discuss possible reasons for deviations from this behaviour. The relevance of attractors and chaos to disordered spatial structures is discussed.

1. INTRODUCTION

So far, in this school we have been concerned with dimensions and fractal shapes of real-space objects. When considering dynamical systems, we move into phase space. In this case too, we shall find that the fractal concept is a useful one. The objects to consider now are the attractors. In particular we shall be concerned with nonlinear (dissipative, driven, non-equilibrium, irreversible) systems for which attractors are strange[1-3] or fractal.

To define the initial state of a dynamical system we may need a large number of independent variables. The initial state is thus represented by a point in a multidimensional phase space. As time evolves, the dynamical behaviour of the system is given by a set of trajectories that constitutes the phase portrait. After a transient period the portrait covers a limited region of phase space called an attractor. One ususally distinguishes three kinds of attractors: a fixed point, a limit cycle and a strange attractor. A fixed point represents a stationary state, a limit cycle or periodic attractor a periodic state. Strange attractors are more complicated phase-space objects and correspond to temporally chaotic states.

A characteristic of chaotic systems is the sensitive dependence on initial conditions. This is reflected in the value of a Lyapunov exponent, which gives the average rate of separation of neighbouring trajectories. A positive exponent implies a quite different temporal behaviour and a divergence of neighbouring initial states. A strange attractor has at least one positive Lyapunov exponent[2-3].

To characterize an attractor one may alternatively use its metric dimension[2]. This dimension is always smaller than the dimension of the phase space it sits in. For a fixed point the dimension is zero, for a limit cycle it is one. A doubly-periodic state (2-torus) has dimension two. Strange attractors have non-integer dimensions, reflecting their topological irregularity which may consist of a fuzzy or many-leaved structure.

The dimension of an attractor gives the number of independent variables in the motion. A knowledge of the dimension may provide valuable physical insight into the problem. Take as an example convection in a two-component fluid, a problem treated theoretically by Velarde and Antoranz[4], using a five-mode Lorenz model. The dimension of 4.05 that they found indicates a weak coupling between two independent attractors, each of dimension close to two.

Strange attractors are connected with chaotic states, i.e. with nonlinear systems. The chaotic state develops when a control parameter of an experiment is increased. Several scenarios have been introduced for this development: period doubling-, intermittency-, and quasiperiodicity- to chaos are those relevant for our review. In the last couple of years experimental data have been analyzed to give the dimensions of strange attractors. The main purpose of this talk is to briefly review these results. Most of the reported dimensions are probably included, but we may have overlooked some cases. We start, however, by describing a method by which fractal exponents are derived from data in nonlinear dynamics[5].

2. TIME SERIES AND THE DERIVATION OF ATTRACTOR DIMENSION

The experimental data on a nonlinear system usually consists of a time-series of a single observable. The corresponding phase portrait is, however, a body in a multidimensional phase space. Still, this high-dimensional object can be reconstructed from such data by a procedure justified by mathematical embedding theorems. The method consists of forming an m-dimensional vector $[A(t_i), A(t_i+T), \ldots A(t_i+(m-1)T)]$ where $A(t_i)$ is a measured point in the time series and T is a time delay. Taking $m = 2$ and letting i vary to include many points in the time series, a two-dimensional projection of the portrait appears. An impressive portrait is shown in Fig. 1, which is from a work of Roux et al.[6] on an oscillating chemical reaction. The shape of the portrait depends on the chosen time delay, but the dimension is little sensitive to it.

Having reconstructed the phase portrait, the fractal dimension may in principle be derived in the same way as for real-space objects, using a box-counting method. The analysis for such a complicated object may be simplified by making a Poincaré cut, i.e. by cutting across the m-dimensional portrait with an (m-1)-dimensional hyperplane. For the attractor, whose two-dimensional projection is shown in Fig. 1, a planar cut through its three dimensional portrait gives an almost linear curve of dimension $\gtrsim 1$, corresponding to a dimension $\gtrsim 2$ for the complete attractor[6].

For a singly periodic time series A(t) the time delay could be

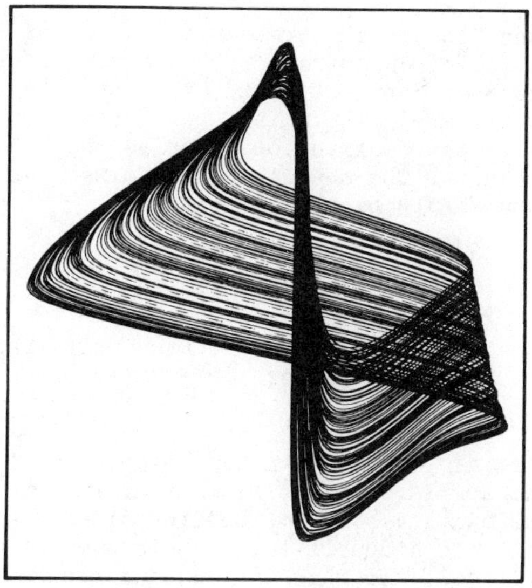

Fig. 1 Phase space portrait of an oscillating chemical reaction. The concentration of one particular chemical in a solution of ~ 30 others at time t and t+T (a fixed delay) are plotted as x and y coordinates, respectively. (From Ref. 6)

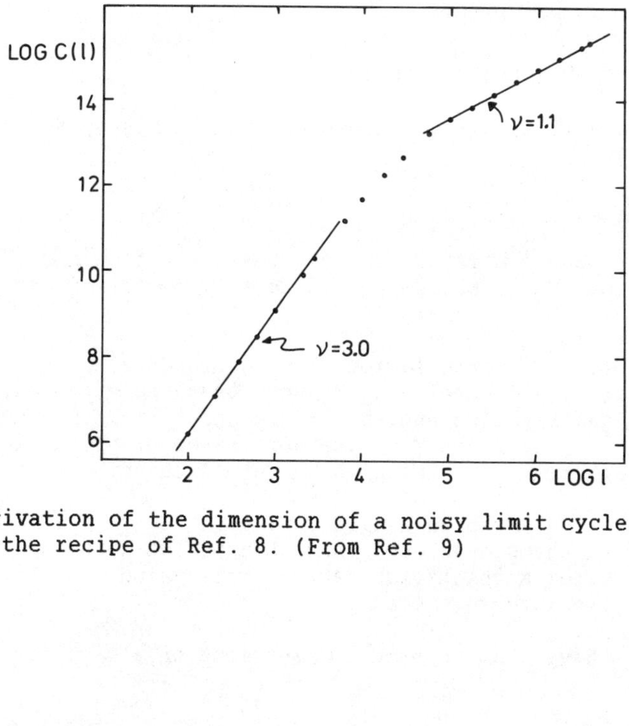

Fig. 2 Derivation of the dimension of a noisy limit cycle, according to the recipe of Ref. 8. (From Ref. 9)

chosen such that A(t+T) corresponds to the derivative $\dot{A}(t)$, and a 2-dimensional plot would simply be a plot for $(A, \dot{A})$. This is a well-known construction in Hamilton-Jacobi mechanics.

Grassberger and Procaccia[7] have devised a faster method for deriving the dimensionality of an attractor from a time series. Their method takes advantage of the fact that the points lie on an attractor and are spatially correlated. The method involves calculation of $C(\ell)$, the number of pairs in the d-dimensional series whose Euclidian distance is smaller than ℓ. The characteristic dimension ν of the attractor is then given simply by the relation

$$C(\ell) \propto \ell^{\nu} \tag{1}$$

for small ℓ's. ν is a lower limit for the Hausdorff fractal dimension D. If all the points on the attractor are equally often visited, then ν and D are identical.

Ben Mizrakhi et al.[8] have shown how noise will affect the analysis. Stochastic noise, as distinct from deterministic noise, will always have the dimensionality of the embedding dimension d. This is demonstrated in Fig. 2[9]. In a double logarithmic plot of $C(\ell)$ vs. ℓ for a time series sin $(2\pi\ t/100)$; to which is added a stochastic noise of amplitude 0.125, the noise has dimension $\nu_1 = d = 3$, the embedding dimension. The second linear slope of $\nu_2 = 1.1$ is close to the expected dimension 1.0 for a limit cycle. A decrease of the noise amplitude would move the knee on the curve closer to the origin and make ν_2 approach 1.0. Changing the embedding dimension d would always change ν_1 correspondingly, whereas ν_2 would stay unchanged. The last statement is true when $d > \nu_2$.

In some cases the observed signal A(t) is noisy enough to make plots like Figs. 1 and 2 difficult. It is then possible to "filter" this noise by smoothing the curve A(t) through a Fourier transform method[9].

3. A SURVEY OF EXPERIMENTAL RESULTS

This survey will mainly include experiments in hydrodynamics and solid state physics.

3.1 Hydrodynamics

Most experiments here are on Couette-Taylor flow and Rayleigh-Bénard convection. The experiments are usually performed near the onset of temporal chaos.

Experiments on Couette-Taylor flow by Brandstäter et al.[10] give Fig. 3, which shows a dimension d_μ deduced by three different methods. R_c is the critical Reynolds number for the onset of Taylor vortex flow. The increase in d_μ coincides with the appearance of broad band noise in the the power spectrum, i.e. with the onset of chaos.

Experiments on Rayleigh-Bénard convection by Malraison et al.[11], on a system in which chaos developed through a period-doubling cascade, gave a dimension of 2.8 for a chaotic attractor. This value was oberved at $R/R_C = 235$, in the inverse cascade.

Giglio et al.[12] have also reported measurement on a Rayleigh-Bénard system

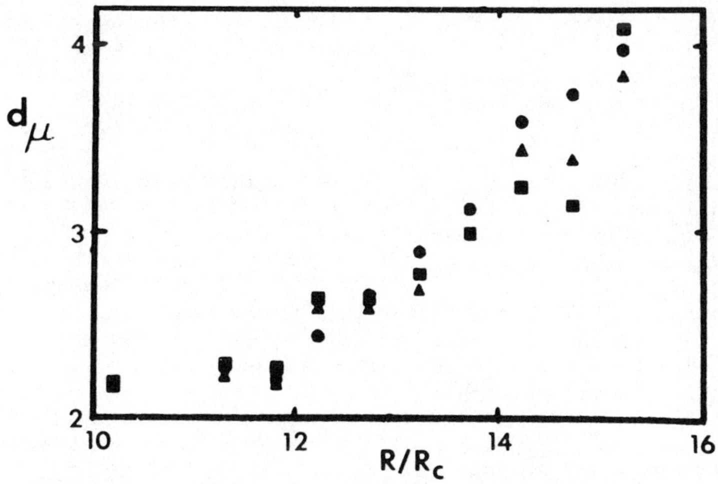

Fig. 3 Attractor dimensions d_μ from experiments of Couette-Taylor flow. Different symbols denote different procedures of data processing. (from Ref. 10)

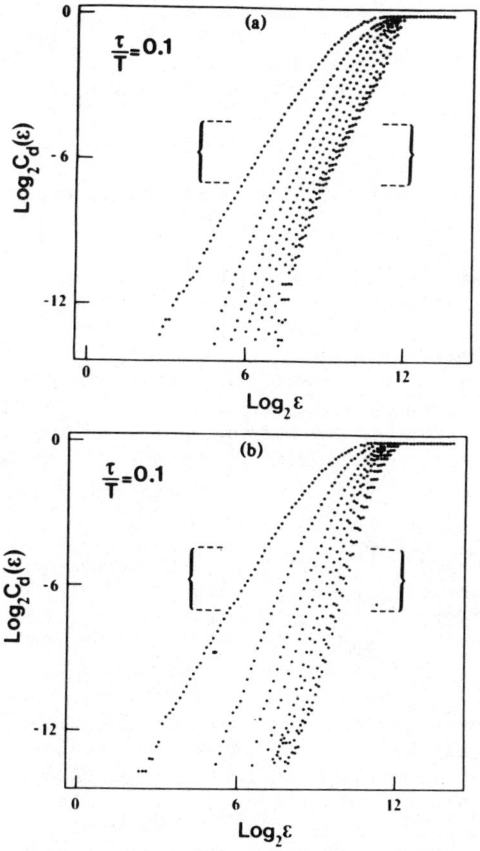

Fig. 4 Plots according to a recipe similar to that of Fig. 2 for convection in a Rayleigh-Bénard system. From left to right curves are for increasing embedding dimension. (From Ref. 12)

that displays a period-doubling scenario. In the chaotic regime, at R/R_C = 71.1 and 80.1, Figs. 4a and b show plots corresponding to eq. (1) for different values of the embedding dimension d. Over the indicated region of the curves the slope becomes independent of d as d increases, and this defines ν = 2.3 and 4.7 for (a) and (b), respectively.

Our own measurements[9,13] on convection in a nematic liquid crystal subject to a temperature gradient and a magnetic field, both in the vertical direction, gave the phase portrait of Fig. 5. This portrait was obtained after smoothing the time series through the Fourier method. The dimensionality of this attractor is seen from Fig. 6 to consist of <u>two</u> indices, 2.1 and 1.25. According to Ruelle[14] one may expect this for a system with two non-interacting attractors of different amplitudes. Actually our system can be classified as a two-component one. The scenario for the development of chaos is not yet known.

A value of ν = 5.1 has been reported by Malraison et al.[11] for the chaotic regime of an EHD instability. The system was a dielectric liquid in which unipolar ions were injected. The control parameter was a vertical d.c. electric field.

Guckenheimer and Buzyna[15] have studied the development of geostrophic turbulence in a rotating, differentially heated annulus of a fluid. The scenario for this development is probably a complicated version of the quasiperiodic scenario of low-dimensional Rayleigh-Bénard convection. The complication is possibly related to the more involved pattern of fluid flow in the geostrophic case, which also could be reflected in the high dimension (between 7 and 12) of the observed attractor.

3.2 Solid State Physics

It is only recently that experiments on deterministic chaos in solids have gained momentum. Most of the studies have been performed by the Berkeley group, and a comprehensive review of their results as of summer 1984 is available[16]. One reason for the interest in these experiments is the greater insight, as compared with liquids, into the microscopic origin of the instabilities.

The Berkeley work includes instabilities in p-n junctions in silicon, resonantly driven by an applied a.c.-voltage. In the case of a single junction the consecutive current maxima display a bifurcation diagram which is a textbook case of a cascade of period doublings to chaos. Just below the period-3 window an attractor dimension of 2.03 was measured[17], see Fig. 7.

For two resistively coupled p-n junction resonators both period doubling and quasiperiodic scenarios are observed. The first period doubling is followed by a Hopf bifurcation with a new incommensurate frequency, frequency locking etc. In the chaotic region the attractor is a folded object with dimension 2.23[17].

The same group studied plasma wave oscillations in a Ge-rod placed in parallel electric and magnetic d.c. fields[18]. At increasing control parameter V_0, a quasiperiodic route to chaos is observed, as seen from the sequence of Poincaré sections of Fig. 8. The deduced dimensions shown in Fig. 9 have steps corresponding to a limit cycle (ν = 1), a 2-torus (ν = 2) and a strange attractor (ν = 2.65).

Martin et al.[19] studied the chaotic behaviour of the d.c.-induced electrical conduction in barium niobate single crystals, a mixed electronic-ionic conductor. A quasiperiodic scenario was observed, invol-

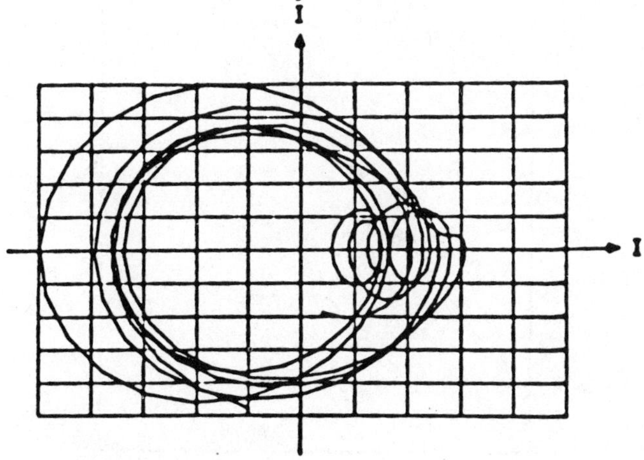

Fig. 5 Phase space portrait from neutron scattering data on a convecting nematic liquid crystal. (From Ref. 9)

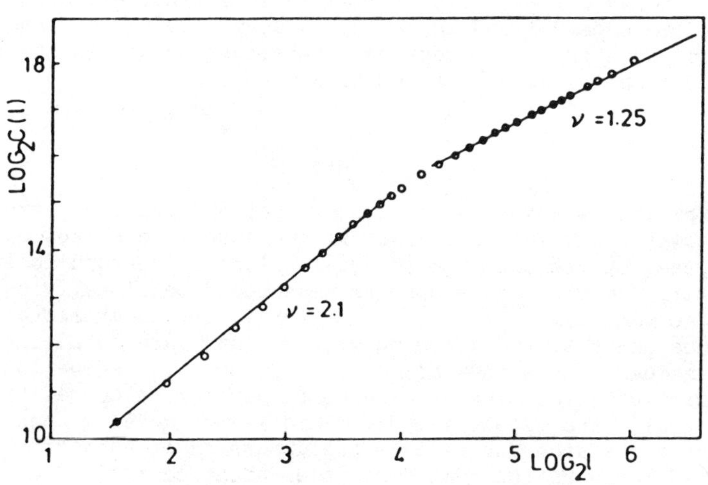

Fig. 6 Derivation of fractal dimension ν from data of Fig. 4. The embedding dimension is four. (From Ref. 13)

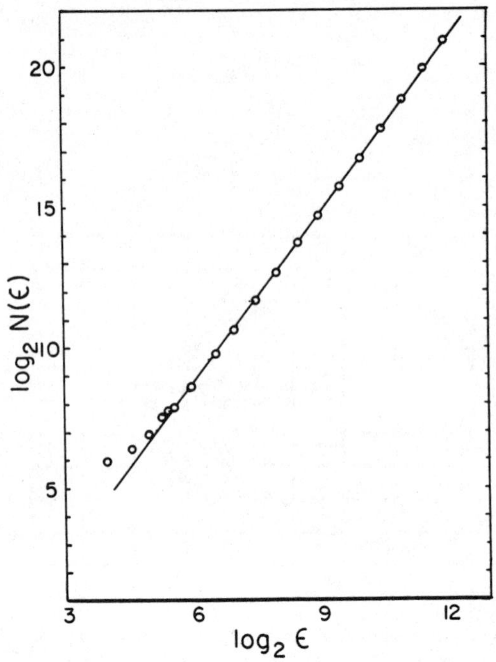

Fig. 7 Plot giving fractal dimension of chaotic attractor for a driven p-n junction in Si. The embedding dimension is six. (From Ref. 17)

ving a 3-torus, leading to a strange attractor with fractal dimension $\nu = 5.3$.

3.3 Other Systems

Chaotic behaviour has been studied quite extensively also in other systems, e.g. in chemical and in optical systems, but to our knowledge no measurement of an attractor dimension has been reported. In the former case a value $\nu \gtrsim 2$ seems relevant[6], as mentioned above.

4. DISCUSSION

The experiments reviewed above show that dissipative systems, when driven away from equilibrium, first condense into states of a few integral degrees of freedom. The integral values correspond to limit cycles and tori. As chaos develops the measured dimensionalities ν are slightly higher and fractal, as expected for a strange attractor. The fractal values are $\gtrsim 1$, in agreement with a conjecture by Kaplan and Yorke[3]. The actual values show no correlation with the scenarios for the development of chaos. The data suggest, however, a correlation between the real-space geometry and ν. This is not totally unexpected. In a hydrodynamic system one expects the dimension of an attractor to be at least as large as the number of independent eddies[3]. The number of eddies is determined by the aspect ratio, and for low-aspect-ratio systems the convection rolls are strongly coupled. The low values of ν

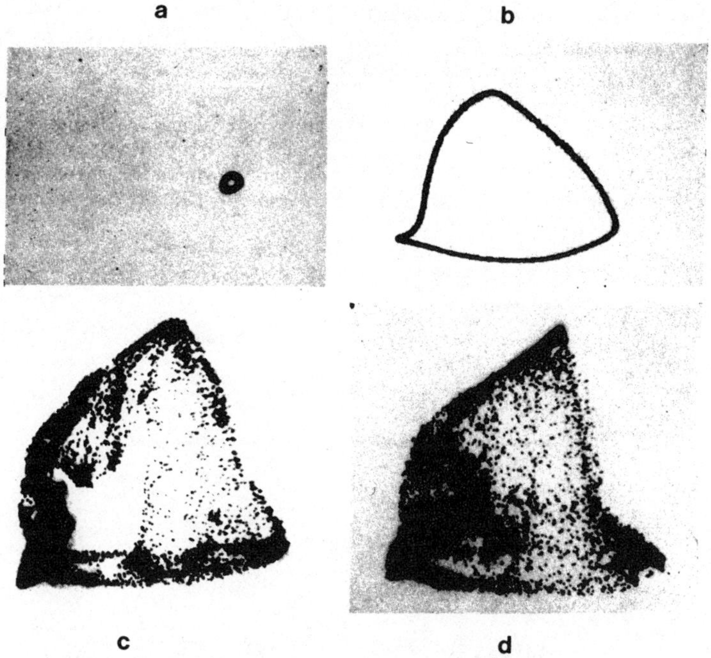

Fig. 8 Sequence of return maps of current maxima I_{a+1} vs. I_n for plasma oscillations in Ge. Applied voltage V_0 increases from (a) to (d). For (d) the measured fractal dimension is 2.65. (From Ref. 16)

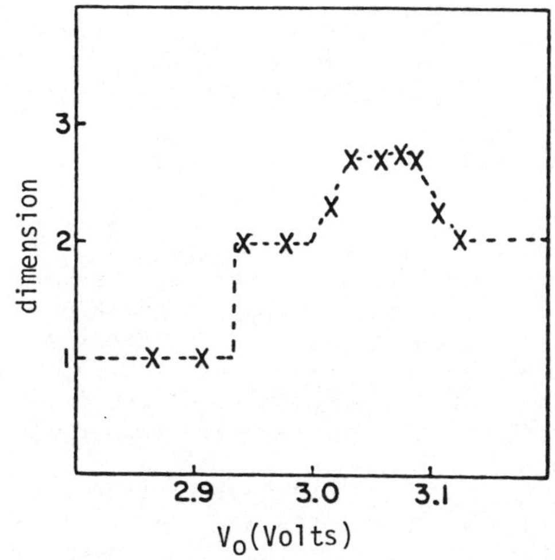

Fig. 9 Attractor dimension for plasma oscaillations in Ge as a function of applied voltage V_0. Some of the points are from data in Fig. 8. (from Ref. 18)

in Table 1 may thus reflect the fact that the systems are "small". In agreement with this an increase of the attractor dimension of the geostrophic system from two to seven was accompanied by a change in the rotational symmetry of the flow.

TABLE 1

COLLECTION OF DATA ON FRACTAL DIMENSION OF CHAOTIC ATTRACTORS

System	Scenario	Fractal dimension	Reference
Couette-Taylor	quasiperiodic	2 - 4	10
Rayleigh-Bénard	period doubling	2.8	11
Rayleigh-Bénard	period doubling	2.3;4.7	12
Rayleigh-Bénard, two component	?	1.25;2.1	9,13
EHD in di-electric liquid	intermittency	5.1	11
geostrophic	quasiperiodic?	7 - 12	15
p-n junction, single	period doubling	2.03	17
p-n junction, coupled	quasiperiodic	2.23	17
solid-state plasma	period doubling and quasiperiodic	2.65	18
electronic-ionic conduction	quasiperiodic	5.3	19
chemical reaction	period doubling	$\gtrsim 2$	6

A sudden change of ν is expected under a condition called crisis, when an attractor collides with a coexisting, unstable attractor of a different dimension[20]. In a cyclic crisis coexisting, coupled attractors suddenly merge into one attractor[21,22]. Experimental indication[13] and a clear evidence[17] for the latter have been reported. Coexisting, uncoupled attractors of different amplitudes may lead to the observation of two coexisting values of ν, as observed in one system[9,13].

An interesting and amusing discussion about the importance of the concept of dimensionality in dynamics was held at a Nobel symposium in Sweden in 1984[5]. Some held the view that dimensions in this case are useless and not worth measuring: for low-dimensional systems we don't care about them, and for high-dimensional systems we can't find them. High-dimensional chaos may exist in fully developed turbulence, although a poll among the experts present indicated that most people doubted that it would. If it exists, it may in any case be difficult to distinguish it from the effects of stochastic noise.

The fractal dimension D reflects the static, geometrical properties of an attractor. ν is to some extent sensitive to the dynamics, since regions of the attractor are weighed by the frequency with which the trajectories are visited. A Lyapunov exponent, λ, is the best measure of the dynamic properties of an attractor[2]. To deduce λ directly from experimental data is a difficult task. To do so one must be able to follow also adjacent trajectories that are not on the attractor[2]. Ideally one should perturb the system from the attractor, and in a welldefined way. A significant progress in data analysis for the derivation of λ has recently been made by Wolf et al.[23].

Up to this point we have discussed strange attractors only in the context of chaos in time. It has, however, been shown that certain chaotic <u>spatial</u> structures can be explained as chaotic orbits in mapping according to mean-field theory[24]. The analogy between chaos or disorder in temporal and spatial structures is particularly striking in a quasi-periodic scenario in which chaos develops as a competition between different modes[25]. Experimentally the development of chaotic structures by this mechanism has been observed in hydrodynamic systems[26] and in solids[27]. Further progress in the understanding of disordered systems from deterministic chaos may be expected.

ACKNOWLEDGEMENT

We are greatly indebted to M. Giglio, C.D. Jeffries, H.L. Swinney and their collaborators for allowing us to reproduce some of their figures.

REFERENCES

1. D. Ruelle, The Mathematical Intelligencer <u>2</u> (1980) 126
2. J.D. Farmer, E. Ott and J. Yorke, Physica <u>7D</u> (1983) 153
3. J.D. Farmer in "Evolution of Ordered and Chaotic Patterns in Systems treated by the Natural Sciences and Mathematics", ed. H. Haken, (Springer, Berlin 1983) p. 228
4. J.C. Antoranz and M.G. Velarde in "Nonequilibrium Cooperative Phenomena in Physics and related Fields", ed. M.G. Velarde, (Plenum 1984) p. 103.
5. Most information in this review, also as regards theory, is contained in papers by other authors in Proceedings of the 59th Nobel symposium, June 1984, published as Physica Scripta <u>T9</u> (1985), ed. S. Lundqvist.
6. J.-C. Roux, R.H. Simoyi and H.L. Swinney, Physica <u>8D</u> (1983) 257
7. P. Grassberger and I. Procaccia, Phys. Rev. Lett. <u>50</u> (1983) 346
8. A. Ben-Mizrakhi, I. Procaccia and P. Grassberger, Phys. Rev. <u>29A</u> (1984) 975
9. T. Riste and K. Otnes in "Condensed Matter Resaerch Using Neutrons", eds. S.W. Lovesey and R. Scherm, (Plenum 1984) p. 227
10. A. Brandstäter, J. Swift, H.L. Swinney and A. Wolf, and J.D. Farmer and E. Jen, and P.J. Crutchfield, Phys. Rev. Lett. <u>51</u> (1983) 1442
11. B. Malraison, P. Atten, P. Bergé and M. Dubois, J. Physique, Lettres <u>44</u> (1983) L. 897
12. M. Giglio, S. Musazzi and U. Perini, Phys. Rev. Lett. <u>53</u> (1984) 2402
13. T. Riste and K. Otnes in Ref. 5, p. 76
14. D. Ruelle in Ref. 5, p. 210
15. J. Guckenheimer and G. Buzyna, Phys. Rev. <u>51</u> (1983) 1438
16. C.D. Jeffries in Ref. 5, p. 11
17. R. Van Buskirk and C.D. Jeffries, Phys. Rev. A (to be published)
18. G.A. Hald, C.D. Jeffries and E.E. Haller in Proceedings of the 17th Internat. Conf. on Physics of Semiconductors, San Fransisco 1984 (Springer Verlag, to be published)

19. S. Martin, H. Leber and W. Martienssen, Phys. Rev. Lett. 53 (1984) p. 303
20. C. Grebogi, E. Ott and J. Yorke, Physica 7D (1983) 181
21. Y. Gu, M. Tung, J.M. Yuan, D.H. Feng and L. Narducci, Phys. Rev. Lett. 52 (1984) 701
22. J. Frøyland, Physica 8D (1983) 423
23. A. Wolf, J.B. Swift, H.L. Swinney and J.A. Vastano, Physica D (to be published)
24. P. Bak, Phys. Rev. Lett. 46 (1981) 791
25. M. Høgh Jensen in Ref. 5, p. 64
26. J.P. Gollub in Ref. 5, p. 79
27. B. Lebech, K. Clausen and O. Vogt, quoted in Ref. 5, p. 68

ERGODIC RENORMALIZATION AND UNIVERSAL STRANGE ATTRACTORS

Indubala I. Satija

Bartol Research Foundation
University of Delaware
Newark, DE 19716
and
Center for Nonlinear Studies
Los Alamos National Lab
Los Alamos, NM 87545

Renormalization has emerged as an important tool in a variety of problems in physics. It was originally invented in quantum field theory. In recent years, it has been used extensively in the theory of critical phenomenon. In the last few years it has been proved to be an important technique in the study of transition to chaos in dynamical systems theory. In simple cases, under renormalization, system converges to a fixed point, i.e., the system is invariant under renormalization. The significance of a fixed point is that all systems attracted to it under successive renormalization will have the same large scale behavior, so that the behavior is universal. In critical phenomenon, the fixed point corresponds to the fact that the renormalized system looks the same at every length scale corresponding to an infinite correlation length. The infinite correlation length is the case of second order phase transition. In dynamical systems theory, renormalization operation corresponds to looking at a system on longer time scales and smaller spatial scales. Universally self-similar period-doubling sequences in 1-d logistic map[1] and 2-d area preserving maps[2] have been explained in terms of fixed points of renormalization. Another important application of renormalization ideas has been the breakdown of invariant circles.[2]

In almost all cases studied in the past, the asymptotic behavior is the convergence to the fixed point of the renormalization. However, there are many cases, where the order in the behavior of the system cannot be explained in terms of simple attractors such as fixed points. In condensed matter physics some of the disordered systems, random field systems and spin glass systems may belong to this class. In dynamical systems, one of the obvious examples is the order in the behavior of the system beyond period doubling transition to chaos in 1-d logistic map as shown in Fig. (1). Problems such as the breakdown of an arbitrary KAM trajectory[2] in 2-d area preserving map or a typical two-torus breakdown[3] cannot be described in terms of fixed points of renormalization. However, some special cases of these problems can be discussed in terms of fixed points.

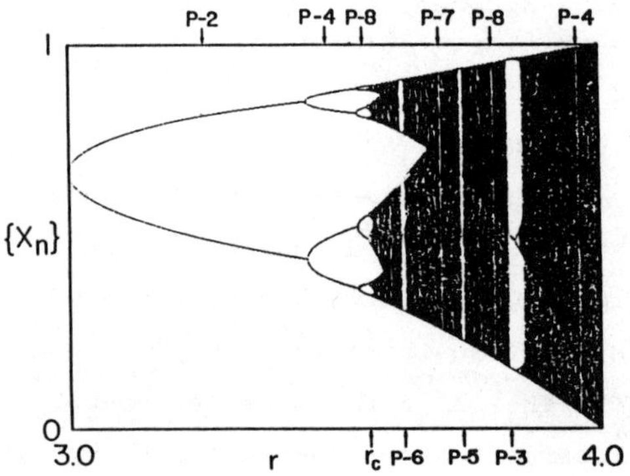

Fig. (1) shows the bifurcation diagram of logistic map $X_{n+1} = rX_n(1-X_n)$. The parameter "r_c" denotes the onset to chaos.

In this paper, I would briefly discuss one example[3] where a renormalization scheme is developed to study the case of asymptotic behavior which is different from a fixed point behavior. However, the ideas described here will have applications in many other problems. This example is the study of quasi periodic transition to chaos with two incommensurate frequencies. In phase space, this corresponds to the breakdown of two-torus.

The model system describing quasiperiodic motion is the circle map,

$$X_{i+1} = X_i + \omega - \frac{K}{2\pi} \operatorname{Sin}(2\pi X_i) = f(X_i)$$

X is taken modulo 1, so that X_i represents the position on the circle at the ith iteration. K is a nonlinearity parameter, and ω is a parameter that determines the rate at which points travel around the circle, called the winding number (or average rotation rate), defined as

$$\rho = \lim_{m\to\infty} \frac{(f^m(X_0) - X_0)}{m} .$$

ρ is independent of X_0. For $K<1$, the map is monotonic and describes periodic or quasiperiodic motion depending upon whether ρ is a rational or an irrational number. For $K>1$, map is nonmonotonic and can exhibit chaotic behavior.[4] Therefore, $K=1$ is a special point describing the onset of transition to chaos.

Rand et al[4] developed a renormalization scheme to obtain universal properties at the onset of transition to chaos. Basic idea behind the renormalization scheme is to approximate a quasiperiodic orbit by a sequence of periodic orbits. This is done by expanding irrational winding number in terms of continued fraction expansion

$$\rho = \cfrac{1}{n_1 + \cfrac{1}{n_2 + \cfrac{1}{n_3 \; \cdots\cdots}}}$$

Rational approximants to the irrational number are obtained by truncating this expansion. Under this scheme, winding number ρ transforms according to the Gauss map

$$\rho' = \frac{1}{\rho} - \left[\frac{1}{\rho}\right]$$

where [X] denotes the integer part of X. The Gauss map converges to a fixed point for quadratic irrationals (one with periodic continued fraction expansion). For a typical winding number, the map is ergodic with invariant probability density

$$P(\rho) = \frac{1}{(\log 2)\ (1+\rho)} .$$

In ref. (4), the universal properties of the transition were studied for quadratic irrationals (set of measure zero) for which renormalization equations converge on a fixed point. However, the universal properties obtained in these cases depend discontinuously on the winding number. For a typical winding number, renormalization trojectories are chaotic. In order to obtain universal features of the transition for a random winding number, one should be able to iterate the renormalization equations indefitely. In previous work,[4] the renormalization is done at a fixed value of the parameter ω while the variable X is scaled. Therefore, these equations cannot be iterated large number of times. I modified the renormalization equations and scaled both X and ω. (The details of the renormalization scheme are discussed in ref. (3)). With this modification, the equations could be iterated indefinitely. The great advantage of formulating the renormalization equations in this manner is that they asymptotically converge on stable attractors unlike the unstable fixed points of ref. (4).

For a typical winding number, the renormalized functions obtained by iterating renormalization equations wander about in the function space of circle maps. It turns out that, asymptotically, the renormalized circle maps settle down on a subset of function space of dimension 1.8 ± .1. It is remarkable that the infinite dimensional function space converges to an attractor of dimension less than two. Fig. (2) shows the 2-d projection of the function space. The attractor seems to have fractal structure along the vertical direction. I believe that the gaps in the attractor are a consequence of the global mode-locking phenomenon in the critical circle map.[5] Fractal dimension of the attractor, interpreted as one plus the fractal dimension of the cantor set along the vertical direction is consistent with this conjecture. It is worth mentioning that the fixed points corresponding to quadratic irrationals lie in an orderly fashion on the attractor. I also computed other critical exponents summarizing the average properties of the strange attractor (see ref. (3)). Unlike the case of quadratic irrationals the universal exponents for random winding number are independent of the winding number.

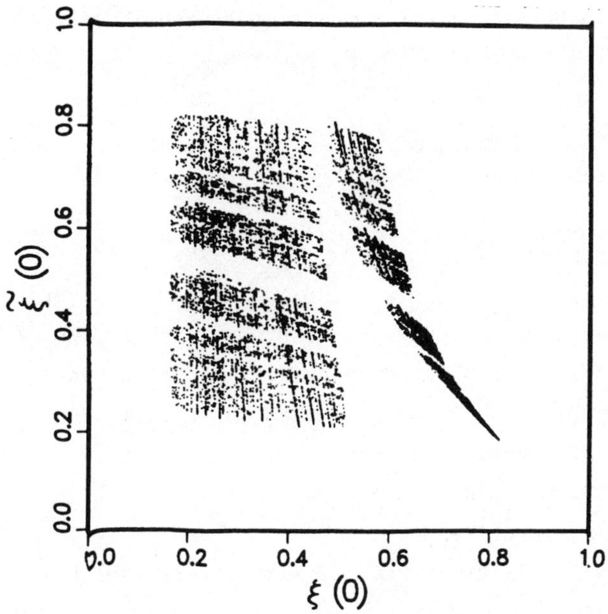

Fig. (2) shows the 2-d projection for the critical circle map. ξ (0) and $\tilde{\xi}$ (0) respectively correspond to two successive renormalized circle maps.

As far as I know, this is the first explicit example where it has been shown that renormalization group transformation has a strange attractor. In principle, all the universal information concerning the transition to chaos is contained in this attractor. I believe that the approach described above has applications in other problems besides circle map. Preliminary study[6] shows that the order in the bifurcation diagram beyond period-doubling transition point can be described in terms of universal strange attractor. The breakdown of an arbitrary KAM trajectory in standard map can be understood in terms of ergodic renormalization scheme. It is possible that the order in spin-glass, random field and disordered systems can be described in terms of universal strange attractors. McKay et al's[7] work on hierarchical Ising models with competing ferro and anti-ferromagnetic interaction already supports this idea.

The work on quasiperiodic transition to chaos was done in collaboration with J. D. Farmer and D. K. Umberger.

REFERENCES

1. M. J. Feigenbaum, J. Stat. Phys., 19 (1978) 25.
2. R. S. MacKay, Ph. D. thesis (1982), Princeton University.
3. J. D. Farmer and I. I. Satija, in press in Phys. Rev. A; J. D. Farmer, I. I. Satija and D. K. Umberger, submitted to Phys. Rev. Lett.
4. D. Rand, S. Ostlund, J. Sethna and E. D. Siggia, Phys. Rev. Lett. 49 (1982) 132; Physica 8D (1983) 303.
5. M. Jensen, P. Bak and T. Bohr. Phys. Rev. Lett. 21 (1983) 1637.
6. J. D. Farmer, unpublished.
7. S. R. McKay, A. N. Berker and S. Kirkpatrick (48) Phys. Rev. Lett. (1982) 767.

COMPUTABLE CHAOTIC ORBITS OF ERGODIC DYNAMICAL SYSTEMS

J. L. McCauley, Jr.

Department of Physics
University of Houston--University Park
Houston, Texas 77004

J. I. Palmore

Department of Mathematics
University of Illinois
Urbana, Illinois 61801

The central theme of this Institute is scaling in disordered systems, especially the use of both regular (periodic) and 'random' fractals to describe the scaling properties. In a related set of lectures, Wolfram[1] has shown how both regular and pseudo-random states can be generated systematically via simple integer maps called cellular automata. The question whether a given problem is 'decidable or intractable'[1-b] was touched upon in a general way. For example, the question whether a given cellular automaton generates a truly random sequence of bits when no pattern can be found may be undecideable: it is always possible that an algorithm generates a pattern that is sufficiently complex that we may fail to decode it, yet it may pass all standard tests for randomness. We regard this as an example of the sort of limitation upon computability that is suggested by the work of Turing.[2,2-b]

In this seminar, we will analyze the limitation imposed upon the computation of orbits of completely chaotic (ergodic) dynamical systems by the use of finite precision arithmetic. The motivation for this study is that when a computer makes an error in the computation of such an orbit, a positive Liapunov exponent will tend to magnify the error exponentially so that it becomes essential to ask what is the meaning of the numbers[3] that are generated by the computation. Toward that end, it is useful to study chaos theory on a finite binary lattice, although in principle arithmetic in any base can be used. Since chaos in a conservative dynamical system can be mapped locally onto a Baker's Transformation,[4] which is equivalent to a two-sided Bernoulli shift,[4-b] our work is about the sort of randomness[5] that is essential for classical statistical mechanics. It is well-known that ergodic motions can be mapped quite generally onto shifts, so that our work outlines the limits on computability of randomness in both conservative and dissipative ergodic systems. A simple example of this behavior is given by the logistic map

$$x_{n+1} = Dx_n(1 - x_n) \tag{1}$$

when D = 4. A simple transformation of coordinates in this case maps (1) into the tent map[6]

$$x_{n+1} = \begin{cases} 2x_n & , \ 0 < x_n < 1/2 \\ -2(x_n - 1) & , \ 1/2 < x_n < 1 \end{cases} , \tag{2}$$

and by writing both x_n and x_{n+1} as binary expansions[7]

$$x_n = \sum_{j=1}^{\infty} \varepsilon_j(n)/2^j \tag{3}$$

with $\varepsilon_j(n) = 0$ or 1, it is easy to see that the recursion formula for the binary coefficients $\varepsilon_j(n)$ is just a shift of the binary point to the right: if $x_n = .\varepsilon_1(n)\,\varepsilon_2(n)\ .\ .\ .$, then $x_{n+1} = .\varepsilon_2(n)\,\varepsilon_3(n)\ .\ .\ .$ or $.(1 - \varepsilon_2(n))(1 - \varepsilon_3(n))\ .\ .\ .\ .$ This rule for constructing $\varepsilon_j(n + 1)$ from $\varepsilon_j(n)$ is an example of a simple cellular automaton.

In what follows, we analyse the limits of computability of Bernoulli shifts on <u>finite</u> binary lattices where (3) is replaced by the expansion

$$x_n = \sum_{j=1}^{N} \varepsilon_j(n)/2^j \tag{3-b}$$

with $N < \infty$. In actual computation, this finite precision limitation is necessary and it is important to recognize its consequences in order to understand how randomness can be predicted computationally in statistical mechanics. We have shown elsewhere[8] that the question whether a non-periodic, truly random, sequence $\{x_n\}$ can be computed in finite N-bit precision can be answered affirmatively and that the result can be formulated on a gedanken computer called a Turing machine.[2,2b] The latter suggests that we have arrived at an absolute limit of computability for Bernoulli shifts, and therefore for ergodic motions. In that analysis[8] and in what follows, central roles are played by a random distribution of bits $\{\varepsilon_j(n)\}$ for almost all numbers x_n in the unit interval, and by a dense set of unstable periodic orbits in phase space. For Bernoulli shifts, the phase space is the unit interval modulo 1, or the circle of unit circumference.

As Wolfram[9] has noted, a Bernoulli shift can generate randomness only if the initial condition x_0 is random (the $\varepsilon_j(0)$ are to be determined by tossing a fair coin). In general, two properties seem necessary for randomness in ergodic systems: a random initial condition and a positive Liapunov exponent. According to Borel,[7] such initial conditions will occur with 'certainty' in a random draw of numbers x_0 from the unit interval. Computationally, this is impossible because we "draw" only N-bit numbers. N-bit numbers, with $N < \infty$, are rational, and therefore have periodic binary expansions. It would therefore seem that randomness (other than the pseudo-randomness that results from unstable periodic orbits) is computationally impossible in classical Hamiltonian dynamics and statistical mechanics. However, this is not the case and we will show below how a truly random orbit can be generated to within at least N-bit precision. We show also that Boltzmann's style of reasoning[11] is quite useful for understanding why unstable periodic orbits are pseudo-random, and we end by describing how discretized physical systems can be transformed into cellular automata, or integer maps.

Consider a prototype statistical mechanical system[10,12] on the circle of unit circumference

$$x_{n+1} = 2x_n \quad \mod 1, \tag{4}$$

whose solution is $x_n = 2^n x_o$ modulo 1, with $x_o \varepsilon [0,1)$. Irrational x_o implies non-periodic behavior with invariant density[12] $p(x) = 1$ describing the attractor, which is the entire 'phase space' (the unit interval). If x_o has the form p/q where p and q are odd integers, the orbits are periodic, unstable, and the corresponding x_o are dense in the phase space. We call such orbits 'pseudo-chaotic.' This property of a dense set of initial data yielding unstable periodic orbits (pseudo-chaotic orbits) in chaotic portions of phase space is fundamental for ergodic dynamical systems. Since the system (4) is nonconservative, it can exhibit limit cycles: if q is even, then (1) contracts onto the limit cycle given by p/q' where $q = 2^m q'$, m is an integer and q' is odd. This system is in principle statistical mechanical in the following sense: if x_o is irrational, or belongs to a very long cycle, then after a "time" $n \sim -\ln \delta x_o / \ln 2$, the system is in statistical equilibrium in the sense that for given uncertainty δx_o in x_o (which is generally present in a physical system) our ignorance is maximal. Since a computer produces only rational x_o, one might expect to see only periodic behavior, but on a binary computer this periodicity is trivial: when (4) is iterated a number of times $n > N$ then $x_n \equiv 0$. The reason why is that an N-bit binary machine represents every initial condition x_o of (4) in the finite form (3-b) where 'single precision' means $N \sim 30$. Hence, $2^n x_o$ is an integer when $n \geqslant N$, and equals 0, modulo 1. Note that for n = 1, 2, 3, . . . , $1 = 2^n x_o$ has a solution $x_o = 1/2^n$, and that linear combinations of these solutions, which are dense in the unit interval,[7] provide the finite binary lattice upon which computations are performed on a digital computer. Instead of (4), consider now the Bernoulli shift

$$x_{n+1} = \mu x_n \quad \mod 1 \tag{4-b}$$

where $\mu > 1$ is an odd integer. With $\mu = 1 + \delta\mu$, $\delta\mu = \sum_{j=1}^{N} \varepsilon_j 2^j$,

and $x_n = \mu^n x_o = [1 + \sum_{\ell=1}^{n} \binom{n}{\ell} \delta\mu^\ell] x_o$. Since $\sum_1^n \binom{n}{\ell} \delta\mu^\ell$ will 'annihilate' (modulo 1) any N-bit binary inital condition x_o, whenever n is even and large enough, it follows that all such x_o yield periodic orbits. Consider the case $\mu = 3$ where $x_n = 3^n x_o$ mod 1 and consider initial conditions x_o in the N-bit form (3-b).The period of x_o is determined by the least significant binary digit $\varepsilon_m = 1$ ($\varepsilon_j = 0$ for $m + 1 \leqslant j \leqslant N$). For $m > 3$, the length of the cycle is 2^{m-2} and since there are 2^{m-1} points with this length of cycle, there are two 2^{m-2}-cycles for each m. As $N \rightarrow \infty$, initial conditions yielding these cycles are dense in [0,1), and Kac[7] has studied the distribution of their binary digits: if we let $\varepsilon_j = 1$ correspond to heads and $\varepsilon_j = 0$ to tails and imagine an infinite number of tosses of a fair coin, then the binary digits of almost all numbers in [0,1] exhibit the randomness implied by this gedanken experiment in the sense of the law of large numbers.[7] Hence, if we toss a fair coin N times to determine N ε_j's, this corresponds to a random choice of one of the 2^N numbers (3-b) on the finite binary lattice. For N large, a random choice of the ε_j will 'draw' an x_0 with one of the largest cycle lengths with high probability: all N-bit initial conditions have the same probability $1/2^N$ of being chosen in a random draw, but it is easy to show that the probability to draw an x_0 with any one of the 2ℓ longest periods (2^{N-2}, 2^{N-3}, . . . , $2^{N-(\ell+1)}$) is $1 - 1/2^{\ell}$. For fixed N, there are 2^N states on the lattice and 2^{N-2} or 50% of them lie upon the (two) 2^{N-2} cycles, 2^{N-3} or 25% of them lie upon the 2^{N-3} cycles, and 98% of all the available states lie upon the eight cycles with the four longest periods. This explains why a random draw of x_0 will almost certainly lead to a 'long Poincaré cycle' on a computer, even in single-precision. The effect of the Liapunov exponent $\lambda = \ell n\, 3$ is to spread out these states pairwise evenly over the unit interval: in fact, the iterations form a tiling of the circle of unit circumference with two tile sizes $1/2^{m-1}$ and $3/2^{m-1}$. Equation (4) with $\mu = 3, 7, 11, 15, \ldots$ has a similar property whereas $\mu = 5, 9, 13, 17, \ldots$ defines a tiling of the circle with tiles of a single size. Hence, the distribution of iterates is either even or pairwise even.

We have produced a Boltzmann-style analysis[11] that explains why long 'Poincaré cycles' look random, or chaotic, for times short compared with the period of the orbit. We can apply this analysis to a specific computation in order to obtain a clearer picture of the effect of 'rounding-off,' say $x_0 = 1/10$, to an N-bit binary initial condition. A typical way of describing the effect of finite precision is as follows. Let the initial condition be given by (3) with n = 0,

$$x_o = \sum_{1}^{\infty} \varepsilon_j(0)/2^j \quad . \tag{3-c}$$

Then, the seemingly innocent choice $x_0 = 1/10$ forces the computer to commit an "error"

$$\delta x_o = \sum_{N+1}^{\infty} \varepsilon_j/2^j \sim 2^{-N-1} \quad . \tag{3-d}$$

The effect of a positive Liapunov exponent λ causes this "error" to grow, and one can say that the resulting loss of a finite number of bits of information per iteration of the map yields a computed orbit that is eventually far from the one that we intended to compute. Let us now see in detail what happens for equation (4-b) with $\mu = 3$ in a typical single precision calculation performed in BASIC. With $\mu = 3$, $x_0 = 1/10$ yields a 4-cycle with iterates 3/10, 9/10, 7/10, 1/10, ... Instead of $x_0 = 1/10$, a binary computer will use as x_0 a nearby lattice point. On the Commodore plus/4, $2^n(1/10) = 0 \pmod 1$ occurs for n = 35, indicating N = 35 for $x_0 = .1$ on that computer. The base 10 printout of the first four iterations on the Commodore reads as follows: .1, .3, .9, .7, .100000001. If we check these numbers for bit size, we find 35, 32, 32, and 30, respectively. 30 bits implies a 2^{28} cycle, and all further iterations fall on this cycle until n = 7464 where three fluctuations of 32, 31, and 27 bits occur and the system settles onto a 2^{25}-cycle. We have not determined the binary output to check for errors that may be confined to a 2^{28}-cycle. For n < 22 the iterations (with increasingly poor accuracy, since the 4-cycle is unstable) approximate the 4-cycle in the sense that the most significant digits .1, .3, .9, .7 are maintained in the base 10 printout. Beyond $n = 22$, the iterations "appear" chaotic, and at 250 iterations/minute, the time required to see recurrence on the 2^{28}-cycle would be about 2 years if "noise" in the form of imprecise arithmetic did not jiggle the system off that cycle. The "noise" problem can be seen more clearly as follows: if we set $x_o = 1/2^m$ with $m > 31$, then after 22 iterations there is a 'noise-driven' transition to a cycle of shorter period, thus violating the exact dynamics. In contrast, if $m = 30$, we have checked to find that the system remains on the 2^{28}-cycle for 22,000 iterations. Presumably, the same will be true if $m < 30$. Benettin et al.[13] have introduced the term 'pseudo-orbit' to denote a "numerically computed orbit . . . [that] can be completely different from the true orbit . . . ." They expected that such an orbit would arise from the accumulation of errors in computation. The example given above for $\mu = 3$ and $x_o = 1/10$ provides an example of a 'pseudo-orbit.' We will now show that by relaxing the constraint that x_o is always fixed at N-bits, (4) can be used to deterministically produce a completely random orbit, but an unlimited supply of Turing tape[2,8] will be required to specify any initial condition x_o that yields x_n to N-bits for arbitrary n. In contrast, any pseudo-chaotic (unstable periodic) orbit can be printed on a finite length of Turing-tape.

Consider a binary Turing machine with an unlimited tape containing an unlimited number of cells where binary digits ($\varepsilon_j = 0$ or 1) can be read, written, or re-written as the tape moves to the left. We introduce our method for computing x_n to N-bit accuracy by noting that there is a way around the 'convergence to zero' problem of (4) in finite binary arithmetic. Using (4), consider the action of 2^n on an N-bit x_o of the form (3-b). Since $x_1 = 2x_o = \sum_{j=1}^{N-1} \varepsilon_{j+1}/2^j \pmod 1$ the bit ε_1 is missing in x_1. Let the tape move one cell to the left and print a new bit ε_{N+1} that is determined by the toss of a fair coin: $\varepsilon_{N+1} = 1$ if heads, $\varepsilon_{N+1} = 0$ if tails. x_1 is then known to N-bits while x_o is known to $(N + 1)$-bits. This proceedure is to be repeated at every iteration for as long as one likes. The result is the deterministic generation of a completely random N-bit state x_n from completely random $(n + n)$-bit initial state x_o. We emphasize that such an x_o <u>cannot</u> be generated by a computer's pseudo-random member generator. According of Kac[7], as $n \to \infty$ such an x_o will be produced with certainty (measure one) in a random draw of numbers form the unit interval. The resulting sequence, if we consider the first N-bits in each iterate $x_o, x_1, . . . , x_n$ is recurrent but non-periodic: the recurrence time will be thw same as for states generated byN tosses of a fair coin. The orbit can be computed as far into the future as one likes. To summarize, we have used an $(N + n)$-bit initial state x_o, randomly chosen, to deterministically generate an N-bit final state that is non-periodic, and it is the action of the positive Liapunov exponent $\lambda = \ell n\ 2$ that guarantees that Maxwell's Demon[11] is paid the required price for the information that we have gained!

Since the Baker's Transformation is equivalent in binary to a two-sided Bernoulli-shift,[4-b] our method can be applied to the Baker's transformation[8] and therefore also to the computation of truly random orbits in conservative dynamical systems. If one could know with certainty that a given cellular automaton[1] generated truly random rather than pseudo-random states, then the initial state x_0 could be generated on a computer. Another way to generate a random x_0 is to interface a computer with radioactive decays.†† We will now discuss via an example how discretized physics can be rewritten as cellular automata.

We return to the logistic map (1) and replace the terms x_n and x_{n+1} by finite binary expansions of the form (3-b). We must also rewrite the control parameter D in binary form, $D = 1 + 2 + \sum_{j=1}^{N} d_j/2^j$ with $d_j = 0$ or 1. This permits us to study the region $3 \leq D \leq 4$ which is interesting, since $D > D_c \sim 3.62$ for chaos.[4,6] In order to solve the model, it is necessary to restrict both x_0 and D to a finite number of bits. If $x_n = .\varepsilon_1(n) \cdot \varepsilon_2(n) \ldots \varepsilon_k(n)$, then $1 - x_n = .(1 - \varepsilon_1(n))(1 - \varepsilon_2(n)) \ldots (1 - \varepsilon_{k-1}(n))1$, so long as $\varepsilon_k(n) = 1$, $\varepsilon_j(n) = 0$ for $j > k + 1$; and since D, x_n, and $(1 - x_n)$ have a finite number of bits, one can obtain a recursive algorithm for $x_{n+1} = .\varepsilon_1(n+1)\ \varepsilon_2(n+1) \ldots \varepsilon_\ell(n+1)$ where, in general, $\ell > k$. Whether one generates an approximation to a strange attractor (Cantor set) such as occurs for $D = D_c \sim 3.62$, or random motion, such as occurs for $D = 4$, is reflected by the asymptotic distribution of the bits $\{\varepsilon_j(n+1)\}$.

In principle, it seems that discretized physical systems can be rewritten quite generally via binary (or ternary, etc.) expansions as cellular automata. In practice, we foresee two stumbling blocks: (1) it may be non-trivial to discover the correct integer-map algorithm, and (2) it is possible that the correct algorithm cannot be written down as a <u>finite</u> number of rules (an example is the logistic map with <u>infinitely many</u> bits in x_0). We encountered neither problem with the Bernoulli-shifts, and it is possible that even the discretized Navier-Stokes equations can be studied as cellular automata.

ACKNOWLEDGEMENT

One of us (Joseph L. McCauley) is grateful to S. Wolfram for conversation and helpful comments and to the students from his chaos theory class and the Turbulence Laboratory at the University of Houston for discussions of computer arithmetic. His part in this work is dedicated to the memory of his father.

††This was pointed out to one of the authors by a member of the audience at the University of Siegen when he (McCauley) presented a similar seminar there.

REFERENCES

1. S. Wolfram, in these proceedings (1985).
1-b. S. Wolfram, Phys. Rev. Lett. <u>54</u>, 735 (1985).
2. A. Turing, Proc. Lond. Math. Soc. (2) <u>42</u>, 230 (1937).
2-b. Andrew Hodges, <u>Alan Turing: The Enigma</u> (Simon and Schuster, 1983), pp. 91-110.
3. J. G. Nash, BYTE <u>10</u>, 177 (1985).
4. A. J. Lichtenberg and M. A. Lieberman, <u>Regular and Stochastic Motion</u>, Springer-Verlag (1983).
4-b. M. V. Berry in <u>Topics in Nonlinear Dynamics</u>, edited by S. Jorna, Vol. 46, AIP (1975).

5. G. J. Chaitin, Sci. Am. 232, 47 (1975).
6. L. P. Kadanoff, Physics Today 36, 46 (1983).
7. M. Kac, Statistical Independence in Probability, Analysis and Number Theory, The Carus Mathematical Monographs No. 12 (1959).
8. J. L. McCauley Jr. and Julian I. Palmore, to be published (1985).
9. S. Wolfram, Origins of Randomness in Physical Systems (1985).
10. J. Ford, Physics Today 36, 40 (1983).
11. S. G. Brush, Kinetic Theory, Vol. 2, Pergamon Press (1966).
12. S. Grossmann and S. Thomae, Z. Naturforsch. 32a, 1353 (1977).
13. G. Benettin, G. M. Casartelli, L. Galgani, A. Giorgilli, and J. M. Strelcyn, Nuovo Cimento 44B, 183 (1978).

FRACTAL STRUCTURE OF SUBHARMONIC STEPS IN DISSIPATIVE SYSTEMS DESCRIBED BY A DRIVEN DAMPED PENDULUM EQUATION

Preben Alstrøm and Mogens T. Levinsen

H.C. Ørsted Institute
Universitetsparken 5
DK-2100 Copenhagen Ø, Denmark

INTRODUCTION

Since it became clear that the return map for the driven damped pendulum equation

$$\ddot{\phi} + G\dot{\phi} + I(\phi) = I + A\sin\omega t \qquad (1)$$

with $I(\phi) = \sin\phi$ at and below the transition to chaos is a circle map for every irrational rotation number W,[1] the connection between the scaling laws for the pendulum system and the structures obtained by iterations of circle maps has been studied extensively. In particular, the universal scaling behavior of the complete devil's staircase obtained by iterations of circle maps[2] like the sine map

$$x_{n+1} = x_n + \mu - \frac{1}{2\pi}\sin 2\pi x_n \qquad (2)$$

with a zero slope third order inflection point has been conjectured also to be found at the transition to chaos in every dissipative dynamical system with two competing frequencies.[3]

It is the aim here to summarize and compare essential results obtained in this connection in order to examine the conjecture above.

CRITICAL LINE

We shall in this note consider three potentials $I(\phi)$ in equation (1), namely $I_1(\phi) = \sin\phi$, which leads to the well-known capacitively and resistively shunted Josephson equation, $I_2(\phi) = \sin(\phi + 0.25\sin\phi)$ in order to look into the effect of a distortion of the potential $I_1(\phi)$ which leaves the zero external force rotation numbers unchanged in the overdamped case, and finally $I_3(\phi) = \frac{6}{7}(\sin\phi + \frac{1}{6}\sin^3\phi)$ to underline that the universal scaling is independent of the inflection point order of the potential. In every case the equation (1) was set up on an analog computer and I-W characteristics were plotted. Typically, the external force parameters A and ω were kept constant while the damping constant G was varied obtaining a series of I-W curves as shown for $I(\phi) = I_1(\phi)$ and $(A,\omega) = (1,1.76)$ in Fig. 1.

It is observed that the subharmonic stepsizes increase when G is lowered, and at the same time the positions of the steps tend against lower values of I. With this in mind the onset of hysteresis can be understood as

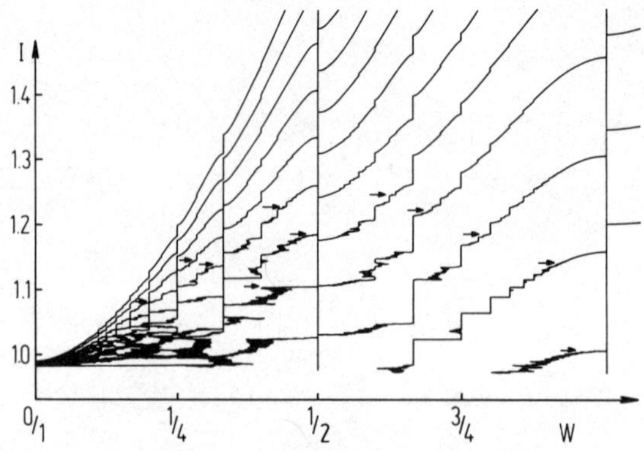

Fig. 1. I-W curves obtained for A=1 and ω=1.76 by analog-computer calculations. The damping factor G is lowered successively by ΔG=0.08 starting with G=1.6. Critical points are indicated by arrows.

appearing due to overlapping of subharmonic steps, and consequently the step-structure will develop a complete devil's staircase along the critical line defined by the points in parameter space where hysteresis sets in. A test of

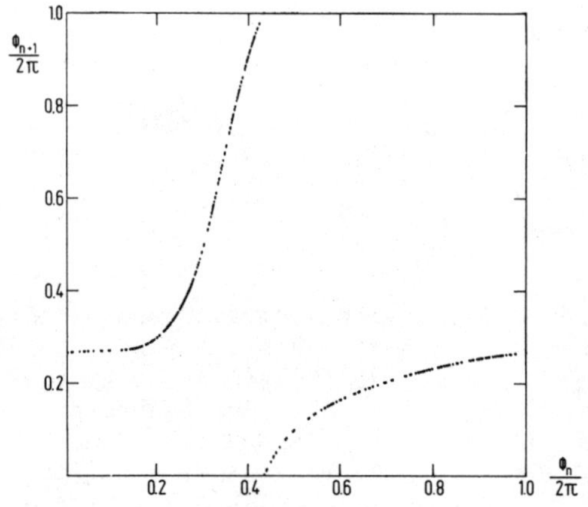

Fig. 2. Return map for A=1, ω=1.76, I=1.16, G=1.36 and W≃0.30.

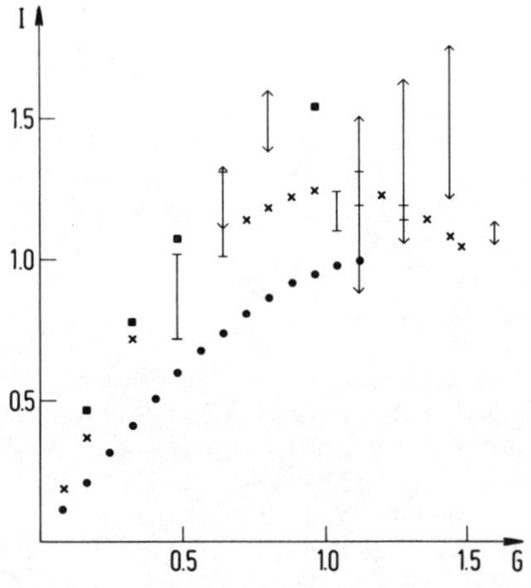

Fig. 3. Critical points in the (I,G) space. A=1, and ω=0.88 (■,↕), 1.76 (×,I), and 3.52 (●). Intervals are given in cases where hysteresis is found below but not above a step.

Table 1. Exponents associated with the complete devil's staircases obtained from different potentials $I(\phi)$ in (1) and from the return map (2).

$I(\phi)$	D	δ	y	$\bar{\delta}$	$\bar{D}$
$\sin\phi$	0.87±0.02	2.95±0.08	2.0±0.2	2.27±0.03	0.882±0.010
$\sin(\phi+0.25\sin\phi)$	0.86±0.02	3.09±0.08	2.3±0.2	2.29±0.03	0.873±0.010
$\frac{6}{7}(\sin\phi+\frac{1}{6}\sin^3\phi)$	0.87±0.02	2.98±0.08	2.1±0.2	2.30±0.03	0.868±0.010
return map (2)	0.8700 ±0.00037	3.0	2.16443 ±0.00002	2.292 ±0.0034	0.873 ±0.0021

this interpretation can be made on a digital computer by obtaining return maps associated with critical points, and then control whether zero slope inflection points are present in these maps. Such a return map related to the G=1.36 curve in Fig. 1 is shown in Fig. 2. This figure supports the picture of the transition to chaos given above.

By changing the external force, i.e. (A,ω), the critical line moves, as seen in Fig. 3, where ω has been varied. As ω tends to infinity the critical line moves down against some limit curve. When A is varied, the limit curve is reached for $A\to 0$.

SCALING LAWS

The stepsizes of the complete devil's staircase obtained along the critical line was measured, and about 140 stepsizes were collected. For these measurements we used rescaling facilities on the analog computer as well as enlarging by means of a projektor. From the collection of stepsizes several scaling parameters can be found, e.g. the fractal dimension which enters the scaling law $N(r)\sim r^{-D}$, where N(r) is the number of steps larger than the scale r.[3] Thus by plotting N(r) versus 1/r in a double-logarithmic plot, D can be determined. For the external force in Fig. 1 such a plot is shown in Fig. 4. It is noticed that the slope obtained is in agreement with the fractal dimension D=0.870 found for the iterated sine map (2).

But also several decay exponents for the stepsizes can be found, and in Table 1 the exponents achieved are given for the different potentials (the external force is chosen as in Fig. 1). The three decay exponents δ, y and $\bar{\delta}$ are defined by the limit behavior of the stepsizes $\Delta(P/Q)\sim Q^{-\alpha}$, where $\alpha=\delta$ for P=1, $\alpha=y$ for $P/Q \in \{F_i\}$, where (F_i) is the Fibonacci sequence, and $\alpha=\bar{\delta}$ when $\Delta(P/Q)$ is replaced by the average $\langle\Delta(P/Q)\rangle_P$ over P. In every caes the results are in agreement with the exponents obtained for the sine map (2).[3,4]

OTHER ASPECTS

Returning to the selected choices of potentials we first note that for $I(\phi)=I_1(\phi)$ it is easy to find the rotation number in the case where no external force is present and the initial term can be neglected. Then simply $W=2\pi/T\omega=(I^2-1)^{\frac{1}{2}}/G\omega$, where

$$T = \int_{-\pi}^{\pi} \frac{G\, d\phi}{I-\sin\phi} . \tag{3}$$

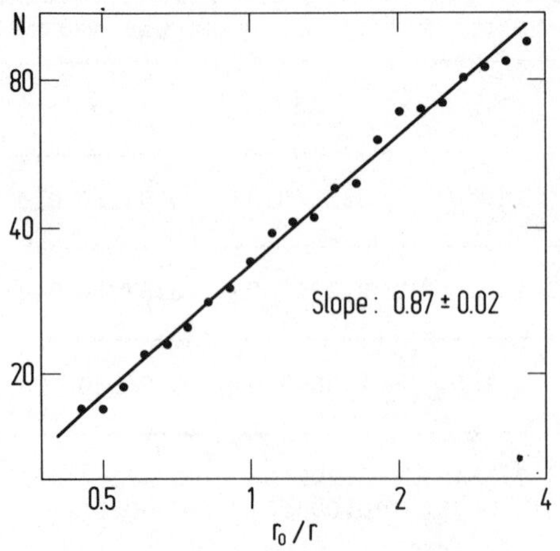

Fig. 4. Number N(r) of steps larger than the scale r plotted in a double-logarithmic plot versus r_o/r, where r_o=0.004.

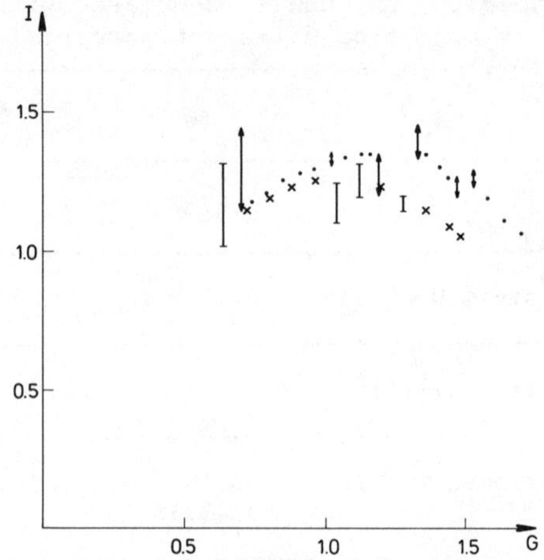

Fig. 5. Critical points in (I,G) space. A=1 and ω=1.76. (×,I) belongs to $I(\phi)=I_1(\phi)$; (•,↕) belongs to $I(\phi)=I_2(\phi)$.

The interesting point about the potential $I_2(\phi)$ is that the period T is unchanged when sinϕ is replaced by $I_2(\phi)$. However, the critical line obtained at finite external forces moves up in parameter I, as seen in Fig. 5 where (A,ω)=(1,1.76) as before. These circumstances are worth noting when the no external force conditions are used to model an overdamped system, as e.g. the charge-density-wave system at certain temperatures.[5]

The purpose to choose the potential $I_2(\varphi)$ is to make clear that higher order inflection points in the potential will not change the scaling laws, in contrast to the changes caused by the introduction of higher order inflection points in the circle map (corresponding to the return map).[3]

Finally we point out the very large degree of self similarity. The stepsizes as a function of the rotation number W is shown in Fig. 6 for $I_3(\phi)$, but similar pictures are found for the potentials $I_1(\phi)$ and $I_2(\phi)$. The pictures are very like those found in iterated circle map cases.[3]

CONCLUSIONS

For the driven damped pendulum equation the critical line indicating the transition to chaos has been identified to be given by the points in parameter space where hysteresis sets in. Along this line a complete devil's staircase is formed, and the scaling laws describing this are found to be consistent with the results obtained for discrete circle maps.

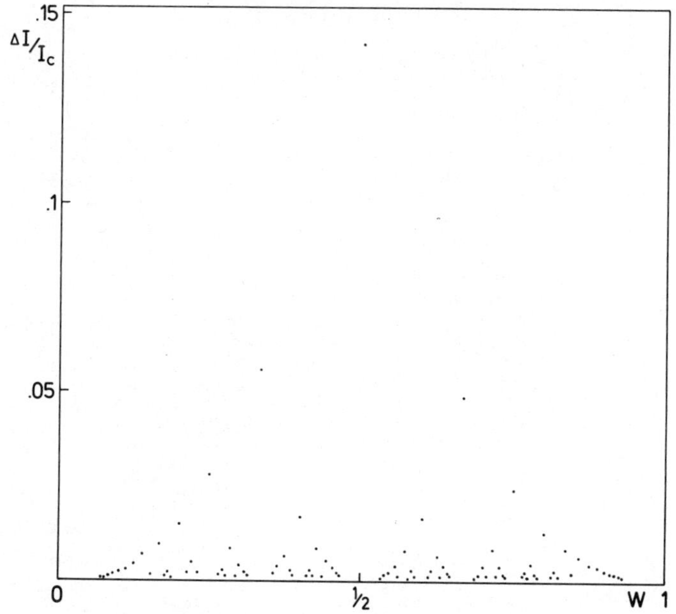

Fig. 6. The sizes of the subharmonic steps along the critical line as a function of the rotation number W=P/Q between 0/1 and 1/1 for $I(\phi)=I_3(\phi)$.

REFERENCES

1. P. Bak, T. Bohr, M. H. Jensen, and P. V. Christiansen, Josephson junctions and circle maps, Solid State Commun. 51:231 (1984).
2. M. H. Jensen, P. Bak, and T. Bohr, Complete devil's staircase, fractal dimension, and universality of mode-locking structure in the circle map, Phys. Rev. Lett. 50:1637 (1983).
3. M. H. Jensen, p. Bak, and T. Bohr, Transition to chaos by interaction of resonances in dissipative systems. I. Circle maps, Phys. Rev. A 30:1960 (1984); II. Josephson junctions, charge-density waves, and standard maps, ibid. 30:1970 (1984).
4. S. J. Shenker, Scaling behavior in a map of a circle onto itself: Empirical results, Physica D 5:405 (1982).
5. S. E. Brown, G. Mozurkewich, and G. Grüner, Subharmonic Shapiro steps and devil's-staircase behavior in driven charge-density-wave systems, Phys. Rev. Lett. 52:2277 (1984).

PARTICIPANTS

Gabriel Aeppli
Bell Laboratories 600 Mountain Avenue MURRAY HILL, N.J. 07974
USA

Amnon Aharony
Dept. of Physics University of Tel-Aviv 69978 TEL-AVIV
Israel

Preben Alstrøm
H.C. Ørsted Institute Universitetsparken 5 DK-2100 COPENHAGEN
Ø Denmark

David Andelman
Exxon Research and Engineering Co. Route 22 East ANNANDALE,
N.J. 08801 USA

Arne F. Andresen
Institute for Energy Technology 2007 KJELLER Norway

Javid Ashraff
Dept. of Physics University of Manitoba WINNIPEG, Manitoba
Canada 3RT 2N2

Jose Antoranz
Dept. of Physics UNED 28080 MADRID Spain

Trond Aukrust
Inst. of Theoretical Physics Norwegian Institute of Technology
7034 TRONDHEIM-NTH Norway

John Axe
Dept. of Physics Brookhaven National Laboratory UPTON, L.I.,
N.Y. 11973 USA

Nils Baas
Mathematical Institute Norwegian Institute of Technology
7034 TRONDHEIM-NTH Norway

Per Bak
Dept. of Physics Brookhaven National Laboratory UPTON, L.I.,
N.Y. 11973 USA

R. Ball
Dept. of Physics University of Cambridge CAMBRIDGE UK

Leigh G. Baxandall
T.C.M. Cavendish Laboratory CAMBRIDGE CB3 OHE UK

Joseph F. Benzoni
Dept. of Physics Imperial College LONDON SW7 UK

Lars Inge Berge
Dept. of Physics University of Oslo, Blindern 0316 OSLO 3
Norway

Tom Egil Berland
Dept. of Physics University of Oslo, Blindern 0316 OSLO 3
Norway

Kurt Binder
Dept. of Physics Johannes Gutenberg University D-6500 MAINZ
W-Germany

Rafael Blumenfeld
Dept. of Physics University of Tel-Aviv 69978 TEL-AVIV
Israel

Himadri Bohidar
Dept. of Physics University of Oslo, Blindern 0316 OSLO 3
Norway

William D. Brown
TCM, Cavendish Laboratory, University of Cambridge CB3 OHE
UK

David Callaway
MS-B285 Los Alamos National Laboratory Los Alamos, N.M.
87545 USA

Maria Carvalho
Dept. of Physics University of Aveiro 3800 AVEIRO Portugal

Elisabeth Charlaix
L.H.M.P., ESPCI 10 rue Vauquelin 75005 PARIS France

Michel Cloitre
L.H.M.P., ESPCI 10 rue Vauquelin 85005 PARIS France

Eric L. Courtens
IBM Research Laboratory CH-8803 RUSCHLIKON Switzerland

Roger A. Cowley
Dept. of Physics Edinburgh University EDINBURGH EH9 3JZ
UK

Mohamed Daoud
LLB-C.E.N. Saclay B.P. No. 2 91191 GIF-SUR-YVETTE France

Silvana De Lillo
Dept. of Physics University of Salerno 84100 SALERNO ITALY

Zoran Djordjevic
Lab. of Solid State Physics Institute Boris Kidric-Vinca 11001
BELGRADE Yugoslavia

Gerald Dolling
Chalk River Nuclear Laboratories CHALK RIVER, ONTARIO
Canada KOJ 1JO

William Dozier
Exxon Research and Engineering Co. Route 22 East ANNANDALE, N.J. 08801 USA

Spiros Evangelou
Research Centre of Crete P.O. Box 517 HERAKLION, Crete Greece

Jens G. Feder
Dept. of Physics University of Oslo, Blindern 0316 OSLO 3 Norway

Shechao Feng
D24 Leverett House Harvard University CAMBRIDGE, MA 02138 USA

Nils Fjeldsø
Dept. of Physics University of Oslo, Blindern 0316 OSLO 3 Norway

Kristian Fossheim
Dept. of Technical Physics Norwegian Institute of Technolgy 7034 TRONDHEIM-NTH Norway

Torsten Freltoft
Dept. of Physics Risø National Laboratory 4000 ROSKILDE Denmark

Walter Goldburg
Dept. of Physics University of Pittsburgh PITTSBURGH, PA 15260 USA

Leonard Gray
Oak Ridge National Laboratory OAK RIDGE, TN 37830 USA

Etienne Guyon
Eccole Superieure de Physique et de Chimie 75231 PARIS France

Christopher K. Harris
Physics Department University of Edinburgh EDINBURGH EH9 3JZ UK

Wilfried M. Hein
Hahn-Meitner-Institut Glienicker Strasse 100 1000 BERLIN 39

Stewart Higgins
Physics Department University of Edinburgh EDINBURGH EH9 3JZ UK

Einar L. Hinrichsen
Dept. of Physics University of Oslo, Blindern, 0316 OSLO 3 Norway

Daniel C. Hong
Dept. of Physics University of Boston BOSTON, MA 02215 USA

John Huang
Exxon Research and Engineering Co. Route 22 East ANNANDALE, N.J. 08801 USA

Janina M. Janik
Inst. of Chemistry, Jagelonian University, Krakow, Poland p.t. Institute for Energy Technology 2007 KJELLER Norway

Jerzy A. Janik
Inst. of Nuclear Physics, Krakow, Poland p.t. Institute for Energy Technology 2007 KJELLER Norway

Henrik J. Jensen
Institute of Physics University of Aarhus 8000 AARHUS C Denmark

Tom H. Johansen
Dept. of Physics University of Oslo, Blindern 0316 OSLO 3 Norway

Giancarlo Jug
Food Research Institute Colney Lane NORWICH NR4 7UA UK

Torstein Jøssang
Dept. of Physics University of Oslo, Blindern 0316 OSLO 3 Norway

Yacov Kantor
Exxon Research and Engineering Co. Route 22 East ANANNDALE, N.J. 08801 USA

Jørgen Kjems
Risø National Laboratory 4000 ROSKILDE Denmark

Max Kolb
Lab. de Physique des Solides Centre de Orsay 91405 ORSAY France

Aglae Magalhaes
Centro Brasileiro de Pesquisas Fisicas p.t. King's College LONDON WC2R 2LS UK

Anthony C. Maggs
Dept. of Theoretical Physics University of Oxford OXFORD OX1 UK

Frank Mallezie
Inst. for Theoretical Physics The Catholic University B-3030 LEUVEN Belgium

Joseph L. McCauley
Physics Department University of Houston HOUSTON, TX 77004 USA

Yigal Meir
Physics Department Tel-Aviv University 69978 TEL-AVIV Israel

Knut J. Måløy
Dept. of Physics University of Oslo, Blindern 0316 OSLO Norway

Toshiya Ohtsuki
Department of Chemistry   Boston University   BOSTON, MA 02215
USA

Raymond Orbach
Dept. of Physics   University of California   LOS ANGELES, CA.
90024   USA

Unni Oxaal
Dept. of Physics   University of Oslo, Blindern   0316   OSLO 3
Norway

Iveta Pimentel
Dept. of Physics   University of Lisbon   1699   LISBON   Portugal

K.V. Rao
Dept. of Solid State Physics   Royal Institute of Technology
S-100 44   STOCKHOLM   Sweden

Finn Ravndal
Dept. of Physics   University of Oslo, Blindern   0316   OSLO 3
Norway

Dieter Richter
IFF - Kerforschungsanlage Julich   D-5170   JULICH   W-Germany

Tormod Riste
Institute for Energy Technology   2007   KJELLER   Norway

Risto K. Ritala
NORDITA   Blegdamsvej 17   2100   Copenhagen Ø   Denmark

Leonard M. Sander
Dept. of Physics   University of Michigan   ANN ARBOR, MI 48109
USA

Indu Satija
CNLS, MSB 258   Los Alamos National Laboratory   LOS ALAMOS,
N.M. 87545   USA

Sushil Satija
Physics Department   University of Delaware   NEWARK, DA 19716
USA

Michael Schick
Dept. of Physics FM-15   University of Washington   SEATTLE, WA 98195
USA

Peter Schofield
Bldg. 521-2, AERE Harwell   OXON OX11 ORA   UK

Pado Sibani
NORDITA   Blegdamsvej 17   2100   COPENHAGEN Ø   Denmark

Byron W. Southern
Dept. of Physics   University of Manitoba   WINNIPEG, Manitoba
Canada R3N 2T2

H. Eugene Stanley
Center for Polymer Studies University of Boston BOSTON, MA 02215 USA

Robin B. Stinchcombe
Dept. of Physics University of Oxford UK OXFORD OX1 3PU

Birger Stølan
Dept. of Technical Physics Norwegian Institute of Technology 7034 TRONDHEIM-NTH Norway

Sunil Sinha
Exxon Reserach and Engineering Co. Route 22 East ANNANDALE, N.J. 08801 USA

Martin Söderberg
Dept. of Theoretical Physics Royal Institute of Technology S-100 44 STOCKHOLM Sweden

Harry Thomas
Dept. of Physics University of Basel CH-4056 BASEL Switzerland

Chris Tigges
Dept. of Physics Yale University NEW HAVEN, Ct.06520 USA

Peder A. Tyvand
Dept. of Physics Norwegian Agricultural College 1432 ÅS-NLH Norway

Werner R. Uhler
Dept. of Physics University of Basel CH-4056 BASEL Switzerland

Jacques Villain
IFF Kerforschungsanlage Julich D5170 JULICH W-Germany

Richard Voss
IBM, T.J. Watson Research Center YORKTOWN HEIGHTS, NY 10598 USA

David A. Weitz
Exxon Research and Engineering Co. Route 22 East ANNANDALE, N.J. 08801 USA

Stephen Wolfram
Institute for Advanced Studies PRINCETON, N.J. 08540 USA

Claude Zeyen
Institute Laue-Langevin 38042 GRENOBLE-CEDEX France

Klaus Ziegler
Dept. of Physics Gesamthochschule D-4300 ESSEN W-Germany

Timothy Ziman
Institute Laue-Langevin 38042 GRENOBLE-CEDEX France

Ignacio Zuniga
Dept. of Physics UNED 28080 MADRID Spain

ORGANIZING COMMITTEE

Director:	Roger Pynn	Institute Laue-Langevin 156X 38042 GRENOBLE France
Co-director:	Arne T. Skjeltorp	Institute for Energy Technology 2007 Kjeller Norway
Secretary:	Gerd Jarrett	Institute for Energy Technology 2007 Kjeller Norway

INDEX